Marine Fisheries Ecology

Marine Fisheries Ecology

Simon Jennings
Centre for Environment, Fisheries and Aquaculture Science
Lowestoft Laboratory
Lowestoft
NR33 0HT
United Kingdom
(www.cefas.co.uk)

Michel J. Kaiser
School of Ocean Sciences
University of Wales
Bangor
LL59 5EY
United Kingdom
(www.sos.bangor.ac.uk)

John D. Reynolds
School of Biological Sciences
University of East Anglia
Norwich
NR4 7TJ
United Kingdom
(www.uea.ac.uk/bio)

Blackwell
Publishing

© 2001 by Blackwell Science Ltd,
a Blackwell Publishing company

350 Main Street, Malden, MA 02148-5020, USA
108 Cowley Road, Oxford OX4 1JF, UK
550 Swanston Street, Carlton, Victoria 3053, Australia

The right of the Authors to be identified as the Authors of this Work has been
asserted in accordance with the UK Copyright, Designs, and Patents Act 1988.

First published 2001
Reprinted 2003, 2004

Library of Congress Cataloging-in-Publication Data

Jennings, Simon.
  Marine fisheries ecology / Simon Jennings, Michel J. Kaiser,
  John D. Reynolds.
     p. cm.
  Includes bibliographical references (p.).
  ISBN 0-632-05098-5
  1. Fishery management. 2. Marine fishes—Ecology.
3. Fisheries—Environmental aspects. I. Kaiser, Michel J.
II. Reynolds, John D. III. Title.
SH328. J46 2000
333.95′6 — dc21   00-021514

A catalogue record for this title is available from the British Library.

Set by Graphicraft Ltd, Hong Kong
Printed and bound in the United Kingdom
by the Alden Press Ltd, Oxford and Northampton

For further information on
Blackwell Publishing, visit our website:
http://www.blackwellpublishing.com

Contents

Preface

Fisheries play a key role in providing food, income and employment in many parts of the world. We might expect fishing to be a profitable and effective way of getting food, since fishers take harvests that they need not sow. Sadly, this is rarely true. Fisheries are often subsidized, wasteful, cause excessive environmental damage and ignite conflicts between otherwise friendly nations. Relatively few fisheries realise their potential benefits to fishers and society.

For many years, the main objective of fishery management was to maximize the yield taken from a fishery without compromising future catches. Intuitively, this seemed a simple and sensible objective, but in practice, there were good biological, sociological and economic reasons why it was rarely achieved. Not only were some fisheries so heavily fished that they collapsed, but they became economically inefficient and threatened species and habitats of conservation concern that were not their intended targets. Clearly, there is much scope to improve management of fisheries and the way we utilize the marine environment.

Fisheries science is now a more exciting and varied field of study than ever before. This is because contemporary management objectives are increasingly diverse, and our attempts to manage and conserve fisheries are based on a much broader scientific understanding of fishers and the fished ecosystem. Indeed, many fishery scientists are now asked to address biological, economic and social concerns. These range from dealing with uncertainty and reducing incidental catches of dolphins to ensuring that fished coral reefs remain attractive to tourists, that fisheries are profitable, and that conflicts between fishers are minimized.

Aims of this book

Effective fisheries management requires clear objectives and a decision-making process supported by the best scientific advice. The aim of this book is to give the reader a broad understanding of biological, economic and social aspects of fisheries science and the interplay between them. The overall emphasis, however, is deliberately ecological. By the time you have read this book we hope that you will appreciate:

• how physical and biological processes drive the production of fished species and why the abundance of these species changes in space and time;
• the scale, social and economic significance of global fisheries, the species that are caught and the gears used to catch them;
• the factors that motivate and limit human fishing activities and why fishers behave as they do;
• the economic, social and biological reasons why fished species tend to be overexploited and why governments and other authorities, including fishers, intervene to encourage sustainable fishing;
• how to make basic quantitative assessments of single and multispecies fisheries and to estimate the parameters needed for these assessments;
• the key strengths and failings of different fisheries assessment methods and the effects of uncertainty on the outputs;
• those aspects of the behaviour and life history of fished species, at individual and population levels, that make them vulnerable to fishing mortality;
• the impacts of fishing on marine ecosystems, birds, mammals, non-target species and habitats and what can be done to mitigate them;
• the objectives of fishery management and how fisheries can be regulated to achieve specific economic, social and biological objectives.

Why we wrote this book

Previous fisheries textbooks have concentrated almost

exclusively on the mechanisms of stock assessment and management with little reference to the ecosystem in which the management practices are applied. However, students have often told us that they wanted to learn more about various fisheries-related issues that traditionally receive scant coverage in fisheries textbooks. These include trawling impacts, coral reef fisheries, discarding, seabirds, marine mammals, and poverty and conflict in fishing communities. They found it difficult to relate 'fisheries' courses that focused purely on quantitative assessment of population dynamics to many of the issues that confronted them in their own lives, the media and even in future employment. Thus, we wanted to write a book that reflected the global diversity of fished species, fisheries and fishing behaviour and showed how useful generalities could be drawn by working across conventional boundaries and treating fisheries as part of an ecosystem that includes the fishers. We also hope to convey our enthusiasm for fisheries ecology through our personal experiences and convince you that it is an exciting subject area that is helping to drive our understanding of marine ecosystems.

Plan of this book

We begin this book by introducing the world's fisheries, their diversity and history, suggesting why we need to conserve fisheries and the marine environment and discussing the main objectives of management. The remaining chapters can be divided into related groups that treat various aspects of fisheries ecology and their relevance to the assessment and management of fisheries. Chapters 2, 3 and 4 cover production processes, the species that are fished, the food chains that support them, and the factors that control the rates and variability of production. Chapters 5 and 6 consider the fishers: how and why people go fishing, the social importance of fisheries and the factors that affect the ways in which people fish. Chapters 7–11 deal with fishery assessment: how to assess the effects of fishing on target species, multispecies communities and ecosystems, bioeconomics, and how to get the data needed for assessments. Chapters 12–15 look at the wider effects of fishing, on community structure, diversity, population genetics, habitats, seabirds, marine mammals and species of conservation concern. In Chapter 16 we ask whether aquaculture can replace fishing as a source of

protein, and in 17 we show how knowledge of production processes, fishers, assessment and fishing effects can be used to reach a range of management and conservation objectives.

We introduce most of the quantitative methods used for fisheries assessment but this is not a recipe book, and our emphasis is on why such models are needed, their assumptions and potential pitfalls. Many ideas are illustrated conceptually rather than mathematically, since several excellent texts already deal with quantitative stock assessment. For those who want to develop expertise in quantitative assessment we strongly recommend the excellent texts by Clark (1985), Hilborn and Walters (1992) and Quinn and Deriso (1999). These are packed with useful ideas and examples of their application. The history of fishery science and its contribution to wider ecological thinking are well described by Smith (1994).

Attention please!

Before you start reading we would like to draw your attention to the structure of this book. The chapters were written with the intention that they would be read in sequence, although each chapter has its own introduction and summary and can be read in isolation by the casual browser. We have referenced the text rather more extensively than many student textbooks. Our aim was to cite classic references from people who developed an idea coupled with contemporary sources that reveal the state of understanding that has now been achieved. These allow an easy lead-in to the research literature on a particular topic. For the reader who would like to know more, each chapter ends with a 'further reading' section. This lists 3–4 key books or reviews that develop ideas introduced in the chapter. Lecturers can download the figures that appear in this book from the Blackwell Science website (**www.blackwell-science. com/jennings**).

All places mentioned in the text are indexed on page 389 and their locations are shown on the accompanying map. Key terms are shown in bold at their first significant use and accompanied by a brief description. In the subject index, the pages on which these terms appear are shown in bold italic so the description can be found. The meanings of all symbols used in equations are given at their first usage and in the table on page 380. Note that biologists and economists often use the same symbols to denote different parameters. Common and species names are given in full at first usage and the taxonomic family to which they belong is shown in parentheses. This can be used to trace the taxonomic affiliation of the species in Tables 3.2 and 3.4. Common and scientific names of all species mentioned are indexed on page 393. The appendices also include a list of fisheries websites that provide a wealth of fisheries information and from which software, data and publications may be downloaded.

We hope you enjoy reading this book and find it useful.

Acknowledgements

We could not have produced this book without the help of many people. Sections of the draft text were read by John Coppock, Isabelle Côté, Chris Darby, Dan Duplisea, Chris Francis, Tim Hammond, Joe Horwood, Rognvaldur Hannesson, Alex Lincoln, Iago Mosquera, Mike Pawson, André Punt, Terry Quinn, Jo Ridley and Bill Sutherland, while Jeremy Collie read everything! Their comments substantially improved the final text. We apologise to the undergraduate students who served as guinea pigs for some of the material in this book and thank them for their feedback! Ian Sherman, Dave Frost and the staff of Blackwell Science were helpful, patient and a pleasure to work with.

Unpublished manuscripts, reprints, materials, data and advice were provided by David Agnew, Geoff Arnold, Mike Beardsell, Andrew Brierley, Martin Collins, John Cotter, Mark Chittenden, Tas Crowe, Chris Darby, Nick Dulvy, Jim Ellis, Karen Field, Nick Goodwin, Ewan Hunter, Diane Kaiser, John Lancaster, Ray Leakey, Chris Mees, Julian Metcalfe, Richard Millner, Jack Musick, John Nichols, Carl O'Brien, Hazel Oxenford, Daniel Pauly, Mike Pawson, Nélida Pérez, Nick Polunin, John Pope, André Punt, Callum Roberts, Rob Robinson, Stuart Rogers, Garry Russ, Ian Russell, Melita Samoilys, John Stevens, Rob Tinch, Geoff Tingley and David Walton.

Photographs were kindly provided by Frank Almeida, Gian Domenico Ardizzone, David Barnes, Quentin Bates, David Berinson, Steve Blaber, Dan Blackwood, Lesley Bradford, Nigel Brothers, Blaise Bullimore, Channel 7 News (Perth, Australia), the Countryside Council for Wales, Derek Eaton, Fishing News International, Fisheries Research Services (Aberdeen), the Food and Agriculture Organization of the United Nations, Bob Furness, Brett Glencross, Thomas Heeger, Cecil Jones, Adam Jory, John LeCras, Berend Mensink, Steve Milligan, North-west Marine Technology, Alfonso Angel Ramos-Esplá, Rudy Reyes, Will Reynolds, Paul Rozario, Joseph Smith, David Solomon, Ian Strutt, the United States National Marine Fisheries Service, Page Valentine, Clem Wardle and Ted Wassenberg. We apologise to those whose photographs could not be included. Martin Collins, Dennis Glasscock, Julian Metcalfe, Graham Pickett and Monty Priede helped us to track down photographic material. Sheila Davies reproduced many photographs, Jim Jennings drew the fish, invertebrates and fishing gears in Figs 3.1, 3.2 and 5.4, and John Nichols provided the larval fish drawings for Fig. 4.9. The staff of the excellent CEFAS library were a great help throughout. Thank you everyone.

Simon would like to say a special thank you to Jess, friends and family for their support and sense of fun and to John Cotter, Mike Pawson, Nick Polunin, John Ryland the late Ray Beverton who, as supervisors and managers, gave him the time, space and encouragement to develop a wide-ranging interest in fisheries and ecological issues. He is also grateful to the University of East Anglia and CEFAS for providing support and facilities and to all at the Marine Studies Programme, University of the South Pacific who did so much to stimulate his interest in tropical fisheries. Michel would like to thank Diane and Holly for their love and inspiration. He would also like to thank his friends and former colleagues at CEFAS whose help and encouragement gave him a privileged insight into the world of fisheries science. He'd also like to thank his PhD supervisor Roger Hughes who gave him the enthusiasm to convey his science to a wider audience and has made writing a pleasure. John would like to thank Isabelle and Geneviève for their support and patience. Many friends and colleagues at the University of East Anglia and at CEFAS provided insights that helped him to bridge his interests in ecology, evolution and conservation. He also thanks five key supervisors from successive stages of his career: Richard Knapton, Jim Rising, Fred Cooke, Mart Gross and Paul Harvey.

We hope to update and improve the book. Please send any comments, suggestions, gifts or abuse to us at 'Marine Fisheries Ecology' c/o Simon Jennings, CEFAS, Lowestoft Laboratory, NR33 0HT, United Kingdom, or e-mail mfe@cefas.co.uk.

Simon Jennings
Michel Kaiser
John Reynolds
July, 2000

1 Marine fisheries ecology: an introduction

1.1 Introduction

Humans have fished since prehistoric times, but in the last 50 years fisheries have expanded faster than ever before. Marine fisheries now yield around 90 million tonnes per year, more than 80% of global fish production. Catches have increased because a growing human population demands more food and because improved technology has simplified capture, processing, distribution and sale. Greater fishing power and increased competition between fishers, vessels or nations has led to the economic collapse of some fisheries that had flourished for centuries. The resultant reductions in fish production, income and employment are usually seen as undesirable by society. This is why governments intervene to regulate fisheries.

Effective fisheries management requires that managers work towards clearly specified objectives. These may be biological, economic and social. Thus, the fisheries scientist has to understand links between different disciplines and the ways in which science can usefully inform the manager. This chapter introduces the history and diversity of the world's fisheries, their current status and the main problems they face. This provides a basis for suggesting why we need to conserve fisheries and the marine environment and identifies the main objectives of fishery management.

1.2 Fisheries of the world

1.2.1 History of fisheries

Fisheries in ancient civilizations
Humans living in coastal areas have always eaten marine organisms. Initially, animals and plants were simply collected by hand from the shore, but more effective fishing methods were soon developed. Fish hooks fashioned from wood and bone have been found

at sites dated 8000 BC, and there are references to fisheries in Greek, Egyptian and Roman texts. In Egypt, nets and spears were in use by 2000 BC. As the Pacific Islands were colonized, fish provided protein on islands with few other animal resources, and the successful migration of Melanesian and Polynesian people often depended on their ability to catch reef fish. Fishers were aware of cycles in the abundance of species they caught, and the Greeks used storage ponds and fish farms to ensure a continuity of supply. Latterly, fish could be preserved by salting and drying, allowing fishers to work further from their home ports and fish products to be traded and exported.

Pre-industrial fisheries
As nations developed their seafaring skills and began to explore the oceans, they discovered abundant fish resources. Explorers reported that huge numbers of cod *Gadus morhua* (Gadidae) could be found off Newfoundland, and by the early 1500s, French and Portuguese fishers were already crossing the north Atlantic Ocean to fish for them. The cod were caught with baited hooks, dried and shipped to the Mediterranean countries and the Antilles where they were known as **bacalao** and fetched high prices. Subsequently, English vessels joined the fishery. The countries that fished for cod were the major sea powers of the time and fought to control trade routes. They were often at war and many fishing vessels were lost. In the Anglo-Spanish war of 1656–1659, 1000 English vessels were sunk. Cod were such a valuable commodity that the vessels were also targeted by pirates. Moroccan pirates would rob vessels returning to Mediterranean ports, and the Sallee rovers, French, Spanish and English pirates under the Turkish flag, attacked fishing boats in so many areas that the fishers eventually sailed in convoy for protection. In later years, the cod fishery on Georges Bank and Grand Bank was increasingly

fished by boats from New England ports. By the mid-nineteenth century, cod were caught with hook and line from fleets of eight or more small dories that transferred their daily catches to a larger schooner for storage in ice and salt. In 1880, some 200 American schooners with eight or more dories were operating (Cushing, 1988a; Kurlansky, 1997).

In Europe, pre-industrial fisheries flourished in France and in countries bordering the North Sea. Sardines *Sardina pilchardus* (Clupeidae) were caught with fine nets off the French Breton coast from the seventeenth century. Initially they were sold fresh but with the development of the oil press, sardine oil was distributed all over Europe. Canneries were opened from 1822. There were around 3000 boats in the fishery at the end of the nineteenth century. A tunny *Thunnus thunnus* (Scombridae) fishery developed in the same area. Five hundred boats would spend up to 2 weeks at sea and catch tunny with trolled lures. The tunny could also be canned successfully. Swedish (Scanian) fishers started to catch Atlantic herring *Clupea harengus* (Clupeidae) from the North Sea in the eleventh century, and by the late sixteenth century, large fleets from Sweden, Holland, England and Scotland all fished for herring. The fish were caught in drift nets and preserved in barrels with salt. The fishing industry would move from port to port as they followed the shoals of herring on their seasonal migrations (Cushing, 1988a).

Industrialization

The power and range of fishing vessels increased rapidly at the time of the Industrial Revolution, as did the demand for fish and fish products. In the 1860s paddle tugs powered by steam were first used instead of sailing boats to drag fishing nets in the North Sea. Calm weather no longer limited fishing activities, and catch rates were four times those of sailing vessels. At much the same time, steamers started to fish for Atlantic menhaden *Brevoortia tyrannus* (Clupeidae) off the east coast of the United States. From the 1840s, there was rapid development of industry and growth of the US economy. A modern farming industry was needed to feed the growing urban population, and fishmeal was a potentially cheap and accessible source of high-protein animal feed. The first **industrial fisheries** began to catch small and abundant shoaling fish that were ground and dried to make fishmeal. In 1840, Charles Mitchell of Halifax Nova Scotia started to preserve fish in hermetically sealed tins. Such canned products could be dis-

tributed and sold throughout the world. The markets for fish became more accessible as rail and road transport improved and consumers were increasingly concentrated in cities where they sought work in factories and service industries.

Similar patterns of development followed elsewhere as farmers demanded fishmeal, and urban populations demanded food. Fishery landings rose rapidly and continued to do so until the present time. Although we have seen that European vessels fished across the Atlantic for several centuries, steel fishing vessels with diesel engines and cold storage facilities made all the oceans accessible to fishers who could remain at sea for months at a time. These were the so-called '**distant water**' or '**high seas**' fleets. Japanese vessels fished the tropical and subtropical oceans for tuna, the former Union of Soviet Socialist Republics (USSR) fished for krill *Euphausia superba* (Euphausiidae) in the Antarctic, and Alaskan pollock *Theragra chalcogramma* (Gadidae) in the North Pacific. The greatest expansion in high-seas fisheries took place after World War II and the fleets, mostly from the former USSR, Japan, Spain, Korea and Poland, were catching between 7 and 8 million tonnes by the 1970s, over 10% of global landings at that time.

Control of the high seas

Traditionally, the oceans were regarded as common property and fishers were free to go where they liked. Although marine tenure systems were in place around many tropical islands, and fishers in temperate waters protected their local fisheries from outsiders, the 'ownership' of marine resources rarely extended more than a few miles from land. From the fifteenth century until the 1970s there was little restriction on where fishing took place in the world's oceans and what was done. It may seem strange that countries fished so freely in oceans many miles from their home ports when hunters could not take terrestrial mammals from countries without invading them and evading or repressing the local people, but the perception that land is owned and that the sea is open to all has persisted in many societies and for many centuries.

The existence of '**freedom of the seas**' had a powerful influence on the development of the world's fisheries. It was formalized in the sixteenth and seventeenth centuries when world powers decided to resolve ongoing and expensive conflicts over trade routes by allowing multinational access. In 1608, Hugo Grotius defended Holland's trading in the Indian Ocean in *Mare Lib-*

erum, and used fisheries to support his arguments for free access. He suggested that fish resources were so abundant that there would be no benefits from ownership and that large areas of national jurisdiction could never be defended. His argument prevailed and freedom of fishing became synonymous with freedom of the seas (FAO, 1993a; OECD, 1997).

However, as human populations grew and fisheries became more intensive, there was increasing conflict over fisheries. Scotland claimed exclusive national rights to inshore waters in the fifteenth century, and other countries followed. However, there was no international agreement on the size of territorial waters. The 1930 Hague Conference on the Codification of International Law decided that the claims to territorial seas were acceptable but did not suggest how large these claims could be. In practice, most countries claimed no more than a few kilometres of inshore waters while the high seas could still be fished by any nation with a suitable vessel.

Following World War II and two United Nations (UN) Conferences on the Law of the Sea, coastal states had greater expectations about the limits of their authority over fisheries. Many states increased their fishery limits to 12 miles (19.3 km) and by 1972, 66 countries had 12-mile limits. Even this did not satisfy countries with important fishing interests who were concerned about the status of 'their' stocks, and countries such as Iceland, where cod were fished by British and German distant water fleets, went to international court in an attempt to impose extended unilateral limits. This claim for jurisdiction was opposed by foreign fleets and led to the 'cod wars' of the 1970s (Burke, 1983).

Most 12-mile claims had little impact on the activities of high-seas fleets, but nations were increasingly concerned that they were not in control of their most accessible fish resources. Further territorial claims were inevitable. As early as 1947, Chile and Peru had claimed 200-mile jurisdiction, and by 1972 another eight countries had joined them. At the 1973 UN Conference on the Law of the Sea, the right to 200-mile limits was formalized. By 1974, 33 countries claimed this **extended jurisdiction** and another 18 joined them in the next 3 years. Since 90% of global fisheries yield was taken within 200 miles of the coast, these changes brought 90% of fisheries under national control. The catches of the high-seas fleets fell and many small island states were suddenly in control of vast fishery resources (Burke, 1994). The tiny Seychelles Islands in the Indian Ocean, for example, with a land area of 455 km^2 and population of 75 000, were now responsible for fisheries in an Exclusive Economic Zone (EEZ) of 1 374 000 km^2. Kiribati in the South Pacific, with a land area of 690 km^2 and population of 72 000, took control of an EEZ of 3 550 000 km^2, which included rich tuna fishing grounds. In subsequent years, small island nations would receive license payments from larger and wealthier nations who wanted to keep fishing in their EEZs. However, extended jurisdiction did not bring an end to battles over fisheries, and they continue to ignite conflicts between otherwise friendly fishers and nations (Fig. 1.1).

Fig. 1.1 Fisheries continue to ignite conflict between individuals and nations.

Recent trends in fisheries

Marine capture fisheries now yield 86 million tonnes per year, more than four times that from freshwater fish production and aquaculture (FAO, 1999). The first sale value of fished marine species is around $US50 billion. Fished species include vertebrates, invertebrates and plants. Fish dominate landings and account for 80% by weight of the total (Box 1.1). Although there are around 17 000 species of marine fish, 50% of fish landings are composed of just 20 species. The other groups that dominate landings are molluscs and crustaceans that are commonly referred to as shellfish.

Few areas of the ocean remain unexploited and economic costs rather than technology now limit fishing power. Around 3.5 million fishing vessels are in use and most shelf seas to depths of 200 m or so are heavily fished. The main fish-producing nations are Peru, Japan, Chile, China and the United States. Target spe-

cies are generally decreasing in abundance, but the high efficiency of modern fishing vessels makes catching them worthwhile. In the developed world, even the smallest inshore fishing vessels carry acoustic devices to provide images of the seabed and locate shoals of fish, accurate navigation systems to pinpoint fishing grounds and powered drums and blocks to haul fishing gear.

Small-scale and artisanal fishers working from the shore, canoes or small boats account for 25% of the global catch and more than 40% of the catch for human consumption. The highly mechanized fisheries of the developed world contrast dramatically with subsistence fisheries in poorer countries where fish are a vital source of dietary protein and fishing may provide the only source of income (Fig. 1.2). In highly populated coastal regions of South-East Asia, Africa and Central America, people turn to the sea as a last possible source of food and income when no alternatives are available. The fishers are usually very poor, have little equipment and are sufficiently desperate to support themselves and their families that they do anything to maintain daily catches. In practice this may mean fishing with dynamite or poisons that destroy the fished habitat.

Despite the collapse of individual fisheries and the demise of some high-seas fleets, new resources have been exploited and total landings have risen steadily since 1950 (Fig. 1.3). Landings have increased because fishing effort has increased in response to growing demands for fish. Global landings are dominated by low-value species, such as anchovies and pilchards, which are mostly converted to fishmeals and oils for

---

Box 1.1

Catches and landings.

The fish and invertebrates that fishers bring ashore are often called **'catches'** or **'landings'**. Strictly speaking, these terms are not synonymous, since much of the catch is discarded at sea and never landed (Chapter 13). Landings statistics compiled by the FAO and other organizations reflect the quantities of fished species brought ashore, and not necessarily the quantities caught.

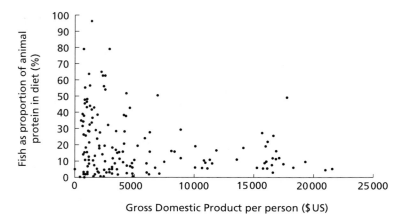

Fig. 1.2 Relationship between the proportion of fish protein in human diets and the relative wealth of nations where those people live. Wealth is measured as the total gross domestic product (GDP) divided by the population size. GDP is the total income of a country from all sources. After Kent (1998).

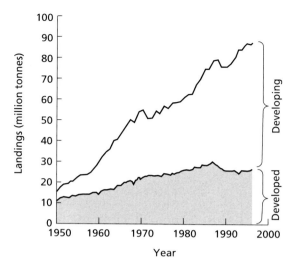

Fig. 1.3 Landings of marine fishes and invertebrates by economically developed countries (shaded) and developing countries (unshaded). Data from OECD (1997) and FAO (1999).

rose and then fell with the imposition of extended jurisdiction (Fig. 1.5). At present, landings in less developed nations are rising faster than those in developed ones. This is partly because most resources in developed countries have already been located and fished as heavily as they can with the available technology. In developing countries, conversely, the availability of new technology and external funding has allowed access to new and lightly exploited resources. Developing countries took 27% of global landings in 1950 but now take over 60% (OECD, 1997).

Since 1950, the effort (e.g. time at sea or fuel consumed) needed to catch a given weight of fish has increased. Technological development and increases in fleet size meant that total yield rose, but for many vessels fishing was no longer cost effective and government subsidies kept vessels at sea and fishers in jobs. Of course, their continued activities further depleted stocks and made fishing even less profitable. In 1989, the former USSR fleet was estimated to have operating costs of $US10–13 billion and yet their landings of around 10 million tonnes of low-value species were worth $US5 billion. This left a deficit of $US5–8 billion, to be met by subsidy (FAO, 1994).

farming and food manufacture (Fig. 1.4). Despite the steady rise in overall landings, landings of individual species fluctuate widely. Catches by high-seas fleets

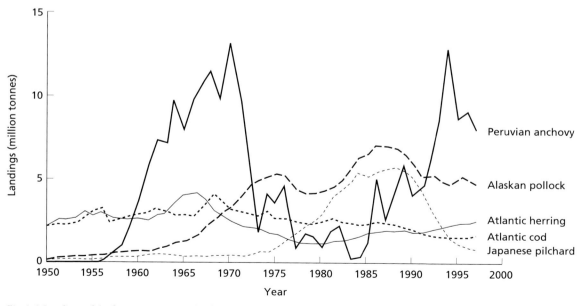

Fig. 1.4 Landings of the five marine species that have dominated global landings since 1950. These species are: Peruvian anchovy, *Engraulis ringens* (Engraulidae); Alaskan pollock, *Theragra chalcogramma* (Gadidae); Atlantic herring, *Clupea harengus* (Clupeidae); Atlantic cod, *Gadus morhua* (Gadidae); and Japanese pilchard, *Sardinops melanostictus* (Clupeidae). Data from FAO (1995a, 1999).

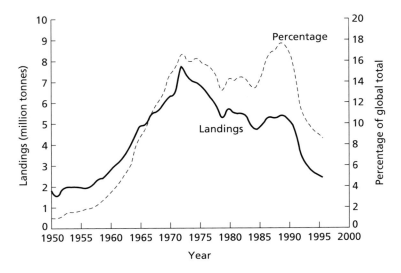

Fig. 1.5 Landings by distant water fishing nations (high-seas fleets) since 1950 and their landings as a proportion of the global total. After FAO (1999).

1.2.2 Fishery science

Fishery science has been recognized as a scientific discipline since the late 1850s, when the Norwegian government hired scientists to find out why catches of Atlantic cod fluctuated from year to year (Smith, 1994). Cod had been caught off the Lofoten Islands in northern Norway for several centuries and the local people relied on them for food and income. As in the Newfoundland fishery, cod were caught with lines and preserved by drying. The money made from exporting dried cod was used to import other foods and to pay traders and bank loans in southern Norway. If the cod fishery failed, then the people of northern Norway would default on their bank loans. For the Norwegian government, fluctuations in the cod fishery caused economic and political strain.

Soon, fishery scientists were also working in Germany, Sweden and Russia, and within 20 years fisheries research was in progress in the United States, Denmark, the Netherlands and the United Kingdom. Scientists were increasingly concerned that fluctuations in abundance could be driven by fishing. At the Great International Fishery Exhibition, held in England during 1883, Thomas Huxley, then president of the Royal Society, declared that 'the cod fishery, the herring fishery and probably all the great sea fisheries are inexhaustible; that is to say, that nothing we do seriously affects the numbers of fish'. Although his statement is often used to suggest that people were generally unaware of the finite limits to natural resources at that time, Huxley himself had qualified many of these remarks, and many fishers in Europe had reported declining catch rates. Moreover, tens of generations of people who depended on the reef fish resources of tropical islands were well aware they could not sustain intensive harvesting and were forced to move to new islands as they depleted fish stocks. There was a growing scientific consensus that research was needed to identify the effects of fishing, and new laboratories were established in Europe and North America (Cushing, 1988a; Smith, 1994).

By the end of the nineteenth century, scientists had developed techniques for ageing fish and tagging them to follow their migrations, but had not shown why catches varied. Work focused on whether killing small fish, before they were able to spawn, would cause catches to fall, whether fishing affected the size and abundance of fish, and what could be learnt from the collection and analysis of catch and effort data from the fishery. The major breakthrough came early in the twentieth century, when a Norwegian scientist, Johan Hjort, showed that the abundance of a year class of fish was established within the first few months of life and that renewal of fish stocks did not take place by a constant annual production of young but by highly irregular annual production with a few fish surviving in most years and many fish surviving in a very few years.

Changes in the rate of survival would cause fluctuations in catches. Soon, fisheries scientists would predict catch rates in some of the larger fisheries by measuring the abundance of young fish in the years before they grew large enough to be caught by fishers.

The number and efficiency of vessels fishing increased rapidly in the years before World War I and the fish caught were getting smaller and less abundant. During World War I, military activity led to a drop in fishing effort, and when the war ended in 1918 fishers caught larger fish and enjoyed higher catch rates. It was clear that the reduction in fishing effort during the war allowed fish populations to recover from the effects of fishing, and fishery science increasingly focused on understanding these effects. By the 1950s, 100 years after fisheries science had begun, models that described the dynamics of fish populations and their responses to fishing were developed by scientists working in the United States, England and Canada. These predicted the yields from fisheries when fish were killed at different rates and provide the background to much of the science we describe in this book. In subsequent years, fishery science would play a leading role in the field of ecology as a whole, describing why populations fluctuate and how population structure changes in space, time and in response to increased rates of death imposed by humans.

Many countries had interests in the same fisheries and there was strong international cooperation between fishery scientists. In 1902, the International Council for the Exploration of the Sea (ICES) was established in Europe and set up a number of international programmes to examine the effects of fishing. Despite the cooperation between scientists, only two fisheries were managed internationally before World War II, the Pacific halibut *Hippoglossus stenolepis* (Pleuronectidae) and Baltic plaice *Pleuronectes platessa* (Pleuronectidae).

After the war, many new scientific and regulatory bodies were established. The International Commission for North-west Atlantic Fisheries (ICNAF) and the Inter-American Tropical Tuna Commission (IATTC) formed in 1949, to be followed by a series of Food and Agriculture Organization (UN) commissions.

In the late twentieth century, concerns about fish stock collapse, the uncertainty that underlies fisheries assessment and the recognition that management objectives do not relate solely to the fished stock, has led to a diversification of fisheries science. There is increasing input from sociologists and economists who consider how different management strategies will affect the lives and incomes of fishers and associated communities. There is also greater emphasis on the effects of fishing on the marine environment and the impacts on species or habitats of conservation concern. Fisheries laboratories are found in many countries around the world, many fisheries are assessed by more than one country and international fisheries statistics are collected by the Food and Agriculture Organization of the United Nations (FAO) (Box 1.2).

1.2.3 Diversity of fisheries

The sheer diversity of fisheries makes fisheries science a fascinating field. Fishers are found on almost any coastline and working in any ocean. At one end of the spectrum a woman will fish daily with hook and line from a coral reef flat in Kiribati to supply food for her family. She winds the line on a simple spool and holds the bait and catch in a bag of woven palm leaves. She collects small crabs from the shore to use as bait and may glean shellfish for food at the same time. Most of her fishing is done within a few hundred metres of her village and total investment in fishing equipment is

---

Box 1.2
International assessment and management of fisheries.

For the purposes of collecting fisheries data, the FAO divides the world's oceans into Statistical areas shown in Fig. B1.2.1 and listed in Table B1.2.1. International fishery management areas are not necessarily the same as the FAO statistical areas. There are many international fisheries organizations that try to develop rational management policies for fish that migrate across national boundaries. Some examples are given in Table B1.2.2.

References: FAO (1994, 1999) OECD (1997).

Continued p. 8

Box 1.2 (*Continued*)

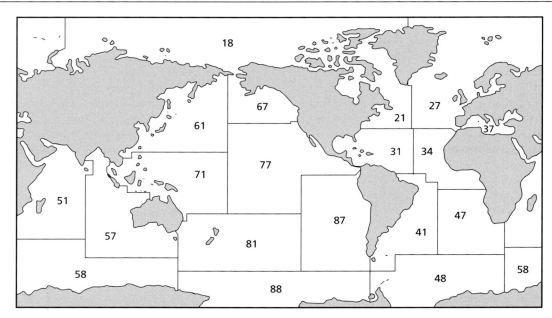

Fig. B1.2.1 Boundaries of FAO statistical areas.

Table B1.2.1 FAO statistical areas.

Area code	Statistical area	Area (million km²)
Atlantic Ocean and adjacent seas		
18	Arctic Sea	7.3
21	North-west Atlantic	5.2
27	North-east Atlantic	16.9
31	Western Central Atlantic	14.7
34	Eastern Central Atlantic	14.0
37	Mediterranean and Black Sea	3.0
41	South-west Atlantic	17.6
47	South-east Atlantic	18.6
48	Atlantic Ocean–Antarctic	12.3
Indian Ocean and adjacent seas		
51	Western Indian Ocean	30.2
57	Eastern Indian Ocean	29.8
58	Indian Ocean–Antarctic	12.6
Pacific Ocean and adjacent seas		
61	North-west Pacific	20.5
67	North-east Pacific	7.5
71	Western Central Pacific	33.2
77	Eastern Central Pacific	48.9
81	South-west Pacific	28.4
87	South-east Pacific	30.0
88	Pacific Ocean–Antarctic	10.4

Continued

Box 1.2 (*Continued*)

Table B1.2.2 Examples of International fisheries organizations.

Organization	Founded	Countries	Fisheries
International Pacific Halibut Commission (IPHC)	1923	United States, Canada	Pacific halibut
North-west Atlantic Fisheries Organization (NAFO)	1979	European countries, United States, Canada	Pelagic and demersal stocks
North-east Atlantic Fisheries Commission (NEAFC)	1963	European countries, Former USSR, Eastern Europe, Iceland, Norway	Pelagic and demersal stocks
Inter-American Tropical Tuna Commission (IATTC)	1949	North, Central and South American countries, France, Japan, Vanuatu	Tuna
Convention for the Conservation of Southern Bluefin Tuna (CCSBT)	1994	Australia, Japan, New Zealand	Bluefin tuna
Commission for the Conservation of Antarctic Living Marine Resources (CCAMLR)	1982	Most economically developed countries	Various

probably less than $US1. At the other extreme, a purse seining vessel pursues schools of tuna over the entire Pacific Ocean using satellite navigation devices. The vessel carries a helicopter to search for fish, a crew of 35 and costs $US40 million. One catch may weigh 30 tonnes and can be frozen to −20°C in a few hours. After several weeks at sea the frozen fish are delivered to a processing plant where they are thawed, cooked and canned for distribution to all the continents of the world. Between these extremes, fishing ranges from trawling for prawns in the muddy waters off Louisiana in the United States to hand-netting fish for the aquarium trade in Sri Lanka.

We will introduce many fishing methods in subsequent chapters, but hope that Figs 1.6 and 1.7 capture some of their variety. It is vital to appreciate this variety because fisheries provide different things for different people and thus the objectives of management differ too. Thus, the provision of a constant food supply may be essential for the small fishing villages in Kiribati while maximum profitability may be essential for a multinational company that catches tuna for canning. Despite the diversity of fisheries, many common principles describe the way in which fished species respond to fishing and the effects of fishing on the environment.

Fisheries scientists realize that research in tropical waters can usefully inform scientists working in temperate waters and vice versa. We have attempted as far as possible to use examples that reflect the global diversity of fisheries.

1.3 Patterns of exploitation

1.3.1 Boom and bust

Fisheries are not static, as any study of the history of exploitation will show. Fisheries develop, some fish stocks collapse, some fishers get rich, other fishers go bankrupt and move elsewhere and into other jobs. An examination of numerous fisheries around the world shows that when a potentially fishable resource is discovered, new fishers start to explore its potential and begin exploitation as soon as possible. Indeed, on an individual basis, the faster they can exploit the new fishery the more income they will receive. Other fishers who find out about the new fishery and think it has more potential than their own will join the race to fish there. Eventually, there will be too many fishers chasing too few fish, the stock will be depleted and catch rates and profits will fall. Once fishing is no longer profitable,

(a)

(b)

(c)

(d)

Fig. 1.6 The diversity of fisheries. (a) Fish traps set in the Lupar Estuary in Sarawak to catch prawns and small fish (see Blaber, 1997), (b) cast-netting for prawns in a small estuary near Mukah, Sarawak, (c) a tuna purse seiner trans-shipping catches to a freezer vessel in the Seychelles, and (d) tuna fishing from a small boat off Cape Verde Islands (d). Photographs copyright S. Blaber (a, b), S. Jennings (c), M. Marzot (FAO photo, d).

the fishers who can will redirect their energies, but those who cannot may continue to fish the depleted stock. To avoid the unemployment and social costs from fishers becoming bankrupt, governments may subsidize fishers to continue exploiting the depleted stock. By this stage, production of the stock is minimal and it will either recover because fishers shift their attentions elsewhere, or collapse, because fishers have no choice but to keep fishing.

Fishers can exploit fisheries at rates exceeding their capacity for replenishment. As this occurs, so the fishery passes through a series of phases (Fig. 1.8). These can be described as fishery development, full exploitation, over-exploitation, collapse and recovery (Hilborn & Walters, 1992). During each of these phases, there are clear trends in the abundance of fished species, fleet size (or fisher number), catch and profit. While the trends

rarely follow the smooth and necessarily stylized curves shown in Fig. 1.8, the general patterns are consistent.

As a new fishery develops so more fishers or boats enter the fishery because it is profitable for them to do so. The fishers who start fishing first, or catch most with least effort, will make the greatest profit. During **fishery development**, the effects of fishing are seen as a slight reduction in catch rates and the size of individuals in the catch. Total catch will rise as fishing effort increases. Prior to industrialization and the human population boom on many tropical coasts, fisheries often stayed in the 'development' phase and replenishment was easily fast enough to counter the effects of fishing. As more fishers enter the fishery, it passes through the fully exploited phase. Here, abundance falls and total catch increases. During the **fully exploited** phase replenishment is sufficient to maintain fish production and hence

Fig. 1.7 More diversity of fisheries. (a) Unloading sardines from a coastal purse-seiner in the Portuguese port of Aveiro, (b) a Japanese squid jigger sailing to fishing grounds off Hokkaido, (c) fishing with dynamite on a reef in the Philippines, and (d) a large catch of demersal (bottom dwelling) fish on a Russian stern trawler. Photographs copyright *Fishing News International* (a, b, d) and T. Heeger (c).

catches. If yet more fishers enter the fishery, their catching capacity will exceed the rate of replenishment (**over exploitation**). This usually leads to a fall in profits because more fishers are competing for a dwindling resource. As profits tend to zero, so fishers stop entering the fishery. Total catch will peak close to the time when the number of fishers reaches a maximum, and subsequently fall as abundance and the capacity for replenishment is reduced. If fishing effort is not reduced then the fishery will ultimately **collapse**, with marked falls in abundance and catch. Following collapse, the fishery will no longer be profitable. If fishers can afford to leave the fishery they usually will, and **recovery**, the rebuilding of stock biomass, may begin. In other cases, the fishers' only option will be to stay because they have to keep fishing to pay loans on boats and equipment.

We will now look at four examples of the many fisheries that have followed this common pattern. They are for different species, in different countries and have been overexploited on different time scales.

The Chilean loco

The loco *Concholepas concholepas* is a muricid gastropod (family Muricidae, subclass Prosobranchia). It is collected from the seabed by divers and is the most economically important shellfish in Chile. Loco are widely eaten in Chile and provide a very valuable export to Asian markets. From 1960 to 1975, 2000–6000 tonnes of loco per year were collected and sold for domestic consumption. The fishery appeared to be sustainable, and divers could make a good living from the fishery. However, in 1976, divers began to collect loco for sale and export. The fishery developed rapidly as more fishers became loco divers and by 1981 annual export revenue

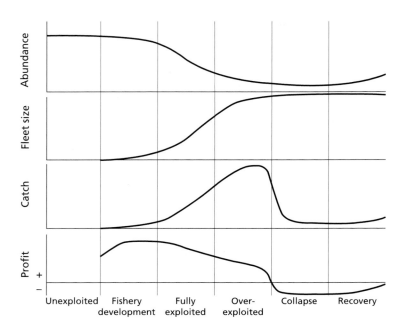

Fig. 1.8 Trends in the abundance of fished species, fleet size or fisher number, catch and profit as a fishery is developed and exploited through to collapse and recovery. Based on Hilborn and Walters (1992).

had exceeded $US20 million. Landings peaked at 25 000 tonnes in 1980, but the abundance of loco and the catch rates of many divers were falling fast. In the mid-1980s the fishery collapsed, with divers losing their incomes or jobs and coastal communities suffering from reduced economic activity and trade. The Chilean government intervened and closed the fishery completely in 1989. When it was reopened, strict catch limits were imposed. The fishery is now recovering slowly (Bustamante & Castilla, 1987; Castilla & Fernandez, 1998).

The Californian soupfin shark
Before 1937, the soupfin shark *Galeorhinus galeus* (order Carcharhiniformes) was caught in small numbers off the Californian coast. The flesh was sold as shark fillet and the fins were dried for export to Asian markets where they were used in shark fin soup. In 1937, nutritionists discovered that the livers of soupfin shark were the richest known source of vitamin A. Vitamin A was in great demand as wealthier Americans began to worry about what they were eating and considered the health benefits of vitamins and minerals. Fishers knew that soupfin shark were abundant, but demand for the livers was so high that the price of sharks rose from $US50 to $US1000 within a few years. The fishery developed rapidly, and many boats switched from

other activities to catch sharks. Hook and line was the favoured fishing method in early years, but latterly nets were used as these increased the catch rates. Landings peaked in 1939 at over 10 times the levels prior to 1937, and 600 boats were operating in the fishery. These catch rates were not sustainable and the fishers found it increasingly hard to catch sharks as abundance fell. By 1944, the fishery had effectively collapsed. Many fishers became bankrupt and others reverted to their former fishing habits (Ellis Ripley, 1946).

The Galápagos sea cucumber
Sea cucumbers (Holothurians) have been exploited throughout the Pacific for several centuries, and provide many examples of boom and bust fisheries. Sea cucumbers, often traded as *bêche-de-mer*, are cooked and dried after they have been collected from the seabed by divers and can be stored dry for many months. They are mostly exported to markets in Asia and the Pacific rim countries, where they are used to thicken and flavour soups and other dishes. Until the 1990s, the Galápagos Islands was one area where the fishers had not been. However, the waters around Galápagos had been increasingly fished by Korean tuna vessels and fishers learnt that the sea cucumber *Isostichopus fuscus* (class Holothurioidea) was abun-

dant. Exporters soon established a market, and in 1992 there was a sudden and massive expansion of the sea-cucumber fishery. Fishers could make more money from a day of sea-cucumber diving than from a month spent catching fin fish for local markets. Local people switched from other types of fishing to dive for sea cucumbers and when people in mainland Ecuador saw how much money could be made they flocked to the islands with their families. The fishers, known locally as pepineros, set up camps for cooking and drying the sea cucumbers on protected islands in the National Park. The camps caused great concern for conservationists who wanted to prevent the introduction of new species and saw pepineros arriving with rats, cats and dogs. The influx of money to the Galápagos changed the lives of many Ecuadorians, and traders became very rich. At one stage a local prostitute was demanding payment in sea cucumbers. The Ecuadorian government attempted to ban the fishery, but the momentum was too strong. Faced with violent protest that threatened the island's stability and the valuable tourist industry, and with the deliberate slaughter of rare tortoises in the National Park, the government revoked the controls and the fishery continued unabated.

The Galápagos fishery grew so fast that the whole cycle of fishery development through to collapse was complete in only 5 years. Sea cucumbers are now so scarce that fishing them is not worthwhile, and the pepineros who moved to the islands have to make money in other ways. This has led to increased fishing pressure on other resources which were formerly seen as sustainable.

The Canadian cod

Our final example shows the cycle from development to collapse over a much longer time scale. In section 1.2.1 we saw that cod, *Gadus morhua*, were caught off Newfoundland in the early 1500s, but in 1992, after almost five centuries of fishing, these were fished to commercial extinction. Cod were largely caught with hook and line until the early 1900s, and although the fishery supported many fishers and provided good economic returns, the replenishment of the stock was fast enough to sustain the fishery. In later years, more effective fishing techniques such as bottom trawling, traps and gill nets were used. Landings of the so-called northern cod increased from 100 000 to 150 000 tonnes per year from 1805 to 1850 and reached 200 000 tonnes

at the end of the nineteenth century. Factory freezer trawlers from Europe started fishing northern cod in the 1950s and landings rapidly increased from 360 000 tonnes in 1959 to a peak of 810 000 tonnes in 1968. In 1977 Canada extended jurisdiction to 200 miles and took over management of the stock. The stock was already overexploited and landings were now 20% of 1968 level. Under Canadian management the stock continued to be fished at low but increasing levels and in 1992 the stock finally collapsed and the Canadian government closed the fishery. The size of the spawning stock had fallen from an estimated 1.6 million tonnes in 1962 to 22 000 tonnes in 1992. The collapse was a disaster for a Canadian fishing industry that had few other resources to turn to, and the government had to support many fishers with welfare payments (Hutchings & Myers, 1994).

Other fishery collapses

Our examples are not unique. Similar stories of development and collapse apply to whales, fur seals, sea-lions, turtles and other species of fish. Marine mammal and turtle populations were hunted to such low levels that they contributed a fraction of 1% to global fisheries yield in the 1990s, and many species are protected by conservation legislation. The indirect impacts of fishing on these species are now the main concern, and we deal with these, rather than the few remaining fisheries directed toward them, in this book. However, the history of whale fishing can provide some important lessons for the fisheries scientist. Sperm and right whales, which were the most accessible to fishers with simple harpoons and rowing boats, were hunted from the twelfth century. As their populations declined, and new technologies such as powered vessels and explosive harpoons were developed, so the whalers could target larger and faster species such as blue, fin and sei whales. These populations were also hunted much faster than they could reproduce and became increasingly scarce. Thus, the whalers would reduce one species as far as possible with the available technology and then move on to another. This is shown very dramatically in Fig. 1.9, which illustrates catches of five species in the Antarctic Ocean (Allen, 1980). These landings data show fisheries for the blue whale developing and collapsing, to be followed by similar development and collapse for fin, sei and minke whales. Later in this book we will see that fishers often maintain yields by

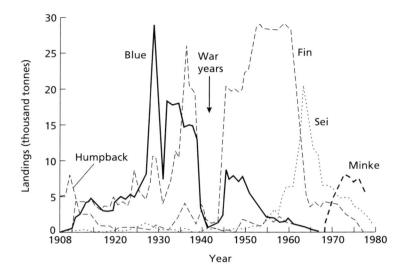

Fig. 1.9 Landings of whales in the Antarctic Ocean. Catches of all species dropped during World War II. After Allen (1980).

shifting from species to species as the most desirable ones are depleted.

1.3.2 Conservation and ecosystem concerns

For many years, fisheries scientists tried to provide advice that could be used to prevent the overexploitation or collapse of fished stocks. However, the increasing intensity of fishing throughout the world has had impacts on the marine ecosystem other than those on target species and these impacts are now the focus of many research and management programmes.

In many fisheries, species of fish, seabirds and marine mammals are caught on lines or in nets when the fishers are pursuing other species. Skippers of tuna fishing boats in the eastern Pacific Ocean would set their nets around dolphins because these often swam above tuna and marked their position. However, when the nets were hauled, many dolphins would be trapped and killed. As a result, populations of spinner *Stenella attenuata* and spotted *Stenella longirostris* dolphins fell to around 20% and 45% of pre-exploitation levels between 1960 and 1975 (Allen, 1985). Baited hooks on long-lines set in the Southern Ocean to catch bluefin tuna *Thunnus maccoyii* (Scombridae) and Patagonian toothfish *Dissostichus eleginoides* (Nototheniidae) would be taken by seabirds such as wandering albatross *Diomedea exulans* before they sank. These long-lived birds cannot reproduce as fast as they are being

killed and are now threatened with extinction. In the Irish Sea, the large common skate *Dipturus batis* (Rajiidae) is caught in towed nets that are used to catch other bottom-living species. Because the target species are usually much smaller than the common skate, the nets have relatively small mesh and any common skate that passes into the net will be caught. The common skate only reaches maturity after 12 years and has a very low reproductive rate. The abundance of these fish has been so reduced by fishing that they are rarely recorded in catches (Brander, 1981). Similar reductions have recently been recorded for other skate species in the same area (Dulvy *et al.*, 2000).

Other concerns about the ecosystem effects of fishing focus on the impacts of towed gears on benthic fauna and habitat, the impact of fishers who use dynamite and poisons on coral reefs, and the effects of fishery waste on populations of scavenging birds and fish. Fishing can also affect food webs by removing predators, such as fish that eat other fish, or prey, such as small shoaling fish that are eaten by seabirds and marine mammals. We consider these effects, what is known about them and what can be done to mitigate them.

1.4 Why manage fisheries?

Fisheries are managed because the consequences of uncontrolled fishing are seen as undesirable. These consequences could include fishery collapse, economic inefficiency, loss of employment, habitat loss or decreases in the abundance of rare species. Management is intended to maximize some specified biological, social or economic benefits from the fishery while minimizing costs. Management is usually imposed by an external regulator rather than the fishers themselves. It may seem surprising that external regulation is needed if the effects of uncontrolled fishing are always undesirable. Surely, fishers would regulate the fishery for their own benefit and would be happy to comply with regulations. By the time you have read this book we hope you will understand why fishers rarely manage their own fisheries and why their desires often conflict with those of the regulators.

1.5 Objectives of management

1.5.1 Range of objectives

If fisheries management is to work, the management objectives must be specified. Without clear objectives it is impossible to judge the success or failure of management or to design a management strategy. Fisheries could, for example, be managed to increase food pro

duction, income or employment, to conserve non-target species and habitats, to placate lobbyists or to encourage fishers to vote for you. For many years the objective of fishery management was to maintain the maximum biologically sustainable yield (in weight) from a fishery, with limited concern for social, economic and environmental factors. In many cases this objective was assumed rather than specified (Clark, 1985; Hilborn & Walters, 1992). Table 1.1 lists some objectives of fishery management. We have grouped them as biological, economic, social and political objectives although many have wider influence.

In order to determine whether objectives are met the manager will need indicators (Hilborn & Walters, 1992). Table 1.2 lists some examples of indicators. When indicators are truly quantitative it is much easier to judge the success of management than when they relate to emotional well-being or a lack of civil strife, although social scientists have increasingly proposed measures for such things. While some of these indicators are relatively easy to measure, it is much harder to set appropriate targets for them. Fisheries scientists do the science that allows managers to set these targets.

Table 1.1 Objectives of fishery management. Based on Clark (1985).

Objective	Biological	Economic	Social	Political
Protect habitat	*			
Increase selectivity	*			
Prevent mortality of rare species	*			
Prevent ecosystem shifts	*			
Rebuild overexploited stock	*			
Reduce discarding	*			
Maximize protein supply	*		*	
Maximize income		*		
Maximize profit		*		
Maximize employment			*	
Keep prices low		*		
Minimize variability in catch			*	*
Minimize variability in income		*	*	
Reduce overcapacity			*	
Raise government revenue		*	*	*
Improve catch quality		*		
Increase exports		*		
Do not upset lobby groups			*	*
Do not upset fishers			*	*
Do not upset conservationists			*	*
Preserve status quo			*	*
Reduce conflicts			*	*
Boost sport fisheries			*	

Objective	Indicator
Protect habitat	Proportion of habitat impacted by fishing gears
Increase selectivity	Proportion of target species in total catch
Prevent mortality of rare species	Mortality rate and abundance of rare species
Prevent ecosystem shifts	Abundance and diversity of indicator species
Rebuild overexploited stock	Abundance of overexploited species
Reduce discarding	Proportion or quantity of material discarded
Maximize protein supply	Weight of landings
Maximize income	Total income from fishery
Maximize profit	Total profit from fishery
Maximize employment	Number of people working in fishery
Keep prices low	Price of fished species
Minimize variability in catch	Variation in yield per unit time
Minimize variability in income	Variation in income per unit time
Reduce overcapacity	Number of fishers or vessels in fishery
Raise government revenue	Value of taxes collected from fishers
Improve catch quality	Price paid per unit weight of catch
Increase exports	Value and volume of exports
Do not upset lobby groups	Person days spent campaigning
Do not upset fishers	Number of complaints to government
Do not upset conservationists	Number of complaints to government
Preserve *status quo*	Rate of change in profits or employment
Reduce conflicts	Number of prosecutions for conflict
Boost sport fisheries	Value of recreational fishery

Table 1.2 Indicators that show whether a management objective has been met.

1.5.2 Balancing objectives

The objectives in Table 1.1 are not independent. Thus, an attempt to meet one objective is likely to compromise one or more of the others. For example, attempts to maximize catches from a trawl fishery that targets many species may mean that some species are overexploited. Should catches, income and employment be sacrificed to save a species of conservation concern? Conservation groups would say 'yes' and fishers may say 'no'. The most simplistic quantitative approach may be to express all objectives using a common unit (such as currency) and select the combination of objectives that maximize 'value'. Alternatively, objectives can be scored in other ways and the objectives set to maximize this score (Hilborn & Walters, 1992).

Most of the biological and economic models for fisheries that we introduce in Chapters 7–11 deal with a single objective, and the outputs are subsequently modified to account for objectives not specified in the model. For example, a government may ask a manager to improve the profitability of a fishery. The manager asks a fisheries scientist to run a bioeconomic analysis and this demonstrates that profitability will increase if 20 large vessels are licensed to operate from a single port and subsidies to the existing fleet of small boats are scrapped. However, the larger vessels will provide fewer jobs than the numerous small boats that currently land catches in small harbours all along the coast, and representatives of fishing communities begin to lobby government. Government, concerned that conflict with the fishers would result in bad publicity and jeopardize their chances of re-election, decides to allow 20 larger vessels to enter the fishery, but also subsidizes fishing communities so the small boats can keep fishing. The new policy is no longer based on the bioeconomic model but attempts to meet non-quantitative economic, social and political objectives. The result is that too many vessels are now chasing too few fish, and overall profitability of the fishery falls even further. Our example is extreme but not unrealistic. Most managers are dealing with multiple objectives and need a framework for making balanced decisions.

A number of **multiple-criteria decision-making** techniques (MCDM) can be used for making trade-offs between objectives. With a large number of possible objectives, a single optimal solution is unlikely, but at least the trade-offs become explicit rather than implicit

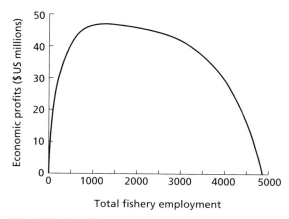

Fig. 1.10 Trade-offs between economic profits and employment in British fisheries of the English Channel. After Pascoe *et al.* (1997).

(Mardle & Pascoe, 1999). For example, an MCDM technique was used in a study of the English Channel fishery where three objectives were identified: to maximize economic profit, to maintain employment and to minimize discarding of quota species (Pascoe *et al.* 1997). **Quota species** are those target species for which catches are limited by managers. When catches exceed these limits, quota species have to be discarded at sea (often dead or dying) even if they are caught. Since the three objectives are measured in different units, they have to be normalized. Profit was expressed as a proportion of the maximum profit that would be made if profit were the sole objective, employment was expressed as a proportion of current levels, and discarding expressed as increases above zero. The analysis allowed managers to see what the optimal fleet configuration in this fishery would be when objectives were given different weightings. One important aspect of this study was that results were presented in a form that would be accessible to managers and fishers, and made the consequences of choices between objectives very clear. Thus, the trade-off between employment and profits was presented in graphical form (Fig. 1.10). Methods of MCDM and examples of their use are discussed in Mardle and Pascoe (1999). In reality, however, trade-offs between objectives are rarely quantitative. Rather, governments make such decisions by 'feel' based on consultation, scientific advice and their own political goals.

1.5.3 From objective to action

The first stage of the management process is to express the objective as a **management strategy**. The strategy should not be mumbo-jumbo that gives no explicit guidance (Hilborn & Walters, 1992). Rather, it should be easy for scientists, fishers and managers to see what the strategy is and whether it is being followed. Examples of clear strategies are those for the cod and Atlantic capelin *Mallotus villosus* (Osmeridae) fisheries around Iceland. For cod, the strategy is to catch 25% of the stock aged 4 years and older each year. For capelin, the strategy is to leave 400 000 tonnes to spawn each year (Jakobsson & Stefánson, 1998). Quantitative strategies may be difficult to set in relation to ecosystem objectives because the complexity of the ecosystem is difficult to describe in quantitative terms.

Management action is taken to implement the strategy. Management actions can be divided into catch controls, effort controls and technical measures. **Catch controls** limit the catches of individual fishers or the fleet as a whole, **effort controls** limit the numbers of fishers in the fishery and what they can do, while **technical measures** control the catch that can be made for a given effort. These would include mesh size restrictions, fishery closures and fishing seasons. There are also a range of other technical measures that may be used to address conservation and ecosystem related objectives.

1.6 Meeting management objectives

The fishery manager has to meet management objectives and the fisheries scientist provides the advice that makes this possible. To meet management objectives, the manager has to know how the fishery responds to different management actions. Most quantitative fisheries science is concerned with calculating the probability that a given outcome will result from a given management action. The scientific methods that have been developed help to answer questions such as 'how will yield and income from the fishery change if 10 more boats are given access?' and 'how many albatrosses can be caught accidentally by a long-line tuna fishery without threatening the population?'.

Historically, most fisheries scientists have concentrated on biological assessment. **Assessment** describes the mathematical approaches that are used to predict how a fished species will respond to different management actions. In providing the scientific advice that helps managers to meet specified objectives, single-species biological assessment is the most quantitatively advanced area of fisheries science, although the answers it provides are always uncertain and depend on the validity of many assumptions and the quality of input data. Biological assessment is, however, just one facet of a process that usually includes other biological, economic, social and political concerns.

The bioeconomic assessment of fisheries is also increasingly quantitative and can provide a basis for meeting specified management objectives such as maximizing profit or optimizing investment in a fluctuating fishery. However, in other areas, such as assessing the impact of fishing on the environment, ecosystem processes or species and habitats of conservation concern, predictive science is in its infancy. In these cases, we are often more concerned with describing the impact and asking whether it matters than with predicting exactly

how it will change in response to different management actions.

1.7 Structure of this book

This book introduces those aspects of fisheries science that should be understood in order to conduct assessments and provide the best scientific advice to managers. We consider fished species, fishers, the fished ecosystem and the links between them. Figure 1.11 summarizes the stages in the assessment and management process where fisheries scientists can inform and advise.

Three chapters deal with fished species, production and variability in the marine environment. We begin by describing how primary production is turned into production of fished species and how environmental factors lead to great variation in production processes (Chapter 2). Knowledge of the interplay between production and environment is essential if we are to explain uncertainty in the abundance of fished species.

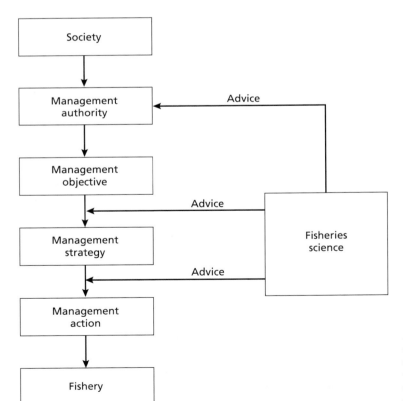

Fig. 1.11 Stages in the management of fisheries where fisheries scientists provide advice. They also receive feedback throughout the management process, especially through monitoring of the fishery.

In Chapter 3 we consider the range of species that are fished, their taxonomic relatedness, life histories and distribution. Such biological knowledge is necessary to understand how they respond to additional mortality imposed by fishing. In Chapter 4 we explain why the population structure of fished species changes in space and time and show that it is very variable, even in the absence of fishing.

Three chapters consider the fishers themselves, how and why they fish and the social and economic roles of fisheries. In Chapter 5 we describe the gears used to catch exploited species, their selectivity and the impacts they have on marine environment. We need to know how fishers are likely to respond to regulation and this is achieved by understanding their behaviour. In Chapter 6 we ask why fishers behave as they do and how they cooperate and compete. The remaining chapter in this group deals with economic issues. Fishing is a major economic activity and can be a major contributor to the wealth of some nations. Fishers are driven by the desire for profit and many of their decisions about when, where and how to fish are governed by economic factors. Fisheries cannot be understood, assessed or managed without an understanding of economics. The bioeconomics chapter (Chapter 11) does not follow Chapter 6 because knowledge of biological assessment methods is needed to understand bioeconomic assessment.

The four chapters on assessment cover the material that dominates existing fisheries texts, but we hope that its role will be better appreciated in the wider context of this book. Chapter 7 describes the application of methods for the assessment of single-species stocks; how stocks respond to fishing and the numerical methods for describing these responses and predicting catches that can be taken. Species interact by eating each other or competing for food and space, so multispecies and ecosystem analyses (Chapter 8) are a logical extension of the single-species approach. The assessment methods all require input data and in Chapters 9 and 10 we describe how such data can be collected. Good data collection requires good knowledge of the study species. In Chapter 9 we consider how to identify the stock and understand its dynamics and in 10, how to estimate the abundance of fished species by direct methods and with catch and effort data from the fishery.

The wider effects of fishing on the marine ecosystem are playing an increasingly significant role in fisheries management. Four chapters deal with these. Few people have not heard of the tuna–dolphin debate, the destruction of coral reefs with dynamite and poisons or the capture of endangered turtles, albatrosses and sharks. Without good science there is a risk that the debate becomes purely emotional and that managers behave as fire-fighters without looking at their overall objectives. Chapter 12 considers the direct and indirect effects of fishing on the structure and diversity of fish populations, their genetic structure and predator–prey relationships. In Chapter 13 we consider the amounts of fish discarded in different fisheries and the effects of discards on populations of scavengers such as birds and invertebrates. Chapters 14 and 15 deal with the ecological impacts of fishing and fishery interactions with birds and mammals, respectively.

The last two chapters of the book look at issues relating to the conservation and management of fisheries. In Chapter 16 we ask whether aquaculture is a viable way of producing a large amount of fish protein and reducing pressure on wild fisheries. In Chapter 17 we see how the results of the scientific studies and analyses we have described can be used to achieve specific management objectives, what management can really achieve and how the demands of different interest groups might be reconciled.

Summary

• Fisheries have expanded rapidly because a growing human population demands more food and because improved technology has simplified capture, processing, distribution and sale. Marine fisheries now yield around 90 million tonnes per year, more than 80% of global fish production.
• Many major fisheries have been overexploited, usually because too many fishers are chasing too few fish. Fisheries are not static: fisheries develop, fishers compete, some fisheries collapse, some fishers become bankrupt and move elsewhere.
• Overexploitation leads to reductions in fish production, income and employment. This is why governments intervene and manage fisheries.

Management is intended to maximize some specified biological, social or economic benefit from the fishery.
• For management to work, the objectives of management must be known. Without clear objectives it is impossible to judge the success or failure of management, or to design a management strategy. Objectives may be biological, economic and social. There is increasing emphasis on objectives that minimize the effects of fishing on the marine environment and reduce impacts on species or habitats of conservation concern.
• The fishery manager has the job of meeting management objectives and the fisheries scientist provides the scientific advice that makes this possible.

Further reading

The development of fisheries, their history and role in society are described by Cushing (1988a). Popular books by Jensen (1972) and Kurlansky (1997) give fascinating accounts of the role of the North Atlantic cod fisheries, the lifestyles of the fishers who crossed the Atlantic in small boats and the wars which they fought to control trade routes and fisheries. Kurlansky's book contains some useful recipes! Smith (1994) has written an interesting account of the history of fishery science that helps to place current work in context. We recommend this to anyone who wishes to pursue a career in fishery science. Hilborn and Walters (1992) contains excellent chapters on the objectives of fishery management and the role of fishery science.

2 Marine ecology and production processes

2.1 Introduction

Fished species are conspicuous and widely studied inhabitants of the marine ecosystem, and their capture provides food and income for millions of people. However, they account for only a small fraction of production and biomass in the marine environment. Production of fished species ultimately depends on the fixation of carbon by marine plants and its transfer along food chains. Marine scientists focus on carbon because it is incorporated into sugars, fats, cellulose and other structural molecules. Consumers break down these molecules to provide energy and create tissues. The food chains leading to the production of fished species may be very short, such as when sea urchins graze algae, or longer, when fishes eat smaller fishes which eat zooplankton which eat phytoplankton. Since the conversion of carbon from prey to predator tissue is inefficient, longer food chains tend to produce biomass less efficiently.

Biological production processes are governed by physical processes that operate on many scales. Both biological and physical processes are often unpredictable. The coupling of physical and biological processes governs the potential yield of fished stocks, the effects of fishing on ecosystem processes and helps explain why recruitment is so variable in space and time. In this chapter we describe sources of primary production in the marine environment, their variability and how primary production is coupled to physical processes on many scales. We proceed to discuss production at different trophic levels and the links between primary production and the production of fished species.

2.2 Primary production: sources and magnitude

Primary production is the photosynthetic fixation of carbon by chlorophyll-containing organisms and is measured as weight of carbon fixed per unit area per unit time. **Gross primary production** is the total carbon fixed while **net primary production** is the carbon that remains after losses due to respiration. There is some uncertainty about the magnitude of gross primary production in the marine environment, but most estimates suggest that photosynthesis by marine organisms produces $30-60 \times 10^9$ tonnes of organic carbon annually: around 40% of global primary production (Melillo *et al.*, 1993; Smith & Hollibaugh, 1993).

Primary producers are found in illuminated seas and coastal margins, where light provides the energy to drive photosynthesis. Light is rapidly absorbed and scattered by water, and most primary production takes place at depths less than 200 m. The **continental shelf**, usually treated as depths from 0 to 200 m, occupies only 7.5% of ocean floor, while the **abyssal plains** and trenches, where depths are typically > 4000 m, account for over 50% of ocean area (Fig. 2.1). Given that the mean depth of oceans is 3700 m, most of the ocean is never illuminated, and production at the surface and on continental margins fuels all other production. This leads to reduced fish production in deeper areas (Merrett & Haedrich, 1997).

The majority of carbon in marine systems is fixed by phytoplankton in the open ocean and coastal waters (Table 2.1). Macroalgae, mangroves, reef algae, seagrasses and marsh plants make a smaller contribution to global production, but their production per unit area is higher. Their contribution to total production is greatest in inshore and coastal systems which are often heavily fished and provide essential nursery habitats for many fishes. In our description of primary producers we focus on phytoplanktonic systems because of their overriding importance in driving the global production of fished species. There is some chemosynthetic primary production at hydrothermal vent sites in the deep sea

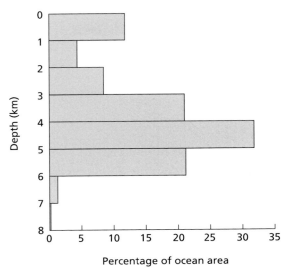

Fig. 2.1 The proportion of ocean area by depth. The mean depth of the oceans is 3.7 km and oceans cover 70% of the globe. Data from Sünderman (1986).

but this makes a minor contribution to global production and will not be considered here (see Felbeck & Somero, 1982; Grassle, 1986; Lutz & Kennish, 1993).

2.3 Phytoplanktonic production

2.3.1 Links between production and physical processes

Phytoplankton, consisting of autotrophic prokaryotes and eukaryotes, are responsible for most carbon fixation in marine systems. **Autotrophs** are organisms that obtain all their energy from inorganic material. The magnitude and variability of their production is deter-

mined by the coupling of physical and biological processes on scales of centimetres and seconds to thousands of kilometres and decades. Thus, sinking rates and diffusion can determine the uptake of nutrients to individual cells and global ocean circulation determines the distribution of upwellings of nutrient-rich water (Mann & Lazier, 1996).

By definition, plankton cannot maintain their distribution against the movement of water masses, and any mobility is generally overriden by the strength of currents. Plankton responsible for primary production range from bacteria to flagellates and diatoms of 20 μm diameter to dinoflagellates of 200 μm or more. They require light, nutrients and carbon dioxide to grow and multiply. Nitrogen (as nitrate), phosphorus (as phosphate) and iron may limit primary production in some circumstances (Martin, 1992). Light is only available close to the surface of the ocean but sensitivity to very high light intensities means that production may be lower at the very surface of intensely illuminated waters.

Light levels fall rapidly with increasing depth. Longer wavelengths are lost first followed by shorter ones (Fig. 2.2). Photosynthesis continues to a depth of around 200 m in the clearest oceanic water but only 40 m on coastal shelves and in upwellings. Since phytoplankton need light to photosynthesize it is important that they remain at depths where light is available. If the density of plankton cells exceeds that of the surrounding water they will sink. The rate of sinking is proportional to their weight and form (Smayda, 1970). If sinking is not countered by upcurrents the cells will fall from the photic zone. However, since phytoplankton rely on diffusion for the uptake of carbon dioxide and nutrients, sinking has advantages because it increases the flow of water over the cell surface. Swimming can

Primary producer	Area (10^6 km^2)	Area (% total)	Total NPP (10^9 t C y^{-1})	NPP (% total)
Oceanic phytoplankton	332.0	88.46	43.0	81.10
Coastal phytoplankton	27.0	7.19	4.5	8.49
Macroalgae	6.8	1.81	2.55	4.81
Mangroves	1.1	0.29	1.1	2.07
Coral reef algae	0.6	0.16	0.6	1.13
Seagrasses	0.6	0.16	0.49	0.92
Marsh plants	0.4	0.11	0.44	0.83
Microphytobenthos	6.8	1.81	0.34	0.64

Table 2.1 Production by marine algae. NPP is net primary production, the carbon which remains after losses due to respiration. From Duarte and Cebrián (1996).

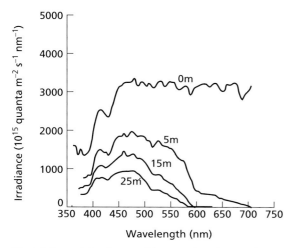

Fig. 2.2 Spectral distribution of downward irradiance at four depths in oceanic water. Based on Kirk (1992) after Tyler and Smith (1970).

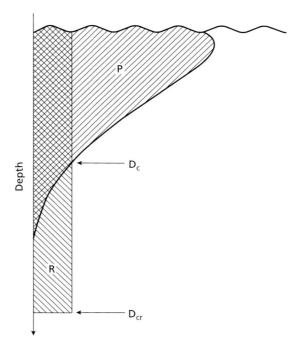

Fig. 2.3 Relationships between depth, phytoplankton photosynthesis (P) and phytoplankton respiration (R). The compensation depth (D_c) and critical depth (D_{cr}) are indicated. After Sverdrup (1953).

also increase water flow and prevent sinking. However, the smallest plankton are predominantly affected by the viscosity of the water so their attempts at swimming are equivalent to a human trying to swim in syrup. As size increases, however, the relative effect of viscosity decreases and swimming helps to increase water flow and diffusion. Models show that populations of large cells have higher growth rates than small cells in conditions of relatively high light, nutrients and upwelling (Parsons & Takahashi, 1973).

Rates of respiration, unlike photosynthesis, are largely independent of depth. Thus, as depth increases, an increasingly large proportion of the carbon fixed by photosynthesis is dissipated by respiration. The depth when fixation balances dissipation is known as the **compensation depth** (Fig. 2.3). This depth will vary in space and time but the light intensity there is usually around 1% of that at the surface. The compensation depth may be 100 m or more in very clear oceanic water but only 1 or 2 m on a turbid continental shelf.

The relationship between the depth of surface mixing and the compensation depth determines rates of production. When phytoplankton are carried below the compensation depth they release rather than fix carbon. Phytoplankton are likely to be carried below the compensation depth in tidally mixed coastal waters and in shelf waters during the months when wind stress

is high. However, when water is heated by solar radiation, and subject to limited tidal or wind stress it will start to stratify. Most heating occurs in the top few metres since light of longer wavelengths is rapidly absorbed (see Fig. 2.2). At the interface of the warm, low-density water and the cool, dense water below, a **thermocline** will form. The warmer water above the thermocline is mixed by wind stress. If the warm, mixed layer is shallower than the compensation depth then phytoplankton will not be carried below the compensation depth. This balance between **mixed layer depth** and compensation depth determines production rates: an association explored in the classic work of Sverdrup (1953).

While stratification may prevent phytoplankton from being carried below the compensation depth, the thermocline prevents nutrient transfer from deep to shallow water. As a result, nutrients in the region above the thermocline are quickly depleted. Vertical migrations of many zooplankton also carry nutrients out of the photic zone: a process known as the **biological**

pump. These vertical migrants graze on phytoplankton near the surface at night and then move to areas below the thermocline in the day. Since the zooplankton can metabolize, excrete waste and be eaten in deeper water, the nutrients which they contain are effectively lost (Longhurst & Harrison, 1989).

Production is divided into two components based on the source of nutrients. **New production** is the proportion of photosynthesis that relies on the supply of inorganic nitrogen from below the photic zone whereas **regenerated production** depends on the recycling of nutrients in the photic zone. For new nutrients to enter the photic zone, they must be carried up from below. This requires turbulence from waves, tidal currents or major ocean currents. The relative stability of the water column in much of the open ocean means that the ratio of new to regenerated production is low, particularly in the stratified central ocean gyres (Dugdale & Goering, 1967).

In many tropical oceans, the water is permanently stratified and production is nutrient limited. In temperate areas, stratification may occur only in the summer months, when wind stress is low and solar radiation is most intense. When stratification first occurs in spring, nutrient concentrations above the thermocline are high. Since the depth of the mixed surface layer is typically less than the compensation depth, phytoplankton multiply rapidly and the spring 'bloom' occurs. Cells can become so dense that the water takes on a green tint and self shading reduces light penetration. The bloom ends when nutrients become depleted, and by mid summer the situation is similar to that in the tropical ocean with regenerated production increasingly significant.

Biological factors also govern phytoplankton production. Zooplankton graze on phytoplankton and in temperate waters this grazing is one of the factors which brings the spring bloom to an end. As zooplankton graze they take energy into food chains and excrete nutrients which can be reused. In temperate waters there is often a second phytoplankton bloom in the autumn when zooplankton biomass falls.

The differences between polar, temperate and tropical production cycles are summarized in Fig. 2.4. The reproduction of many species of fish is closely linked to these cycles since the larval fish rely on phytoplankton, and to a greater extent, zooplankton, for food. In temperate waters, the spawning and larval development of many fishes coincides with the spring phytoplankton

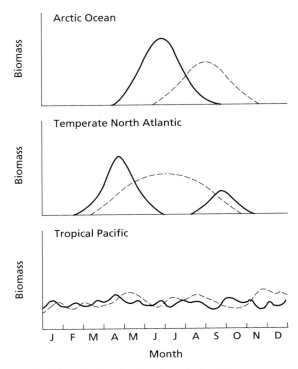

Fig. 2.4 Phytoplankton (——) and zooplankton (– – –) production cycles in polar, temperate and tropical environments. After Cushing (1975).

bloom. The spatial and temporal links between fish larval distributions and the distribution of their planktonic foods can be critical to the survival, growth and year class strength of many fished species (section 4.2.2).

2.3.2 Upwellings and fronts

Nutrients can also be supplied to the photic zone by upwelling. **Upwellings** are currents that carry nutrient-rich water towards the surface. The increased availability of nutrients in upwelling regions means that there are high levels of phytoplankton, zooplankton and fish production. Indeed, around 25% of total global marine fish catches come from five upwellings occupying 5% of ocean area. Upwellings can be induced when islands, ridges or seamounts force deeper waters to the surface, but the upwellings that have the greatest impact on global fishery yields are driven by diverging open ocean or coastal currents.

Local and intermittent upwellings occur close to offshore islands that rise from deep water, leading to higher levels of phytoplankton production 'downstream'. Such upwellings in the Seychelles and Galápagos Islands are the focus for pelagic fisheries. The major upwellings in the open ocean are associated with the divergence of currents that draw deeper, colder and nutrient-rich water to the surface. The main example is the equatorial divergence. Here, Coriolis force causes currents induced by westerly trade winds to be deflected north in the northern hemisphere and south in the southern hemisphere. This sets up a divergence with upwelling water spreading north and south from the equator. The **equatorial upwelling** leads to a biological community that parallels the lines of latitude and determines patterns in the fisheries of this region (Vinogradov, 1981). Primary producers are found closest to upwellings since these have fast growth and short generation times. Further to the north and south, zooplankton biomass maxima and then pelagic fish biomass maxima are encountered. These provide productive fisheries for tuna and other pelagic fishes, and those seabirds which feed on shoaling fish are particularly abundant. Moving further north or south, the effects of the additional nutrients become less apparent, and reef fishes rather than pelagic fishes start to dominate catches. Mean production in the equatorial upwellings is lower than in many coastal regions, but their vast extent means that they are major contributors to global primary production.

Wind-induced coastal upwellings are sites of very high production and occur in regions where the prevailing winds parallel the shore. The currents they induce turn left in the southern hemisphere and right in the northern hemisphere as a result of Coriolis force, and continue to turn until they run at right angles to the wind direction (Mann & Lazier, 1996). If the resulting current is offshore, then deeper water is drawn to the surface close to the coast. There are five major coastal currents associated with upwelling areas, known as the California, Peru, Canary, Benguela and Somali currents (Fig. 2.5). All support major fisheries. The global distribution of mean upwelling intensity is shown in Fig. 2.6.

Upwelling intensity depends on wind strength, and in some areas upwelling is highly seasonal, leading to episodes of production similar to the spring blooms in coastal waters. In the Peruvian upwelling, which supports one of the world's largest marine fisheries for anchovies and sardines, wind stress is relatively low but the upwelling continues for most of the year. As the intensity of upwelling changes, phytoplankton respond quickly, but zooplankton and fish are less responsive because they have longer generation times. Long-term changes in the intensity of upwelling, and hence nutrient supply, can have massive impacts on primary production and subsequent fish production.

Other oceanographic features can produce areas of high production. Spatial differences in tidal flow, and hence turbulence, produce mixed and stratified zones. Where these zones meet, a **front** will form. Fronts also form at the shelf edge where the warm and saline water of the continental shelves meets stable and colder offshore water. At the front, light penetration and nutrient levels are both high and blooms may develop (Pingree *et al.*, 1975; Simpson, 1981). Fronts are intensively fished and support a rich benthic fauna. While phytoplankton production increases at fronts, much of the apparent productivity may result from the sinking of organic matter when tidal turbulence no longer mixes the water column. Thus fisheries at fronts may be supported by aggregations of feeding fish and invertebrates rather than *in situ* production.

Differential solar heating of the globe and the earth's rotation cause trade winds to develop and these drive **gyres** in the ocean basins (see Fig. 2.5). Gyres are areas of low production. The mixed layer depth does change with season and there is evidence for transient upwelling, but otherwise they are permanently stratified. Within the gyres, water is heated by solar energy and then carried polewards by boundary currents such as the Gulf Stream (see Fig. 2.5). As boundary currents flow, they release small gyres of warm and cold water that can persist for several months. These are commonly known as **rings** and can be sites of high production.

2.3.3 Rates of phytoplanktonic production

Productive waters are well known to the fisher and seafarer. They have a green cast by day and will often produce a strong phosphorescent glow at night when zooplankton are disturbed by the boat's propeller and wake. Conversely, the deep, stratified and nutrient-poor waters of the gyres have immense clarity and a blue cast.

Measurements of chlorophyll concentration are routinely used as measures of algal biomass since it would be highly impractical to count individual cells on a large

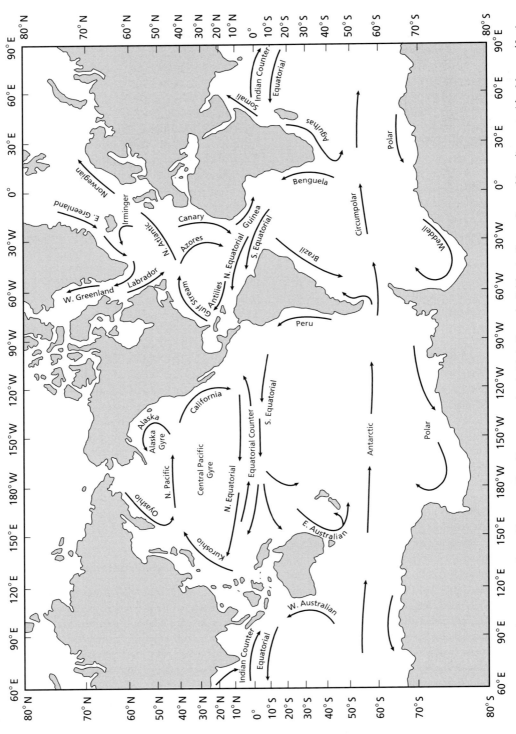

Fig. 2.5 Principal surface currents in the world's oceans. Upwellings are associated with the Benguela, California, Canary, Peru and Somali currents. After Mann and Lazier (1996).

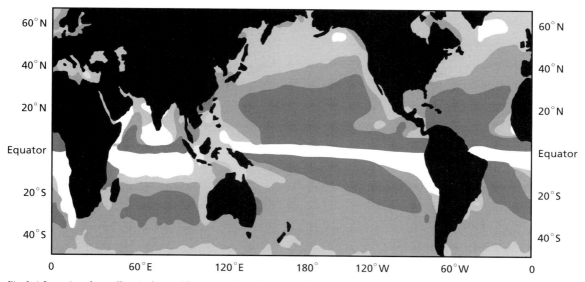

Fig. 2.6 Intensity of upwelling in the world's oceans. Upwelling strength is shown as the annual mean vertical velocity (cm day^{-1}). Dark shade: under –10; medium shade: –10 to 0; light shade: 0 to 10; white: over 10. After Xie and Hsieh (1995).

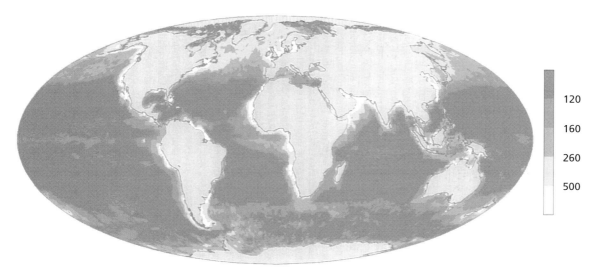

Fig. 2.7 Global distribution of phytoplankton production (g C m^{-2} y^{-1}) as estimated from satellite chlorophyll measurements. From Longhurst *et al.* (1995). Reproduced with permission.

scale! Similarly, primary production is not measured on a cell-by-cell basis but as the uptake rate of radioactive CO_2 by growing organisms under natural conditions. There are many concerns about the accuracy and reliability of this methodology (Li & Maestrini, 1993) but it is generally accepted that production rates range from around 25 g C m^{-2} y^{-1} in subtropical gyres to over 250 g C m^{-2} y^{-1} in upwellings and coastal regions. The areas of highest production are very localized (Fig. 2.7). Approximately 25% of phytoplanktonic production occurs in 8% of total ocean area, 50% in 24% of area and 75% in 54% area (Fig. 2.8) (Longhurst *et al.*,

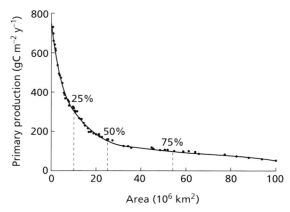

Fig. 2.8 Annual marine primary production in the world's oceans in relation to the area where production occurs. The data are based on the assumption that 100% of the water-leaving radiation detected by satellite is due to chlorophyll. If only 50% of water-leaving radiation is due to chlorophyll in turbid coastal regions then 25% of primary production occurs in 10% of ocean area, 50% in 27% and 75% in 57%. Data from Longhurst *et al.* (1995).

Fig. 2.9 A *Macrocystis* kelp forest off the coast of southern California. Photograph copyright E. Hanauer, by courtesy of M. Tegner.

1995). The distribution of primary production explains why so much fish production comes from a small proportion of ocean area. Satellite chlorophyll measurements suggest that total net primary production is 45–50 Gt C y^{-1} (1 Gt = 10^9 tonnes). The 5 Gt C y^{-1} range in this estimate allowed for the assumption that 25–100% of water-leaving radiation may represent chlorophyll in coastal areas where particles other than algae cause turbidity (Longhurst *et al.*, 1995). Other estimates of global marine primary production fall between 30 and 60 Gt C y^{-1}.

2.4 Non-phytoplanktonic production

While phytoplankton account for the majority of primary production, other autotrophs such as macroalgae, mangroves and seagrasses contribute around 10% of global marine primary production (Table 2.1). These species need to remain in air or shallow water where light is available and so they must expend energy building structural support to stop currents or gravity carrying them away.

2.4.1 Macroalgae

Macroalgae, the seaweeds of tidal and shallow seas, form important habitats for fishes and invertebrates and

are a locally important source of primary production. In shallow water, light can penetrate to the seabed, and currents or wave action increase water flow over the photosynthesizing cells. However, while increased water flow over macroalgae increases productivity, it has to be countered by changes in morphology to withstand the flow. In areas with strong currents, plants tend to be more streamlined, with increased blade thickness and less convoluted blade structures. Photosynthetic rates are highest in shallow, well-lit water, but the benefits of increased light are countered by tidal exposure, which leads to drying and reduced photosynthetic rates. Species differ in their tolerance to drying and this is one determinant of their distribution on tidal shores.

There are many species of macroalgae, from those that live on tidal shores to the giant kelp that grow from depths of over 20 m to the surface (Fig. 2.9). Three genera of kelp form the majority of the world's kelp forests. *Laminaria* is dominant in the North Atlantic and north-west Pacific, *Macrocystis* in the eastern Pacific and south-east Atlantic, and *Ecklonia* in southern Africa and Australasia. They are usually found in temperate waters and as with phytoplankton, the regions close to nutrient-rich upwellings are particularly productive. Rates of production can be very high, with *Laminaria* from 120 to 1900 g C m^{-2} y^{-1}, *Macrocystis* 800–1000 g C m^{-2} y^{-1} and *Ecklonia* 600–1000 g C m^{-2} y^{-1}. Material from kelp and other seaweed beds passes into food webs by the erosion of the blades, which produces detritus, and the production of dissolved organic matter. Blades are also grazed directly by herbivores and fished species such as sea urchins (Mann, 1982).

2.4.2 Mangroves

Mangrove trees grow in shallow coastal lagoons and estuaries in the tropics and subtropics. They have evolved salt tolerance and breathing roots (pneumatophores) because they often grow in anaerobic sediments. Mangroves reduce water flow and increase sedimen-

tation, leading to gradual development of mangals. **Mangals** (forests of mangrove trees) are typically found from the upper shore to mean sea level and their roots provide important habitats for many species of juvenile fish and crustaceans.

Although terrestrial rather than truly marine plants, the detritus that mangroves produce has an important role in coastal food chains and may even fuel production in the deep sea. The majority of mangrove production enters food chains as leaf litter. Leaf litter production is dependent on wind speed and season. Estimates of net primary production by mangroves are typically 1000–2000 g C m^{-2} y^{-1} although some exceed 5000 g C m^{-2} y^{-1} (Clough *et al.*, 1997; Wafar *et al.*, 1997).

2.4.3 Coral reef algae

Coral reefs are formed by reef-building (**hermatypic**) corals, and support high levels of production (Polunin, 1996). Corals are usually found in tropical areas where water temperatures range from 25 to 30°C. The diversity, colour and accessibility of reefs makes them a focus for naturalists and conservationists (Fig. 2.10). Production is attributable to phytoplankton in the

Fig. 2.10 A coral reef slope on the Great Astrolabe Reef in Fiji. Most reef-building (hermatypic) corals are found at depths less than 30 m. Photograph copyright S. Jennings.

overlying water, corals, and epibenthic algae on sediment and coral rubble. Phytoplankton make a small contribution on an area-specific basis, but if currents flow over the reef then much of their production may pass into reef food webs.

Coral production takes place in a thin veneer of living coral tissue. The primary producers are unicellular algae (**zooxanthellae**) living in symbiosis with the coral polyps. Rates of production increase with illumination and water flow, and corals may be the dominant primary producers in shallow reef habitats (Carpenter *et al.*, 1991; Hatcher, 1997). Less than half the carbon fixed is released to consumers as mucus, wax esters and dissolved organics, while the rest is respired, recycled or accumulated within the coral colony (Muscatine, 1990). As a result, a major proportion of primary production on reefs is never available to the food web that drives the production of fished species (Muscatine & Weis, 1991). Only a few species, such as the gastropod *Drupella*, the crown-of-thorns starfish *Acanthaster* and some parrotfishes (Scaridae) or butterflyfishes (Chaetodontidae), graze corals directly.

The shallow and well-lit reefs created by hermatypic corals provide an ideal substrate for algal growth. Most reef areas not covered with corals are colonized by turfs of filamentous and small, fleshy algae that are grazed directly by fished species such as sea urchins, parrotfishes and surgeonfishes (Acanthuridae). The algae and photosynthetic bacteria that account for most reef production rely on the surrounding sea water to supply organic nutrients and disperse the organic compounds they leak or excrete. Turf algae are constantly grazed by herbivores that keep them in the fastest phase of growth and increase productivity (Hatcher, 1988). There are many algal species with many morphologies: from encrusting corallines to stands of sargassum (South & Whittick, 1987). Rates of gross primary production by reef algae range from 1 to 40 g C m^{-2} d^{-1}. Twenty to 90% of algal production is grazed and much of this leads directly to the production of fished species (Hatcher, 1981; Polovina, 1984; Polunin, 1996). Coral reefs are amongst the most productive ecosystems known. However, since autotrophs recycle much of the carbon, the excess production may only just exceed zero and is comparable with new production in the surrounding ocean (Polunin, 1996; Hatcher, 1997).

2.4.4 Seagrasses and marsh plants

Salt-marsh communities develop intertidally where silt and mud accumulates (Mann, 1982). The plants reduce water flow and encourage further sedimentation. The networks of pools and convoluted channels that form in salt marshes on estuaries and sheltered coasts are important habitats for juvenile fishes and invertebrates. Species of *Spartina* are among the most common plants that form and colonize salt marshes (Chapman, 1977).

Seagrasses are subtidal plants, such as *Zostera* and *Posidonia* (temperate) and *Thalassia* (tropical). These form large submarine meadows in sheltered localities, binding sediments with their roots and rhizomes and causing further sediment deposition. In the Mediterranean Sea, seagrass meadows are usually found in shallow, well-lit bays while in the tropics they are often found in sheltered lagoons behind reefs. Seagrass meadows extend to a depth of 20 m or more in clear waters. **Epiphytic algae** are algae that grow on many seagrasses and also contribute to production.

Most salt-marsh production enters food chains as detritus. Estimates of salt-marsh production vary with species, latitude, exposure and distance from the shore. Gross primary production tends to be 2000–4000 g C m^{-2} y^{-1} and net primary production one-tenth of this. Seagrasses are often grazed directly by fish, gastropods or urchins, and many fishes feed directly on the epiphytic algae which colonize seagrasses. On a *Posidonia* meadow in the Mediterranean, leaf production ranged from 68 to 147 g C m^{-2} y^{-1} and rhizome production from 8 to 18 g C m^{-2} y^{-1}. Of the carbon fixed, 3–10% was consumed directly by herbivores, 23–34% was decomposed *in situ* by detritivores and 27–35% exported as litter (Pergent *et al.*, 1997). The remaining material was stored in the sheaths and rhizomes. In Australia, two *Posidonia* species produced 600–1100 g C m^{-2} y^{-1} with epiphytes on the leaves adding 130–160 g C m^{-2} y^{-1} (Cambridge & Hocking, 1997). Some salt-marsh and seagrass production is also lost through burial. Seagrasses such as *Posidonia* in the Mediterranean bury 20–200 g C m^{-2} y^{-1} into sediments. This may account for 5–10% of total production (Romero *et al.*, 1994).

2.4.5 Microphytobenthos

Microphytobenthos are microscopic, photosynthetic eukaryotic algae and cyanobacteria that grow on or close to sediment surfaces in illuminated inshore areas. They are found on sandy exposed beaches, estuaries, mudflats, salt marshes and lagoons and appear as a thin green or brown film or sheen on the sediment. When anoxic sediments are present in the euphotic zone, anaerobic and chemosynthetic bacteria may also contribute to microphytobenthic production. Because they are found at the sediment interface, the microphytobenthos also modulate the exchange of nutrients between the sediment and water column. Their production and role in food webs were well described by MacIntyre *et al.* (1996) and Miller *et al.* (1996).

Photosynthesis by microphytobenthos relies on the penetration of light onto and into the sediment. This depends upon incident light level, wavelength, sediment particle size and organic enrichment. One per cent of incident light remains at sediment depths of 0.2–13.2 mm with a mean depth among studies of 3.5 mm (MacIntyre *et al.*, 1996). In calm conditions, phytoplankton sink and mix with the microphytobenthos. However, in turbulent conditions, microphytobenthos may be resuspended and enter the water column. Many microphytobenthos and phytoplankton are close taxonomic relatives, and distinctions are often based on morphology and distribution. In exposed habitats the microphytobenthos are often attached to particles, but in sheltered regions they can form surface mats. Some species make vertical migrations, moving to the surface when sediments are exposed and returning as the tide floods. This behaviour may stop them from being resuspended and swept away by currents. There may be 10^5–10^7 cells cm^{-3} and so it is most convenient to estimate biomass from chlorophyll concentration! Gross primary productivity of microphytobenthos is often higher in sheltered, muddy habitats than sandy habitats but muddy habitats have higher rates of respiration and net primary production may be lower. In subtidal habitats the benthic primary productivity is typically less than that in the water column but the converse is true in intertidal areas. Estimates of daily gross primary production reach over 2 g C m^{-2} but most studies have reported values of 0.1–1.0 g C m^{-2} (MacIntyre *et al.*, 1996; Miller *et al.*, 1996). Production by microphytobenthos and resuspended microphytobenthos can account for almost 50% of total productivity in some inshore areas (Admiraal, 1984; de Jonge & van Beusekom, 1992; de Jonge & Colijn, 1994).

2.5 Heterotrophic production

2.5.1 The fate of primary production

Primary production can be grazed by herbivores, decomposed or stored. **Herbivory** is the consumption of living plant material by heterotrophs, those organisms that depend on an external source of organic compounds to obtain energy. **Decomposition** refers to the use of dissolved organic carbon (DOC) or particulate organic carbon (POC) by consumers. **Stored carbon** is not decomposed but accumulates in the sediment (from which it may ultimately be recycled). Carbon may also be exported from, or imported to, an ecosystem. Measures of export depend on system boundaries.

The relative importance of herbivory, decomposition and storage (Table 2.2) reflects the properties of producers and the physical dynamics of the environment. Thus, herbivores tend to eat fewer large plants because these contain a greater proportion of unpalatable and nutrient-poor structural material. Eighty-eight per cent of herbivory takes place in the open ocean and accounts for a greater proportion of open ocean net primary production than coastal net primary production. In open ocean and upwelling systems, a large and variable part of production is lost from the pelagic system and exported to deeper waters, the seafloor and sediment. The amount of export is positively correlated with productivity, and in productive coastal areas, may exceed 50% of fixed carbon (Berger *et al.*, 1987). The integrity of the autotrophic community can only be maintained if the export of organically bound nutrients from the system does not exceed the rate of inorganic carbon input.

Export is generally higher and more variable in productive regions, so the proportion of production taken into food webs that drive the production of fished species is also more variable. The variability of export largely depends on the relative abundance of organisms in the food chain. When autotrophs and heterotrophs are in dynamic balance, the transfer is efficient. In this case, recycling within the water column will occur and the organic matter produced by autotrophs is largely eaten by abundant heterotrophs waiting for food

Table 2.2 The fate of marine primary production. From Duarte and Cebrián (1996).

Primary producer	Herbivory (10^{15} g C y^{-1})	%	Decomposition (10^{15} g C y^{-1})	%	Storage (10^{15} g C y^{-1})	%
Oceanic phytoplankton	24.40	88.0	14.70	77.5	0.17	26.5
Coastal phytoplankton	1.80	6.5	1.80	9.8	0.18	27.0
Microphytobenthos	0.15	0.5	0.09	0.4	0.02	3.1
Coral reef algae	0.18	0.6	0.45	2.0	0.00	0.7
Macroalgae	0.86	3.1	0.95	4.2	0.01	1.6
Seagrasses	0.09	0.3	0.25	1.1	0.08	12.0
Marsh plants	0.14	0.5	0.23	1.0	0.07	11.3
Mangroves	0.10	0.3	0.44	1.9	0.11	17.6

within the photic zone. The waste produced by the heterotrophs is then recycled by bacteria and these small particles tend to settle very slowly, if at all, leading to very low leakage from the system. In productive and variable environments, however, the dynamics of autotrophs, heterotrophs and decomposers rarely reach any form of equilibrium. When light and nutrients reach their seasonal peak, the autotrophs proliferate and the heterotroph populations take some time to respond (see Fig. 2.4). During this period, the abundance of autotrophs can increase almost exponentially and many die through senescence and nutrient shortage rather than predation. Dead autotrophs, along with the faecal pellets of heterotrophs, sink from the system. There may also be a large downward flux of DOC released by growing cells. Export as DOC appears to approach particle flux in some regions of the ocean but the results are still controversial (Carlson *et al.*, 1994; Thomas *et al.*, 1995; Murray *et al.*, 1996).

In the North Atlantic, phytoplankton aggregations fall to the bottom as the spring bloom ends, forming a blanket of material on the seabed (Billet *et al.*, 1983). In the course of a few weeks this enters benthic food chains that fuel fish and shellfish production. Sediment traps beneath the photic zone indicate that the downward flux of matter is very variable between seasons and years.

Some primary production is grazed directly. Anchovies eat larger phytoplankton in upwellings, mullet (Mugilidae) graze microphytobenthos in shallow estuaries or creeks, seabreams (Sparidae) graze seagrasses and parrotfishes or surgeonfishes graze algae on tropical reefs. Sea urchins and abalone also graze kelp and algae in temperate and tropical waters. On shallow reefs, direct grazing by fishes and urchins removes 50–100% of total algal production (Hay, 1991; Hatcher, 1997).

Most higher level production in unvegetated inshore habitats is supported by microphytobenthos. Benthic particle and deposit feeders, such as gastropods, bivalves and polychaete worms, feed on microphytobenthos. They, in turn, are eaten by juvenile fish and fished directly by humans. Small ciliate protozoans also graze microphytobenthos (Mallin *et al.*, 1992). Most primary production from macroalgal, mangal and saltmarsh systems enters food webs as detritus. Detritus is broken down by microbial activity and invertebrate 'shredders' and consumed by deposit and filter feeders in sediment communities. In inshore areas, 100–300 g C m^{-2} y^{-1} of algal or seagrass detritus may reach sediments. Further from land the inputs fall, but even Caribbean deep-sea food chains are fed by seagrass detritus (Suchanek *et al.*, 1985).

2.5.2 Transfer along the food chain

Once carbon is fixed by autotrophs, a proportion is assimilated by heterotrophs. So begins a food chain in which each consumer becomes a potential resource for another consumer. Each step in this chain is inefficient, and only a small proportion of prey carbon is incorporated into predator tissues. The life histories and population dynamics of organisms change in predictable ways as we ascend food chains. Thus, organisms higher in pelagic food chains tend to have longer life spans, larger size, lower production to biomass ratios, and slower responses to environmental change. As humans, our perception of organisms also changes as we move up food chains. Hence, we aggregate microbe or phytoplankton species, refer to genera or species of larger zooplankton, and frequently consider edible fishes or crustaceans at the individual or population level. This approach partly results from the intractibility of working with individual phytoplankton, but is also due to the influence of commercial factors such as the species-based marketing and management of fished species.

Each step in the food chain may be referred to as a **trophic level**. Autotrophs are usually assigned a trophic level of one. The trophic level of a consumer is one above that of the species it consumes. Most heterotrophs feed at many trophic levels. For example, an anchovy will consume a mixed diet of phytoplankton (trophic level 1) and zooplankton that have fed on phytoplankton (trophic level 2). This would give the anchovy a fractional trophic level of between 2 and 3. In reality it is very difficult to determine trophic level from studies of diet because individuals eat many other organisms that also have complex diets. Thus, even a herbivore such as a sea urchin is unlikely to have a trophic level of two, because microbes and invertebrates will live on the algae that it eats. Trophic levels are usually determined as the outputs of ecosystem models (section 8.7) or from chemical studies of nitrogen isotope ratios in structural tissues (Owens, 1987).

Size usually determines what eats what in the pelagic environment. Therefore the size distribution of the biota usually follows regular patterns that reflect size constraints on predator–prey relationships and energy transfer. Production, respiration and production to biomass ratios are closely related to body size and are reflected in relationships between biomass density and body size. These relationships suggest that energy flow within ecosystems can be studied through analysis of body size distributions (Borgmann, 1987; Dickie *et al.*, 1987; Fry & Quinones, 1994) and we explore this later in the book (sections 8.6 and 12.4.3). Size-based models are particularly appealing when a single species of fish may be a prey organism, a predator on other fishes or a cannibal during its life history, and when its feeding strategy may depend on the relative abundance of prey species which vary in space and time.

Growth and transfer efficiency

There is carbon loss at each step in a food chain because the consumer cannot digest and assimilate everything that it eats and because it uses energy to respire, reproduce and feed. The proportion of prey carbon converted to predator carbon is known as **gross growth efficiency** (GGE). For copepods, the most abundant grazers in many pelagic ecosystems, the mean GGE is about 25%, but can exceed 60%. The GGE of copepods is positively correlated with food concentration. Mean GGE for other taxa such as nano- and microflagellates, dinoflagellates, ciliates, rotifers and cladocerans is 25–30% (Straile, 1997). Heterotrophic bacteria are amongst the most efficient consumers with a mean GGE of around 40% (Cole *et al.*, 1988). The GGE of fishes rarely exceeds 20% (Hoar *et al.*, 1979). Growth efficiency of cephalopods can be very high. It is 40–60% for the sedentary octopus and 20–30% for more active species of squid in the late juvenile and adult stages (Wells & Clarke, 1996). Indeed, lifetime food requirements may be only 2.5–3.0 times the final body weight of octopus and 4–8 times for more active squid (O'Dor & Wells, 1987). Fisheries for species with such high GGE have the potential to be very productive.

Transfer efficiency (TE) is the product of GGE and predation efficiency: the proportion of prey production taken by the predator. TE will always be less than GGE because no predators are 100% efficient. Zooplankton, for example, have feeding thresholds and stop feeding altogether at low phytoplankton densities (Parsons *et al.*, 1969). Transfer efficiencies tend to be lower at higher trophic levels (Iverson, 1990). In Chapter 8, we see that TE estimates are used to predict unknown fluxes in production models. Errors in estimated TE can seriously affect the output (Baumann, 1995). Given their significance, there are remarkably few estimates of TE in marine food chains.

Since carbon transfer is inefficient, the production of fished species per unit primary production will fall as the number of links in the food chain rises. Thus, krill or anchovies, which feed directly on primary and secondary production, have high rates of production and can provide more animal protein for human consumption than tunas or groupers, which feed at the apex of long food chains. However, the acceptability of marine protein to human consumers is often linked to trophic levels. Hence, tunas or groupers are eaten directly, anchovies and sandeels are used for animal and fish feeds, and jellyfish or zooplankton are rarely caught.

Role of bacteria

Bacterial production can equal 20% of primary production, and bacteria have a key role in cycling excreted nutrients and returning them to food chains (Azam *et al.*, 1983; Cole *et al.*, 1988). Protozoa or heterotrophic microflagellates consume heterotrophic bacteria, cyanobacteria and smaller eukaryotes, and are then consumed by ciliates. Ciliates and protozoa are eaten by zooplankters. This food chain is known as the **microbial food loop** and is distinct from the short food chains described in section 2.5.2 where large (> 5 μm) phytoplankton are consumed directly by zooplankton. Transfer in the longer chain, which involves the microbial food loop, is much less efficient. In upwelling and coastal areas where seasonal blooms occur, the shorter chain is dominant, but the longer chain dominates in **oligotrophic** (nutrient poor) regions of the ocean and during stratification in the temperate summer.

Zooplankton production

Because larger heterotrophs are usually studied on a species-by-species basis, there are few studies of secondary and higher level production. Groups that divide by binary division have proved the easiest to study, and are also amongst the most productive (Pierce & Turner, 1992). The main group of microzooplankton are the ciliates, which consume microbes and the smaller phytoplankton. Despite difficulties with measuring their production, a number of production estimates are now available (Leakey *et al.*, 1992, 1994). In the Kattegat of northern Europe, for example, ciliates consumed 143 g C m^{-2} y^{-1} which was equivalent to 49% of annual primary production in this region. Ciliate production was 57 g C m^{-2} y^{-1} (Nielsen & Kiørboe, 1994). Typically ciliates consume 40–60% of primary production during summer when the very small nanoplankton dominate and 10–60% on an annual basis (Pierce & Turner, 1992). Larger zooplankton, such as the copepods, are increasingly dealt with as generic or functional groups. Copepod production is often calculated based on the study of their population dynamics. Such approaches are data intensive and thus gross estimates of total copepod production are very rare (Hay, 1995). In the Kattegat, copepod production was 12 g C m^{-2} y^{-1}, less than one-quarter of ciliate production. Annual copepod production in the Benguela upwelling was 17–150 g C m^{-2} y^{-1} (Hutchings *et al.*, 1995).

Krill are large zooplankton which inhabit polar waters (*Euphausia superba* is the most abundant in the southern oceans). They are an important source of food for fish and mammals and may be fished directly by humans. Acoustic biomass estimates (section 10.2.3) coupled with experimental studies of production to biomass ratios suggest that production in the southwest Atlantic is $28.6–96.7 \times 10^6$ t y^{-1}. The higher estimate exceeds the current global landings of marine species (Trathan *et al.*, 1995). On an aerial basis, krill production is 60.8–205.2 g wet weight m^{-2} y^{-1}, equivalent to 6.1–20.5 g C m^{-2} y^{-1} if krill wet weight is 10% carbon (Morris *et al.*, 1988; Ikeda & Kirkwood, 1989). Krill consume both autotrophic and heterotrophic production; one estimate suggests they removed 10–59% of daily phytoplankton production. Krill also meet a very significant part of their dietary requirements by consuming smaller heterotrophs (Pakhomov *et al.*, 1997). Krill abundance varies massively from year to year, a factor which affects the behaviour and population structure of krill consumers in the polar oceans. Off South Georgia, for example, there are 20-fold annual differences in krill abundance (Brierley *et al.*, 1997).

2.5.3 Production of fished species

Fish production results from fish growth. To grow, a fish must feed effectively and convert food into tissue. Food is also metabolized to meet energy requirements. The remainder is excreted.

Feeding strategies

Within constraints such as minimising risks of being killed by predators, natural selection leads to feeding strategies that maximise energy intake relative to the energy expended per unit time (Stephens & Krebs, 1986). This leads to faster growth and provides more energy for reproduction. The main components of the **feeding strategy** are the size and type of prey to consume, the order in which to eat them, and how to balance this against the costs of searching for food. This may involve **prey switching** to feed on organisms of a size and abundance that give the greatest energetic rewards for the energy expended on capture. Prey switching matters to the fisheries ecologist because models that are based on fixed feeding relationships, and used to predict the effects of predators on prey populations, may not be valid when prey abundance changes (sections 8.4, 12.4.4 and 15.3.2).

Food such as zooplankters are patchily distributed. A consumer needs to decide whether it is better to stay in a patch as it is depleted or to move to a new patch. The movement of predators between patches is predicted by the **marginal value theorem** (Charnov, 1976). This predicts that the optimal time to stay should be greater in more productive patches than less productive ones. Moreover, the time an animal remains in a patch will be greater when the patches are further apart. Scale is very important here. For example, small fish larvae have low energy reserves and swimming speeds relative to the distance between patches and they rely on the spatial and temporal coincidence of patches and their hatching locations to survive (section 4.2.2). Conversely, large fishes, such as pelagic tunas, search over hundreds of kilometres for schools of prey fish. For smaller species, the highest prey densities may be found in areas where they risk higher predation. As such, they have to balance the risk of mortality against the potential for growth (Werner & Gilliam, 1984; Gilliam & Fraser, 1987).

Predators often have a favoured range of prey sizes. Larger prey may contain more energy, but more energy may be needed to catch them and they may take longer to find and process. Small prey, conversely, may be easy to catch and quick to process, but so many may be needed to maintain energy intake that feeding upon them is inefficient. The result of such trade-offs is that every organism has an optimal size range for prey. In general, larger individuals eat larger prey, although this idea does not always apply across species, where the largest fishes, such as the whale and basking sharks, feed on highly productive zooplankton. Scavenging species, which feed on carrion, also tear pieces from larger dead organisms. Small crustaceans, for example, are major scavengers of dead fish dumped by trawlers (section 13.2).

Swimming and feeding have metabolic costs and studies of the plankton feeding Cape anchovy *Engraulis capensis* (Engraulidae) have demonstrated that different levels of energy investment are used to obtain different food sources (James & Findlay, 1989; James & Probyn, 1989). Plankton filter feeding was shown to be more costly than feeding on particles, presumably due to the drag caused by swimming with an open mouth and opercula, and this method was only used if plankton were sufficiently abundant to maximize the ratio between energy intake and energy costs of feeding.

Hungry fish swim more slowly during routine swimming but faster after they begin feeding and they change swimming strategies for different food sources in order to maximize energy gain (Videler, 1993).

Uses of assimilated energy

Energy is needed for metabolism. This includes the costs of sustaining bodily functions, digestion and respiration. For fishes and motile invertebrates the major cost above the **standard metabolic rate** (energy cost at rest, SMR) is locomotion. The **active metabolic rates** (AMR) of fish are directly proportional to mass, contrary to the idea that larger fish are relatively more efficient. This may be a result of their higher muscle mass, which results in higher running costs in larger fish (Brett, 1979; Brett & Groves, 1979; Videler, 1993). AMR increases with speed but there is a point where the amount of work per unit distance is minimized. This is the optimum speed which, measured as body lengths per unit time, is higher for smaller fish. Absolute swimming speed, measured as distance per unit time, is slower. At optimum speed, energy expenditure is typically two to three times SMR. At maximum swimming speed, requirements may approach 10 times SMR.

When energy is not limiting, growth rate is predominantly determined by temperature (Brett & Groves, 1979). Fish have an optimum temperature for growth that is determined by phenotypic and genotypic factors (section 3.4.2). Fish expend a lot of energy synthesizing new body tissue. The cost of synthesis can represent 40% of the energy content of the synthesized tissue (Pederson, 1997). Fished invertebrates such as urchins and abalone have much lower metabolic requirements than fish. This means that when they compete with fish for the same food sources, such as algae on reefs, the invertebrates can subsist at lower levels of algal production than fish (section 14.7).

Production estimates

Estimates of fish production in different regions depend on the magnitude of primary production, the length of food chains and transfer efficiency. Variance in fish production within regions may increase with increasing primary production (Box 2.1). In temperate shelf waters, fish production rates of 20–50 g m^{-2} y^{-1} are typical (Iversen, 1990; Sparholt, 1990). Fish production rates of 14–35 g m^{-2} y^{-1} have been recorded on tropical reefs (Polunin, 1996). Currents flowing over reefs carry plankton that augment reef production. The extent of plankton use by reef fishes was dramatically illustrated by Hamner *et al.* (1988). As water flowed over the reef slope, virtually all copepods and other zooplankters were grazed from the water by planktivorous fishes (Fig. 2.11). Although production of waters surrounding reefs is low on a per unit area basis, the continuous flow of water over many reefs means that reef planktivores can eat plankton from a very large area of ocean (Polunin, 1996).

Fish production is much lower in the deep sea than in shallow surface waters because the deep sea relies on exports from the shallow and well-lit surface waters where primary production takes place (Merrett & Haedrich, 1997). Biomass and production for demersal (bottom dwelling) fish communities in the eastern North Atlantic (Fig. 2.12) show that production rates are an order of magnitude lower than in the shelf areas described above. Despite the great extent of the deep sea, fish are relatively scarce and production is low (Merrett & Haedrich, 1997).

Gross fishery production estimates based on landings may not be reliable indicators of differences in production since there are major differences in the accessibility of fishes to fishers. The remarkably high yields in upwellings can be attributed to the behaviour of the clupeoid fishes which shoal in surface waters and are very accessible to fishers. Conversely, the total biomass of small pelagic fishes (family Myctophidae) in the open

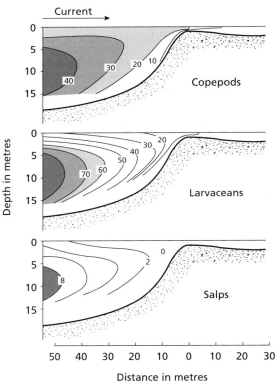

Fig. 2.11 Changes in the abundance (numbers m^{-3}) of copepod, larvaceans and salp zooplankton as water flows across a reef on the Great Barrier Reef, Australia. Larvaceans are a class of Urochordates, also known as Appendicularia. After Hamner *et al.* (1988).

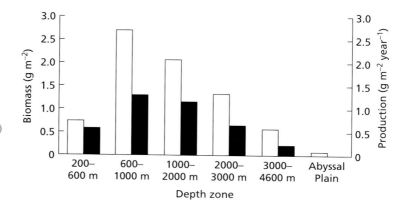

Fig. 2.12 Estimates of production (shaded) and biomass (unshaded) of demersal fishes at different depths in the eastern North Atlantic. The areas at depths of 200–2000 m are on the continental slope. After Haedrich and Merrett (1992).

ocean is also high, but they are not fished commercially because they do not form tight shoals. The expense of towing vast midwater trawls to catch them would be greater than the value of catches.

2.5.4 Linking primary production and landings

It would be useful if estimates of primary production could be used to assess the potential fishery yields of ecosystems. This may help to establish limits for global fish production. To calculate fish production from primary production it is necessary to know transfer efficiencies between predator and prey and the number of transfers involved (Lasker, 1988).

Ryther (1969) predicted global fish production from primary production. He divided the world's oceans into open ocean, coastal and upwelling regions and assumed a set number of steps in the food chain in each region: from 5 in the open ocean to 1.5 in upwellings. Transfer efficiency was assumed to be 10%. The results suggested that global fish production was 24×10^7 t wet weight year^{-1} (fish wet weight is approximately 10% carbon (Vinogradov, 1953)) and that global fish catches could not exceed 10^8 t y^{-1} given primary production of 20×10^9 t C y^{-1}. Ryther's assumptions about the lengths of food chains and transfer efficiency were probably not unreasonable, although better estimates now exist (Parsons & Lee Chen, 1994). However, the global estimate of marine primary production was less than half the current estimate (section 2.3.3), suggesting that the estimate of potential fish production might be doubled.

In ecosystems with similar food chains and sources

of primary production there are good correlations between fish yields and primary production. Phytoplankton production is closely correlated with carnivorous fish and squid production in a number of seasonally stratified marine ecosystems (Iverson, 1990) (Fig. 2.13).

Estimates of trophic levels, transfer efficiencies and landings can be used to estimate the proportion of primary production required to sustain existing global

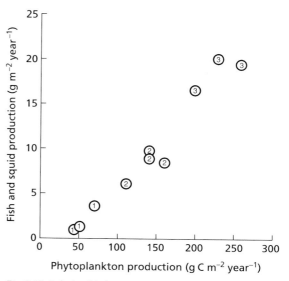

Fig. 2.13 Relationship between annual phytoplankton production and the production of carnivorous fishes in open ocean and coastal environments. (1) Sites in open ocean and gyres, (2) sites in Baltic and adjacent seas, (3) sites in north-west Atlantic shelf waters. After Iverson (1990).

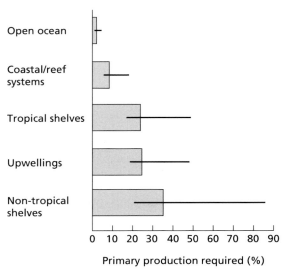

Fig. 2.14 Global estimates of the primary production required to sustain fisheries in five marine ecosystems (± 95% C.I.). Estimates were calculated from the size and trophic composition of the global fish catch and by-catch from 1989 to 1991. Data from Pauly and Christensen (1995).

fisheries. If fish landings by species groups are assigned **fractional trophic levels** (effectively weighted averages of the trophic levels at which they feed, plus one) and a transfer efficiency of 10% between trophic levels is assumed, then the amount of primary production needed to sustain landings in each group can be calculated (Pauly & Christensen, 1995). The results show that over 20% of primary production is required to sustain fisheries in many intensively fished coastal ecosystems (Fig. 2.14).

Linking primary and fish production is not easy, but it is an important area of research. The understanding of fish production processes provides a basis for understanding fluctuations in fisheries and the ecosystem effects of fishing. There has been renewed interest in fish production processes in recent years, particularly with the development of relatively simple and accessible models which provide descriptions of trophic structure in marine ecosystems. We describe these models and their role for description and prediction in section 8.7.

Summary

• Phytoplankton are responsible for around 90% of global marine primary production, and macroalgae, mangroves and coral reef algae for a further 8%.
• Production is limited by light and nutrient supply and is highest in shallow, well-lit coastal waters and in upwellings. Rates of phytoplankton production are governed by physical processes that operate on many scales and are highly variable in space and time.
• Production that is eaten (as opposed to stored or decomposed) enters a food chain where each consumer is a potential resource for another consumer. The food

chain is inefficient, so the production of fished species per unit phytoplankton production falls as the number of links increases.
• The production of fished species is highest in coastal shelf waters, upwellings and around coral reefs, broadly reflecting the high levels of primary production in those areas. Production is lower in the deep sea where fished species rely on carbon exported from shallow water.
• In many coastal areas over 20% of primary production is required to sustain current fisheries yields.

Further reading

The book by Mann and Lazier (1996) is useful reading for any fisheries scientist. It emphasizes the physical forcing of biological processes on many scales and the natural variation in marine ecosystems. The book should convince any sceptic that most events

in the marine environment are unpredictable and that the future is uncertain. Barnes and Hughes (1999) describe production processes in a range of marine ecosystems, and Cushing (1975, 1982) discusses links between climatic events and fisheries. Duarte and Cebrián (1996) review the fate of marine autotrophic production.

3 Fished species, life histories and distribution

3.1 Introduction

The aim of this chapter is to describe fished species, their contribution to global catches, life histories and distribution. Understanding the life cycles and distribution of target species is central to understanding how they are affected by fishing and the environment. We discuss how fished species reproduce, the evolutionary advantages of their reproductive strategies and the links between life-history traits such as growth, fecundity and age at maturity. We show that there are fundamental relationships between life-history traits and that these apply whether the fish is a pelagic tuna in the central Pacific, a cod in the north Atlantic or a parrot fish in the tropical Indian Ocean. In the final sections of the chapter we describe latitudinal trends in life histories, why fishes have different geographical distributions, why they migrate and how larvae are transported between spawning and nursery areas.

3.2 Fishes

Fishes exhibit great diversity in ecology, life history patterns and morphology. Adult fishes range in size by over three orders of magnitude from Indian Ocean gobies which mature at less than 10 mm to whale sharks of 10 m or more. Fish live in waters with average temperatures from less than 0°C to over 30°C. Over half of the recognized vertebrate species are fishes and more are described every year. Currently, there are around 25 000 valid species and Nelson (1994) has predicted that there are some 28 500 extant species in total. These species have been classified into 482 families and about 60% of them live permanently in the sea (Table 3.1). There are also a small number of **catadromous** species such as eels which live in fresh water but enter the sea to breed and **anadromous** species such as salmons and sturgeons which live in the sea but enter fresh water to breed. This book focuses on wholly marine fishes.

Table 3.1 Extant fishes of the world in taxonomic order. From Nelson (1994).

Class and order	Families	Species	% use marine	% fully marine
Class Myxini				
Myxiniformes	1	43	100	100
Class Cephalaspidomorphi				
Petromyzontiformes	1	41	22.0	0
Class Chondrichthys				
Chimaeriformes	3	31	100	100
Heterodontiformes	1	8	100	100
Orectolobiformes	7	31	100	100
Carcharhiniformes	7	208	99.5	96.2
Lamniformes	7	16	100	100
Hexanchiformes	2	5	100	100
Squaliformes	4	74	100	100
Squatiniformes	1	12	100	100
Pristiophoriformes	1	5	100	100
Rajiformes	12	456	94.7	93.9

Continued p. 40

Table 3.1 (*cont'd*)

Order	Families	Species	% use marine	% fully marine
Class Sarcopterygii				
Coelacanthiformes	1	1	100	100
Ceratodontifomes	1	1	0	0
Lepidosireniformes	2	5	0	0
Class Actinopterygii				
Polypteriformes	1	10	0	0
Acipenseriformes	2	26	46.1	0
Semionontiformes	1	7	14.3	0
Amiiformes	1	1	0	0
Osteoglossiformes	6	217	0	0
Elopiformes	2	8	100	12.5
Albuliformes	3	29	100	100
Anguilliformes	15	738	99.2	96.5
Saccopharyngiformes	4	26	100	100
Clupeiformes	5	357	79.8	77.6
Gonorhynchiformes	4	28	20.0	17.1
Cypriniformes	5	2 662	0	0
Characiformes	10	1 343	0	0
Siluriformes	34	2 405	5.2	4.9
Gymnotiformes	6	62	0	0
Esociformes	2	10	0	0
Osmeriformes	13	236	82.2	69.9
Salmoniformes	1	66	31.8	0
Stomiiformes	4	321	100	100
Ateleopodiformes	1	12	100	100
Aulopiformes	13	219	100	100
Myctophiformes	2	241	100	100
Lampridiformes	7	19	100	100
Polymixiiformes	1	5	100	100
Percopsiformes	3	9	0	0
Ophidiiformes	5	355	98.6	98.3
Gadiformes	12	482	99.8	99.6
Batrachoidiformes	1	69	92.8	91.3
Lophiiformes	16	297	100	100
Mugiliformes	1	66	98.5	89.4
Atheriniformes	8	285	48.8	40.0
Beloniformes	5	191	73.3	70.7
Cyprinodontiformes	8	807	1.6	0.2
Stephanoberyciformes	9	86	100	100
Beryciformes	7	123	100	100
Zeiformes	6	39	100	100
Gasterosteiformes	11	257	92.6	84.0
Synbranchiformes	3	87	3.4	0
Scorpaeniformes	25	1 271	95.9	95.1
Perciformes	148	9 293	79.3	76.2
Pleuronectiformes	11	570	99.3	96.5
Tetraodontiformes	9	339	96.5	94.1
Totals	482	24 618	61.5	59.5

Taxonomy of fishes

There are five main classes of extant fishes, and those in the Chondrichthys and Actinopterygii dominate global landings. The orders in class Chondrichthys, excluding Chimaeriformes, are collectively known as the elasmobranchs and have cartilaginous skeletons. All the Actinopterygyiian fishes, with the exception of orders Polypteriformes, Acipenseriformes, Semionontiformes and Amiiformes, collectively belong to the division Teleostei and have bony skeletons. Ninety-six per cent of extant fishes are teleosts.

Of wholly marine fishes, approximately 11 300 species are found in coastal and littoral waters to depths of 200 m. The diversity of marine fishes is highest in the Indo-West Pacific, particularly in the areas between Papua New Guinea and Queensland, Australia. Coral reefs have the highest fish diversity on an area-specific basis and 4500 species are known from coral reefs and associated habitats. Approximately 130 marine species have circumglobal distributions in tropical oceans and many of these, such as the tunas, are very important in fisheries.

Of the 14 650 marine fishes, a tiny proportion are fished. Families which include regularly fished species are indicated in Table 3.2 and Fig. 3.1. This list of families is by no means exhaustive: a huge range of species are caught as by-catch (Chapter 13) or in specialist fisheries supplying the marine aquarium or curio trade. Listings of most species recorded in world fisheries are maintained by the Food and Agricultural Organization (FAO) of the United Nations (UN). The FAO records landings statistics for fishes in 163 of the 482 extant families (FAO, 1997a).

In 1997, global landings were 93.3 million tonnes and included 85.6 million tonnes of marine species (FAO, 1999). Twenty species accounted for almost 40% of global landings in 1997, and they are all marine (Table 3.3). The landings reported by the FAO are not equivalent to the weight of fish caught or killed by fishing activities since many unwanted individuals are discarded at sea (Box 1.1; Chapter 13). Discarding accounts for over 25% of reported landings (Alverson et al., 1994). Of the total global marine landings from 1950 to 1997, 33% consisted of herrings, anchovies and sardines of order Clupeiformes and 19% of cod, hakes and haddocks of order Gadiformes. Thus, two orders accounted for over 50% of the global marine landings.

Global fisheries yield has increased steadily over the period from 1950 to 1997, but this steady increase masks quite dramatic fluctuations in the catches of individual species (Fig. 1.4). Such fluctuations are characteristic of many fisheries and are often driven by unpredictable climatic processes (e.g. see section 2.3.2). They have important consequences for fishers, economies and marine ecosystems. We return to these fluctuations in Chapter 4. Species that make a tiny contribution to global catches by weight may be important because they are valuable, support fisheries of local significance, or are recognized as endangered. Fisheries scientists have sometimes overlooked the concerns of conservationists (Reynolds et al., 2001). Thus, a number of endangered skates are listed only as 'skates and rays' in FAO statistics and the demise of individual species would not be obvious from an analysis of these data (Dulvy et al., 2000).

3.3 Invertebrates

Invertebrates contribute less to global landings than vertebrates, but are often valuable and sustain many significant fisheries. The term **shellfish** is generally used to refer to edible marine invertebrates including shellless animals such as sea cucumbers, octopus or jellyfish, while the term **commercial invertebrates** includes forms captured for uses other than food, such as ornamental shells and corals (Orensanz & Jamieson, 1998). Molluscs comprised 7.3% of total landings in 1997 and crustaceans 5.7% (FAO, 1999). The mollusc species that dominated landed weight were the Argentine shortfin squid *Illex argentinus* (Ommastrephidae) and Japanese flying squid *Todarodes pacificus* (Ommastrephidae). These species accounted for 1.7% of global landings (Table 3.3). The crustacean species that dominated landed weight were the Akiami paste shrimp *Acetes japonicus* (Sergestidae, subfamily Penaeoidea) and the northern prawn *Pandalus borealis* (Pandalidae, infraorder Caridea). They accounted for 0.5% and 0.3% of global landings, respectively.

The main fished invertebrate groups are listed in Table 3.4 and illustrated in Fig. 3.2. These lists cover the bulk of species removed from the sea for food although others are harvested for jewellery making, fishing bait and building materials. Some other phyla are also fished for food. These include jellyfish (phylum Cnidaria) that are eaten as sashimi (uncooked) and

Table 3.2 Fished families of commercial significance in taxonomic order. Examples of species from each family as shown in Fig. 3.1. The size of the largest individuals seen in catches is usually 50–75% of the maximum recorded size.

| Order/families | Common names | Species example | | Range | Maximum size (cm) |
| | | Example | Scientific name | Common name | | |

Order/families	Common names	Example	Scientific name	Common name	Range	Maximum size (cm)
Carcharhiniformes						
Carcharhinidae	requiem sharks	1	*Carcharhinus albimarginatus*	silvertip shark	Indian and Pacific, tropical	300
Lamnidae	sharks	2	*Isurus oxyrinchus*	shortfin mako shark	Circumglobal, tropical and temperate	400
Squaliformes						
Squalidae	dogfishes	3	*Squalus acanthias*	spurdog/ spiny dogfish	North Atlantic, temperate and cold	120
Rajiformes						
Rajiidae	skates and rays	4	*Raja undulata*	undulate ray	North-east Atlantic, temperate	120
Elopiformes						
Elopidae	ladyfishes	5	*Elops machnata*	tenpounder	Indian, tropical and warm temperate	90
Megalopidae	tarpons	6	*Megalops cyprinoides*	oxeye	Indian and West Pacific, tropical	60
Albuliformes						
Albulidae	bonefishes	7	*Albula nemoptera*	bonefish	West Atlantic, tropical and warm temperate	51
Anguilliformes						
Anguillidae	eels	8	*Anguilla anguilla*	common eel	North Atlantic, temperate	100
Congridae	conger eels	9	*Conger conger*	conger eel	North-east Atlantic, temperate	280
Clupeiformes						
Clupeidae	herrings and pilchards	10	*Clupea harengus*	Atlantic herring	North Atlantic, temperate and cold	43
Engraulidae	anchovies	11	*Engraulis encrasicolus*	European anchovy	East Atlantic, temperate	20
Osmeriformes						
Osmeridae	smelts and capelins	12	*Mallotus villosus*	capelin	North Atlantic and Pacific, temperate and cold	23
Salmoniformes						
Salmonidae	salmons and trouts	13	*Salmo salar*	Atlantic salmon	North Atlantic, temperate	150
Gadiformes						
Gadidae	cods	14	*Gadus morhua*	Atlantic cod	North Atlantic	180
Macrouridae	rat-tails and grenadiers	15	*Macrourus carinatus*	ridge-scaled rat tail	Southern oceans, temperate	65
Merlucciidae	hakes	16	*Merluccius merluccius*	hake	North-east Atlantic, temperate	180

Order / Family	Common name	No.	Scientific name	Common name	Distribution	Max length
Lophiiformes						
Lophiidae	angler fishes	17	*Lophius budegassa*	black-bellied angler	East Atlantic, temperate	90
Mugiliformes						
Mugilidae	mullets	18	*Moolgarda pedaraki*	longfin mullet	Indian and West Pacific, tropical	60
Beryciformes						
Trachichthyidae	roughies	19	*Hoplostethus atlanticus*	orange roughy	Atlantic and Pacific, temperate	45
Scorpaeniformes						
Scorpaenidae	rockfishes	20	*Sebastes ruberrimus*	yelloweye rockfish	West Pacific, temperate	91
Triglidae	gurnards	21	*Eutrigla gurnardus*	grey gurnard	North-east Atlantic, temperate	45
Perciformes						
Acanthuridae	surgeonfishes	22	*Naso unicornis*	bluespine unicornfish	Indian and Pacific, tropical	70
Ammodytidae	sandlances	23	*Ammodytes tobianus*	sandeel	North-east Atlantic, temperate	20
Anarhichadidae	wolffishes	24	*Anarhichas lupus*	wolffish	North Atlantic, temperate and cold	125
Carangidae	jacks	25	*Seriola dumerili*	amberjack	Indian, Pacific and Atlantic, tropical	190
Centropomidae	snooks	26	*Centropomus nigrescens*	black snook	East Pacific, tropical and warm temperate	55
Coryphaenidae	dolphinfishes	27	*Coryphaena hippurus*	common dolphin fish	Circumglobal, tropical	200
Haemulidae	grunts	28	*Plectorhinchus gibbosus*	brown sweetlips	Indian and West Pacific, tropical	60
Istiophoridae	marlins	29	*Tetrapterus audax*	striped marlin	Indian and Pacific, tropical and temperate	420
Lethrinidae	emperors	30	*Lethrinus nebulosus*	spangled emperor	Indian and West Pacific, tropical	86
Lutjanidae	snappers	31	*Lutjanus bohar*	red bass	Indian and West Pacific, tropical	75
Mullidae	goatfishes	32	*Parupeneus ciliatus*	cardinal goatfish	Indian and West Pacific, tropical	38
Nototheniidae	icefishes	33	*Notothenia rossii*	icefish	Southern Oceans, cold temperate and cold	60
Percichthyidae	temperate perches	34	*Dicentrarchus labrax*	seabass	North-east Atlantic, temperate	100
Pomatomidae	bluefishes	35	*Pomatomus saltatrix*	bluefish	Atlantic, Pacific and Indian, warm temperate	130
Scaridae	parrotfishes	36	*Scarus ghobban*	bluebarred parrotfish	Indian and Pacific, tropical	75
Sciaenidae	drums	37	*Protonibea diacantha*	spotted croaker	Indian, tropical and warm temperate	150
Scombridae	mackerels and tunas	38	*Thunnus thynnus*	bluefin tuna/tunny	Atlantic, tropical, temperate and cold	400
Serranidae	seabasses	39	*Plectropomus leopardus*	coral trout	Western Pacific, Western Indian, tropical	75
Siganidae	rabbitfishes	40	*Siganus argenteus*	forktail rabbitfish	Indian and Pacific, tropical	37
Sparidae	porgies	41	*Sparus aurata*	gilthead	North-east Atlantic, temperate	70
Trichiuridae	cutlassfishes	42	*Lepidopus caudatus*	cutlassfish	Circumglobal, tropical and temperate	200
Xiphiidae	billfishes	43	*Xiphias gladius*	swordfish	Circumglobal, tropical and temperate	450
Pleuronectiformes						
Bothidae	left eye flounder	44	*Lepidorhombus boscii*	four spot megrim	North-east Atlantic, temperate	41
Pleuronectidae	right eye flounder	45	*Pleuronectes platessa*	plaice	North-east Atlantic, temperate	90
Scophthalmidae	turbots	46	*Scophthalmus maximus*	turbot	North-east Atlantic, temperate	100
Soleidae	soles	47	*Solea solea*	common sole	North-east Atlantic, temperate	60

Fig. 3.1 Common representatives of the fished families listed in Table 3.2. Drawings copyright S.R. Jennings.

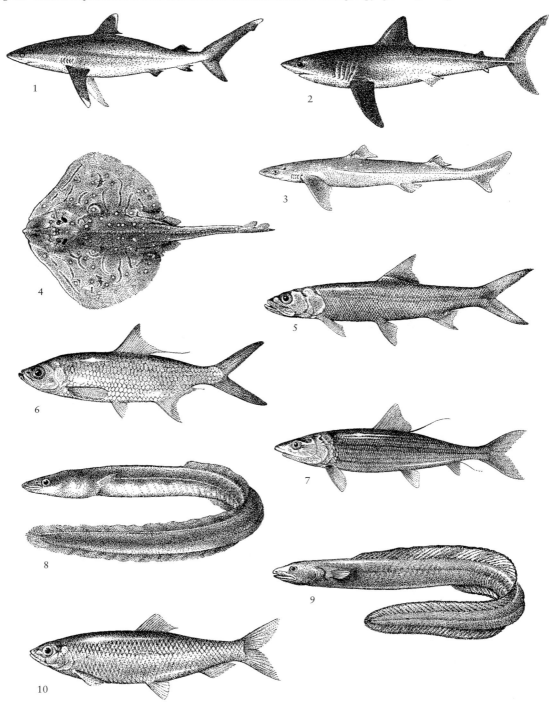

Continued

Fig. 3.1 *(Continued)*

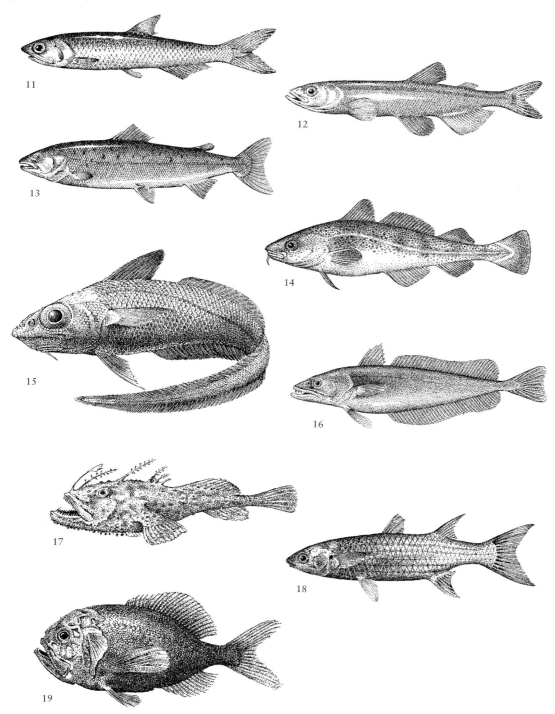

Continued

Fig. 3.1 *(Continued)*

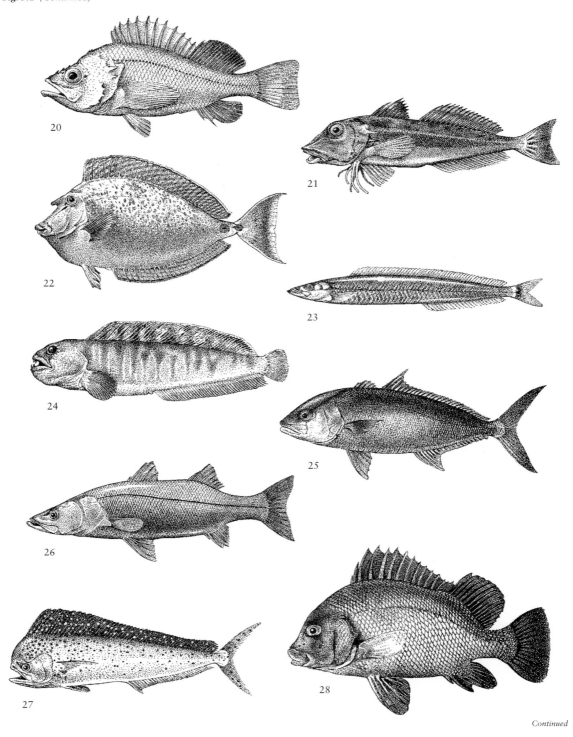

Continued

Fig. 3.1 *(Continued)*

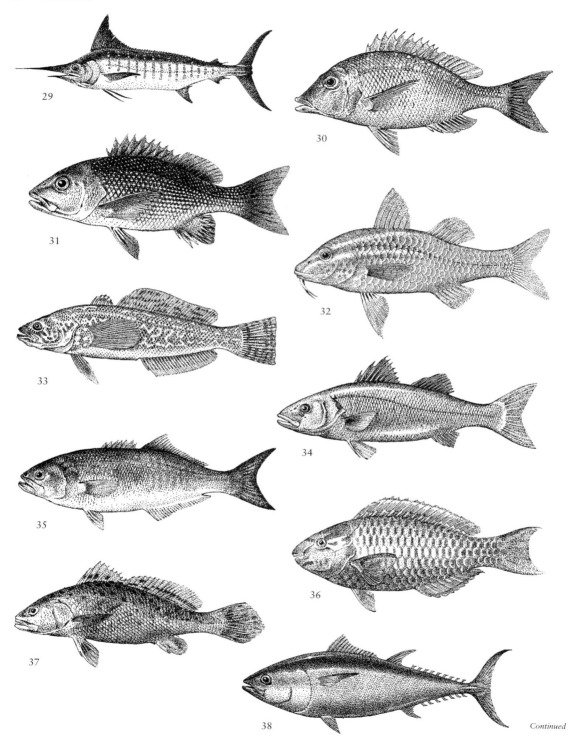

Continued

Fig. 3.1 *(Continued)*

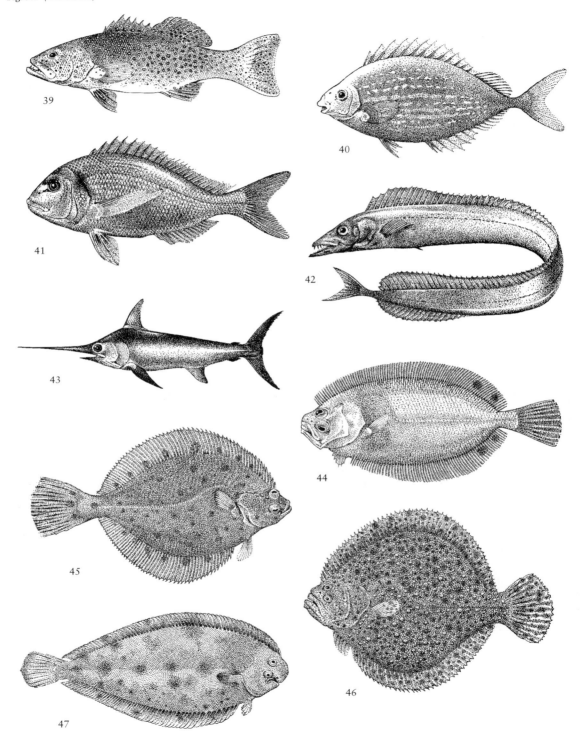

Table 3.3 Landings of marine species in 1997. The 20 species that contributed most to global landings are listed. Percentage is the percentage of global landings. From FAO (1999).

Common name	Species	Family	Catch (t × 10^6)	%
Peruvian anchovy	*Engraulis ringens*	Engraulidae	7.69	8.23
Alaska pollock	*Theragra chalcogramma*	Gadidae	4.37	4.68
Chilean jack mackerel	*Trachurus murphyi*	Carangidae	3.60	3.85
Atlantic herring	*Clupea harengus*	Clupeidae	2.53	2.71
Chub mackerel	*Scomber japonicus*	Scombridae	2.42	2.60
Japanese anchovy	*Engraulis japonicus*	Engraulidae	1.67	1.79
Capelin	*Mallotus villosus*	Osmeridae	1.60	1.71
Skipjack tuna	*Katsuwonus pelamis*	Scombridae	1.43	1.53
Atlantic cod	*Gadus morhua*	Gadidae	1.36	1.45
Largehead hairtail	*Trichiurus lepturus*	Trichiuridae	1.20	1.29
Yellowfin tuna	*Thunnus albacares*	Scombridae	1.13	1.21
European pilchard	*Sardina pilchardus*	Clupeidae	1.03	1.11
Argentine shortfin squid	*Illex argentinus*	Ommastrephidae	0.96	1.03
South American pilchard	*Sardinops sagax*	Engraulidae	0.72	0.77
European sprat	*Sprattus sprattus*	Clupeidae	0.70	0.75
Blue whiting	*Micromesistius poutassou*	Gadidae	0.70	0.75
Round sardinella	*Sardinella aurita*	Clupeidae	0.65	0.69
Argentine hake	*Merluccius hubbsi*	Merluccidae	0.63	0.68
Japanese flying squid	*Todarodes pacificus*	Ommastrephidae	0.60	0.65
Gulf menhaden	*Brevoortia patronus*	Clupeidae	0.60	0.64

the polychaete palolo worm (phylum Annelida). The gamete-filled tails of palolo worms are netted when the tails rise from some tropical reefs in a writhing mass once each year.

Phylum crustacea contains 35 000 species. Many are zooplankton and spend their entire life in the pelagic phase. Most zooplankton are not fished, with the exception of krill *Euphausa superba*, which form dense swarms in South Atlantic waters. The most familiar fished crustaceans are the decapods, which also account for most of the value of fishery landings in this group (**landed value**). Commercially important decapods are classified in three main groups: the shrimps (Penaeidae, Pandalidae and Crangonidae), crabs (Brachyura and Anomura) and lobsters (Nephropidae and Paniluridae). Most species have benthic adults and pelagic larvae. Of the shrimps and prawns, penaeids are generally limited to tropical and warm temperate waters while pandalids have a boreal distribution. Clawed lobsters are found in cold temperate regions and spiny lobsters are found in the tropics and warm temperate regions. Crabs have a worldwide distribution. Crustacea inhabit environments ranging from tidal shores to deep-sea abyssal

plains, but most fisheries take place at depths < 200 m. Life spans range from 1 year for tropical penaeid prawns to over 50 years for lobsters such as the clawed European lobster *Homarus gammarus* (infraorder Astacura) in the north-east Atlantic.

Phylum Mollusca also contains many fished species. Class Polyplacophora, the chitons, includes 600 species from a few mm to 30 cm in length (*Cryptochiton stelleri* of the north Pacific coast). Many of these species are gleaned by fishers from rocky shores. The largest molluscan class is the Gastropoda, containing approximately 40 000 species, some 80% of extant molluscs. Marine gastropods are distributed throughout tropical, temperate and polar seas. Most have shells. A range of gastropods such as abalone, snails, whelks and winkles are fished. Gastropods generally crawl or glide on the substrate while grazing algae, but usually within a very limited home range. Their larvae are the dispersive phase in the life cycle. As well as providing meat, some species are targeted for their nacreous (mother of pearl) shells. Two of the largest gastropod fisheries are for the green snail *Turbo marmoratus* (Turbinidae) and the trochus *Trochus niloticus* (Trochidae). Both of

Table 3.4 Invertebrate groups of commercial significance in taxonomic order. Different taxonomic levels have been used to describe groups most appropriately. Species examples from each family as shown in Fig. 3.2. The size of the largest individuals seen in catches is usually 50–75% of the maximum recorded size.

Taxonomic group	Common name	Example	Species example Scientific name	Common name	Range	Maximum size (cm)
phylum Crustacea **class Malacostraca**						
order Stomatopoda family Squillidae	mantis shrimps	1	*Squilla mantis*	spot-tail mantis	East Atlantic and Mediterranean, temperate	20 (TL)
order Euphausiacea family Euphausiidae	krill	2	*Euphausia superba*	krill	Southern oceans and Antarctic, cold	7 (TL)
order Decapoda **suborder Dendrobranchiata** Superfamily Penaeoidea	shrimps and prawns	3	*Metapenaeus ensis*	red endeavour prawn	Indian and west Pacific, tropical	16 (TL)
Suborder Pleocyemata infraorder Caridea	shrimps and prawns	4	*Crangon crangon*	brown shrimp	North-east Atlantic, temperate	9 (TL)
infraorder Astacidea	clawed and Norway lobsters	5	*Nephrops norvegicus*	Norway lobster	North-east Atlantic, temperate	8 (CL)
infraorder Palinura	spiny and rock lobsters	6	*Panulirus cygnus*	Australian spiny lobster	East Indian (West Australia), temperate	14 (CL)
infraorder Brachyura	crabs	7	*Scylla serrata*	mud crab	Indian and west Pacific, tropical	24 (CW)
phylum Mollusca						
class Polyplacophora	chitons	8	*Chiton squamosus*	squamose chiton	West Atlantic, Caribbean, tropical	7 (TL)
class Gastropoda **subclass Prosobranchia** family Haliotidae	abalones and ormers	9	*Haliotis rufescens*	red abalone	East Pacific (US west coast), temperate	27 (SL)
family Trochidae	trochus shells	10	*Trochus niloticus*	trochus	East Indian and west Pacific, tropical	16 (SH)
family Turbinidae	turban shells	11	*Turbo marmoratus*	green snail	Indian and west Pacific, tropical	22 (SH)
family Buccinidae	whelks	12	*Buccinum undatum*	common whelk	North-east Atlantic, temperate	11 (SH)
family Littorinidae	winkles	13	*Littorina littorea*	edible winkle	North-east Atlantic, temperate and cold	3 (SH)

class Bivalvia

family Ostreidae	oysters	14	*Ostrea edulis*	flat oyster	North-east Atlantic, temperate	10 (SL)
family Pectinidae	scallops	15	*Pecten maximus*	great scallop	North-east Atlantic, temperate	16 (SW)
family Mytilidae	mussels	16	*Mytilus edulis*	common mussel	North Atlantic, temperate and cold	12 (SL)
family Arcticidae	quahogs	17	*Arctica islandica*	ocean quahog	North Atlantic, temperate and cold	12 (SW)
family Cardiidae	cockles and clams	18	*Cerastoderma edule*	cockle	North and east Atlantic, temperate	5 (SW)
family Mactridae	trough shells	19	*Spisula elliptica*	elliptical trough shell	North Atlantic, temperate and cold	4 (SL)
family Veneridae	carpet shells	20	*Tapes decussatus*	chequered carpet shell	East Atlantic, temperate	8 (SL)
family Tridacnidae	clams	21	*Tridacna gigas*	giant clam	Pacific, tropical	130 (SL)
family Solenidae	razor clams	22	*Ensis siliqua*	pod razor clam	North Atlantic, temperate	20 (SL)
family Myidae	sand gaper	23	*Mya arenaria*	gaper clam	North Atlantic and Pacific, temperate	15 (SL)

class Cephalopoda

subclass nautiloidea	nautilus	24	*Nautilus pompilius*	emperor nautilus	East Indian, west Pacific, tropical	22 (SD)
subclass coleoidea	squids and octopuses	25	*Todarodes pacificus*	Japanese flying squid	Southern Oceans, temperate and cold	55 (ML)

phylum Echinodermata

class Echinoidea	sea urchins	26	*Tripneustes ventricosus*	Caribbean edible urchin	Atlantic, tropical	15 (TD)
class Holothurioidea	sea cucumbers	27	*Stichopus fuscus*	brown sea cucumber	East Pacific, tropical	25 (TL)

Key to size measurements: CL, carapace length; CW, carapace width; ML, mantle length; SH, shell height; SL, shell length; SW, shell width; TD, test diameter; TL, total length. These measurements are described in Fig. 9.10.

Fig. 3.2 Common representatives of fished groups of marine invertebrates listed in Table 3.4. Drawings copyright S.R. Jennings.

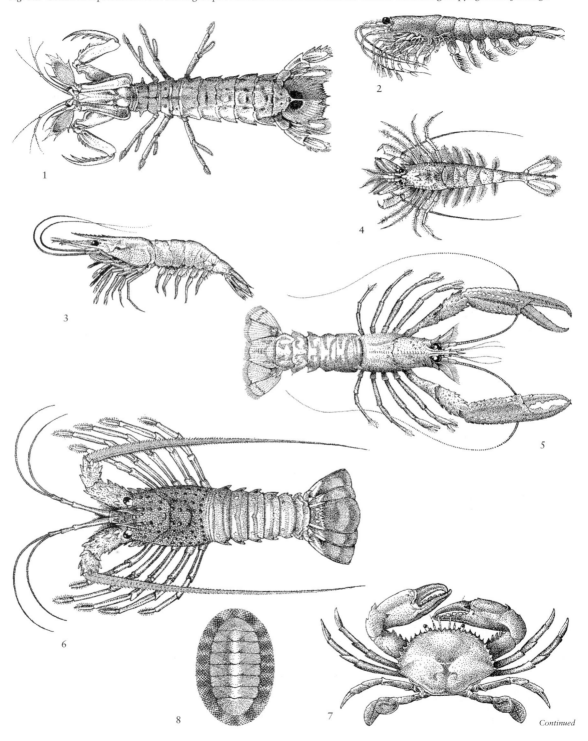

Continued

Fig. 3.2 *(Continued)*

Continued

Fig. 3.2 *(Continued)*

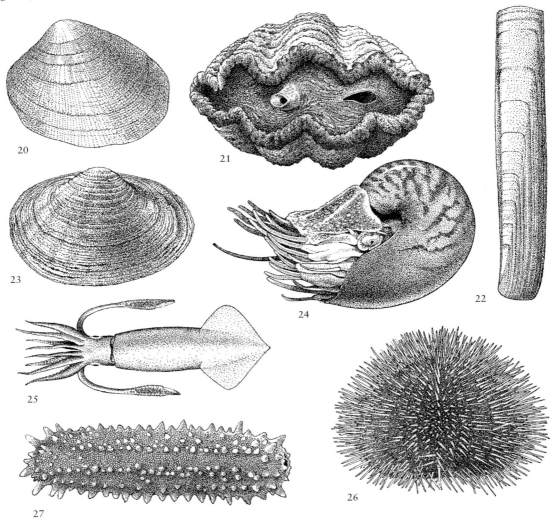

20
21
22
23
24
25
26
27

these species are widely distributed in tropical seas. The meat is eaten and their nacreous shells are used for inlays in furniture, jewellery and button manufacture.

The molluscan class Bivalvia contains 7500 mostly marine species including the scallops and oysters. These support fisheries in many temperate coastal regions. Class Cephalopoda includes 650 species. In the sub-class Nautiloidea there is a single family Nautilidae, containing the shelled *Nautilus* species which are caught in the tropics with baited traps. The larger subclass is Coleoidea with 46 families in five orders.

Squids and octopus of the families Loliginidae and Sepiidae are coleoids.

The remaining species we consider are sea cucumbers and sea urchins of phylum Echinodermata. Landings of echinoderms are small on global basis but can be very valuable. Sea cucumbers are fished to produce a high value dried product known as *bêche-de-mer* or trepang, and sea urchins are fished for their gonads. These fisheries may also have ecological effects disproportionate to catch rates since urchins act as keystone species in some ecosystems.

Humans have a virtually unlimited capacity to utilize or eat products from the marine environment. As well as the fish and invertebrate groups we have highlighted, there are many smaller fisheries for plant, reptile and mammal species. These fisheries have been reviewed by Bustard (1972), Allen (1980), Chapman and Chapman (1980), Bjorndal (1982), Beddington *et al.* (1985), Horwood (1987b), Groombridge and Luxmoore (1989), Hirth (1993) and South (1993). We refer to some of these fisheries when they have conservation significance or lead to shifts in the structure and function of marine ecosystems.

3.4 Life histories

In this section we describe the life histories of fished species, discuss why these have evolved and the ways in which they may affect the responses of species to exploitation. These basics are often overlooked, but a seemingly insignificant factor such as the development rate of a larva can have a major impact on the ability of a species to recolonize areas where it has been extirpated by fishing and the survival time of a sperm will determine the probability that invertebrate eggs are fertilized when the population is fragmented by fishing.

Fished species display a remarkable range of life histories, from squids (Loliginidae) which mature after 1 year and all die after spawning to rockfishes *Sebastes* spp. (Scorpaenidae) which attain maturity after 10 or more years and continue to spawn every year for three decades or more. Why have these life histories evolved and why are they so diverse? Natural selection favours individuals that make the greatest contribution to future generations. The perfect organism, be it a fish or lobster, would mature early at a large size and produce numerous large offspring over a long reproductive life span. However, in reality, resources are limited and when individuals allocate resources to improve one aspect of fitness there will be a cost elsewhere (Leggett & Carscadden, 1978; Roff, 1992; Stearns, 1992). There are many potential ways of combining growth, maturity, fecundity and egg size to maximize fitness, depending on physical and ancestral constraints and costs and benefits set by the environment. This accounts for the diversity of life histories that we see. In this section we consider sex ratios, growth, maturity, longevity, egg size, fecundity and the links between these life history parameters.

3.4.1 Sex, sex reversal and sex ratios

Fished species may be **dioecious** (separate males and females that do not change sex in the course of their lives) or **hermaphroditic** (individuals function as both sexes either simultaneously or sequentially). Sexual reproduction can provide an advantage because it creates and perpetuates genetic diversity and allows the formation of novel genetic combinations that provide adaptations to ever changing environments (Williams, 1975; Bell, 1982; Michod & Levin, 1988). All elasmobranchs have separate sexes throughout their lives and population sex ratios are usually close to 1 : 1. Most temperate teleost fishes are dioecious, but almost 50% of fished families in the tropics contain hermaphroditic species (Sadovy, 1996). Sequential hermaphrodites may be **protogynous** in which adult females change to adult males or **protandrous** in which adult males change to adult females (Fig. 3.3). The fished families Lethrinidae, Scaridae, Labridae, Sparidae and Serranidae (see Table 3.2) contain protogynous species. Protandrous hermaphroditism has been observed in Sparidae, Centropomidae and Platycephalidae. Sex ratios of hermaphroditic species are affected by fishing because they may be size related. In protogynous species, for example, males are larger than females so size-selective fishing leads to a female-dominated population. We return to this in section 12.3.2.

Why does sex change occur? Selection should favour sex change with increasing size or age when one sex gains a relatively greater reproductive advantage with size or age (Warner, 1975, 1988; Charnov, 1993; Reynolds, 1996). Thus, if male reproductive advantage

Species	Sex	
Simultaneous hermaphrodite	♂♀ ⟶ ♂♀	
Protogynous (sequential hermaphrodite)	♀ ⟶ ♂	
Protandrous (sequential hermaphrodite)	♂ ⟶ ♀	

Fig. 3.3 Patterns of sex change.

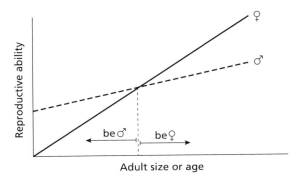

Fig. 3.4 Sex change in sequential hermaphrodites. If males and females gain a reproductive advantage with size or age, but females gain an advantage more rapidly, then protandry is favoured. After Charnov (1993).

increases faster than that of females, then female to male change will be favoured. Conversely, if female reproductive advantage increases faster than that of males, then male to female change will be favoured (Fig. 3.4.). For example, in the blue-head wrasse, *Thalassoma bifasciatum* (Labridae), only the largest males are able to defend spawning sites and attract females. Individuals therefore start as females and switch to males when they can capitalize on their size advantage for sexual selection. The converse is true for species that change sex from male to female: here there is lower disparity between the sexes in sexual selection due to dispersed breeding resources, which cannot easily be monopolized by a single male. The rate of gain with size or age is therefore shallower for males and is exceeded by gains that females receive per unit size due to fecundity advantages (Fig. 3.4). The first sex is always more abundant in fishes that change sex, thus giving a male bias for protandrous species and a female bias for protogynous ones (Choat & Robertson, 1975).

Fished invertebrates also exhibit a range of reproductive strategies. Within the crustacea, most malacostracans are dioecious, but some are protandrous hermaphrodites. Stomatopods and the fished penaeid prawns also have separate sexes. Caridean shrimps of the widely fished family Pandalidae are protandrous hermaphrodites. Reptantian decapod crustaceans, the crabs and lobsters, have separate sexes.

Molluscs may be dioecious or hermaphrodite. Gastropod prosobranch molluscs such as abalone, trochus shell and green snail are mostly dioecious but there are some examples of protandric hermaphrodites.

The widely fished green snail *Turbo marmoratus* and the trochus *Trochus niloticus* have separate sexes (Nash, 1993; Yamaguchi, 1993). In class Bivalvia, oysters of family Ostreidae are mostly sequential hermaphrodites while some scallops and clams (families Pectinidae and Cardiidae) are simultaneous hermaphrodites. Giant clams *Tridacna* spp. are protandrous hermaphrodites which can become simultaneous (Munro, 1993). Cephalopods are dioecious. In phylum Echinodermata, class Echinoidea and class Holothuroidea are dioecious with external fertilization, relying on simultaneous spawning with an external chemical trigger (Conand, 1989).

3.4.2 Growth, maturity and longevity

Organisms grow because greater body size confers a number of advantages that can ultimately result in higher lifetime reproductive output. Larger individuals are subject to lower predation mortality and the faster they grow the more rapidly this mortality decreases. Larger individuals can store more energy so they are less susceptible to fluctuations in food supply. They also show increased tolerance of environmental extremes (Sogard, 1997). Fecundity and ability to compete for mates and resources also increase with size. Large size can be attained by hatching at a large size, growing fast or growing for a long time. Delaying the age at maturity provides more energy for growth.

Indeterminate growth
Most fished species have **indeterminate growth**, that is, they begin reproduction before attaining maximum body size. We have seen in Chapter 2 how energy is allocated to growth and other functions. It is worth investing spare energy in growth as long as this increases expected future reproductive output. Growth in length of marine organisms is widely described using the von Bertalanffy growth equation (see also section 9.3.3). This fits the growth trajectory of many species rather well and since it has been used for many years there are many compilations of parameters over long time periods that are useful for comparative studies. It is given as:

$$L_t = L_\infty (1 - e^{-K(t-t_0)}) \qquad (3.1)$$

Where L_t is the length at age t, L_∞ is the asymptotic length (the length at which growth rate is theoretically zero), K is the Brody growth coefficient and t_0 is the

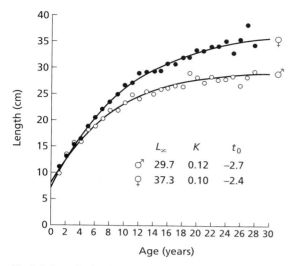

reproductive success is countered by a corresponding decrease in fitness resulting from compensatory changes in other traits (Roff, 1984, 1992; Stearns & Crandall, 1984; Stearns, 1992). The resulting balance between traits is expected to maximize lifetime reproductive output. The result of compensatory trade-offs is that while fish exhibit many life histories, the parameters that describe life histories are related in consistent ways (Beverton & Holt, 1959; Beverton, 1963, 1987, 1992; Charnov, 1993). These relationships have practical value because relatively simple measures of life history, such as growth rate and age at maturity, can be used as surrogates for parameters such as natural mortality which are needed in studies of population dynamics but are much harder to measure (section 9.3.6).

The relationships between life-history parameters suggest that the evolution of fishes is governed by some very general features of life history trade-offs (Charnov, 1993). These features may result from the evolutionary advantage of completing as much potential growth as possible within a life span that is constrained by the forcing effects of temperature on metabolic processes and growth (Beverton, 1963). Thus age at maturity and reproductive output would be adjusted to life span and lifetime reproductive output would be maximized. Where on the growth trajectory should a species mature to maximize lifetime reproductive output? Deferred maturity increases reproductive value at maturity because reproductive output increases with size and age (section 3.4.3). However, fish consistently mature at 65–80% of the maximum size they can attain (Beverton & Holt, 1959; Beverton, 1963). At this size, the costs of delaying maturity (risk of mortality) exceed those of the slower growth that results from allocating resources to reproduction.

Fig. 3.5 Growth of male and female greenstripe rockfish in the southern California Bight. Fitted von Bertalanffy growth curves and associated parameters are shown. After Love *et al.* (1990).

time at which length is zero on the modelled growth trajectory (Beverton & Holt, 1957). t_0 can be seen as a parameter that shifts a growth trajectory defined by K and L_∞ left or right along the x-axis. Examples of the von Bertalanffy curve fitted to length at age data for the male and female greenstriped rockfish *Sebastes elongatus* (Scorpaenidae) of the southern California Bight are shown in Fig. 3.5. Methods for fitting these curves are described in section 9.3.3. For squid that die immediately after spawning, apparently asymptotic growth curves are not necessarily meaningful. Since the fastest growing individuals tend to mature first the mean size is increasingly biased by the slower growing animals that are still alive. As a result, the mean curve is asymptotic even though no individual follows this trajectory.

Parameters of the von Bertalanffy growth equation are widely used to describe fish life histories. Across species there are close relationships between these parameters and others that describe other aspects of the life history, such as age and size at maturity. In general, high K is associated with low age and size at maturity, high reproductive output, short life span and low asymptotic length. Conversely, species with low K have greater age and size at maturity, lower reproductive output, longer life spans and greater asymptotic length. The relationships are consistent with **trade-offs**, such that change in a single trait which may increase lifetime

Determinant growth

Many crustaceans have exoskeletons that do not grow and they must moult to increase body size. The measured growth rate of species with exoskeletons has two components, size increase at moult and frequency of moult (Fig. 3.6). For short-lived species, moulting may be frequent, and continuous growth models may be used. Length increments with moulting tend to decrease as size increases: ranging from 20 to 40% in small individuals to 5% or less in large adults. It is difficult to age crustaceans since hard skeletal materials that grow throughout the life span are usually needed

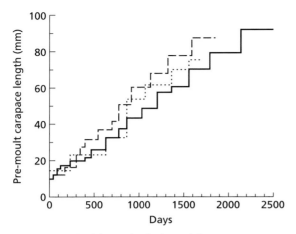

Fig. 3.6 Growth of three individual spiny lobsters. M.J. Fogarty unpublished, from Cobb and Caddy (1989).

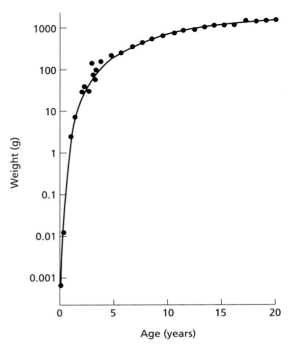

Fig. 3.7 Growth of the plaice *Pleuronectes platessa* from larval metamorphosis to an age of 27 years. After Cushing (1975).

(section 9.3.2). Some crustaceans exhibit a terminal moult at maturity, after which no further growth occurs and more resources are allocated to reproduction. These species have **determinant growth**. In other species male and female growth rates differ, female rates often lagging those of males because they moult less frequently (Cobb & Caddy, 1989).

Extent of growth

Most species with planktonic larvae grow in weight by several orders of magnitude during their life cycle. This has important ecological implications because they may switch from being the prey of larger organisms to their predators within a few months. Fish routinely grow across five orders of magnitude in 5–10 years (Fig. 3.7). Large clams such as *Tridacna gigas* can reach 200 kg in weight and increase in size by at least six orders of magnitude in 10 years or more. The fastest growth is probably shown by some cephalopods. For example, the Argentine shortfin squid *Illex argentius*, a widely fished species in the south-west Atlantic, increases in size by over five orders of magnitude, from 0.01 g to 600 g, in 50 weeks. Fast growth means that cephalopods switch rapidly from consumed to consumers. For example, juvenile Argentine shortfin squid are important prey for juvenile hake *Merluccius hubbsi* (Merlucciidae) in the south-west Atlantic, but within a few weeks the larger squid have grown sufficiently to prey on the hake (Boyle & Bolettzky, 1996).

Longevity

Maximum life spans of fished species are very variable and rarely realized because there is an increased probability of capture with age. Fished teleosts range in longevity from 2 years for some anchovies to 100 years or more for rockfishes (*Sebastes* spp.). Several elasmobranchs do not mature for 20 years or more (Compagno, 1984a, b). Life spans of molluscs range from 1 year for many squids to over 100 years for giant clams such as *Tridacna gigas* (Tridacnidae) and the quahog *Arctica islandica* (Arcticidae). Humans are the only major predators of adult giant clams. The Nautiloidea mature at 5–10 years, grow slowly, reproduce many times and often live for 10–20 years. Many squids and cuttlefishes in families Loliginidae and Sepiidae have life spans of 1 year (Boyle, 1990; Boyle & Bolettzky, 1996; Clarke, 1996). Sea urchins may be long lived and slow growing (Ebert & Russell, 1992; Russell *et al.*, 1998). For example, the urchin *Strongylocentrotus droebachiensis* (class Echinoidea) may live for over 50 years (Russell *et al.*, 1998).

3.4.3 Egg size, fecundity and reproduction

The limited energy available for reproduction can be allocated in a number of ways. Animals may produce many small eggs, a few large eggs or live-born young. Spawning may take place over many years or a single year. Many strategies of allocation are observed in fished species, but the production of thousands to millions of small eggs tends to dominate.

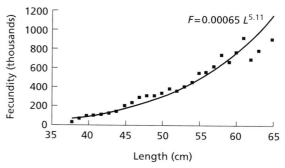

Fig. 3.8 The fecundity–length relationship for blue whiting. After Trella (1998).

Fecundity of fishes

Elasmobranchs include species with various reproductive modes including egg laying, live-bearing, and live-bearing with maternal input. Egg laying is the ancestral reproductive state from which various forms of live-bearing have evolved, although there is evidence of at least one evolutionary reversal to egg laying (Dulvy & Reynolds, 1997). Fecundity is usually low, with high parental input to each egg or offspring. Many fished elasmobranchs produce tens and occasionally hundreds of offspring each year. For live-bearers, gestation may last from less than 6 months to over 2 years. Offspring in egg cases may hatch in 2–12 months (Compagno, 1990). Elasmobranchs have internal fertilization.

Most fished marine teleosts have high fecundities of thousands to millions per female per year. A few species, such as rockfishes of genus *Sebastes*, are live-bearing (viviparous) (Breder & Rosen, 1966). Oviparous species may broadcast their eggs into the pelagic environment, release them on or near the substrate, or deposit them on rocks, shells or in nests. They are fertilized when the male releases sperm into the water. Most fished species produce pelagic eggs. The primary exceptions are the herrings (genus *Clupea*) which lay demersal eggs, the triggerfishes (Balistidae) that tend demersal eggs and the rabbitfishes (Siganidae) that scatter adhesive demersal eggs. Pelagic eggs may drift tens or hundreds of kilometres before hatching. Most fish have defined spawning seasons although these are often extended in the tropics (Sadovy, 1996).

Reproductive output increases with body size. An example is the relationship between **annual fecundity**, the number of eggs produced per year, and length for the blue whiting *Micromesistius australis* (Gadidae) that is fished south of the Falkland Islands (Fig. 3.8). The relationships between fecundity and length is given by:

$$F = aL^b \tag{3.2}$$

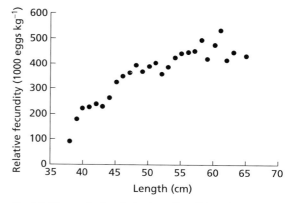

Fig. 3.9 Changes in the relative fecundity of blue whiting with length. Data from Trella (1998).

Where F is fecundity, L is length and a and b are constants. Fecundity increases with length faster than weight increases with length. As a result, **relative fecundity** (fecundity per unit body mass) increases appreciably with size (Fig. 3.9) and cannot be treated as a constant for individual species. Ratios between somatic (body) weight and reproductive output (the product of egg size and fecundity) are often used as measures of reproductive effort, because they are more convenient to measure than the entire energy budgets of a fish. Although biases exist in this approach, increases in reproductive output with age suggest that fish put more energy into reproduction as their life span progresses (Hirshfield, 1980). What is the evolutionary basis for this? As with age and size at maturity, investment in reproduction has costs in terms of mortality, because energy is diverted from growth. The amount of effort put into reproduction at a given time is determined by the difference between present gains in fitness resulting from reproduction to future gains from surviving to reproduce in subsequent years. When potential future gains exceed present gains, low reproductive output is favoured. When they do not, high reproductive output is favoured. Fish therefore exhibit a range of reproductive strategies according to their environments and ancestry, from a single spawning to multiple spawnings over 10 or more years.

Within the energy available for reproduction, fecundity can only be increased at the expense of egg size. We have seen that most species of fish produce many small eggs and those that lay a few eggs, or give birth to live young, are relatively rare. Why do many species have high fecundity and small eggs? Winemiller and Rose (1993) modelled the relative survival of larval cohorts in environments with food resources that varied in patchiness and density. Each cohort of larvae had the same total biomass but was composed of different numbers of individuals. In prey-rich and patchy environments, the best strategy was to produce many small larvae, but as prey abundance and the size of prey patches was reduced, so fewer larger larvae led to higher survival rates. The simulations indicated that reproductive investments should be towards larger numbers of smaller eggs when resources were patchy on a relatively large spatial scale. In patches that were large in relation to swimming ability, the larvae would encounter the same mean prey availability over many days. However, larvae that were not in prey patches would starve and die. Given the patchiness of planktonic food in the marine environment it is not surprising that many species have evolved high fecundities and small egg size.

A few species spawn once and then die (**semelparous**) but most spawn over many successive seasons (**iteroparous**). Within a spawning season some species, especially pelagic fishes such as sardines and anchovies, mature and release eggs in batches over periods of days, weeks or months (**batch spawners**), while others spawn all their eggs in a single episode. To calculate the annual fecundity of batch spawners it is necessary to know the number of eggs in each batch and the number of batches spawned (section 9.3.5). The Norwegian coastal cod *Gadus morhus*, for example, may spawn from 50 000 to 250 000 eggs in each batch, and produce batches at 3-day intervals over a spawning season of 50 days (Kjesbu, 1989).

Fecundity of invertebrates

Fecundity of the decapoda is very variable, from hundreds in lobsters and pandalids to millions in crabs (Cobb & Caddy, 1989). Most species mate, and the male passes spermatophores to the female. These are stored for periods of days (some crabs and prawns) to over 1 year (lobsters). The penaeids lay demersal eggs, but other species carry their eggs on the ventral surface of the abdomen. *Homarus* may carry eggs for 10–12 months, but other species, such as *Panulirus*, may carry eggs for a month or less. Many eggs incubated by the female may not hatch. Larval stages are pelagic and they complete metamorphosis into a post larva with similar morphology to the adult. This settles to the benthic habitat. Of course, if populations are stable, high fecundities must be correlated with large numbers of larvae dying before they reach adulthood. In a study of South African decapods, the number of eggs that produced one adult was 1.2×10^6 for the mangrove crab *Scylla serrata* (infraorder Brachyura), 0.22×10^6 for the West African red crab *Chaceon maritae* (infraorder Brachyura), 1.0×10^6 for the west African rock lobster *Jasus islandii* (infraorder Palinura) and 0.5×10^6 for spiny lobster *Panulirus gilchristi* (infraorder Palinura) (Pollock & Melville-Smith, 1993). Given that decapods have planktonic larvae that feed and develop in pelagic habitats it is likely that high egg production has also evolved for reasons outlined by Winemiller and Rose (1993). Many warm-water decapods may spawn more than once each year and this must be considered when calculating annual fecundity. Population fecundity is highest in intermediate size classes which produce fewer eggs per female, but the females spawn more frequently and are more numerous (Cobb & Caddy, 1989).

Other invertebrate groups with pelagic larvae also produce numerous small eggs. Some species, such as abalone (Haliotidae), have external fertilization and the presence of male sperm in the water stimulates egg release. Internal fertilization is also common in this group and eggs are released in gelatinous strings or cases. The gastropod *Trochus niloticus* produces 2 million eggs when 10 cm in diameter and 6 million eggs at 14 cm diameter (Nash, 1993). The eggs hatch to a trochophore stage at 10–12 h after fertilization and 20 h after fertilization the larvae have a shell. They settle to the benthic habitat after 50–60 h. The green snail *Turbo marmoratus* also has high fecundity, around 7 million eggs are produced by a 2-kg female. Fertilized eggs hatch to trochophores after 12 h, form a pediveliger stage (the larva develops a foot) after 3 days and settle on day 4 (Yamaguchi, 1993). These short larval life spans limit dispersal and ensure the larvae settle on appropriate substrates.

The giant clams such as *Tridacna* spp. (Tridacnidae) are amongst the most fecund fished species. A female of 70–80 cm in length will produce 240×10^6 eggs. The larvae form swimming trochophores 12 h after fertilization and a shelled veliger after 36 h. Larval life is usually 5–15 days and shortly after reaching the pediveliger stage the larva settles and metamorphoses. The newly settled larvae are known as spat (Munro, 1993).

Oysters (family Ostreidae) release eggs and sperm and the eggs are fertilized in the water column. Fecundity ranges from hundreds to several million. A few species brood eggs which are fertilized by sperm drawn in from the open water through an inhalent aperture. Larvae hatch as ciliated trochophores which pass through shelled (veliger) stages in the plankton before settling on the substrate. Brooders release larvae as veligers. Dispersal occurs in the planktonic phase. Adult bivalves are sedentary and spawning is induced by the presence of male sperm and other chemical cues. As such, they have to be in close proximity for successful fertilization. In the case of cockles *Cerastoderma edule* (Cardiidae), for example, close means a few centimetres (Andre & Lindegarth, 1995).

Cephalopod fecundity varies from tens to tens of thousands. *Nautilus belauensis* (subclass Nautiloidea), which is fished in Palau and other Pacific Islands, may produce 10–20 eggs per year. Fecundity is 200–60 000 in loginids and from 50 000 to over 1 million in the ommastrephids. The size of cephalopods is generally not related to fecundity within species. Cephalopods usually mate shortly before spawning, but in some species males may mate with immature females which store the sperm. This permits opportunistic mating. In octopods, an interval of 100 days or more has been recorded between mating and spawning. Pelagic squids produce gelatinous egg masses which are neutrally bouyant or float. Species found on the continental shelves may lay demersal eggs in strings on the seabed. Hatchlings resemble the adult in most features, although the juveniles of some bottom-dwelling species may live in the pelagic environment for a few weeks. Hatchlings are generally much larger than fish larvae, often exceeding 15 mm at hatch. They feed immediately on zooplankton. The annual life cycles of squids account for the marked interannual fluctuations in

catches and apparent instability of populations (Boyle, 1990). Unlike teleost fishes, there is virtually no overlap of generations, so biomass depends almost entirely on the success of recruitment that year. Most species are semelparous, with their lifetime fecundity realized in a spawning period of hours or days, usually followed by mass mortality. The most complex population structures occur when species have protracted spawning seasons. For example, *Loligo forbesi*, which is widely fished in the north-east Atlantic, may have breeding seasons of several months and produce successive waves of recruitment (Boyle, 1990).

The sea urchins (class Echinoidea) and sea cucumbers (class Holothuroidea) have separate sexes. Sperm and eggs are released into the water where they are fertilized. In echinoideans, the fertilized eggs develop and hatch to form a planktonic **echinopluteus** larva that settles after periods of weeks or months. In holothurians, the eggs hatch and pass through two planktonic larval stages, the **auricularia** and **dodiolaria**, before metamorphosing and recruiting to the seabed at a length of around 1 mm. The auricularia is ciliated while the doliolaria is barrel shaped and free swimming. Holothurians have fecundities of millions to tens of millions. Relatively high densities of sea urchins and sea cucumbers are often required for successful fertilization.

In marine invertebrates there is considerable intraspecific variation in the egg size. Levitan (1993) considered why variations in egg size evolved based on a theoretical and empirical study of sea urchins in the genus *Strongylocentrotus* (class Echinoidea). He was curious to know why eggs were not as small as possible in all species and thought that this may be due to sperm limitation if larger eggs had a greater probability of being fertilized. Levitan suggested this because he had previously observed very low fertilization successes in *Strongylocentrotus franciscanus* at a shallow site in British Columbia, Canada. At this site, he recorded fertilization success of 0–82%. Individual reproductive success was highly dependent upon population parameters and environmental conditions. Increases in group size and aggregation, decreases in current velocity and location within an aggregation (central to downstream) all led to increased fertilization success (Levitan *et al.*, 1992). Given these results, it is reasonable to assume that sperm availability often limits reproduction, as it does for many other invertebrates in the marine environment (Levitan, 1991; 1995). Levitan's (1993) study confirmed that egg size could influence fertilization success in the sea urchin and that this would provide powerful selection for egg sizes which were fertilized more frequently. Using a mathematical model he demonstrated that fertilization success was sensitive to egg size and that the cost of producing fewer eggs was offset by an increase in fertilization success. The results were supported by his experimental studies that showed higher fertilization rates with larger eggs.

3.5 Distribution in space and time

3.5.1 Geographical ranges and stock structures

There are marked variations in the geographical ranges inhabited by fished species. A number of larger pelagic fishes such as tunas and dolphinfishes (Scombridae and Coryphaenidae) have circumglobal distributions throughout tropical and warm temperate waters while some reef fishes such as grunts (Haemulidae) are endemic to single island groups such as the Galápagos. However, some useful generalizations can be made about the ranges of species and the factors that determine them. Latitudinal ranges of species within given families and genera are greater when the midpoints of their ranges are near the equator (Rohde *et al.*, 1993). Tropical species generally have the largest latitudinal ranges. This may be due to the more stable temperatures in warm seas, where habitats are relatively uniform over large areas and facilitate widespread dispersal by planktonic eggs and larvae. Mean minimum and maximum surface temperatures differ by only 7.8°C throughout the tropics (46° of latitude) whereas a similar temperature range is found in a 4° latitude band between 40 and 44°N (Rohde *et al.*, 1993).

The suitability of an environment for a fish population is probably determined by density-independent factors, as well as by physiological response curves. The most favourable combination of these factors occurs close to the midpoint of the species range. The factors determining suitability may be different at opposite ends of the species range, but the suitabilities themselves may be quite similar (MacCall, 1990).

Across the latitudinal range of a species, individuals rarely mate randomly with conspecifics. Rather, they form populations that are known to fishery biologists as **stocks**. In the short term, stocks usually respond independently to the effects of exploitation and are

therefore used as management units (Chapter 9). This is because birth and death rates within stocks, rather than immigration or emigration, drive their dynamics (Carvahlo & Hauser, 1994; Pawson & Jennings, 1996). If genetic exchange between stocks is extremely low they may become genetically distinct. Different stocks of the same species often have different life-history traits. This variation is predominantly a phenotypic response to the environment.

The growth rate of individuals in a stock is largely determined by temperature and, as we saw in section 3.4.2, other life-history parameters will undergo compensatory adjustments. As a result, faster-growing stocks in warmer environments have shorter reproductive life spans but earlier ages at maturity and higher annual reproductive output. An example of these effects is shown in Fig. 3.10. In this case, Atlantic herring *Clupea harengus* in northerly stocks (such as the Norwegian spring spawning herring) live at low environmental temperatures and have potentially long reproductive lifespans. Conversely, southern stocks (such as the herring of the southern Irish (Celtic) Sea) have high reproductive output and short lifespans (Jennings & Beverton, 1991). The net effect of this **plasticity** in life histories is that lifetime reproductive output remains relatively constant across the latitudinal range. Similar variation in life-history traits also occurs across the latitudinal range of invertebrate species. For example, southern populations of the soft-shell clam *Mya arenaria* (Myidae), found on the north-east coast of North America, grow faster and have shorter lifespans than northern populations. Moreover, the southern populations spawn twice each year and those in the north only once (Appeldoorn, 1995). Differences in life histories are important to the fishery manager because they mean that the stocks will respond differently to the same rates of fishing mortality.

3.5.2 Migration

Some fished species remain in one location throughout their adult life, but these sedentary species, such as the giant clams, are exceptions rather than the rule. Sedentary species must rely on a dispersive larval phase to colonize new habitats. Most fished species make movements on different scales throughout their life. These may be local and linked to short-term activities contributing to growth, survival and reproduction.

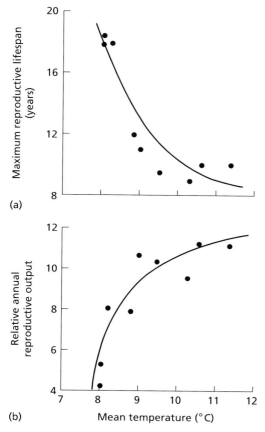

Fig. 3.10 The relationship between (a) maximum reproductive lifespan and (b) annual reproductive output of Atlantic herring *Clupea harengus* stocks and the mean temperature at which they live. Annual reproductive output is an index of gonad mass as a proportion of somatic mass. Data from Jennings and Beverton (1991).

Others are longer and encompass new environments. These movements and migrations are important in the fisheries context: they may take fishes in to and out of a protected area, lead to their crossing national and international boundaries, or make them particularly accessible to fishers as they congregate to spawn or migrate along narrow routes.

The smallest scale movements made by fished species are termed **station keeping** movements. These would, for example, keep fish in position as they wait to feed on zooplankton that are swept past in a current. Over longer time and space scales most fished species

make foraging movements in search of resources. An example would be the diel commuting of goatfishes *Mulloides flavolineatus* (Mullidae) between coral reefs and other habitats (Holland *et al.*, 1993) or the vertical migration of plankton (Longhurst & Harrison, 1989). Individuals will also make territorial movements where agonistic behaviour is directed towards intruders and ranging movements where an area is explored in search of resources (Dingle, 1996).

Migrations differ from station keeping, foraging and territorial movements because responses to local resources may be temporarily suppressed. An example of a migration would be repeated movement between spawning and feeding grounds. Migration involves persistent movements of greater duration and more consistent direction than occurs during station keeping or foraging. There may also be activity patterns particular to arrival and departure, and specific patterns of energy allocation to support movement.

There must be good reasons for migration when the energy costs can be high and the energy used for migration cannot be used for reproduction. In the long term, migration may reduce the environmental heterogeneity experienced by the organism, or place it in optimal conditions for feeding, growth or reproduction for a greater period (Dingle, 1996).

Many of the most important fished species migrate. Perhaps the best known migrations are those of anadramous fishes that leave the sea to spawn in rivers (e.g. Salmonidae) and catadramous fishes that leave rivers to spawn in estuaries and the open sea (e.g. Anguillidae). Many other marine species make return journeys between spawning, feeding or wintering grounds (Fig. 3.11). Many migrations rely on assistance from currents, and migrating organisms may navigate using a variety of chemical, visual and physical cues (Harden-Jones, 1981). Most pelagic cephalopods make large-scale migrations in ocean currents, and these may exceed several thousand kilometres. Cephalopods of the continental shelf also move hundreds of kilometres from offshore feeding grounds to inshore spawning areas (Boyle, 1990).

Distributions of fish and fisheries show pronounced seasonal changes as a result of migrations (Harden-Jones, 1968, 1981). Conventional tagging, using sequentially coded tags attached to fishes suggests the general migrations of fished species (section 10.2.6). However, recent studies with electronic tags that allow

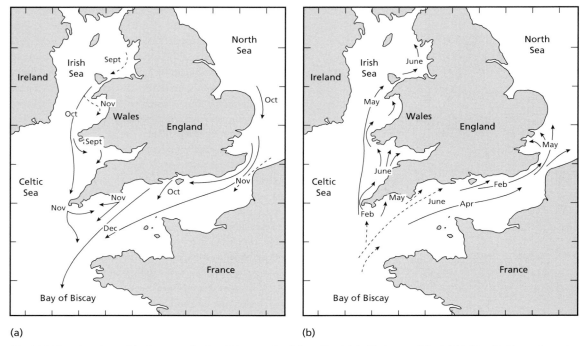

Fig. 3.11 The migrations of the European Sea bass *Dicentrarchus labrax* (Percichthyidae) around the British Isles. (a) Shows the paths of autumn migrations as the sea temperature falls and the bass move to overwintering grounds, (b) shows the path of spring migrations as the fish move north and spawn. After Pickett and Pawson (1994).

the actual movements of fish to be followed (and may provide information on depth, temperature and orientation) have demonstrated that rates of movement are greater than predicted in most conventional studies.

We have already seen that swimming costs are high, often three to four times SMR at the optimal swimming speed (section 2.5.3). Fish exhibit a number of strategies to reduce energy costs during migration. Some species have been shown to migrate using **selective tidal stream transport** (Arnold *et al.*, 1994). Fish, including flatfishes, leave the seabed at, or shortly after, slack water and swim downstream with the tide before returning to the seabed at the next slack water (Fig. 3.12). These vertical movements explain why fishers sometimes report marked changes in catch rates with bottom trawling gears as the tide starts to flow. Plaice, *Pleuronectes platessa*, in the North Sea, for example, use one tidal direction to reach their spawning grounds and the opposing direction to return to their feeding grounds (Metcalfe & Arnold, 1997). A 35-cm plaice could potentially reduce the metabolic cost of a round

trip migration of 560 km by consistently orienting downtide, equivalent to approximately 30% of the energy content of eggs spawned each year (Metcalfe *et al.*, 1993).

Migrating fish often maintain a consistent heading in the open sea for long periods frequently without any apparent access to direct visual or tactile cues. This led to the assumption that fish had an internal compass (Arnold & Metcalfe, 1989). To investigate how plaice navigated in the North Sea, fish were tagged with transponding acoustic compass tags that allowed a research vessel to determine their location and orientation. Random walk models were used to simulate the change in heading that would be expected if the plaice were not actively orienting and the results were compared with the movements of migrating fish in the open sea. By daytime, plaice swimming in midwater did not maintain a heading that was consistent with active orientation to an external directional cue but at night, however, they did. Since there were no obvious visual and tactile clues (the southern North Sea where

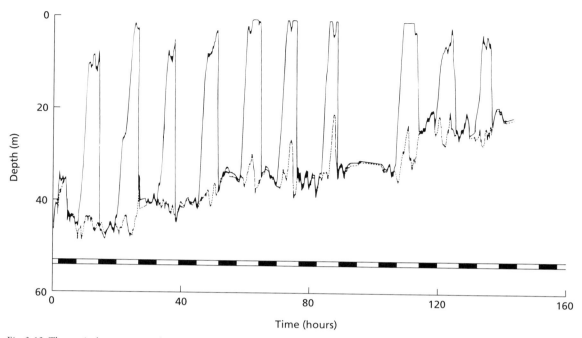

Fig. 3.12 The vertical movements of a migrating plaice *Pleuronectes platessa* using tidal stream transport, recorded by an electronic tag. The continuous line shows the depth of the fish and the broken line the depth of the seabed. The bar along the time axis shows whether the tide is going north (unshaded) or south (shaded). This fish is migrating north and swims up in the water column when the tide is flowing in that direction. After Metcalfe *et al.* (1991).

the study was conducted is very turbid), the external reference appeared to be geophysical. Possible cues could be the electrical field generated by flowing sea water or the earth's magnetic field. The former is more likely since many species of fish are sensitive to the low voltages generated (Metcalfe *et al.*, 1993).

Techniques such as electronic tagging and acoustic observation techniques (sections 10.2.3 and 10.2.6) have improved understanding of fish migration and its relation to oceanographic features. For example, movements of Atlantic cod *Gadus morhua* off New-foundland were investigated with acoustic methods and experimental trawling (Rose, 1993). The cod were observed spawning on the shelf edge in spring. Here, they formed a shoal shaped like an inverted saucer, with a radius of 1.2 km and a central height of 40–45 m. Fish were tightly packed in this shoal, each occupying only 1 m³ on average. After spawning, the cod began to migrate towards the coast. A group consisting of hundreds of millions of fish, some 80% of the total

stock, migrated south-west along a deep 'highway' of warm oceanic water which was 2–3°C warmer than the surrounding shelf waters. On these migrations the cod fed on capelin and shrimp. When the cod arrived in shallower water in summer they dispersed north before returning in autumn to wintering areas on offshore margins of the highway. The migration of the cod could provide a number of advantages. First, it kept them in relatively warm water during the winter when temperatures inshore were around 0°C and provided a spawning area that led to the eggs hatching in a beneficial environment for growth and development. Second, the migration allowed cod to feed successfully in relatively warm and productive inshore waters during summer and autumn.

3.5.3 Larval transport, retention and dispersal

Successful survival, growth and development during the early life history usually rely on a larva getting somewhere (such as an estuary or sandy inshore nursery area) or staying somewhere, such as in the vicinity of an isolated oceanic atoll. Larval transport or retention requires that spawning and larval development coincide, in time and space, with appropriate oceanographic conditions and that the larvae show behavioural responses to environmental stimuli. The coincidence of spawning and favourable environmental conditions for larval transport and prey production has led to the evolution of specific spawning times and spawning grounds. However, year-to-year fluctuations in oceanographic factors and production cycles still account for much of the variation in recruitment success (Chapter 4). The dispersal of some larvae away from the main areas of transport or retention has evolutionary significance because new environments may be colonized. Larval transport is often responsible for sustaining fish or invertebrate populations in intensively fished areas, since the larvae may recruit there from less heavily fished areas. Thus, the links between these sinks and sources need to be understood if larval supply is to be assured. Moreover, larval transport needs to be understood in order to predict the effects of estuarine and coastal development on the use of nursery grounds by juvenile fish (Weinstein, 1988).

Most fished species have pelagic larvae. For example, of 99 teleost families with representatives on coral reefs, only four plus a single species in a fifth family lack a pelagic larval stage (Leis, 1991). The duration of larval life varies among species and with temperature. Some gastropods have a pelagic phase that lasts only 2–3 days, but teleost larvae may be pelagic for 2–3 months. Metamorphosis may be delayed in larvae that do not reach an appropriate substrate for settlement, but longer larval life will increase predation risk. The larval life spans of many fishes have been determined from counts of daily growth rings on larval otoliths (section 9.3.2).

It is remarkable that larvae of a few millimetres in length, often hatching hundreds of kilometres offshore, can recruit to, and stay within, estuaries and other areas with strong tidal flushing. Equally, on isolated islands such as Christmas Island in the Pacific Ocean, that are surrounded by deep ocean in all directions, many larvae still reach suitable inshore settlement sites. Larval behaviour mediates transport and retention processes. The ability of larvae to mediate transport increases with age and development stage. This appears to be due to their rapid development of sensory capability

(Blaxter, 1986, 1987). In an interesting study of ontogenetic changes in larval distribution, Rowe and Epifanio (1994) measured the flux of larval weakfish *Cynoscion regalis* (Sciaenidae) in and out of an American estuary. For yolk-sac larvae there was no significant difference in flood and ebb flux, but for older larvae mean flux was significantly greater as the tide flooded. This suggested that older larvae used behavioural mechanisms to remain in the estuarine nursery.

Behavioural traits, such as vertical migration, could allow larvae to enter water masses moving in different directions at different rates, and control their rate and direction of transport (Weinstein, 1988). Larval Atlantic menhaden *Brevoortia tyrannus*, which recruit to American estuaries from offshore spawning grounds, showed patterns of behaviour that were consistent with such regulation. When placed in water columns that contained offshore or estuarine water of the same salinity and temperature, young menhaden larvae (< 8 mm total length) moved to the surface of the column in estuarine water and towards the base in offshore water. Older larvae of 14–16 mm showed the opposite response. Water type appeared to induce positional changes in the larvae which reflected the vertical positions necessary for horizontal transport in offshore and estuarine areas by larvae of different ages. Thus, young larvae showing this response would be flushed from the estuary back into pelagic environment while larger larvae would be retained (Forward *et al.*, 1996). Such a mechanism could account for patterns of movement to nursery areas in many other species. Table 3.5 summarizes the potential role of physical processes, environmental stimuli and behaviour as larval fish recruit to estuarine nursery areas.

Not surprisingly, the coincidence of spawning time and oceanographic features that ensures larval retention or transport will have a dramatic effect on the survival of larvae and hence on recruitment success. Indeed, many of the key hypotheses for recruitment variation (Chapter 4) emphasize the importance of larvae being retained in appropriate areas for feeding, growth or settlement. Larvae are often retained in gyres. For example, the Tortugas Gyre in the Caribbean retains penaeid shrimp larvae and allows them to recruit to nursery grounds in Florida Bay (Griales & Lee, 1995). A seasonal gyre in the Irish Sea retains fish and decapod larvae. It forms in spring when reduced wind stress and increased solar radiation cause the Irish Sea to stratify,

Table 3.5 The role of physical processes, environmental stimuli and behaviour as larval fish recruit to estuarine nursery areas. Based on Boehlert and Mundy (1988).

Environment	Pelagic	Nearshore	Estuary
Development stage	Yolk-sac and young larvae	Older and metamorphosing larvae	Post-larvae and juveniles
Physical process	Oceanographic (upwelling, surface drift)	Longshore currents	Tidal flux
Activity transport	Drift	Inshore movement	Tidal stream transport
Behaviour	Vertical movements	Circadian activity rhythms	Circatidal activity rhythms
Potential stimuli	Light, temperature, prey distribution	Diel cue (e.g. light)	Tidal cue (e.g. temperature, salinity)

and this isolates a dome of cold bottom water in the deep basin of 100 m depth and more. The resultant differences in density drive the gyre (Dickey-Collas *et al.*, 1997).

The larvae of reef-associated fishes may also be retained around coral reefs (Jones *et al.*, 1999; Swearer *et al.*, 1999). Retention must have evolved around isolated atolls to ensure successful recruitment to the local environment. While wrasse (Labridae) larvae have been caught more than 1000 km from source populations, levels of transport over this distance will be very low and such dispersal will not maintain fished populations (Leis, 1983). Even in species with relatively long larval life, transport over tens and hundreds of kilometres may occur so infrequently that gene flow is restricted. In a study of the convict surgeonfish *Acanthurus triostegus* (Acanthuridae) in French Polynesia, Planes (1993) compared populations at different scales from within reefs to between archipelagos. Despite a larval duration of 2 months, there were differences in the allele frequencies of fishes collected on different islands.

Patterns of spawning on reefs were once viewed as adaptations to maximize the advection of eggs away from areas of very high predation (Figure 2.11) to pelagic habitats. However, even if eggs are spawned in areas such as reef passes and carried seaward it is likely that oceanographic processes retain them in the vicinity, even if they do not pass across the reef (Boehlert & Mundy, 1993; Boehlert, 1996). Spawning sites could equally have evolved because they facilitate mate choice and reduce the predation hazards for spawners (Warner, 1988).

Many reef fish larvae settle in habitats such as mangrove swamps and back-reef rubble areas that are adjacent to the reef slopes where adults are found. Temperate demersal species, conversely, may use nursery areas hundreds of kilometres from the area occupied by the adult stock. Since the survival of larvae that do not reach nurseries may be lower, the prevalence of oceanographic features that make transport possible can have a key effect on recruitment success. Relationships between transport and oceanographic features are well understood for a number of species. For example, larvae of walleye pollock *Theragra chalcogramma* spawned in the western Gulf of Alaska travel southwest towards a nursery area 200 km away. This appears to be an example of passive transport since the larvae followed the same trajectories as drifting satellite bouys (Hinckley *et al.*, 1991). Bluefish *Pomatomus saltatrix* (Pomatomidae) larvae are transported from spawning grounds in the South Atlantic Bight to nurseries in the Middle Atlantic Bight. Gulf Stream-associated flow moves the larvae north-east from spawning grounds and then transports them to the shelf edge on the Middle Atlantic Bight in warm core ring streamers (section 2.3.2). The juveniles then swim actively across the shelf edge, a behaviour which appears to be initiated by the dissipation of the shelf slope temperature front in late spring. The number of recruits reaching nurseries could depend on warm core ring activity and the timing may depend on the timing of front dissipation (Hare & Cowen, 1996). Fronts may act as a barrier to larval transport to inshore nurseries because they constrain cross frontal flows. As a result, currents develop parallel to the fronts and these transport larvae which collect at the front to a greater range of settlement locations: a so-called 'larval conduit'. Such a larval conduit transports Dungeness crab *Cancer magister* (infraorder Brachyura) larvae on the coast of the United States (Eggleston *et al.*, 1998). The direction of transport along the front is strongly influenced by prevailing winds and this could have a significant impact on post-larval supply to particular estuaries. We discuss links between larval transport and recruitment success in section 4.2.2.

3.5.4 Metapopulations

Most marine invertebrates and some relatively site-attached fishes, such as those on coral reefs, exist as **metapopulations**: sedentary or relatively sedentary subpopulations connected by dispersing larval stages (Hanski, 1991, 1999). Metapopulations take various forms, from total connectance, where all populations act as sources and sinks for the dispersing larvae, to zero connectance, where individual components are self sustaining. More frequently, there is simply unbal-

anced connectance, where components act as sources and sinks to various degrees (Fig. 3.13). Patterns of connectance are often well established and consistent, since beds of oysters and other bivalves may persist in the same locations for many years. In some cases there can be asymmetry, where one population is a sink for recruits but does not supply other locations. From a fisheries viewpoint this is very significant since such a component can be fished intensively with no effect on the other populations (Botsford *et al.*, 1998; Orensanz & Jamieson, 1998). There are examples of this for northern shrimp *Pandalus borealis* populations in some fjords (Parsons & Frechétte, 1989) and Pacific oysters *Crassostrea gigas* (Ostreidae) at sites in British Colombia, Canada (Jamieson & Francis, 1986). An understanding of metapopulation structure is key to the siting of protected areas (section 17.5.3) since some populations may only be sustained by larval inputs from a few sources and extirpation of these sources would lead to their demise. Thus, management decisions that affect 'upstream' populations will have impacts on the components 'downstream'.

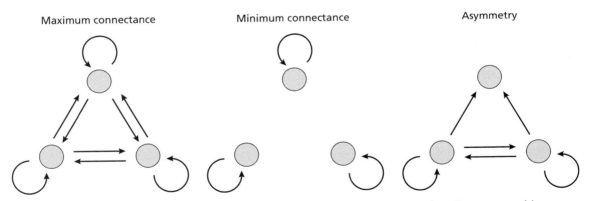

Fig. 3.13 Patterns of connectance and asymmetry in metapopulations. Maximum connectance is where all components of the population are connected, minimum connectance is where all components are isolated but self sustaining, and asymmetry is where one component is sustained solely by larval inputs from another.

Summary

• Landings of marine species are currently 85.6 million $t\,y^{-1}$ and account for 92% (by weight) of global fisheries landings. Twenty marine species account for almost 40% of the total landed weight.

• Fished species have many life histories and these determine their accessibility to fishers and responses to exploitation. Those with high fecundity and small eggs may grow in weight by five to six orders of magnitude during their life cycle.

• Growth is indeterminate in most fished species and maturity occurs when the benefits of reproduction outweigh the disadvantages of growing more slowly. There are close links between growth and other life-history traits. In general, fast growth is associated with low age and size at maturity, high reproductive output, short life span and small maximum size.

• Fished species may form a number of stocks across their geographical range. Each stock will have a characteristic life-history strategy that is predominantly a phenotypic response to the environment.

• Many fished species migrate and this may allow individuals to remain in optimal conditions for feeding, growth or reproduction for the longest possible period. Migrating fish often cross boundaries of marine reserves, nations or international management areas and fishing mortality in one area can influence catches in another.

• The pelagic larvae of many fished species have to be retained in, or transported to, appropriate nursery habitats in order to survive and grow. Larval behaviour mediates retention and transport.

• Some fished species form metapopulations where sedentary or relatively sedentary adult subpopulations are connected by dispersing larval stages.

Further reading

The FAO publish annual reports that summarize global landings by species, countries and regions and the same information is also available from their website (FAO, 1999; www.fao.org). The general biology of fishes, cephalopods and other invertebrates is well described by Bone *et al.* (1995), Boyle (1990) and Caddy (1989), respectively. Charnov (1993) has written an excellent review of relationships between life-history traits and their evolutionary significance. Chambers and Trippel (1997) contains many chapters on aspects of larval biology, transport and recruitment.

4 Population structure in space and time

4.1 Introduction

The abundance of marine species fluctuates in space and time. These fluctuations are driven by:

1 physical and biological processes that affect the production and survival of eggs and larvae;

2 growth and mortality during the juvenile and adult phases; and

3 behavioural processes such as migration or density dependent habitat use.

In this chapter we concentrate on (1), which is thought by many people to be the most important. Fluctuations in abundance have many implications for fishers, fishery managers and the marine ecosystem. For example, changes in the abundance and distribution of pelagic fishes such as the Pacific sardine *Sardinops sagax* (Clupeidae) and Peruvian anchovy *Engraulis ringens* (Engraulidae) led to the proliferation and demise of industries that employed thousands of people and produced hundreds of thousands of tonnes of fish products and fishmeal.

Fish populations fluctuated in abundance long before they were fished. Thus, counts of Pacific sardine and northern anchovy *Engraulis mordax* (Engraulidae) scales in cores taken from anaerobic sediments off California showed that sardine and anchovy abundance had changed dramatically over 2000 years (Fig. 4.1) (Soutar & Isaacs, 1974; Baumgartner *et al.*, 1992). The collapse of the sardine stock that occurred in the latter half of this century, and led to the loss of Californian canning industries described in *Cannery Row* and other novels by John Steinbeck, appears to be just one of a series of such collapses. Many other records of fisheries that extend back to the fourteenth century show cycles of collapse and recovery when fishing pressure was low by contemporary standards (Cushing, 1982).

The aim of this chapter is to show why fished populations fluctuate dramatically in space and time. Fluctua-tions are often unpredictable and create uncertainty for the fisheries scientist and everyone else who depends on the fishery. Thus, recognition of the dynamic beha-viour of fished stocks is central to appreciating the realistic goals and expectations of management.

4.2 Recruitment

When scientists developed techniques for ageing fishes and looked at the age structure of their populations they showed that only a few year classes accounted for most of the biomass even though many year classes were present. Fish in the abundant year classes would often dominate catches for many years before new abundant year classes recruited to the fishery. The first convincing evidence for these patterns was published by Hjort (1914) following his study of Atlantic herring *Clupea harengus* and Atlantic cod *Gadus morhua* populations off Norway. He showed that one year class would make the greatest contribution to catches for several years. For example, the 1904 year class of her-ring dominated catches from 1908 to 1914 (Fig. 4.2). As fishery scientists followed year classes in more stocks, it was increasingly clear that year class success or fail-ure was largely determined during the first few months of life. Interest in the early life history, and the reasons for fluctuations in year-class strength, have preoccu-pied fisheries scientists since (Blaxter, 1974; Lasker, 1978; Blaxter *et al.*, 1989; Chambers & Trippel, 1997).

Recruitment is defined as the number of individuals that reach a specified stage of the life cycle (e.g. meta-morphosis, settlement or joining the fishery). Reef fish ecologists often equate recruitment with the time when larvae first settle on a reef and can be counted by under-water visual census techniques (section 10.2.2), while fishery biologists working on deep-sea species with juve-niles that are inaccessible to sampling gears may meas-ure recruitment as the abundance of the youngest year

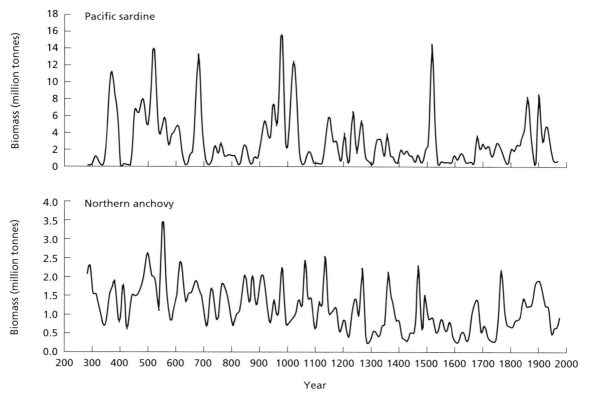

Fig. 4.1 Historical estimates of Pacific sardine and northern anchovy biomass off California. Biomass was estimated from scale deposition rates (density) in two sediment cores from the anoxic Santa Barbara basin. Scale deposition rate was assumed proportional to stock biomass because such relationships can be demonstrated using contemporary data. After Baumgartner *et al.* (1992).

class entering the fishery. The reef fishes may only be a few months old, while the deep-sea species may be 3–5 years old. Recruitment often varies by a factor of 20 or more between years, and the relative effect of each recruitment event on total stock size will depend on the age structure of existing stock (Cushing, 1982, 1996; Rothschild, 1986). Figure 4.3 gives some examples of recruitment time-series for exploited populations, showing the ubiquity of recruitment fluctuation.

Recruitment depends on many factors such as the abundance and distribution of the mature adult population, the number and viability of eggs produced, and the subsequent survival of eggs and larvae. Most fished species release many small eggs (section 3.4.3) and over 99% of mortality occurs between egg fertilization and the settlement or recruitment of juveniles. This is an important period in the life history since small variations in mortality rates have profound effects on sub-

sequent abundance. Higher fecundity is associated with greater recruitment variability (Rickman *et al.*, 2000). The variable supply of recruits determines population structure, the strength of interactions in multispecies communities, and the biomass of fishes that can be caught. Fisheries based on one or two year classes are highly dependent on successful recruitment, and one poor recruitment event when fishing effort is very high may cause collapse.

4.2.1 Spawner and recruit relationships

More spawners should produce more eggs and so one would expect a positive relationship between spawner and recruit abundance. Any such relationships are obscured by considerable variance (Fig. 4.4). Relationships between spawner and recruit abundance (commonly known as **stock-recruitment** or **spawner-recruit**

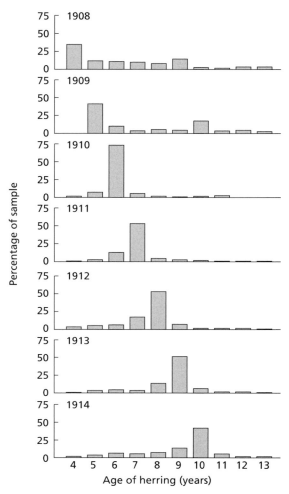

Fig. 4.2 The age composition of Norwegian herring catch samples from 1908 to 1914. Data from Hjort (1914).

larvae. While such a relationship may exist across a range of stock sizes, there must be ultimate limits to the resources available. Decrease in mean recruitment per spawner almost always occurs at high stock densities, so some density-dependent mechanism must operate. For example, competition between larvae could lead to starvation or adult spawners could consume their own eggs and larvae. The effects of spawner abundance on recruitment cannot be discounted, and in those cases where no relationships have been shown this is usually because a sufficiently wide range of spawner abundance has not been considered. Sadly, as stocks are over-fished, the relationships have become more apparent (Myers *et al.*, 1995a).

Spawner–recruit relationships are also important because they provide a basis for predicting the range of recruitment that is expected for a given size of spawning stock and, for semelparous species such as squids and Pacific salmon, *Oncorhynchus* spp. (Salmonidae), they provide a basis for predicting the stock size necessary to ensure that the stock is replaced in the next generation (section 7.8.1). A range of models can be fitted to spawner and recruit data of the type shown in Fig. 4.4. The models describe the average situation rather than what happens in any given year.

Spawner abundance is often used as an index of total egg production in spawner–recruit relationships because egg production is difficult and expensive to measure. However, spawner abundance is not directly proportional to egg production. This is because fishing changes the size and age structure of stocks, leaving fewer age classes with smaller individuals. Smaller individuals have lower relative fecundity (egg production per unit body weight, section 3.4.3). Thus, egg production per unit of stock biomass may decrease as fishing intensity increases (Solemdal, 1997). A population of smaller individuals may also have a shorter spawning season, and this may reduce the probability that larvae hatch when oceanographic conditions favour growth and survival (Trippel *et al.*, 1997). Larger females may also produce larger eggs (Kjesbu, 1989) and hatching success may increase with repeat spawning. In one study, Atlantic cod *Gadus morhua* spawning for the first time had a 40% hatch rate that increased to 70% in the second and third spawning year (Solemdal *et al.*, 1995). The reduced egg production that results from decreasing size and age of spawners may be countered by plasticity in life histories that reduces age at

relationships) are important because they describe the ability of a stock to maintain abundance in response to intensive fishing pressure. If this compensation did not occur, stocks would tend to extinction or infinite population size and neither feature is observed in nature on biological time scales.

If mortality in the early life stages were density independent then factors such as predation, starvation and oceanographic processes would act independently of the number of eggs spawned (spawner abundance). Thus, spawner and recruit abundance would be directly proportional, with variability around the relationship due to interannual fluctuations in the factors affecting

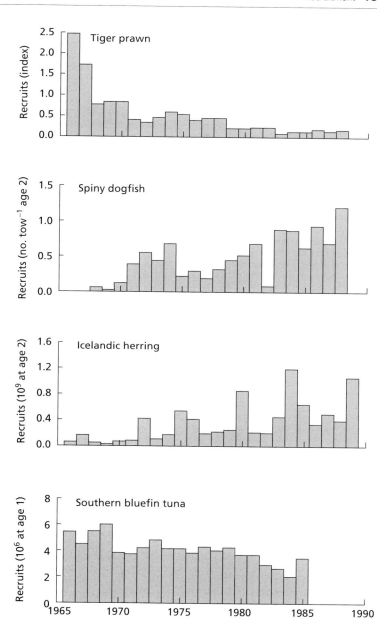

Fig. 4.3 Recruitment variation in fished stocks. Time series of recruitment data for: the tiger prawn *Penaeus esculentus* (Penaeidae) of the Exmouth Gulf, Western Australia; the Spiny dogfish (spurdog) *Squalus acanthias* (Squalidae) of the north-west Atlantic; the Icelandic summer spawning herring *Clupea harengus*; and the southern bluefin tuna *Thunnus maccoyii* of the Pacific Ocean. Note the different measures of recruit abundance and that values of zero indicate that no data were available. Data from the compilation of Myers *et al.* (1995b).

maturity. The effects of stock structure on egg production, and the possibility for relating recruitment to stock structure and egg production, are likely to provide a fertile area for further research. For now, we need to remember that spawning stock biomass is not linearly related to egg production.

Several theoretical relationships have been used to link spawner and recruit abundance. An ideal spawner–recruit relationship should provide an acceptable fit to empirical data, pass through the origin (no spawners produce no recruits!), not fall to the spawner axis at high stock densities (all reproduction cannot be

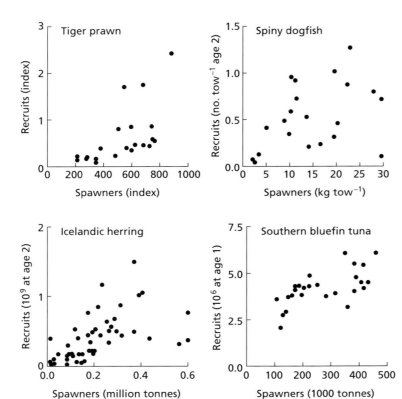

Fig. 4.4 Examples of relationships between spawner and recruit abundance for: the tiger prawn of the Exmouth Gulf (1965–88); the spiny dogfish (spurdog) of the north-west Atlantic (1968–88); the Icelandic summer spawning herring (1947–95); and the southern bluefin tuna (1960–85). Note the different measures of spawner and recruit abundance. Data from the compilation of Myers *et al.* (1995b).

eliminated), and the rate of recruitment should change continuously and smoothly as parent stock increases (Ricker, 1975; Hilborn & Walters, 1992; Quinn & Deriso, 1999).

In estimating one spawner–recruit relationship from several years' data we have to assume that the spawner–recruit relationship is constant over time. This means that historical data points are equivalent to contemporary data points and that historical data can be used to produce a predictive curve. Clearly, this is unrealistic when there are multispecies interactions between stocks, and abiotic features of the environment can change dramatically with time (e.g. O'Brien *et al.*, 2000). Moreover, time series of stock abundance data do not provide independent data points because abundance at time $t + 1$ depends on abundance at time t, a process called autocorrelation. Good recruitment at low stock sizes will lead to large stocks in subsequent years and poor recruitment produces small stocks. If fishing led to rapid positive and negative changes in spawning stock size then bias would be reduced, but

most fisheries are subject to rising effort and proceed on a one-way trip to low abundance. The result is that mean recruitment rates are biased upward at low stock sizes and biased down at high stock sizes (Hilborn & Walter, 1992). Experimental manipulation of fishing effort yields better information (Walters & Hilborn, 1976) but this approach is impractical or politically untenable in most major fisheries.

A further complication is that a stock that has been designated for management purposes may actually consist of several true stocks (largely reproductively isolated subunits, section 3.5.1) each of which is responding differently to exploitation. If the spawner–recruit relationship aggregates across these stocks, it reveals nothing of their true dynamics. The spawner–recruit relationship will change with stock size, habitat and distribution (Walters, 1987). Spawner–recruit relationships help us to understand the impact of exploitation on populations, but they have been misinterpreted in ways that their originators never intended. Spawner–recruit relationships show mean relationships, but the

Box 4.1

Alternative formulations for spawner–recruit relationships

Spawner–recruit relationships can be written in many ways. The meaning of parameters in each formulation is different, but the curves they describe are the same. These alternative formulations simplify some calculations and will be encountered in the fisheries or ecological literature. Some are used elsewhere in this book.

The Beverton and Holt (1957) model (equation 4.1), for example, that describes an asymptotic curve with consistently high mean recruitment at high spawner abundance, can also be written as:

$$R = \frac{S}{a^* + b^* S} \tag{1}$$

where $a^* = 1/a$ and $b^* = b/a$ (a and b from equation 4.1). So, for example, the relationship shown in Fig. 4.5 when $a = 6$ and $b = 0.2$, will also be produced using equation 4.1 when $a^* = 0.167$ and $b^* = 0.0333$. The curve plotted using equation B4.1(1) approaches asymptotic recruitment at $1/b^*$.

Another formulation of the Beverton and Holt (1957) model is:

$$R = \frac{\alpha S}{\beta + S} \tag{2}$$

where α is the maximum number of recruits produced ($= 1/b^*$) and β is the abundance of spawners that gives mean recruitment of $\alpha/2$ ($= a^*/b^*$).

Other parameterizations of the Beverton and Holt, Ricker, and Shepherd models are described in Ricker (1975), Hilborn and Walters (1992), and Quinn and Deriso (1999).

---

large and significant variance around the mean drives short-term stock dynamics. The fishery manager needs to consider variances and not just means; in much the same way that an engineer designs bridges to survive 100 years of storms rather than mean windspeed.

We now present three density-dependent models that describe relationships between spawner (S) and recruit (R) abundance. Each can be written in several different ways (Box 4.1; Cushing, 1988b; Hilborn & Walters, 1992; Elliott, 1994) but we introduce them all in a form where the parameter a is recruits per unit spawners at low stock sizes and b captures the density dependence of the relationship. The same models of density dependence have been developed and used by terrestrial ecologists (Elliott, 1994). We describe the two-parameter models of Beverton and Holt (1957) and Ricker (1954) and the three-parameter model of Shepherd (1982). Other models have been suggested by Cushing (1971, 1973), Deriso (1980) and Paulik (1973); see Quinn and Deriso (1999).

The **Beverton and Holt model** (1957) is:

$$R = \frac{aS}{1 + bS} \tag{4.1}$$

This equation describes an asymptotic curve with consistently high mean recruitment at high spawner abundance (Fig. 4.5). Parameter a increases the height of the

asymptote and reduces curvature while b increases the rate of approach to the asymptote (Fig. 4.5). Maximum mean recruitment is calculated as a/b and is produced at an infinitely large spawning stock sizes.

The **Ricker model** (1954) is dome shaped (Fig. 4.6) and is based on increasing density dependent mortality at high spawner abundance:

$$R = aSe^{-bs} \tag{4.2}$$

High a, the density-independent parameter, leads to a higher steeper peak in recruitment at a fixed level of spawner abundance, while high b, the density-dependent parameter, decreases the height of the peak and reduces the level of spawner abundance at which the peak occurs (Fig. 4.6). Maximum mean recruitment occurs at spawner abundance $S = 1/b$.

The **Shepherd model** (1982) includes an additional parameter and can take both the Beverton and Holt (1957) and Ricker (1954) forms (Fig. 4.7):

$$R = \frac{aS}{1 + (bS)^c} \tag{4.3}$$

Parameter a is the slope of the spawner recruit curve at low stock sizes (Fig. 4.7a). Increases in parameter b, the density-dependent parameter, decrease the height of the recruitment peak and reduce the level of spawner abundance at which the peak occurs (Fig. 4.7b).

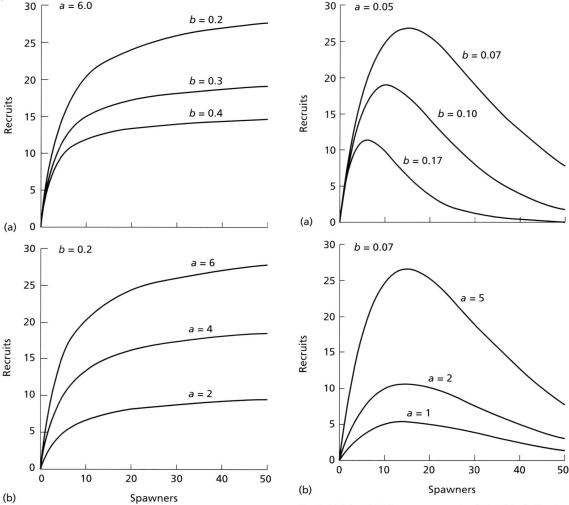

Fig. 4.5 Beverton and Holt (1957) spawner–recruit relationships indicating the effects of (a) changing parameter b when $a = 6$, and (b) changing a when $b = 0.2$.

Fig. 4.6 Ricker (1954) spawner–recruit relationships indicating the effects of (a) changing parameter b when $a = 0.05$, and (b) changing a when $b = 0.07$.

Parameter c determines the shape of the curve. When $c < 1$ the curve rises indefinitely, when $c = 1$, the model takes the Beverton and Holt form, and when $c > 1$, the model is dome shaped like the Ricker form (Fig. 4.7c).

Examples of the three relationships fitted to spawner–recruit data for the spiny dogfish *Squalus acanthias* from the north-west Atlantic and red king crab *Paralithodes camischatica* (infraorder Brachyura) from Alaska are shown in Fig. 4.8. Note the enormous variation around the mean spawner–recruit relationship. This is not unusual. Variation will be the result of errors

in abundance estimates for spawners and recruits (Chapter 10), interannual variation in egg, larval and juvenile survival and the other biases we have already described. It is generally assumed that the variance in recruitment not explained by the relationship has a log normal distribution and this limits the use of linear transformations for estimating parameters because they only meet this assumption in the Ricker case. Non-linear methods of fitting spawner–recruit relationships are reviewed in Hilborn and Walters (1992) and Quinn and Deriso (1999). The residuals from fitted spawner–

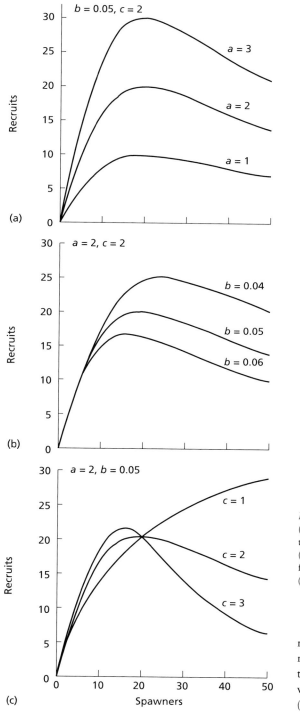

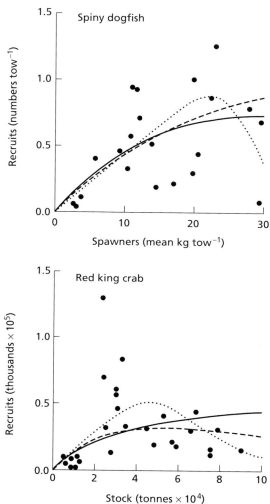

Fig. 4.8 Spawner–recruit relationships of Beverton and Holt (——), Ricker (– – –) and Shepherd (. . . .) fitted to data for the spiny dogfish *Squalus acanthias* of the north-west Atlantic (1968–88), and for the red king crab *Paralithodes camischatica* from Bristol Bay, Alaska (1968–88). Data from Myers *et al.* (1995b).

Fig. 4.7 Shepherd (1982) spawner–recruit relationships indicating the effects of (a) changing parameter a when $b = 0.05$ and $c = 2$; (b) changing b when $a = 2$ and $c = 2$; and (c) changing c when $a = 2$ and $b = 0.05$.

recruit relationships have often been used to look at relationships between recruitment and other aspects of the life history, independent of changes in stock size, as well as environmental and multispecies relationships (e.g. Rickman *et al.*, 2000).

With spawner–recruit models providing such poor fits to empirical data, are there any useful generalizations we can make? Ransom Myers has maintained an

Recruits (millions) (i)								
97.5	0.00	0.05	0.05	0.05	0.05	0.05	0.05	0.05
82.5	0.00	0.05	0.10	0.10	0.15	0.20	0.25	0.25
67.5	0.05	0.10	0.15	0.25	0.30	0.35	0.35	0.35
52.5	0.15	0.20	0.35	0.30	0.25	0.20	0.20	0.20
37.5	0.15	0.40	0.25	0.25	0.20	0.15	0.10	0.10
22.5	0.45	0.15	0.10	0.05	0.05	0.05	0.05	0.05
7.5	0.20	0.05	0.00	0.00	0.00	0.00	0.00	0.00
	12.5	37.5	62.5	87.5	112.5	137.5	162.5	187.5

Stock biomass (tonnes $\times 10^6$) (j)

Table 4.1 A spawner–recruits probability transition matrix for south-west African (Cape) anchovy *Engraulis capensis* (Engraulidae). The transition matrix elements indicate the probability that a spawner biomass in subdivision with midpoint *j* will result in the number of recruits in subdivision with midpoint *i*. Based on Getz and Swartzman (1981).

impressive database of spawner–recruit data for many stocks (www.mscs.dal.ca/~myers) and his analyses of many spawner–recruit relationships have demonstrated that recruitment does fall at low stock sizes. In cases where recruitment appears to be independent of spawner abundance, the stock has simply not been fished to very low levels (Myers *et al.*, 1995a).

Other methods of describing spawner–recruit relationships do not rely on fitting average curves. Rather, they break down the range of potential stocks and recruitments into intervals and indicate the probability that a given spawning stock produces a given recruitment (Getz & Swartzman, 1981) (Table 4.1). This method is not widely used, mainly because data requirements are formidable and long data series over a range of stock sizes are needed to estimate the parameters (Walters, 1975). However, the method has many advantages over the traditonal two- and three-parameter relationships because it explicitly incorporates the type of variation seen in the data and is not constrained by a specific relationship (Hilborn & Walters, 1992).

4.2.2 Mortality during the early life history

Much of the variance around spawner–recruit relationships is due to massive and largely unpredictable variation in egg and larval survival. In this section we focus on the factors that determine mortality rates during the early life history and how these affect recruitment variability and year-class strength. We begin by looking at the development of eggs and larvae, the processes that determine developmental rates and how ontogenetic changes affect the response of larvae to predators or starvation. We then consider the roles of starvation and predation mortality and consider theories that link

changes in production cycles and oceanographic features to variations in recruitment.

Development

Development and mortality rates are closely linked. During development, every larva is challenged daily by circumstances that demand certain levels of performance if it is to survive. Behavioural and physiological performance are the key to survival and subsequent recruitment (Fuiman & Higgs, 1997). **Development** can be divided into growth and ontogeny. **Growth** leads to changes in the size or abundance of existing features, and **ontogeny** leads to the appearance of new features and reorganization or loss of existing ones. Most major ontogenetic changes occur prior to metamorphosis, when the larva adopts characteristics that are essentially fixed until maturity. An example of developmental changes during the early life history of a sole *Solea solea* from the north-east Atlantic is shown in Fig. 4.9. The pelagic yolk sac larva develops into a metamorphosed benthic juvenile in the course of a few weeks. Rates of egg and larval development depend upon the genetic characteristics of a species and water temperature. Figure 4.10 shows how the time from fertilization to hatch (a) and hatch to mouth opening (b, shortly before first feeding) changes with temperature for the European sea bass *Dicentrarchus labrax*. Across its latitudinal range this species spawns at temperatures from 9°C to 20°C. Larvae hatching at higher temperatures have the potential to grow faster so they are less accessible over time to a given size class of predator. However, they have higher daily food requirements and can starve more quickly (Blaxter, 1992).

The relationship between the environment in which a larva hatches and the timing of major transitions

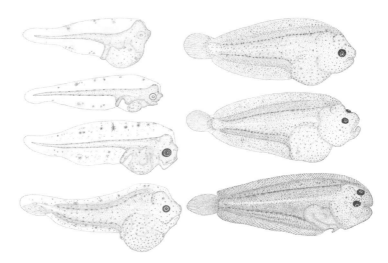

Fig. 4.9 Development of a European sole *Solea solea* (Soleidae) larva. From Nichols (1976).

in body form, muscle and fin development, the feeding apparatus, visual mechanisms, physiology and behaviour, are critical to survival. As Atlantic herring *Clupea harengus* larvae grow from 20 mm to 30 mm in total length (TL) their ability to evade predators improves because lateral line canals and auditory bullae develop (Blaxter & Fuiman, 1990). As larvae grow, their resistance to starvation increases. This starvation threshold is usually measured as the **point of no return** (PNR), the time at which starving larvae become too weak to feed and recover. The PNR depends on temperature and species. For herring larvae the PNR increases from 5 or 6 days immediately after hatching when the larvae are 10–11 mm in length, to 15 days 90 days after hatching (Blaxter & Ehrlich, 1974). Older larvae also feed more effectively. The velocity of feeding strikes in herring larvae increased from 6 cm s^{-1} at first feeding (10–11 mm TL) to 25 cm s^{-1} at 15 mm. The ability to strike at prey more rapidly greatly increased the success in prey capture from 1% at first feed to 25% at 15 mm (Rosenthal & Hempel, 1970).

Starvation and predation

Changes in rates of larval starvation or predation can account for annual variations in recruitment. In his classic study of the factors determining fluctuations in fish stocks (section 1.2.2), Hjort (1914) proposed that the transport of larvae away from nursery areas and the mass starvation of larvae at the time of first feeding were potential causes of recruitment variation.

His ideas were subsequently developed and tested by Cushing (1975, 1996), Lasker (1981) and Sinclair (1988). Lasker (1981) showed that northern anchovy *Engraulis mordax* larvae hatching off the Californian coast would starve during periods of unstable oceanographic conditions. His **ocean stability hypothesis** was based on the observation that patches of high food concentration would develop when the ocean was calm, and that larvae in the patches could feed effectively. However, when the ocean was rough, prey would be dispersed and their density would become too low to support the larvae (Lasker, 1975, 1978). Thus, aggregations of food, rather than total integrated food, were more important to larval survival. The size of available prey was also important, and prey concentration had to exceed a threshold for successful feeding. Small anchovy larvae in the early stages of development had very high rates of feeding failure, and high prey densities were needed to compensate for this. As they grew, the larvae fed more successfully and hunted over larger areas.

Production cycles in temperate latitudes are characterized by a spring phytoplankton bloom that is followed by the proliferation of zooplankton (section 2.3.1). The timing of this production cycle can vary by 6 weeks from year to year, while the date of peak spawning is quite consistent (even though the spawning season often lasted 2–3 months). Cushing (1975) suggested that interannual variation in larval survival could be explained by the match or mismatch between

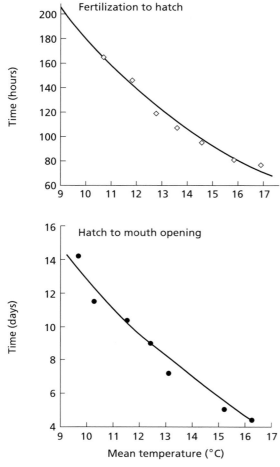

Fig. 4.10 Development rates of European sea bass *Dicentrarchus labrax* eggs from fertilization to hatch, and from hatch to 50% mouth opening at different temperatures. Data from Jennings and Pawson (1991, unpublished).

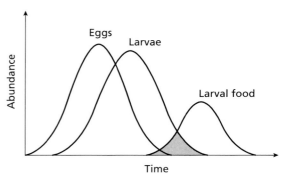

Fig. 4.11 The match–mismatch hypothesis suggests that the timing of larval production in relation to the timing of the production cycle will determine larval survival. The larger the shaded area, the higher survival will be. After Cushing (1982).

the timing of the production cycle and the peak spawning time. The strength of the match would determine the quantity of food available to the larvae (Fig. 4.11). This hypothesis is known as the **match–mismatch hypothesis**. If there is a mismatch in space or time between larval food production and larval hatching time then the larvae may not encounter sufficient food and reach the 'point of no return'.

Testing the match–mismatch hypothesis has been difficult. Mertz and Myers (1994) tested whether recruitment variability in cod stocks increased as spawning duration decreased, because short spawning duration would increase the probability of a mismatch. Their analysis supports the hypothesis, but in the absence of corresponding data on variability in the timing of the production peak, we might expect extended spawning to evolve in response to variable timing of production.

The match–mismatch hypothesis was extended by Pope *et al.* (1994) who expressed the idea in terms of the larvae having to surf on a wave of seasonal production. To maximize recruitment success, spawning would have to occur shortly before the prey biomass wave is propagated by increased solar radiation and decreased wind stress. The hatching larvae would then 'surf' this wave, growing and feeding as the prey themselves grew. Riding the wave at an appropriate time was the key to survival.

Another hypothesis for recruitment variation was proposed by Sinclair (1988) who emphasized the importance of the relationship between spawning time and stable oceanographic features which retain larvae in favourable environments. This **member–vagrant hypothesis** emphasized the role of physical rather than biological factors in governing spawning or year-class success. In reality, both physical and biological processes will interact and both can be important (Heath, 1992).

Initial study of the factors influencing year-class strength focused on starvation, but in recent years there has been much more interest in the effects of predation. Fish eggs and larvae are consumed by many predators whose feeding strategies are often size rather than species related (Bailey & Houde, 1989). In the zooplankton these include copepods, euphausiids, ctenophores and medusae. Many filter-feeding fishes such as engraulids and clupeids will also consume eggs and young larvae, and demersal predators will eat the eggs of demersal spawning fishes. This may lead to density-dependent regulation (MacCall, 1990). Egg and newly hatched larval densities of 100 m^{-3} are not unusual, and may exceed 1000 m^{-3} (Bailey & Houde, 1989). In one study the total mortality of Peruvian anchovy *Engraulis ringens* eggs was 68% d^{-1} and 21.9% of this was attributed to cannibalism (Alheit, 1987). Northern anchovy *Engraulis mordax* may consume 17.2% of their daily egg production, and given that total natural mortality was estimated to be 53% d^{-1}, cannibalism may be responsible for 32% of natural egg mortality (Hunter & Kimbrell, 1980).

Total mortality rates of eggs and larvae are size dependent, and mortality rates fall rapidly from around 50% d^{-1} to 10% d^{-1} in the first few weeks of life (Fig. 4.12). Predation has been estimated from ocean studies of the composition and abundance of predator communities and their consumption rates, or from experiments in large moored enclosures known as mesocosms.

Vulnerability depends on the size, type and abundance of predators and their capacity for ingestion. These factors are modulated by environmental conditions such as temperature, turbidity and light (Bailey & Houde, 1989; Mann, 1993). Temperature determines development rates and hence the rates at which former predators are lost and new predators gained. Density-dependent growth of larvae may lead to density-dependent mortality, since slow-growing larvae are subject to higher predation rates for longer. Density-dependent growth could result from competition for food between larvae. However, it is widely thought that densities of larval fish are usually too low for such competition to occur (Bailey & Houde, 1989; Heath, 1992; Houde, 1997).

Given that mortality rates decline with size during the early life history, it might be expected that getting big quickly will minimize mortality rates. This is known as the **bigger is better hypothesis** (Houde, 1987). However, high growth rates have costs that can lead to increased mortality, and actual growth rate evolves to balance the costs and benefits. Bigger may

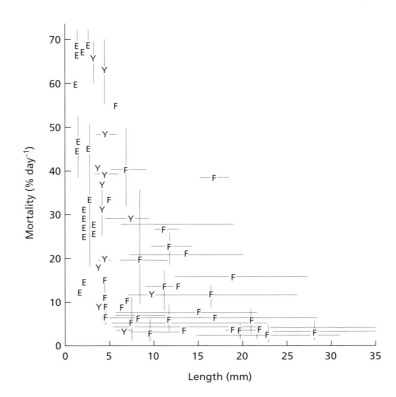

Fig. 4.12 Relationships between mortality rates of fish eggs (*E*), yolk sac larvae (*Y*) and feeding larvae (*F*) and size. Each letter represents a different species or study and the horizontal and vertical lines indicate the ranges in lengths and mortality rates recorded. After Bailey and Houde (1989).

be better but it is not necessarily the best strategy to get big quickly. If it were, then natural selection would drive the genetic capacity for growth to the maximum permitted by physiological and phylogenetic constraints. In a test of the 'bigger is better' hypothesis, Conover and Present (1990) reared Atlantic silverside *Menidia menidia* (order Mugiliformes) larvae through several generations under identical environmental conditions. The silverside has a 1-year life cycle and is found on the Atlantic coast of the United States. In the wild, larvae from northern populations grow for a much shorter period of the year than those in the south, because temperatures in the north are below the growth threshold for longer. Since southern populations are found in warmer water, they would be expected to evolve the fastest possible growth if selection favoured this. However, when northern and southern fish were reared under identical conditions the northern fish grew faster than those from the south. These differences in growth were maintained across several generations, showing they had a genetic basis. This **counter-gradient variation** in growth rate showed that bigger is not always better, a pattern that has since been reported for other species (Conover & Schultz, 1995). In section 2.5.3 we saw that high growth rates also mean high energy expenditure on the synthesis of new body tissue, so growth has a high cost. This may constrain other activities such as swimming, thereby leading to high predation mortality. As such, faster-growing larvae may be less able to escape predators and suffer higher mortality. In the case of the silversides, northern fish must grow rapidly in the summer, regardless of predation risk, because winter mortality is size dependent and small fish cannot store sufficient energy to survive the winter. The southern populations have a longer growing season, and may adopt a more conservative strategy that reserves a larger proportion of metabolic scope for maintenance and activity. This may increase their ability to evade predators. The results of Conover and Present (1990) are not surprising in the context of life-history evolution which is dominated by trade-offs such as those between growth and maturity (section 3.4.2) or egg size and fecundity (section 3.4.3).

Interaction of physical and biological processes
Physical processes have profound impacts on the production of plankton that provide food for larvae,

the distribution of that plankton, and the transport or retention of larvae in suitable areas for development. Variation in production is manifest on many scales since it may be driven by everything from local wind induced turbulence to shifts in ocean currents.

Perhaps the most dramatic effects of climatic influences on recruitment processes come from the study of upwelling ecosystems (section 2.3.2), where high levels of primary and fish production are the consequence of upwellings of nutrient-rich water. When these cease, so does most phytoplanktonic, zooplanktonic and ultimately fish production. Pelagic fisheries off the coast of Peru are affected dramatically by variations in upwelling intensity. These variations are due to the **El Niño–Southern Oscillation** (ENSO), a phenomenon driven by changes in atmospheric pressure systems (Mann & Lazier, 1996). ENSOs occur at intervals of a few years. They lead to changes in current patterns in the Pacific Ocean and affect the source of upwelling water (Graham & White, 1988). During an ENSO, warm, nutrient-poor water rather than cool, nutrient-rich upwelling water is present near the shore, and productivity falls dramatically. Stocks of short-lived species such as anchovies crash, because there is inadequate production to support successful recruitment. The El Niño of 1972–73 had particularly marked consequences on the Peruvian anchovy fishery (section 7.3.4). The yields from the fishery in 1970 was 13 million tonnes (> 25% of global marine landings), but fell to 2–3 million tonnes in subsequent years and 1 million tonnes (< 2%) in 1980. Initially, sardines dominated the fishery, but by 1995, after another fall in production during the 1982–83 ENSO, anchovies recovered and yielded 9 million tonnes (> 9%). There are many other examples of fish stocks affected by large-scale climatic events, and these events often act in synergy with fishing effects (Chapter 12). Physical processes consistently play a significant role in controlling the dynamics of fished stocks (Cushing, 1975, 1982; Mann, 1993; Mann & Lazier, 1996).

The production of plankton and zooplankton is very patchy on a range of scales in both space and time. Patches of zooplankton have particular significance for fish larvae because patches may harbour prey densities adequate for larval growth when adjacent areas do not. We have already seen that northern anchovy larvae needed a specific prey density to survive and that storms broke down patches and led to starvation of larvae.

Patches are affected by the activities of grazers, turbulence and, especially in the coastal zone, surface winds. Patches larger than 1 km or so tend to be formed by turbulence, but increasing turbulence can lead to their break down (Steele, 1978).

On a larger scale, oceanographic processes determine the success or failure of recruitment through their links with larval transport. For example, when haddock *Melanogrammus aeglefinus* (Gadidae) larvae were transported from spawning grounds on Georges Bank into the middle Atlantic Bight, recruitment was unusually successful. The transport was the result of a strong along-shelf surface flow that pushed larvae some 400 km beyond their usual limits of distribution into an area where their survival rates were much higher (Polacheck *et al.*, 1992).

On the Georges Bank, cod larvae need to be retained if they are to recruit to appropriate nursery areas. Frontal currents that are associated with frontal boundaries govern the intrusion of external water masses (originating from Gulf Stream rings) onto the Georges Bank. When intrusions occur, larvae are less likely to be retained and recruitment falls (Townsend & Pettigrew, 1996). Hydrodynamic processes also affect the recruitment and settlement of sea urchin *Strongylocentrotus franciscanus* larvae in southern Oregon, USA (Miller & Emlet, 1997). Peak settlement was associated with warm-water events when the water column was mixed and winds blew from the north. Conversely, settlement rates were low when the water column was stratified and wind stress was low.

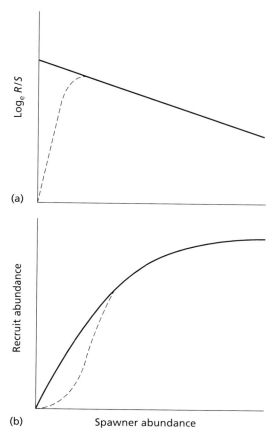

Fig. 4.13 Examples of (a) the relationship between recruits per spawner and spawner abundance when there is compensation (——) or depensation (– – –) and (b) the effects of depensation (– – –) on the mean spawner–recruit relationship.

4.2.3 Depensation

When the spawner–recruit relationship follows one of the models described in section 4.2.1, there is compensation at levels of spawner abundance that satisfy the condition:

$$\frac{d(R/S)}{dS} < 0 \qquad (4.4)$$

That is, all models show **compensation** at lower levels of *S* and recruits per spawner (*R/S*) increases as spawner abundance decreases (Quinn & Deriso, 1999). This means that the intrinsic rate of increase for the stock is highest at the lowest spawner abundance. Compensation is a key part of the regulatory mechanism that allows a stock to withstand the additional mor-

tality imposed by fishing and to recover from low abundance.

However, if *R/S* falls at low abundance (Fig. 4.13), the stock is said to show depensation, and recovery will be slower than predicted by the model or not take place at all. Depensation occurs when:

$$\frac{d(R/S)}{dS} > 0 \qquad (4.5)$$

Depensation leads to lower than expected recruitment success at low population levels. Depensation could occur if predators ate larvae at a relatively constant rate so that high larval abundance swamps them, and proportional mortality goes down. Depensation could also occur if some females fail to find mates when stock size is low or if the fertilization rates of broadcast spawners are dependent upon sperm concentration. Depensation can be thought of as inverse density dependence at low population densities, an effect often referred to by terrestrial ecologists as the **Allee effect** (Allee, 1931). If depensation exists, fishery managers should be extremely nervous because fished stocks may not recover after being fished to very low abundance, even when fishing is stopped.

Spawner–recruit relationships can be modified to allow for the existence of depensation by the addition of a power function. For example, the Beverton and Holt (1957) function of Box 4.1 can be modified to give:

$$R = \frac{\alpha S^d}{\beta^d + S^d} \tag{4.6}$$

where d descibes the level of depensation. When $d = 1$, the model displays the Beverton and Holt stock–recruitment relationship but when $d > 1$ there is depensation (reduced recruitment at low spawner abundance). However, the depensation parameter in this equation has no clear biological interpretation since the extent of depensation for a given value of d will depend on the other parameters of the model (Liermann & Hilborn, 1997). As such, d is not appropriate for making comparisons among stocks. Liermann and Hilborn (1997) reparameterized the model and developed an alternative depensation parameter q. Their measure of depensation is defined as:

$$q = \frac{\left[\dfrac{\alpha(0.1S^*)^d}{\beta^d + (0.1S^*)^d}\right]}{\left[\dfrac{\alpha'(0.1S^*)}{\beta' + (0.1S^*)}\right]} \tag{4.7}$$

where, S^* is the maximum observed spawner abundance, and α' and β' are the parameters from the standard Beverton and Holt model that agree with the depensatory model at $0.5S^*$ and S^*. The models agree at R^*, when spawner abundance is S^*, and at zR^* when spawner abundance is $0.5S^*$, where:

$$z = \frac{\left[\dfrac{\alpha(0.5S^*)^d}{\beta^d + (0.5S^*)^d}\right]}{R^*} \tag{4.8}$$

The full derivation is shown in Liermann and Hilborn (1997) and summarized graphically in Fig. 4.14. q is

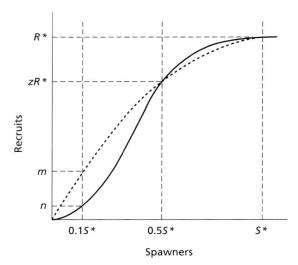

Fig. 4.14 The graphical definition of R^*, z and q, where $q = n/m$. The broken line shows a Beverton–Holt spawner–recruit relationship and the continuous line shows the relationship with depensation. After Liermann and Hilborn (1997).

effectively the recruitment ratio of depensatory and standard models at $0.1S^*$. When $q < 1$ the model is depensatory and when $q > 1$ there is hypercompensation. Liermann and Hilborn (1997) calculated maximum likelihood distributions for q for a variety of codfish (Gadiformes), flatfish (Pleuronectiformes), and herring and anchovy (Clupeiformes) stocks that had been depressed to low population levels. The advantage of this approach is that it emphasizes the considerable uncertainty as to whether depensation exists in stocks where it had previously been disregarded, and provides an indication of the relative likelihood of competing hypotheses. Thus, the tails of the distributions for the taxa examined extend well into depensatory range, and for Pleuronectiformes (Fig. 4.15) the modal value of q was < 1. Depensation may occur, and if it does, stocks pushed to low levels may not recover. Invertebrate stocks often provide stronger evidence for depensation because the adults may not be mobile and low densities prevent successful reproduction (Orensanz & Jamieson, 1998). By way of caution we should also note that egg production per unit biomass will often fall at low population levels (section 4.2.1) and that most spawner–recruit relationships are based on spawner abundance rather than egg production. As such, depensation may exist in spawner–recruit relationships based on spawner

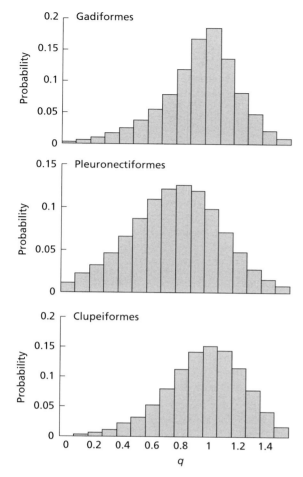

Fig. 4.15 Maximum likelihood distributions of the depensation parameter q in three orders of fishes. Values of $q < 1$ represent depensation and values of $q > 1$ represent hyper-compensation (stronger compensation than predicted by the model). After Liermann and Hilborn (1997).

abundance but not in relationships based on egg-production measures.

4.2.4 Regulation in fish populations

Regulation refers to the tendency of a population to decrease in size above a certain level and to increase in size below that level. Some density-dependent process must regulate populations since fished species have persisted for thousands of years without going extinct or reaching infinite biomass (A.R.E. Sinclair, 1988; M. Sinclair, 1989). Moreover, many fished populations

now experience fishing mortalities at least twice natural mortality with little sign of a consistent decrease in population size (Shepherd & Cushing, 1982). The environment may have density-independent effects on mortality which are so large that they make the observation of density-dependent factors very difficult. Fluctuation does not imply a lack of regulation because the population level may be well regulated while total abundance simply tracks resources or environmental change. Marine systems are not constrained like experimental ponds and lakes where food supply is fixed by the investigator. In most marine fishes, density-dependent regulation must occur at the egg, larval or juvenile stages because this is when most mortality occurs and when the relative size of the year class is fixed. There may also be feedbacks from adults to the early life history through decreases in size at age and fecundity when high adult densities constrain growth.

Some species of fish show strong density-dependent regulation of population size at the juvenile stage. This leads to relatively low recruitment variability. For example, Beverton and Iles (1992) and Beverton (1995) showed that the North Sea plaice *Pleuronectes platessa* had unusually low recruitment variation because density-dependent regulation of population size occurred on the shallow, sandy nursery grounds used by newly metamorphosed juveniles. These nursery grounds were relatively limited in extent, and during years when larval survival was good and juvenile densities were high, juvenile mortality rates rose. This dampened recruitment fluctuations of the plaice stock, even though variations in egg and larval mortality were comparable with those in other species.

Density-dependent growth, fecundity or juvenile mortality could regulate at high stock size, but more powerful regulation would be required to prevent population collapse at high fishing effort and low stock size. An alternative is that density-dependent processes are quite weak but that stochastic processes may provide regulation in practice (Shepherd & Cushing, 1990). Stock size at a given time is approximately dependent on the arithmetic mean of the preceding recruitment events. If regulatory processes determine the median or geometric mean of recruitment variability then there will be larger arithmetic mean recruitment to spawner abundance ratios at lower stock sizes. This would give strong regulation in mean abundance as the mean would increase in relation to the median at low stock sizes due to increasingly large, but increasingly infrequent, outstanding year classes. Simulations in Shepherd and Cushing (1990) showed that this mechanism would prevent stocks being fished to extinction over a wide range of fishing mortality rates, and that the stock increased from near extinction now and again simply by chance events. Even though their model included no climatic shifts, the stocks sometimes grew rapidly, after many centuries at low levels, despite being subject to high mortality.

In invertebrate populations, the evidence for density-dependent control of populations has been rather stronger than for fish. Density dependence can mean that recruitment is reduced at high population densities even though fertilization is often more effective. If density increases, competition for resources may reduce the surplus energy available for reproduction, and reproductive output will fall. Density-dependent processes that act after fertilization include the inhibition of settlement by high densities of conspecifics and cannabilism in crustaceans.

Density dependence in invertebrates can lead to some interesting trade-offs. Thus, although individuals compete for food or space they may benefit from proximity to others due to the increased probability of reproductive success. Indeed, very low numbers of competitors may create ideal conditions for feeding and growth but lead to depensation in the spawner–recruit relationship. These effects are particularly applicable to free spawning marine invertebrates such as sea urchins which have to maintain a clumped distribution to ensure reproductive success, but rapidly suffer from density-dependent reductions in reproductive output as a result of competing for algal resources (section 3.4.3).

4.3 Density-dependent habitat use

As the density of individuals using a habitat increases, so competition for food or space often reduces the area's suitability. As organisms occupy a new area they pick habitats of highest suitability, and then, as the realized suitability of these habitats falls, they switch to previously unoccupied habitats that have lower inherent quality but where resources are not stretched by as many occupants. This process creates the so-called '**ideal free distribution**' where realized suitability is equal in all habitats and no individual can go to a better habitat (Fretwell & Lucas, 1970; Fretwell, 1972). The distribution is called the ideal free because individuals are assumed to be ideal in their judgement of suitability and 'free' because they can move freely to a location of choice (Sutherland, 1996).

Expansion of the range of fished species at high density, and contraction at low density, has been recognized in a number of species (Myers & Stokes, 1989; MacCall, 1990). Why is this important in fisheries? Catchability is inversely proportional to density. Thus, if abundance and geographical range are positively correlated, catchability should not fall as population size decreases. This will maintain relative fishing mortality at low stock levels. It also means that the area where a stock can be caught changes with abundance. For example, the distribution of Atlantic cod *Gadus morhua* in the southern Gulf of St Lawrence was density dependent for all age groups from 3 to 8+. Variation in abundance explained 63–94% of variation in the area occupied by the stock (Fig. 4.16). As abundance increased, density increased slowly in areas where cod were already present and more rapidly in surrounding regions (Swain & Wade, 1993). This meant that catches by fishers on the margins of the cod's range showed a proportionately greater increase than those taken by fishers targeting the hotspots.

There were clear changes in the distribution of the northern cod stock as abundance fell and the stock eventually collapsed (Hutchings, 1996). These changes were such that the catchability of cod increased with declining biomass, and there were rapid increases in fishing mortality as biomass fell. The area inhabited by the northern cod stock is covered by a groundfish survey. The survey showed patterns of change in the distribution and abundance of cod as the size of the population fell. Hutchings grouped the tows into those where catch rates of cod were low (0–100 kg tow^{-1}), medium (100–500 kg tow^{-1}) and high (> 500 kg tow^{-1}). The percentages of tows by catch rates showed consistent trends over time. The number of low-density tows increased steadily as the stock collapsed, the number of medium-density tows fell steadily, but the number of high-density tows remained relatively stable until the stock had collapsed (Fig. 4.17). Between 1989 and 1991, 50–60% of the estimated survey biomass of the northern cod was in high-density areas (assumed to be favoured habitats: catch rates > 500 kg tow^{-1}) whereas the average was 21% from 1981 to 1985.

Despite high trawling effort in the commercial fishery and a reduction in the total biomass of cod, high

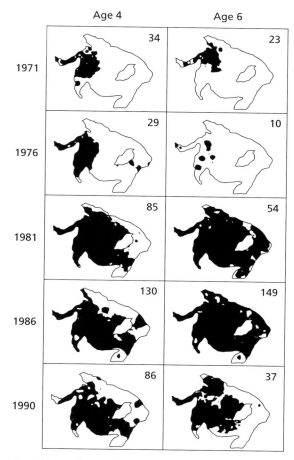

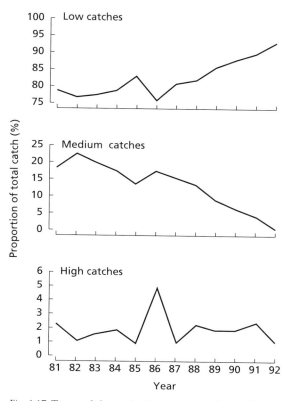

Age 4   Age 6

Fig. 4.16 Distribution of 4- and 6-year-old cod in the southern Gulf of St Lawrence from 1971 to 1990. Numbers inside panels are the abundance of the age group in millions. After Swain and Wade (1993).

Fig. 4.17 Temporal changes in the proportion of low catches, 0–100 kg; medium catches, 100–500 kg; and high catches, > 500 kg, in research vessel surveys of the northern cod stock. After Hutchings (1996).

densities of cod were maintained in the favoured habitats by the relocation of cod from neighbouring areas of lower suitability. Catch rates of trawlers fishing in the hotspots remained high as long as cod were sufficiently abundant to form high-density aggregations there. Stock collapse was only seen when this became impossible, at a time when total stock biomass was very low. Outside the trawled area, however, fishers on the periphery of the cod's range saw catches decrease as the cod relocated to the hotspots.

Research trawl surveys are used to estimate the abundance of many demersal stocks (section 10.2.4) but the results may not indicate abundance trends unless the

effects of density-dependent habitat use are considered. This was demonstrated during the northern cod collapse in the early 1990s (Hutchings, 1996). Simple estimates of northern cod abundance based on arithmetic mean catch rates from trawl survey tows did not demonstrate a declining trend in abundance until the stock collapsed. Annual frequency distributions of biomass per tow approximated to an extended poisson distribution, with a long positive skew. Arithmetic mean catch rates (density estimates) were calculated because they provided an unbiased estimate of biomass. However, this overlooked design bias in the survey (Hutchings, 1996), and the arithmetic mean of all catch rates did not provide a valid estimate of the population mean because it was strongly influenced by a few tows in which very high catch rates were recorded. As we saw from our discussion of density-dependent habitat

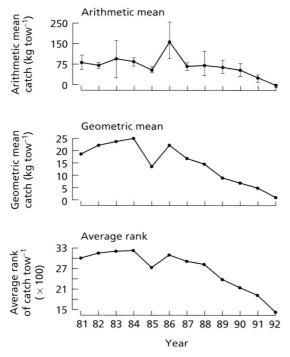

Fig. 4.18 Changes in arithmetic mean; geometric mean; and average rank of catches of northern cod from 1981 to 1992. After Hutchings (1996).

use, a few hotspots continued to attract high densities of cod, even though the size and range of the stock had shrunk. A geometric mean, calculated as the mean of log transformed abundance data, revealed very different temporal trends in population size as it downweighted the effect of the few high-density tows (Fig. 4.18). The rapid decline in arithmetic mean abundance in 1991 and 1992 prompted the comment that the stock had declined rapidly. Using the geometric mean as an estimator of temporal trends in abundance, the decline was apparent from 1986—consistent with reports of fishers operating at the edges of the cod's range who detected rapid decreases in catch rates as the range contracted. A rank-based approach was also robust to consistently large survey catches in areas favoured by the cod stock (Fig. 4.18). In this case, all catch per tow data for all tows in all years were ranked from high to low and the mean rank of catch per tow for each year was calculated. The results from this approach mimicked those indicated by the geometric mean, showing a steady decline in abundance from 1986 rather than a rapid decline in 1991 and 1992.

The density-dependent habitat use described above underscores the importance of using many ways of estimating abundance. Similarity among trends would give the scientist some confidence in the results. Disparity suggests the need for further examination of data and a cautious management strategy.

Summary

• The abundance of fished species fluctuates in space and time. Fluctuations are often driven by physical processes and are largely unpredictable. This creates uncertainty for the fisher, fisheries scientist and everyone else who depends on the fishery.

• The abundance of recruits is determined by egg production and egg and larval mortality during the first few months of life.

• Spawner–recruit (stock–recruitment) models describe mean relationships between spawner and recruit abundance. Much of the variance around fitted models results from variations in egg and larval survival.

• Rates of larval survival are determined by prey availability and the impacts of predators. Both are influenced by coupled physical and biological processes, such as the stability of the water column and timing of the production cycle.

• Spawner–recruit relationships are compensatory, because the abundance of recruits per spawner increases at low stock sizes. Depensation occurs when recruitment success at low stock size is lower than predicted.

• Depensation may prevent stocks recovering from overexploitation and is particularly likely to affect sedentary invertebrate species that release gametes into the water.

• The abundance of fished species may determine ranges. Such density-dependent habitat use has to be considered when interpreting fishery and survey data.

Further reading

Several chapters in Chambers and Trippel (1997) provide up-to-date summaries of the factors affecting the survival of eggs and larvae and the role of biological and physical processes in determining year class strength. Hilborn and Walters (1992) and Quinn and Deriso (1999) describe the application of several spawner and recruit models and methods for estimating parameters. Ransom Myers maintains a website that allows users to view spawner–recruit relationships for many fished species (www.mscs.dal.ca/~myers). A study of the northern anchovy by MacCall (1990) provides many insights into density dependent processes in fish populations. Kawasaki *et al.* (1991) and Schwartzlose *et al.* (1999) review global fluctuations in pelagic fish populations.

5 Fishing gears and techniques

5.1 Introduction

Humans have hunted and gathered marine species for many thousands of years. Early evidence of shoreline gathering and fishing comes from the middens of shell debris around the remains of ancient human settlements (Poiner & Catterall, 1988). Primitive fishing hooks made from bone, cave paintings of fishers using bows and arrows and fishing spears (Yellen *et al.*, 1995), and Stone-Age fish traps that remain intact today provide evidence of early human attempts to improve their ability to hunt marine organisms (Fig. 5.1). Marine species were sought both as a rich source of protein and for other uses. The Romans extracted a rich purple dye for imperial robes from the gastropod *Bolinus brandaris*, South Pacific islanders have traditionally used cowrie shells as currency, and Chinese fishers hunt and collect a wide variety of marine organisms for their reputed medicinal value. At first, fishers were driven by the need

to meet their daily requirements for protein; catching more than was required would have been pointless without any means of preservation. However, once preservation by drying, smoking or salting was accomplished, this enabled marine products to be traded in exchange for other goods or commodities. Indeed, Mark Kurlansky's book *Cod* provides an excellent historical account of the importance of wind-dried and then salted cod for world exploration and trade (Kurlansky, 1997). It is quite likely that fish preservation started the evolution of techniques that enabled humans to exploit marine organisms to a level in excess of artisanal fishers.

5.2 From shoreline gathering to satellites

As technology has developed, so humans have been able to expand the range of species exploited and the extent of the hunting activities that they use. The earliest fishing methods date back to pre-hominid times and involved hand gathering on the shore, although it could be argued that this is not true fishing. The first active fishing techniques can be dated back 85 000 years and involved the use of spears (Yellen *et al.*, 1995), arrows or stones to impale or stun fish. Traps, which fish passively, are relatively easy to construct from wood, vines or stone, and are among those current fishing methods with the most primitive origins (Von Brandt, 1984). The sophistication of fishing devices and techniques parallels the evolution of technological innovations that humans have then used to improve their fishing efficiency. The advent of woven material no doubt permitted the use of finer and stronger lines for setting hooks and producing netting with which to catch or trap fish and crustacea. The advent of sail power would have greatly increased the range over which fishers could pursue their prey.

Fig. 5.1 An ancient fish trap in its present condition at Llygwy Bay, Anglesey in north Wales that is located near to a settlement thought to date from around AD 300 (Bannerman & Jones, 1999). Fish that swim over the walls of the trap at high tide are trapped as the tide recedes, whereupon they are easily collected by hand at low-water. Photograph copyright C. Jones.

However, the size and amount of fishing gear deployed from such fishing vessels would have been constrained by the towing power of the vessel and the physical capability to haul the net and catch aboard. Thus, the advent of steam power marked one of the most significant advances in fishing methodology (Table 5.1). Steam-powered vessels were able to tow large fishing gears at relatively high speed over long distances. They were also able to tow fishing gears independently of the direction or speed of the wind. This permitted much greater safety when towing gears in close proximity to navigational hazards such as sandbanks.

Until the advent of steam, most towed forms of fishing gear had a rigid frame at the mouth of the net that held it open. These were the forerunners of modern-day beam trawls. However, a rigid frame is awkward to handle when it exceeds a certain size, and even modern fishing vessels only operate these gears up to a maximum width of 20 m. The greater power of steam vessels meant that fishers could utilize the hydrodynamic forces acting upon a flat surface to force the mouth of a fishing net open. This is the mechanism by which otter boards maintain the mouth of otter trawl nets in the open fishing position (Box 5.1). The advent of steam power eliminated the constraint of vessel power with respect to gear size. As one constraint has been overcome, so another takes its place. In modern pelagic fisheries it is the size of the potential catch that is the main constraint. Catches of pelagic species can be so large that it is physically impossible to haul the net aboard the vessel without destroying the gear or endangering the safety of the ship. As a result, other methods had to be devised to retrieve the catch from the net. Purse seiners initially used pan-shaped nets called **brailers** (Box 5.1) to empty fish from the main net. This process was very time consuming but it was speeded up greatly by the advent of fish pumps that pump the fish into the fish-hold. These same pumps are then used to unload the fish directly once in the harbour or onto larger factory ships for processing at sea.

The introduction of steam powered and then diesel engines has also reduced the time spent travelling to and from fishing grounds. Consequently, fishing trips across the Atlantic that used to be measured in weeks are now achieved in days. Modern freezer and cooling technology ensures that catches arrive at the market in pristine condition and thus attain the highest possible prices. Distance is no longer an impediment in a world fishery that is becoming dominated by fleets of super-trawlers and seiners.

The advent of power-assisted winches lifted many of the constraints that had previously limited the extent of certain fishing practices. In towed net fisheries, powered winches enabled larger catches to be hauled on-board ship and thereby permitted fishers to tow their gears for longer periods without the need to haul the net in frequently. This has improved the efficiency and safety of fishing practices by increasing the time spent fishing and reducing the frequency of hazardous net hauling operations. Powered winches have revolutionized static gear fisheries. Set nets, long-lines and traps were traditionally set and retrieved by hand, hence it was the physical strength of the fishers that limited the amount of gear that could be hauled complete with its catch. Fishers were much less likely to retrieve gear frequently because of the extra effort that this entailed. Nowadays, much larger fleets of gear can be operated and they are retrieved as frequently as once per tidal cycle. Static gears become less efficient after a certain soak time (Box 5.2), hence fishers have been able to improve their catch rates by increasing their retrieval rate. In addition, they are able to retrieve their gear more quickly if poor weather conditions are forecast, thereby reducing the amount of gear that is lost and left on the seabed 'ghost fishing' (Chapter 13).

A quick perusal of any of the fishing press publications will give the reader a taste of the mind-boggling array of technology that is currently used to enable fishers to catch their quarry. Automation has enabled quick, safer and efficient retrieval and deployment of gear. For example, long-lining involves setting a line from which many hundreds or thousands of baited hooks are suspended. Prior to automation, each of these hooks had to be baited by hand prior to deployment, then on retrieval, fish were removed by hand and the lines carefully coiled to ensure smooth resetting of the gear. Nowadays, the entire operation is done automatically. As the long-line is hauled aboard, fish are removed through rollers and sent down a chute into the fish hold, the line is automatically baited and coiled ready for immediate redeployment ('shooting', Box 5.1).

The advent of World War II coincided with the invention of the sonar to detect the presence of submarines. It was not long before fishers applied this principle to the detection of shoals of fish. The acoustic

Table 5.1 Technological and social advances that have had significant impacts on the fishing industry in the North Sea since 1870. The advent of technological advances has opened up new opportunities such as the industrial fisheries that supply the fish meal trade. After 1950, it is clear that more and more conservation measures have been required to prevent us from over-exploiting populations of some fish species. Based on ICES (1995).

Decade	Events and advances in certain fisheries
1870	Small sailing vessels (smacks) were fishing with beam trawls. Towards the end of the decade the first steam trawlers came into operation which increased towing power.
1880	Otter boards were introduced that enabled larger nets to be deployed. Mechanical ice production enabled vessels to remain at sea for longer periods as the catch stayed fresh for longer.
1890	Steam powered drift netters were introduced that targeted the pelagic stocks of fishes such as herring.
1900	Smack fishing fleet had dwindled in numbers and was confined to sole fishing in the southern North Sea. Fish products were considered useful as animal feeds, prompting fishers to make bulk catches of smaller species. The International Council for the Exploration of the Sea was instigated.
1910	World War I causes reduction in fishing pressure.
1920	Post-Russian revolution sees the emergence of the Soviet fishing fleets. Trawling for herring introduced. Bridles and bobbins introduced to otter trawling gear enabling bottom species to be targeted in rougher grounds. Anchor and fly seining introduced for bottom dwelling species. Fish meal production begins, leading to the more intensive fishing of small pelagic species.
1930	Radio telephones introduced. Echo sounders introduced. Powered line and gill-net haulers were introduced in set gear fisheries and enabled larger fleets of gear to be handled.
1940	World War II caused a 70% reduction in catches. Stern trawling with otter trawls introduced. Mid water pair trawling for pelagic species introduced. Late 1940s DECCA navigation system implemented.
1950	Echo-sounders used as fish finders. Synthetic fibres used for net and rope manufacture. Power blocks for hauling nets. Net drums for hauling. Ship-board freezing facilities. Icelandic cod wars. Danish industrial fisheries for sandeel and Norway pout begin.
1960	North East Atlantic Fisheries Commission instigated. UK distant water fleet diminishes. Double beam-trawling introduced. Single vessel mid-water trawling commences. Quotas on herring catches introduced. Norwegian sandeel industrial fishery commences.
1970	Icelandic cod wars in 1975–1976. UK, Denmark and Ireland join the European Economic Community (EEC). 200 nautical mile fishery limits imposed. Catch quotas introduced on gadoid fish species. Total Allowable Catch recommended for sole. Total Allowable Catch agreed for plaice. Total ban on North Sea herring fishery. Pout box introduced as a conservation measure to exclude this fishery from a large area of the sea.
1980	EEC Common Fisheries Policy signed. Spain and Portugal join the EEC. EEC agrees to instigate cod box, plaice box, and sprat box as conservation measures. Total Allowable Catch agreed for sole. Herring ban continues in central and northern North Sea.

Box 5.1

Jargon that relates to the use or construction of fishing gears.

Brailer: a pan shaped net used to unload a catch from a purse-seine net or from the vessel's fish hold onto the quayside.

Bridle: the cable or rope attached between the net and an otter board or common attachment point.

Codend: the portion at the rear of a towed fishing gear in which the fish collect when the gear is hauled.

Foot-rope: the weighted rope that runs along the bottom of the mouth of a net.

Hauling: the act of retrieving the fishing gear.

Otter trawl: a net towed behind a fishing vessel that is held open by two boards attached to the warps between the net and the vessel.

Shooting: the act of deploying or setting the fishing gear into the water, just before it begins to fish.

Warp: the cable or rope attached to the fishing gear and used to retrieve it.

---

signals sent out from the ship are reflected by substances with different density from water. Thus, the air-filled swimbladders of fish provide strong signals. The signals received can be used to measure accurately the size of the shoal and its depth. Different species of fish form shoals with a characteristic shape. Hence, fishers are now able to decide whether or not to shoot their gear depending on the size of shoal and the species they want to catch. This information is extremely important when one considers that it can take several hours to shoot and retrieve some of the larger pelagic nets. The acoustic signals from a sonar system can be projected in a given direction that enables fishers to track shoals of fish as they move ahead of the vessel. Sonar can also be used to survey the contours and texture of the seabed that can be used to construct detailed maps of different habitats on the seabed. This information can help fishers locate certain types of ground or avoid obstructions that could foul their gear.

Echo-sounders provide fishers with information about the depth of water and type of ground over which they are fishing (Box 5.3). An echo-sounder only provides information about the seabed immediately beneath the vessel, in contrast to sonar which produces a beam of sound that provides information about a wide area of the seabed. Recent developments have produced data-processing systems that are able to interpret the echoes received from the seabed in such a way that the user can differentiate types of seabed. As sound waves strike the seabed they are reflected at different strengths and angles depending on the hardness and the surface topography of the seabed (Box 5.3). The characteristics of the returned signals are interpreted in conjunction with ground-truthing of the seabed. **Ground-truthing** is achieved through a combination of benthic sampling techniques such as sediment grabbing or coring and underwater photography. These systems, such as RoxAnn™ or QTC™, are able to differentiate between much more subtle differences in seabed characteristics than could be achieved using

---

Box 5.2
Soak time.

Soak time is a term that refers specifically to static gears. This is defined as the length of time that the gear remains fishing in the water. With static net gears such as gill and trammel nets, catch rate decreases rapidly 12–24 h after deployment depending upon the catch and prevailing environmental conditions. Other static gears, such as traps and pots, require longer soak times to achieve their best catch rates as these gears tend to use baits to attract the target species. The odours of the bait are dispersed on water currents attracting those animals that are able to detect them. Often, the feeding activities of animals caught within the trap will release more odours and enhance the attractiveness of the bait.

Box 5.3

Echo-sounders.

An echo-sounder works by emitting an acoustic signal from a transducer located beneath the hull of the fishing vessel. The sound signals are reflected by the seabed or air-filled cavities. These echoes of the original signal are received by instruments located on the hull and are translated into images on a video monitor. The strength of the echo is related to the acoustic density of the material, thus it is possible to tell whether the seabed is hard (e.g. rock substratum) or soft (e.g. mud). Fish with swimbladders stand out well because air has very different acoustically reflective properties compared with the surrounding water. Conversely, bottom-dwelling species that tend to have either no or a reduced swimbladder are not easily differentiated against the background of the seabed. The use of echo-sounders for fishery surveying is described in section 10.2.3.

---

the unprocessed sounder data (Sotheran *et al.*, 1997; Kaiser *et al.*, 1998a). For example, while fishers can tell whether they are over a hard or soft seabed using an echo-sounder, once this information is processed by RoxAnn they can discriminate between muddy sand and sand or a similar seabed covered with brittlestars. This information is of little use for finding pelagic species. However, benthic species, such as scallops, crabs, lobsters and flatfishes, have a very strong affinity for certain types of seabed. Hence, this system enables fishers to identify areas of the seabed where they are more likely to find the benthic species they are trying to catch (Kaiser *et al.*, 1998a).

Many modern-day fishing grounds have been fished for many years. Knowledge of those grounds that consistently yield good catches is passed from one generation to another and between fishers within communities. However, the ability to relocate a particular area of the seabed depends upon the ability of fishers to navigate their vessels accurately. Prior to the introduction of electronic navigation systems, fishers relied upon using land-based reference points to determine their position. Thus, accurate navigation was mostly limited to coastal waters with lines of sight to prominent landmarks. The advent of Decca and subsequently satellite navigation systems has allowed fishers to plot the positions of their tows and return to them time after time (Fig. 5.2). Military technology has led to the development of global positioning systems (GPS) that are accurate to within a few metres. This means that fishers are able to navigate safely between areas of seabed with hazardous obstructions, whereas previously it would have been too dangerous to fish in these areas. However, this advance has brought some disadvantages. Obstructions such as wrecks, rock pinnacles and rough ground were relatively undisturbed by fishing gears in the past as fishers gave them a wide berth. The increased accuracy of positioning systems has meant that trawlers can fish very close to these obstructions with little risk of coming fast. Whereas seabed obstructions previously protected the surrounding seabed from disturbance by fishing gears, they are no longer such an effective deterrent to fishers using towed fishing gear.

5.3 Modern commercial fishing gears

Fishing gears are conveniently divided into two categories based on their mechanism of capture. Thus, fishing gears are considered to be either **active gears**, i.e. when they are propelled or towed in pursuit of the target species, or **passive gears**, i.e. when the target species move into or towards the gear (Fig. 5.3). Most active commercial fishing gears involve a vessel towing a net or dredge through the water column or across the seabed. There are a multitude of different designs of nets and dredges, each developed to meet the demands of catching a particular species or fishing in a certain environment. Nevertheless, most of these gears are based on several basic patterns which we will describe here. The greatest diversity of fishing gear design probably occurs amongst passive fishing gears, many of which are based on primitive designs that still work as effectively in modern fisheries. It is impossible for us to cover the vast range of different fishing techniques and gear designs that are currently in use, but the summary below can be augmented by Von Brandt (1984) and Sainsbury (1996).

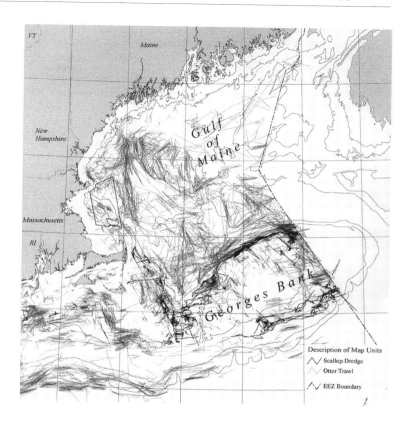

Fig. 5.2 A track plot of 2 years of fishing tows made by trawlers fishing off New England, USA. These plots illustrate the patchy distribution of fishing effort on the seabed and illustrate that fishers are highly selective about where they choose to fish. From Auster and Langton (1999).

5.3.1 Towed fishing gear

The majority of towed fishing gears can be described as either trawls or dredges. We consider these to be distinct from seine or purse nets that are towed into position around shoals of fish prior to being drawn closed and hauled in. Trawls are fished either in mid-water, just off or in direct contact with the seabed. In contrast dredges are exclusively used to capture species that live or feed in benthic habitats, and thus they have been designed to maximize their contact with the seabed. As commercial stocks have diminished, so fishers have adapted their gears as they tried to maintain yield. Some designs have proved more successful than others. For example, in the late 1960s beam trawling began to become popular in the North Sea. Some fishers that switched from otter trawling to beam trawling thought that they could catch both sole *Solea solea* and cod *Gadus morhua* by fishing an otter trawl directly above a beam trawl. This design proved so complicated that it did not fish efficiently and it was soon discarded.

Hence, fishing gear design has undergone a process of economic evolution, as only the most cost-effective designs have survived.

Beam trawls

Beam trawls derive their name from the rigid beam supported by the two shoes at either end (Fig. 5.3d). The net is attached to the beam, shoes and ground rope that runs between the base of the shoes. Thus, the mouth of the net is held open regardless of the speed at which the net is towed through the water. This means that beam trawls can be towed at any speed and still catch organisms, hence they were probably among the first towed fishing gears to be designed by fishers. The shoes act as skis that glide across the surface of the seabed and spread the load of the gear and prevent it from sinking into soft substrata. In some cases, these shoes have been enhanced by the addition of wheels that reduce drag as the gear moves across the seabed. Beam trawls are specifically designed to catch animals such as shrimps, prawns and flatfish that live on or buried in the top few

Fig. 5.3 Examples of modern fishing gears. The purse seine (a), midwater trawl (b), demersal trawl (c), flatfish beam trawl (d) and scallop dredges (e) are examples of active gears. The drift and set gill nets (f and g), long-line (h), fish trap (i) and pots (j) are examples of passive gears. Drawings copyright S.R. Jennings.

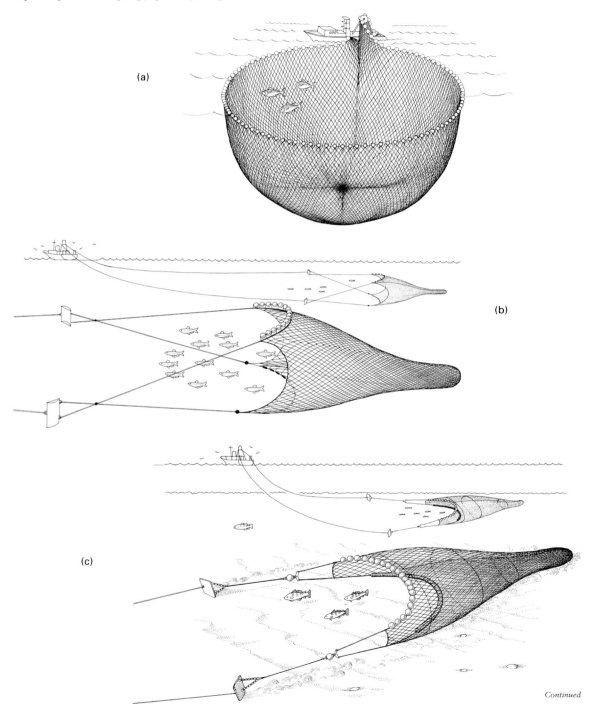

(a)

(b)

(c)

Continued

Fig. 5.3 *(Continued)*

(d)

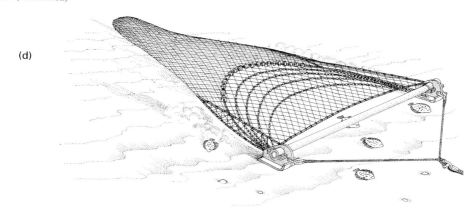

(e)

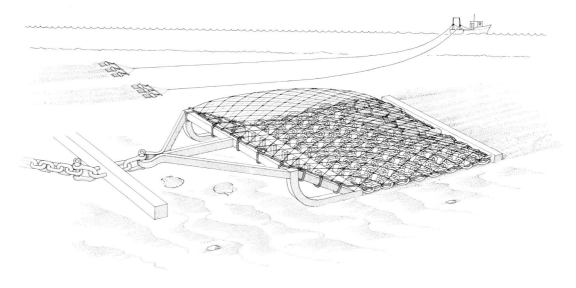

(f)

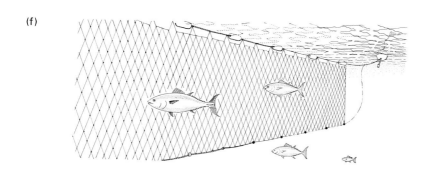

Continued

Fig. 5.3 *(Continued)*

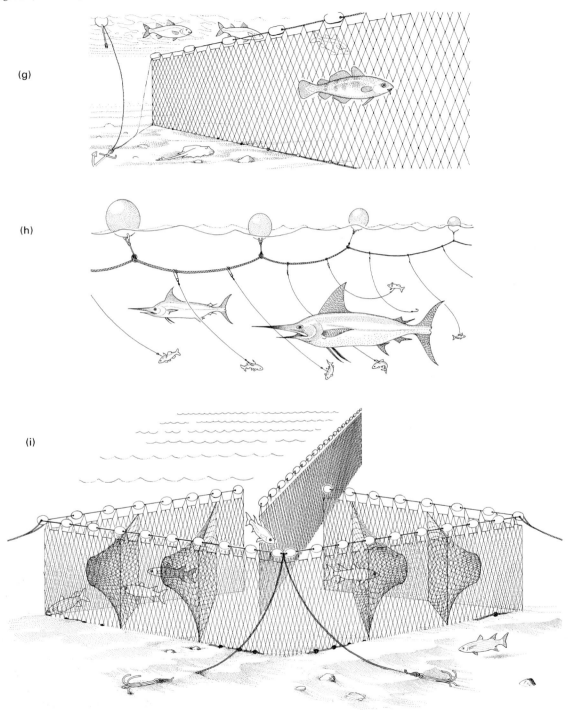

(g)

(h)

(i)

Fig. 5.3 *(Continued)*

(j)

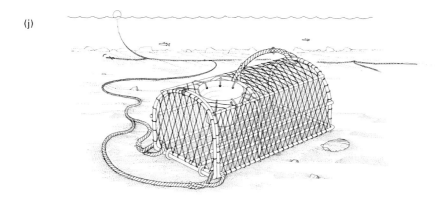

centimetres of the sediment. Various configurations of chains are attached between the beam shoes. These chains, called **tickler chains**, are designed to disrupt the surface of the seabed and disturb or dig out the target species. Small inshore vessels use shrimp beam trawls that are relatively light and rarely have more than one chain fitted between the shoes. This single tickler chain disturbs the sandy substratum sufficiently to cause the shrimp to flee into the water column whereupon they are caught in the net. The addition of extra tickler chains would increase the bycatch of non-target species (Cruetzberg *et al.*, 1987) and would decrease the efficiency of the operation by increasing the time spent sorting the shrimp. However, extra tickler chains are added when fishers seek species that are buried more deeply than shrimp, e.g. flatfish such as sole *Solea solea*. Cruetzberg *et al.* (1987) demonstrated that the catch of sole rose linearly with each extra tickler chain added to their beam trawl (Fig. 5.4). Why should the addition of extra chains increase catch rate? Firstly, extra chains increase the weight of the gear and thereby ensure that it is in close contact with the seabed. Increasing the weight of the gear also means that it can be towed faster as the extra weight compensates for the additional lift generated by the towing forces (Fig. 5.5). Secondly, as each chain passes over the sediment, it fluidizes the sediment making it easier for the following chains to penetrate deeper into the substratum. Large beam trawls can be fitted with over 20 tickler chains and can penetrate soft sand to a depth of over 6 cm.

Beam trawls fitted with tickler chains tend to be fished over clean ground that has few rocks or obstruc-

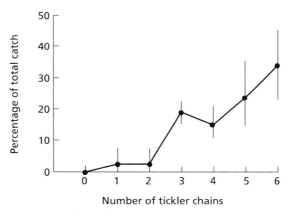

Fig. 5.4 The effect of increasing the number of tickler chains on the proportion of flatfishes in the catch (mean ± S.D.). After Cruetzberg *et al.* (1987).

tions on the seabed. The possibility of snagging the gear on the bottom is one of the greatest threats to the safety of a beam trawler and its crew. Most of these vessels fish two gears simultaneously, one either side (Fig. 5.6). The largest trawlers fish at speeds of 7 knots which leaves little time for the crew to react if one of the trawls comes fast on the bottom and this can result in the vessel capsizing. Despite these risks, beam trawlers target valuable flatfish species such as turbot and brill, *Scophthalmus maximus* and *S. rhombus* (Scophthalmidae) which inhabit areas of rougher ground. A beam trawl fitted with standard tickler chains would soon fill the net with rocks that would destroy the gear and ruin the catch. Fishers have overcome this problem by adding

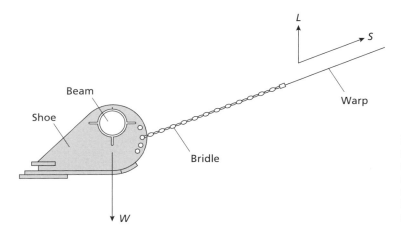

Fig. 5.5 The efficiency of beam trawls is affected by a combination of towing speed and weight of the gear. Increasing towing speed (S) increases the lift on the gear (L) and has to be compensated for by increasing the weight of the gear on the seabed (W). After Sainsbury (1996).

Fig. 5.6 A 12-m flatfish beam trawl is hauled alongside a beam trawler. A similar net will also be hauled on the other side of the vessel. Photograph copyright Q. Bates.

Fig. 5.7 An underwater photograph of an otter board being towed along the seabed (from right to left). The otter boards are set at an oblique angle to the direction of travel causing them to move apart and hold the mouth of the net in the open position. Photograph Crown copyright, reproduced with permission of Fisheries Research Services, Aberdeen.

longitudinal chains across the tickler chains to form a **chain mat**. In addition, a flip-up gear is fitted to the ground rope. The chain mat prevents large boulders from entering the net, while the **flip-up gear** forms a barrier to smaller rocks and debris. Despite these innovations, beam trawls tend to catch large amounts of inert or nontarget benthic species that can rapidly fill the net and reduce the value of the catch by causing damage to the fish in the codend. Fishers reduce this problem by cutting some of the meshes in the belly of the trawl so that the material tends to fall out of the net as it moves back towards the codend.

Otter trawls

Otter trawls derive their name from the two otter boards or doors that are fixed between the warps and

bridles (Fig. 5.3b, c; Box 5.1). **Otter boards** (also known as **trawl doors**) are hydrodynamically designed so that as they are pulled at an oblique angle through the water they plane in opposite directions (Figs 5.7 and 5.8). This action holds the wings of the net open. The otter boards have to be towed at a certain speed (depending on their size) for this effect to be achieved. Modern designs have arisen to meet a variety of different requirements and have been developed by gear technologists using models of otter doors and flume tanks. The net is held open vertically by a series of buoys attached to the headline and a weighted foot-rope (Box 5.1). Factors such as towing speed and the length of warp between the otter boards and the vessel will determine the depth

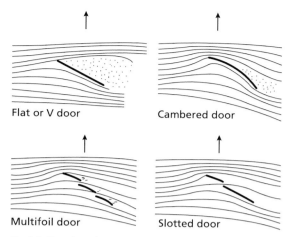

Flat or V door

Cambered door

Multifoil door

Slotted door

Fig. 5.8 Otter boards are sometimes designed like hydrofoils to improve their effectiveness when used in conjunction with large nets. These doors are being towed from right to left and are viewed from above. Arrows show the direction in which the boards are forced as they are towed. From Sainsbury (1996).

and aperture of the net. Nowadays, large otter trawls are fitted with acoustic devices that can be interrogated from the ship and provide information on the depth at which the net is fishing and the distances between the otter boards and between the headline and foot-rope. The otter doors, the plumes of sediment that they create, and the warps attached to the net also have a herding effect and cause fish to aggregate directly in front of the mouth of the net.

Otter trawls are either fished on the bottom for demersal species such as cod, whiting *Merlangius merlangus* and prawns (Fig. 5.3c), or in midwater for pelagic species such as herring *Clupea* spp and mackerel *Scomber scombrus* (Fig. 5.3b). When rigged for prawns or flatfishes such as plaice, tickler chains are added between the otter boards. However, it is more typical for the foot-rope of otter trawls to be fitted with rubber **bobbins** that bounce over obstructions and avoid catching benthic invertebrates. Consequently, the catches of otter trawls generally contain far less bycatch per unit of commercial catch than those of beam trawls. Otter trawls adapted for fishing over rocky ground are known as **rockhopper trawls**. In these, the ground-rope is fitted with rubber discs over 50 cm in diameter and metal bobbins that can weigh more than 10 kg each. The rubber discs are held in position by a wire threaded through their rear half that runs

the length of the ground-rope. As the discs come fast against a snag, they partially rotate against the tension imposed by the wire and then 'spring' clear, allowing the gear to hop over solid obstructions. Another variant on this design is the so-called **'street-sweeper' gear** that replaces the large bobbins with round brushes as used by street-cleaning vehicles. This gear is apparently as effective fishing over rough ground as the rock-hopper configuration (Carr & Milliken, 1998).

The size of beam trawls is limited by their manoeuvrability when they are hauled alongside the vessel. In contrast, otter trawls have no rigid structures other than the otter boards; hence, it is possible to deploy extremely large gears. Otter trawl size is largely dictated by the size of the vessel, net hauling winches and the power of the engine. Otter trawls are occasionally fished by two separate vessels which permits a much bigger net to be fished; this is known as **pair-trawling**. Alternatively, some trawlers now fish two bottom trawl nets side by side from the same vessel, a method known as **twin-rigging**. As inshore fisheries have come under increasing pressure and stocks have dwindled, fishers have begun to exploit the large stocks of deep-sea fishes that occur at depths > 1000 m. Most of the species targeted, such as orange roughy, form dense shoals around seabed structures such as sea mounts. Otter trawls are currently the only suitable gear for this type of fishery as they ensure a catch is taken that is large enough to compensate for the long shooting and hauling time.

Dredges

Dredges fall into two main categories: they are either mechanical or hydraulic dredges. All dredges are used to capture sedentary species such as scallops, clams and gastropods, that live either on the surface of the seabed or within the sediment down to depths of 100 cm. **Mechanical dredges** are operated by the largest range of vessel sizes and tend to have a simple design based on that of the beam trawl. It is often difficult to tell the difference between what some would define as a beam trawl and others a dredge. So, why not use beam trawls that can be towed at faster speeds to catch scallops? Surprisingly, in comparative trials, scallop dredges out-fished a similar sized beam trawl by approximately 100 : 1 (Kaiser *et al.*, 1996b). Either the beam trawl did not dig as deeply into the sediment as the tooth bar on the dredges, or it was being towed too rapidly.

Fig. 5.9 A gang of three scallop dredges. Although scallop dredges can be particularly destructive to seabed habitats, bycatches of non-target species are limited by the use of rigid steel rings instead of netting. Photograph copyright A. Jory.

A typical **scallop dredge** incorporates a heavy-duty bag or net attached to a rigid metal frame to which tooth bars or cutting blades of various designs are fitted (Fig. 5.3e). The rigid rings have the advantage that they do not collapse when the gear is towed and they permit bycatches of other organisms to escape. Thus scallop dredge catches can be very clean. In scallop fisheries, **tooth bars** bearing 11-cm long teeth are fitted to the base of the rigid frame and are designed to disturb scallops that lie slightly recessed in the sediment. These teeth can become snagged on rocks or other seabed obstructions and result in the loss of the dredge. Fishers have overcome this problem by fitting springs and hinges to the tooth bar so that it bends back and springs clear of snags. These modifications have enabled scallop dredgers to access much rougher ground than would otherwise be possible. On snag-free sandy grounds, larger dredges are deployed that have fixed tooth bars and diving vanes to improve the penetration of the teeth into the seabed. Most scallop dredges are between 0.75 and 2 m wide and are fished in gangs with up to six individual dredges attached to a beam (Fig. 5.3e, Fig. 5.9). The largest dredgers fish up to 20 dredges on either side of the vessel.

Hydraulic dredges use jets of water or air to create a venturi effect, which lifts the dredgings up a pipe and onto the operating vessel for further processing on fixed or mechanical **riddles** (grids that separate target species from the sediment). Some of these devices also use jets of water to fluidize the sediment directly in front of the dredge head. Hydraulic dredging barges are used to harvest lugworms, *Arenicola marina*, in the Dutch Wadden Sea. These worms are sold commercially to meet the demand for bait from recreational anglers. These barges operate on intertidal areas at high tide and create furrows 1 m wide and 40 cm deep (Beukema, 1995). Hydraulic dredges operated from boats or mechanical dredges towed behind tractors are used to harvest cockles *Cerastoderma edule* (Cardiidae) and Manila clams *Tapes philippinarum* (Veneridae) at mid to high tide on sandflats in northern Europe (Hall & Harding, 1997). On a smaller scale, divers use hand-held suction dredges to remove razor clams *Ensis siliqua* (Solenidae). Although the area excavated is relatively small, pits can be up to 60 cm deep (Hall *et al.*, 1990a).

Encircling nets

Encircling nets tend to be used for those species whose schooling behaviour means that they are found in dense aggregations. Nets are either set from the shore or deployed by boat at sea, but in all cases the net is set around the fish and drawn closed (Fig. 5.10). Seine nets are commonly used from the shore, when the net is either set from a boat or is walked out from the shore and set by hand. The net is then gradually pulled ashore maintaining the lead line in contact with the seabed, and the headline is supported at the water's surface by floats. However, an extra panel of net that lies flat on the surface of the water may be used to prevent species such as mullet (Mugilidae) from jumping out of the net. The seine net may consist of a simple panel of netting or

(a)

(b)

Fig. 5.10 Two different types of surrounding nets. (a) Two seine boats about to surround a 27 t school of Atlantic menhaden *Brevoortia tyrannus* off North Carolina, USA; and (b) beach seine netting for mullet on the Ivory Coast, Africa. Photographs copyright US National Marine Fisheries Service (a) and M. Marzot, FAO (b).

may incorporate a codend at its centre. The latter design is used in **Danish seining** when the net is deployed at sea from a vessel. In this case, one end of the net is anchored and buoyed while the vessel steams away paying out the net in a circle, eventually returning to pick up the buoyed end before hauling. The fish captured using this technique are usually landed in excellent condition, because they spend little time in the codend, and command high prices at market.

Purse seines can be extremely large and take entire schools of fish (Fig. 5.3a). This method is normally

targeted at pelagic species such as anchovies, tunas and mackerel. Tuna are located using a variety of techniques, either sonar, by spotting schools from helicopters, or from the feeding activities of seabirds that are attracted to smaller prey fish that have been driven to the surface by the feeding tuna below. Modern tuna purse seiners are some of the largest and most expensive boats engaged in marine fishing. Purse seines are set in the same manner as seine nets often using two vessels to deploy the net. The term 'purse' comes from the mechanism by which the net is closed, as the lead line is drawn closed by the purse wire that runs through a series of loops at the bottom of the net. This method is so efficient that the catches are usually too heavy to drag aboard in the net, hence the fish are either scooped up using pan nets, brailers or more usually pumped aboard the vessel. Dolphins are often found in association with schools of tuna and are sometimes accidentally captured during purse seining operations. This led to adverse publicity and the introduction of good fishing practices to ensure that the purse wire was only drawn after dolphins were clear of the net.

5.3.2 Static fishing gear

The gears previously described are all actively fished, i.e. they require manipulation towards the target species by fishers or their vessels. In contrast, static gears are not worked as such, rather they operate passively and entangle or trap the target species that move towards or into them. Fishers improve their capture success by orientating static gear across migration routes, either across or with tidal currents and in close proximity to the refuges used by the target species. For most gears, there is an optimum soak time (Box 5.2) after which the catch rate decreases considerably. Fishers also need to consider the quality of their catch, because the longer fish remain in the gear the more decomposed they become, and they are at risk from damage by seals and crustacean scavengers. Voracious isopod and amphipod scavengers are able to strip all the flesh from a fish within 24 h. Hence, the frequency with which the gear is hauled will depend upon a combination of the cost of retrieval, catch rate and losses to catch degradation.

Set net fisheries have benefited greatly from the development of man-made materials such as monofilament nylon. **Nylon nets** are virtually invisible in water and the strength of the knots of each mesh increases as the material swells when immersed. Nylon is highly resistant to abrasion, hence the netting has the potential to last for many years. This is also one of the less attractive aspects of set netting. Set nets and pots are occasionally lost due to bad weather conditions or when towed away by other fishing vessels or commercial shipping. When lost, the static nets and pots can continue to fish for many years catching hundreds of organisms during this time (see section 13.7 for a fuller discussion).

There are four main types of static gears: gill nets, traps, pots and long-lines (Fig. 5.3f–j).

Gill, trammel and tangle nets
Gill nets derive their name from their main method of capture. They are among the most selective fishing gears with respect to the size range of target species captured (Box 5.4). As fish attempt to swim through the meshes of the net, they become snagged by their gill

---

Box 5.4
Selectivity of gill nets

Gill nets are highly selective for size classes of the target species of fish provided the net is well serviced and tended regularly. For example, Fig. B5.4.1

shows (a) the size distribution of a European sea bass *Dicentrarchus labrax* population, and (b) the size distribution of bass caught in a gill net of 89 mm mesh (Potter & Pawson, 1991). Despite the selectivity of gill nets for target fish, they are responsible for bycatches of marine mammals, reptiles and birds.

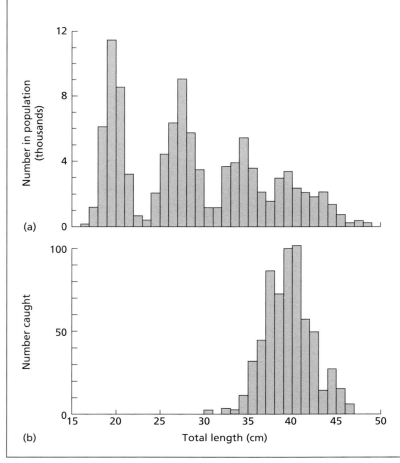

Fig. B5.4.1 Selectivity of gill nets.

operculi, fins or by their scales. The meshes of a gill net are uniform in size and shape, hence they are highly selective for a particular size class of fish. Small, usually undersized, fish are able to swim through the mesh unharmed, whereas excessively large fish are unable to penetrate the mesh sufficiently to become trapped. Gill nets are basically a series of panels of meshes with a lead foot-rope and a headline with floats. These 'fleets' of net are buoyed and anchored at either end to form a barrier. They can be set from the bottom to the surface of the water column (Fig. 5.3f, g). Gill nets are shot either across or with the tide depending on local tidal conditions and the target species. When set on the bottom across the tide, the net will tend to lie flat when the tide is running at its fastest, and will be fully extended at slack water. Hence, catch rate often varies according to the state of the tide. Surface-set gill nets were fished in fleets of up to 50 km in length until recently. These nets were responsible for the entanglement of many non-target organisms such as turtles, seabirds and dolphins. International legislation now limits nets to a maximum length of approximately 2 km. Despite this, set nets continue to catch cetaceans that are unable to detect the presence of the netting. Several attempts have been made to develop acoustic devices to alert or deter cetaceans in close proximity to set nets, but these are not yet fully effective.

Gill nets are relatively cheap to produce, which makes them affordable for small fishing operations. However, this also means that gill nets are a favoured method for individuals fishing illegally and makes the ecological and population impacts of the gill net fishery difficult to assess accurately.

Trammel nets are similar in many ways to gill nets, but they are set mainly on the seabed. They incorporate three layers of netting: an inner small meshed net sandwiched between layers of large meshed net. As the target species swims through the large-meshed layer it meets the small-meshed layer. Swimming forward, the fish pushes the fine-meshed layer through the next layer of large-meshed netting and becomes trapped within a pocket of netting. These nets work in all states of the tide and are particularly effective for catching flatfishes, rays and crustacea.

Tangle nets have much larger meshes than either gill or trammel nets. They are designed so that the meshes hang loose between the foot-rope and headline. As fish or crustaceans move over the net they become snagged on the loose mesh and can become totally rolled up in the netting. Tangle nets work particularly well for spiny organisms, e.g. fishes such as monkfish *Lophius* spp., elasmobranchs, lobsters and spider crabs.

Traps and pots

Traps are among the most primitive of fishing techniques and have remained little changed. Generally, traps take advantage of the movements of fishes along a tidal gradient or migration route. The operational principle of most traps is the same around the world. There is usually a guiding mechanism (e.g. a wall of net or sticks) that directs the fish to the entrance of the trap from which there are a number of non-return chambers or a maze of passageways from which the fish are unable to return (Fig. 5.3i). In South-East Asia, large '**corrals**' consisting of arrow-shaped fences are found along the coast. A similar trap is used in northern Europe to catch salmon *Salmo salar* migrating up estuaries. The salmon tend to swim up the estuary close to the shore, hence the leading wall of the trap extends from the high-water mark out to an enclosure called the 'bunt'. **Fyke nets** tend to have two walls of netting that lead to the main entrance of the net. They are normally set into the current to intercept fish such as eels as they move against tidal or estuarine flow. Many of the earliest traps were simply circular stone walls that were inundated at the high tide and retained fish as the tide receded (Fig. 5.1).

Other designs of trap known as **pots** are most commonly used in crustacean fisheries, although they are also used to capture predatory fishes and molluscs. Most pots are similar in design: they are made of a rigid frame with a mesh covering in which one or several entrances are inserted (Fig. 5.3j). The entrances are designed to prevent animals from escaping, although video observations indicate that in some simple designs the same crab will enter and leave the pot several times. **Parlour pots** are slightly more sophisticated in design as they have a separate internal chamber containing bait. This chamber incorporates a non-return type of entrance and is much more effective at retaining the animals once they are inside.

As in gill nets and long-lines, pots are usually deployed in fleets anchored at both ends and marked by surface buoys. Small-scale fishing operations or artisanal fishers use single pots, but these are less efficient to set and retrieve. Most pot fisheries use bait, but some, such as the octopus *Octopus vulgaris* (subclass Coleoidea) fishery in the Mediterranean, take advantage of the refuge-seeking behaviour of the fished species. In this particular fishery, empty amphorae are set on the seabed and left for several days during which time octupi begin to occupy the empty vessels.

Pots tend to be set for longer than other gears, as it takes time for the bait within the pot to begin to attract the target species. Catch rate increases over several days as the feeding activities of animals consuming the bait increases the dispersion of attractive odours. The most potent attractants released by the bait are amino acids and adenosine triphosphate (ATP) (Zimmer-Faust, 1993). Thus, pots work best for those animals that find food using chemosensory stimuli. Both crustaceans (crabs, lobsters and crayfish) and gastropods (whelks) follow odour trails borne by water currents. The shape and extent of the plume of attractants emitted from the bait will vary according to the prevailing water currents and will affect the number of predators attracted to the pot (Saint-Marie & Hargrave, 1987). Fishers have learnt with experience that different baits attract certain species more than others. For example, fish is a good attractant for crabs, but dead crab works best for whelks. Sometimes repellents are also placed in the pot. For example, green (shore) crabs *Carcinus maenas* are regarded as a pest in edible crab *C. pagurus* and whelk *Buccinum undatum* fisheries. The addition of dead green crabs to a fish-baited pot greatly reduces the number caught without unduly affecting catches of the desired species (Moore & Howarth, 1996; Ramsay *et al.*, 1997).

Long-lines

Long-lines are deployed to catch either demersal or pelagic species (Fig. 5.11). The gear consists of a length of line, wire or rope to which baited hooks are attached via shorter lengths of line (Fig. 5.3h). Long-lines are often set in fleets that may be thousands of metres long with hooks spaced ten metres apart. Bottom-set long-lines are anchored at each end and are marked using surface buoys. Subsurface long-lines may remain attached to the vessel at one end while they are fishing. The line is maintained at the required depth by a series of surface buoys and weights added along the long-line. Automation of the deployment and retrieval of long-lines has greatly improved fishing efficiency and safety. Each vessel may fish several thousand hooks, each of

Fig. 5.11 Long-lines are frequently used to catch pelagic species such as the sharks being hauled aboard this Japanese vessel (Copyright TVW Channel 7 Perth, Australia. Reproduced with permission).

which need to be baited and from which the catch needs to be unhooked. Long-lines are highly selective as a result of hook size, and bycatches of invertebrates are virtually non-existent. However, subsurface long-lines are known to catch diving seabirds (section 13.5.1). Current recommendations suggest that these long-lines should be set below the depth to which these birds dive, which can vary from 1 to > 20 m.

5.4 Other fishing techniques

All of the fishing gears described so far are used extensively around the world in one form or another on a commercial scale. In addition to these commonly used techniques, artisanal fishers have developed a variety of other methods of harvesting marine species. The range of examples is vast and is covered extensively by Von Brandt (1984).

Spears

The use of spears to catch marine species is among the most ancient forms of hunting that is still used to this day. Artisanal fishers use a variety of single- or multiple-tipped spear designs to pursue fish in shallow lagoons and coastal waters. **Spears** can be propelled mechanically underwater using spear guns that are charged with compressed gas or powered by rubber. On the high seas, individual bluefin tuna (tunny) *Thunnus thynnus* are chased by high-powered speed boats until a skilled fisher is able to harpoon the fish which can weigh in excess of 500 kg. Whales and basking shark are captured using harpoons fired from a cannon on the ship's prow. Once these have penetrated the animal's body, an explosive charge in the harpoon head detonates, or alternatively an electrical current is passed down the cable to which the harpoon is attached and causes paralysis and death.

Lures, pole fishing and fish aggregation devices

Lures are fashioned from a variety of materials and are designed to represent the body shape or movement of prey species (Fig. 5.12). Lures can be used individually as in the case of angling or trolling, or they can be fished simultaneously in large numbers from a boat. In all cases, the lures are moved actively through the water, their coloration and movement stimulates a predatory response from the target species. Powerful lights are used to attract fish or squid towards vessels where they form dense congregations. These are then caught on jigs fished on hand reels or automated fishing devices (Fig. 1.7b). Jigs, or rippers as they are sometimes known, are normally fitted with a several hooks or numerous sharp spikes that snag the fish or squid as they lunge at the lure.

Ocean fronts are associated with pelagic species such as tuna that gather near the surface of the water to feed on prey species (section 2.3.2). Fishers have known for many years that floating debris such as logs often become entrained in these areas and are closely associated with target fish species. Fishers have taken advantage of this fish behaviour by making their own debris, usually a raft of logs, that attracts fish beneath it. Schools of tuna are also indicated by the feeding activities of birds, cetaceans and sharks (section 13.5.4). Once located, the school of tuna is often attracted towards the fishing boat using live baitfish whereupon water jets are sprayed onto the water surface to mimic the effect of bait fish escaping the feeding tuna. This induces a feeding frenzy and the tuna will strike any brightly coloured and moving object. At this point, the fishers lower bare or baited hooks into the mass of tuna. As the fishers strike into fish, the fish are simultaneously propelled out of the water and behind the fisher onto the deck of the vessel.

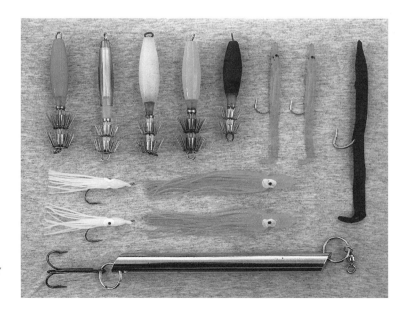

Fig. 5.12 Different designs of squid jigs (top left) and lures used to catch fish (top right and bottom). Lures are designed to imitate the body shape and coloration of prey species or a specific feature of the prey such as movement or reflection of light. Photograph copyright S. Jennings.

Blast fishing

Blast fishing is particularly prevalent in poorer regions of the world (Fig. 1.7c). Blast fishing occurs illegally in the Indian Ocean region, in South-East Asia and in the central and western Pacific. Bombs are made from explosives such as dynamite and detonated using simple fuses. The completed bombs are often placed in glass bottles. The fisher lights the fuse and throws the bomb into the sea, in the hope that it will explode in mid-water and stun or kill fish. Needless to say, many blast fishers have been injured when bombs exploded prematurely. Catch rates can be very high, but this method is indiscriminate, killing all sizes of fishes. In addition, explosives kill invertebrates and structural organisms such as corals and can lead to the total destruction of local habitats (Chapter 14).

Stupefacients

Both naturally derived and fabricated products are used as **stupefacients**. Liquid detergents, plant derivatives and sodium cyanide are all used to catch organisms either for the aquarium trade or for human consumption. One estimate suggests that 150 tonnes of sodium cyanide is used annually on Philippine reefs to catch aquarium fishes (McAllister, 1988). Currently, little is known of the effects of these chemicals on the various life-history stages of reef organisms. Although the concentrations of stupefacients have an acute effect on organisms, they are dispersed quickly. We can only speculate about their chronic effects.

Fishing with other animals

Probably the best known example of humans enlisting the help of other creatures to help catch fish is the Japanese practice of **cormorant fishing**. This ancient fishing method is first recorded between the eighth and ninth century AD. This technique spread to other areas of Asia and eventually to Europe where cormorant fishing ranked alongside falconry in the English courts of James I and Charles I. The cormorants *Phalacrocorax* spp. are either taken from the wild as chicks or reared from eggs and trained to catch fish and return to their keeper's boat. The birds can fish either at night or during the day and one bird can catch up to 150 fish per hour (Von Brandt, 1984). In Finland and Sweden, fishers have been able to exploit the fishing activities of merganser ducks (*Mergus merganser*). Traps constructed of sticks are placed in lakes. The gap between the adjacent sticks allows fish to pass through but excludes birds as the latter pursue their natural prey. When sufficient fish are concentrated within the trap, it is encircled with netting and the fish retrieved. Although the ducks are not trained as such, fishers have exploited the hunting behaviour of these birds and the escape responses of the fish to improve their fishing efficiency.

Diving

Diving is labour intensive and probably one of the most dangerous fishing practices currently used to harvest marine organisms. Generally, the species targeted have a high individual value, e.g. pearl oysters, sponges, seahorses or scallops. The depth to which divers can operate is limited by the time an individual can hold their breath (up to 2 min for female pearl fishers) or the physiological constraints imposed by decompression sickness. There are no bycatches associated with diving, although divers can damage corals and other fragile organisms. Diving is a highly efficient method of removing almost entire populations of organisms as has happened with the white abalone *Haliotis sorenseni* (Haliotidae) off the western coast of the United States.

Drive-netting techniques are used to catch a range of reef-associated fishes which shelter within the reef matrix or shoal above the reef. These techniques are extensively used on coral reefs, and may range from small-scale village-based operations involving four or five fishers, to large commercial operations which target offshore reefs in the Philippines and South China Sea and involve hundreds of divers (McManus, 1996). The process involves fishers scaring reef-associated fishes towards an encircling net or trap, using scaring devices such as weighted lines or poles that are struck against the seabed. In deeper water, the **Kayakas** and **muro-ami** drive-netting techniques involve teams of swimmers that repeatedly drop weighted scarelines onto the reef in order to drive fishes towards a bag net.

5.5 Conservation methods

While some fishing gears are very species specific, others are less discriminate. We have already indicated some of the potential negative side-effects of using certain gear configurations in terms of the resultant number of non-target organisms caught. The issue of bycatches will be dealt with in more detail in Chapter 13. Few

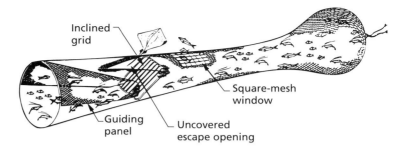

Fig. 5.13 Trawl efficiency devices (TEDs) have been developed to exclude large organisms from catches of smaller species. These devices are based on grids through which the smaller target species can pass, whereas larger organisms such as turtles and skates are deflected out of the net through an escape opening. From Brewer *et al.*, 1998.

target species exist in the marine environment without associated non-target organisms. Demersal fishes live in close proximity to benthic organisms, hence when a towed fishing gear passes through this habitat a bycatch of non-target bottom living species is inevitable. In midwater and pelagic fisheries, nets may be towed for many hours. As a result, any cetaceans, marine reptiles or birds that are caught incidentally are usually drowned by the time the net is hauled. Different fish species share the same environment, thus it is almost impossible to catch cod or haddock with an otter trawl and not catch other smaller gadoid species. Set nets and long-lines are considered to be amongst the most selective fishing gears in terms of the size and identity of the target organisms captured. However, there are numerous examples of dolphins, whales, seals, seabirds and shark species that have been entangled or trapped in static gear. These incidental catches undoubtedly give fishers bad publicity. This situation is unfortunate as most fishers would gladly avoid these accidental catches if it was possible. Regardless of the ethical considerations surrounding bycatch, bycatches often have financial implications for fishing operations. For example, a shark entangled in a set net will effectively render the net useless. Bycatches of invertebrates such as starfishes and crabs are abrasive and can severely decrease the financial value of fish such as sole by causing scale loss and bruising to the tissues. Thus, the pressure to minimize incidental and bycatches of non-target organisms has come from fishers, environmentalists and the general public whose motives may be either financial considerations or conservation interests. Although fisheries managers have argued for many years that there is a need to reduce bycatches of non-target organisms, it is only relatively recently that this has become an issue of public concern. Consequently, research to reduce bycatches only began approximately 20 years ago.

Fig. 5.14 An underwater photograph of a codend as an otter trawl net is towed through the water. Note that the diamond meshes have become constricted under tension which reduces the ease with which bycatch and undersized species can escape through them. Photograph Crown copyright, Fisheries Research Services, Aberdeen.

Large bycatch species such as cetaceans, turtles, sharks and skates can be filtered out of towed fishing nets relatively effectively using **trawl efficiency devices** (TEDs). These devices are relatively simple, consisting of a rigid grid placed towards the codend that deflects large organisms out of the net through an escape panel (Fig. 5.13). The smaller target species continue to pass through the grid and catches are not affected significantly (Robins-Troeger *et al.*, 1995; Brewer *et al.*, 1998).

Mesh-size restrictions have been used as a fish stock conservation tool for many years. However, it is only in recent years that scientists realized that traditionally designed meshes retained much smaller size classes of fish than predicted. Meshes in nets used in many fisheries are typically diamond shaped. On deck, the mesh may be open. However, once the net is fishing, the drag and weight in the codend causes the meshes to close (Fig. 5.14). The meshes then become blocked with debris and bycatch organisms, further reducing the efficiency with which the net filters the catch. A number of recent studies have shown that panels of square-shaped meshes (known as **square-mesh panels**) do not close up when the gear is fishing, and fish are able to escape more effectively through these compared with traditional meshes (e.g. Briggs, 1992). Underwater video observations of the behaviour of fish in towed nets has shown that fish are more likely to swim up and through escape panels if they are used in conjunction with a dark tunnel of netting (Glass & Wardle, 1995; Glass *et al.*, 1995). Presumably, the dark area invokes an escape response. It is difficult to describe in words the difference in the fishes' behaviour in response to these areas of darkened netting, but they emerge through the escape panel far more energetically than if it were not used (Fig. 5.15).

As discussed later in Chapter 13, bycatches of cetaceans are problematic in set-net fisheries, and considerable effort has been made to overcome this problem. The fine materials used to produce set nets are virtually invisible in water and not detected sufficiently well by dolphins and porpoises to alert them to their presence. Gear technologists have tried making the nets more visible by dying the nylon netting with fluorescent or dark colours. These have had limited success, but often they are used in water so turbid that the visibility is less than 1 m. A variety of acoustic devices have been developed, some emit sounds at frequencies either designed to deter

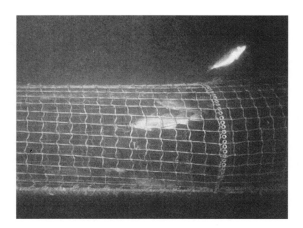

Fig. 5.15 The escape of fish through square-meshed panels can be greatly improved by placing a section of black material immediately behind the escape panel. As the fish approach the darkened area of the net they become startled and elicit a vigorous escape response by swimming at right angles to the direction of tow. Photograph Crown Copyright, Fisheries Research Services, Aberdeen (Glass & Wardle, 1995).

or alert cetaceans to their presence. None of the techniques developed to date seems to work consistently.

Static gears are vulnerable to loss through bad weather conditions or when fishing vessels tow them away. When this occurs, the lost gear can continue to fish indefinitely in some cases (section 13.7). Some fishing and government organizations encourage set-gear fishers to tag their gear to make it easier to find in the event that the surface marker buoys are lost. Other innovations include the use of biodegradable panels in pots that will eventually rot and permit trapped animals to escape (Guillory, 1993; Polovina, 1994). These measures would seem to be common sense, yet they are not yet universally adopted by fishers.

Methods for reducing the damage that bottom-fishing trawls and dredges inflict upon seabed habitats and communities are only just beginning to be studied. At one time, it was suggested that traditional tickler chains could be replaced by use of an **electrified tickler cable**. The tickler cable emitted a charge that shocked or stunned fish as it was towed towards them. In principal, it should be possible to target certain size classes of fish according to the charge that passes through the tickler cable. In practice, the gear was too indiscriminate and the system was never implemented. However, if this system could be improved, damage to the seabed caused by beam and otter trawls could be reduced considerably.

It is difficult to conceive how bycatches of sharks and other large non-target organisms could be avoided in long-line fisheries. Predators such as sharks and seals are attracted to the catch on the hooks. Diving seabirds are also caught by surface-set long-lines. This problem has been overcome by deploying the line deeper in the water, i.e. below the depth to which these birds can dive. In addition, the baited line can be protected by a guard that extends from the boat into the water which prevents birds from attacking the bait as it is shot away, or by deploying a scaring device above the line (Fig. 5.16) (Løkkeborg, 1998).

One of the best ways to reduce bycatches is the use of common sense. Many fisheries research organizations now produce guides on the use of 'good-practices' that are distributed to fishing organizations. These encourage fishers to avoid the use of set-net gear in close proximity to seabird breeding colonies or feeding grounds. It is unavoidable that some of the best fishing grounds will coincide with the presence of birds or cetaceans, as this is often why the latter are found in these localities.

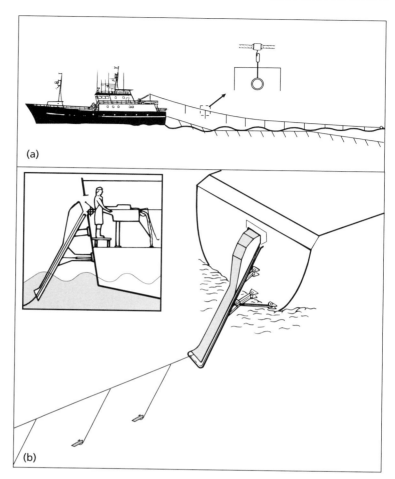

Fig. 5.16 (a) A bird scaring device which is deployed from a line streamed away from the stern of a fishing vessel. The scarers are made of brightly-coloured material and flutter in the wind as the boat rolls. This fluttering action scares the birds from the baited long-line. (b) A setting funnel that deploys the baited line immediately beneath the water and helps to prevent the birds from seeing the baited hooks. Both devices reduce bird bycatches during the setting process, but the investment required to purchase the devices and their vulnerability to damage in the marine environment may make them unpopular with fishers. After Løkkeborg (1998).

Summary

• Fishing gears and techniques are in a constant state of evolution. Fishery and environmental managers will always have to keep pace with these technological changes.

• Fishers are highly innovative and will always strive to improve their catching efficiency. As conservation legislation is introduced, so fishers may alter their behaviour and gear to overcome any short-fall in income.

• Fishing gears can be categorized as active or passive.

• Active gears include nets and dredges that are towed through the water column or across the seabed.

• Passive gears entangle or trap target species that come to them. Examples are set nets, traps, long-lines and pots.

• There is increasing interest in the development of highly selective gears that allow non-target species to escape.

Further reading

Von Brant (1984) and Sainsbury (1996) provide detailed descriptions of fishing gears in use throughout the world.

6 Fishers: socioeconomics and human ecology

6.1 Introduction

The assessment and management of a fishery requires an understanding of the motivations, behaviour and attitudes of fishers. The social aspects of fishery management are often neglected, even though management is rarely effective without some support from fishers. Moreover, social management objectives such as maximizing employment or maintaining the integrity of coastal communities may be as important as biological or economic ones. In this chapter we explore the motivations that encourage fishers to fish as they do, the conflicts that arise between fishers and the ways in which fishers have sought to resolve them.

As predators, humans should hunt or forage like other organisms, hunting in the most energetically cost-effective manner. What will become clear in this chapter is that humans rarely hunt in this way. This is because the currency in modern societies has changed from energy (the food needed to survive another day) to financial reward. In the first part of this chapter, we discuss examples of traditional cultures that hunt according to energetic models. We proceed to discuss the mechanisms that have caused humans to disregard these rules and fish for species that require a disproportionate amount of effort to catch, and how social and religious constraints modify fishers' behaviour. This discussion provides some background to the economic models we describe in Chapter 11. These often assume that fishers are striving to maximize profits and that the overriding objective of fisheries management is profitability. In some societies, these assumptions are wrong.

Since marine resources are finite, fishers often compete for the same resource. In the latter part of this chapter we describe how conflicts over fish have led to war and how some societies have developed management systems that allow fishers to coexist without conflict. These were often in use centuries before the development of modern fisheries science and, in some cases, led to the sustainable use of marine resources. It is ironic that many of these traditional management systems were based on fisheries ownership and limited entry, ideas that are favoured by contemporary managers who have witnessed the collapse of open-access fisheries (sections 1.3.1 and 17.3.3).

6.2 Motivations for fishing

6.2.1 Food

All animals feed selectively, regardless of whether they are herbivores, omnivores or carnivores. How and what organisms decide to eat has been the focus of numerous studies in the disciplines of physiology and behavioural ecology. These different approaches to studying the mechanisms of food acquisition have been united to a significant extent by foraging theory. **Foraging theory** became recognizable after the publication of two papers by MacArthur and Pianka (1966) and Emlen (1966) who drew on economic theories of cost–benefit analysis to construct models of foraging behaviour. The theory is simple. They assumed that an animal would maximize its fitness, in terms of growth, survival and reproductive output, by foraging in ways that maximize the net rate of energy gain over time. This premise of energy maximization underpins all components of foraging theory including searching behaviour, exploitation of food resources, and selection among food types (Stephens & Krebs, 1986). Within the framework of foraging theory is a theory devoted to food selection called **optimal diet theory** (ODT). As in other such models, ODT makes the following simplifying assumptions.

• Foragers can evaluate the profitability, in units of energetic yield per unit time, of all available food types, ranking them in order of profitability.

- Foragers can remember the average profitability of food types encountered and measure encounter rates with different food types.
- This information is used to decide which food items to accept or reject.

This model predicts that the forager should always accept the most profitable food type, and should only accept less profitable types as higher-ranked food types are encountered at a rate below a critical level (Hughes, 1993). Later in this chapter we see that the choices made by fishers in many modern fisheries do not meet these predictions.

In the introduction to his book *Diet Selection*, Roger Hughes relates several studies that illustrate selective human foraging behaviour and how it has altered according to changes in technology (Hughes, 1993). Valuable insights into human foraging behaviour come from the reports of anthropologists who have lived among hunter–gatherers. North American Cree Indians were able to increase their search efficiency as they progressed from paddled canoes, to motorized canoes and then to snowmobiles. The Cree became more selective in their choice of prey as the search time between consecutive encounters decreased with improved technology. Eventually, small prey types, which were hunted previously, were ignored entirely as it was no longer profitable to pursue them (Winterhalder, 1981). Another excellent example of how technological advances affect prey choice is Hill and Hawkes' (1983) study of Ache Indians in Paraguay who use either bows and arrows or occasionally shotguns. The use of a shotgun was far more efficient than hunting with bow and arrows. Hence, ODT predicted that only birds weighing in excess of 1 kg should be hunted with shotguns, whereas birds as small as 0.4 kg were acceptable when hunting with bows and arrows. These predictions were confirmed by observations of the quarry shot by the Ache.

The apparent profitability of a particular food item is subject to a complex array of factors that include the physiological state of the predator and the risks and complications associated with acquiring that food item. Imagine subsistence fishers harvesting marine organisms from the intertidal area. When they first come down to the shore they are probably hungry, they encounter some seaweed that is low in energy content but nevertheless eat it as their physiological state dictates that they should eat the first food items they come across. Moving down the shore they spend a long time searching for high-energy molluscs that live in crevices. Our imaginary fishers are now more selective having already consumed sufficient food to improve their energy balance. If our fishers were to search among the surf, they might find some sea cucumbers that yield high-energy returns, but they decide against this as the risk of being eaten by a shark outweighs the possible rewards from this exercise. You might smile at our simplistic example, but these are the sort of cost–benefit analyses that underlie foraging behaviour.

Subsistence fishers rank the desirability of prey according to energetic or nutritional value, and foraging decisions will be based on a cost–benefit balance in which the currency of the benefits is energetic or nutritional reward. Thus, if we examined the catches from subsistence fisheries we would expect to see a mixed catch that would probably be composed of large numbers of small fish species and fewer larger specimens. However, the catches landed by commercial fisheries are rarely like this, more often they are composed of catches of only one or two species or a certain size range of fishes. In confirmation of the predictions of ODT, abundant, easily captured species are often ignored in preference to lengthy and risky pursuit of rare, large-bodied individuals which are more valuable.

6.2.2 Income

Marine organisms are a valuable source of food. However, at some stage they became valuable as a means of earning money or as a commodity that could be traded for other goods. While local communities used to guard local marine resources as a source of food, Exclusive Economic Zones can now extend to 200 nautical miles from the coast and protect national fisheries resources. Furthermore, while the economic value of a fishery might be relatively small, the wealth created through local support industries and services can be considerable. The decommissioning of one fishing vessel will have obvious repercussions for the boat crew, but will also affect net suppliers, ship repair yards and other support industries. Consequently, while fisheries are managed with the sustainability of the stock foremost, the economic and social consequences of management decisions also play a part in the selection of appropriate management measures (Chapters 11, 17).

The behaviour of fishers often changes during the

transition from a primitive to market economy because fishers realize greater overall benefits from selling fish than from eating them. As ice plants, fish marketing facilities and improved boat services have developed in the Fijian Islands of the South Pacific Ocean, money has played an increasingly important role in island society. On the islands in the traditional fishing ground of Ko Ono, some 70 km south of the major fish markets on the largest island of Vanua Levu, 47% of the fish caught by villagers is now sold. However, on the island of Totoya, over 100 km east of Ko Ono and not within easy range of the fish markets, only 13% of the catch is sold. The islands within Ko Ono use the income from fishing to pay loans on boats, for fuel, for village improvement projects and to purchase a range of foodstuffs and some consumer goods. The villagers of Totoya have a much smaller income that pays for little more than some new boats and fuel.

The relative importance of the market economy affects fishers' behaviour within their own fishing grounds and explains differences in behaviour among separate grounds. Within fishing grounds, those catching fish for sale selectively target species of higher value than when they catch fish for food (Fig. 6.1). Moreover, even though there are no major differences in the structure of the fish communities among grounds, fishers in Ko Ono target high value species, and most of the catch caught for sale consists of the 20 most valuable species. On Totoya, the fishers are less selective and still target many species that they fished traditionally rather than dramatically changing their fishing behaviour to suit the whims of the fish markets (Jennings & Polunin, 1996a).

If the currency of reward gained from fishing were expressed only in terms of net energy gains, fishers would limit their effort as predicted by optimal foraging theory. The energetic value of one average southern bluefin tuna *Thunnus maccoyii* does not alter dramatically whether there are 100 000 or only 100 left in the oceans. Moreover, it is unlikely that such low numbers would ever be reached if energetic returns were the main criterion for foraging choices. Sadly, the financial value of such resources tends to be inversely related to their abundance, based on the laws of supply and demand (Chapter 11). Thus, while 100 000 southern bluefin tuna might be worth $US650 million, the last few tuna might be worth the same amount. Indeed, prices on the Japanese markets range from $US6500 to $US11 000 per fish depending on the quality of its flesh (Fig. 6.2).

The extent of the transition from a primitive to market economy will also determine the effort that fishers

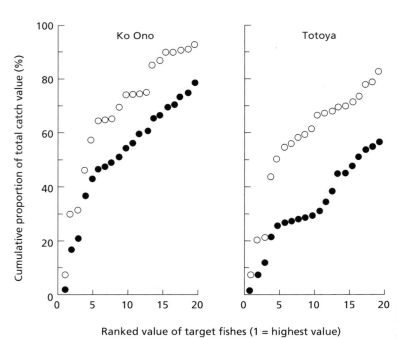

Fig. 6.1 Composition (by value) of catches from subsistence (●) and commercial (○) fishing in two Fijian fishing grounds. From Jennings and Polunin (1996a).

Fig. 6.2 Southern bluefin tuna laid out for sale in a Japanese fish market. These fish are highly prized and a top quality specimen can be worth in excess of $US10 000. Such high individual prices mean it is worthwhile pursuing these species even when their numbers are low. Photograph copyright B. Glencross.

will invest in targeting a particular species. In energetic terms, extirpation should never occur because the costs associated with searching for the dwindling stock would exceed the energetic returns. This is not the case when the currency of cost–benefits is based on financial gain. Rarity, gastronomic culture and certain traditional medical practices, create an aura of mystique and desirability that surrounds certain species. Hence, market forces push the price of such rarities and luxuries so high that it becomes economically viable to expend the extra effort required to harvest them.

Since many fisheries are now exploited for financial gain rather than food, the economic analysis of fisheries has played an increasing role in the assessment and management process. Indeed, the purest forms of economic analysis treat profitability as the overriding goal of fisheries management, with little or no regard for social factors such as employment and quality of life. We address economic analysis, the economic value of fisheries and the problems with pure economic analyses in Chapter 11.

6.3 Modifications to fishing behaviour

Desires for food and income drive the behaviour of fishers, but social and religious factors can also have marked effects. Even in the most primitive economies, relationships between the work input to fishing activities

and the food or income produced cannot be explained if food and income are treated as the only motivations.

6.3.1 Social

Ontong Java is a coral atoll in the Solomon Islands. It consists of 122 small islands with a total land area of around 7.8 km^2 that are inhabited by about 1000 people. The islands are scattered around a large sandy and coral lagoon and the islands are vegetated with coconut palms, bananas and breadfruit trees. The villagers also cultivate a range of root crops such as sweet potatoes and taro. On average, men spend around 23 h each week gathering food and fishing while women spend about 14 h. The foraging and fishing activities of these islanders have been studied in detail (Bayliss-Smith, 1990) (Table 6.1).

The analysis shows that pond-field cultivation, coconut collecting and fishing provide most of the energy in the diet. Coconut collecting yields a particularly large proportion of the energy in the diet for the work input, but energy is a rather simplified predictor of nutritional value since many of the high-energy foods lack protein. Fish yields less energy per unit of work input, but will provide most of the dietary protein that the islanders need to survive. Shellfish diving and turtle hunting together take more time than coconut collecting but yield only 0.4% of dietary energy. There are social motivations for shellfish diving and turtle hunting which are an important part of cultural life for women and men, respectively, even though they are not an important food source for the village. These social motivations can thus confound analyses of fisher behaviour that are based solely on optimal diet theory as discussed earlier.

There are many other examples of social modification to fishing activity that can confound fisheries assessment. For example, when estimating fish abundance from catch and effort data we assume that catch-per-unit-effort is proportional to abundance (Chapter 10). This requires that the catchability of the fish is constant and that fishers put the same levels of effort per time unit into trying to catch fish as fast as possible. However, if fishers only treat fish capture as one aim of fishing and do not consistently choose to go to the best fishing grounds because they do not want to risk their lives, boats or nets in treacherous but productive waters, then a simple measure of effort such as hours

Table 6.1 Annual estimates of inputs to and outputs from the main subsistence activities on Ontong Java Atoll. These activities account for a total of 70% of the energy in the diet. From Bayliss-Smith (1990).

Activity	Work input (h)	% total input	Total food yield	% energy in diet
Pond-field cultivation	86 500	51.5	69 t *Cyrtosperma* corms 40 t *Colocasia* corms 1.3 t *Colocasia* leaves Turmeric	29.1
Coconut collecting	8 800	5.2	121 000 ripe nuts 272 000 unripe nuts 104 000 germinating nuts	20.5
Fishing (nets, lines, spears)	59 300	35.3	81.2 t fish	19.0
Shellfish diving	3 800	2.2	1.3 t shellfish meat	0.2
Turtle hunting	9 700	5.8	60 turtles 60 clutches of eggs	0.2

fished may not reflect their true fishing effort. In both subsistence and commercial fisheries, fishers often make decisions about where and how to fish that do not have the sole aim of maximizing catch. Of course, this information may be completely unknown to the analyst who only sees catch and effort data on a computer screen. This demonstrates why fishers, assessment scientists and managers should always discuss their work.

As an extreme example of fishers choosing not to maximize catch rates, we can look at reef shore fisheries in the Fijian reef fishing grounds of Ko Ono, Moala and Totoya. Here, fishers do not choose to use the most efficient fishing techniques available. Figure 6.3 shows that catch rates are often higher when spears or nets are used for fishing, but that lines are used more frequently. This is not because nets are not available in fishing villages, nor because there are too few fishers to deploy the nets nor because more desirable species are caught using lines. Rather, fishers only try to maximize catch rates when fish are in short supply and at most other times they see no need to hurry with their fishing. Time rarely restricts the potential for food gathering since villagers can easily complete their fishing and agricultural tasks in one day and there are rarely facilities for storing excess catch. Fishers catch enough fish for immediate consumption and treat the prolonged periods of fishing as a recreational and social activity that provides an

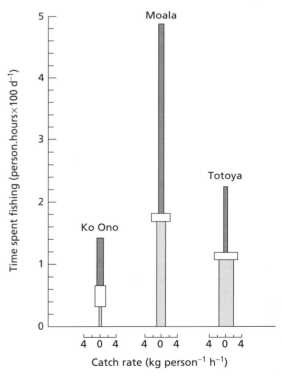

Fig. 6.3 Mean catch rates of reef associated fishes and time spent fishing in three Fijian fishing grounds. Dark shade is hand-line fishing, unshaded is spear fishing and light shade is net fishing. Data from Jennings and Polunin (1995a).

important chance to spend time away from the village (Jennings & Polunin, 1995a).

6.3.2 Religion

Throughout history, fishers have worshipped the sea and sought favour from mythical gods that supposedly ruled the watery domain. Fishing remains one of the most life-threatening jobs and coastal communities live with the omnipresent risk of losing loved ones at sea. Not surprisingly, fishing communities are often closely linked to one form of religion or another and this in turn affects their fishing behaviour.

A good example of the influence of religion is in the rose shrimp *Aristeus antennatus* (order Caridea) fishery off the Catalan coast in the north-western Mediterranean. The fishery is targeted by vessels that leave and return to their home-port every day. In the Catalan region, it is the tradition that fishers do not work during weekends because this is a religious festival. Landings of rose shrimp vary through the week

and the fishers have known for many years that landings are always highest on Fridays (Fig. 6.4a). While this sounds like a superstitious tale, it is true, but why? Sardà and Maynou (1998) used landings per unit effort data, field data and the maximum daily prices paid at auction for rose shrimp to investigate possible causes of this phenomenon. They tested three hypotheses that could lead to an increase in catches. The first hypothesis was that sediment stirred up by the bottom trawling activities increased the food available to the shrimp towards the end of the week and caused them to aggregate in greater numbers. The second theory was that as fishing activity progresses through the week, predators and competitors of shrimp were removed at a greater rate than shrimp. The third hypothesis was that market system dynamics act as an incentive for fishers to improve their catch efficiency towards the end of the week.

Analyses of stomach contents of rose shrimp sampled on each day of the week revealed no differences in diet or the amount of food eaten, hence the first

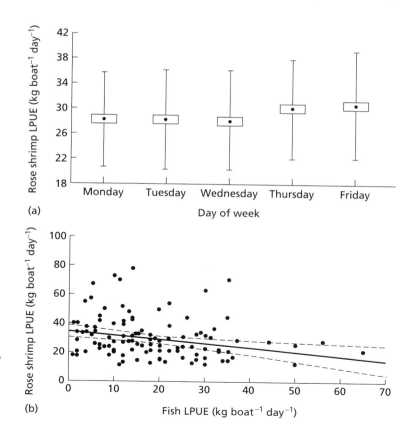

Fig. 6.4 (a) Box and whisker plots of the average shrimp landings per unit effort (LPUE) from Monday to Friday for the data in Sardà and Maynou's study of the Catalan rose shrimp fishery. The limits of the boxes depict the standard errors of the mean while the whiskers denote the standard deviations. (b) The linear relationship between fish and rose shrimp landings for a 122-day period of study showing that shrimp catches rise as fish catches fall. After Sardà and Maynou (1998).

hypothesis was rejected. However, Sardà and Maynou (1998) found that as fish landings fell, so those of shrimp increased (Fig. 6.4b). More impressively, landings were most strongly linked to market price. The highest landings were always made on Friday when demand was highest prior to the weekend when there were no landings. Rose shrimp tend to be confined to a narrow depth band within the water column, hence it takes the fishers several days to locate schools of these shrimp and to pinpoint the optimum depth at which to set their nets. During the week, the fishers improve their fishing efficiency based on the experience gained from the previous day's fishing activity. Hence they are most efficient by Friday. In addition, the first fisher to return to port will gain the highest price for their shrimp at the market. However, after 2 days with no fishing, the schools of shrimp have usually moved to a different locality and will have altered their depth, therefore they take longer to find. Although prices were high on Mondays due to the lack of fishing over the weekend, fishers were unable to meet this demand because their ability to find the shrimp shoals is greatly reduced after 2 days of inactivity. This is an interesting example of how traditional practices, market forces and the behaviour of the target species interact to influence catches.

Religious beliefs also play an important role in regulating fishing activities in other nations. In the Pacific Islands, many forms of fishing are taboo and there is no Sunday fishing by Christian Methodists. The North Sea Dutch beam trawl fleet from Urk stays in harbour for 1 week each year for religious reasons. This fleet contains around 30% of the larger Dutch beam-trawling vessels. In the absence of these large beamers, other boats compete more effectively in the area they usually fish and the revenue of the vessels that continue fishing increases by 15% (Rijnsdorp *et al.*, 2000a, b).

6.4 Conflicts and conflict resolution

6.4.1 Competing for fish

Why do fishers compete?
Many fisheries are viewed as common property and some are entirely open-access. As such, there is little to be gained by one fisher trying to conserve fish because the fish will simply be caught by someone else! In the simplest terms, the race to fish begins because it is better to catch a fish today than to leave it in the water. We will look in detail at the reasons why fishers compete and overfish resources in Chapter 11. Newspaper headlines from around the world demonstrate the strength of competitive feeling between fishers and the nature of local and national conflicts between them (Fig. 6.5).

The consequences of the race to fish have been particularly dramatic in countries such as the Philippines, where management systems and resources have collapsed. For many poor people, the sea is the only potential source of food and income. As more and more people enter the fisheries, so they are faced with declining catches. Lacking any alternative source of protein

Fig. 6.5 Conflicts over fish.

or income, the fishers initiate wholesale destruction of the resource base to maintain their livelihood. This may involve fishing with explosives and poisons that damage the reefs, kill most of the non-target species and compromise any possibility of sustaining yields in future. Such fishing was termed **Malthusian overfishing** (Pauly *et al.*, 1989) after the Reverend I.R. Malthus (1766–1834) the 'prophet of doom', who described problems of feeding an exponentially growing human population.

Competition between fishers using different methods

In areas permanently closed to certain fishing practices stocks may become more abundant than in areas open to that practice (Schmidt, 1997; section 17.5.3). Regardless of whether this is the case, this is often the perception of fishers denied access to those grounds. Fishers are then tempted to contravene the rules of the closure to obtain a higher catch-rate and short-term monetary rewards. These incursions tend to destabilize the effect of the management measure until a free-for-all situation develops. A good example of this situation is given by Hart (1998) who describes fishing grounds used equally by fishers that operate either static or towed gear. Hart (1998) describes an area of seabed off

the south Devon coast in England that is fished with pots for crab and lobster, with dredges for scallops and beam trawls for flatfishes. Pot fishermen set their gear on the seabed where it is left unattended for 24 h or more. Scallops and flatfishes are also found on this fishing ground and these are caught with towed dredge and trawl gears. The density of crab pots in these areas is so high that it is almost impossible for scallop dredgers or beam trawlers to work this ground without damaging the gear of the crab potters. This potential conflict led to the establishment of areas voluntarily closed to trawling in 1978 (Fig. 6.6). Note this is not a closure to protect the target species, rather it is a solution to minimize conflict between different groups of fishers. Hart (1998) emphasized several key points that indicate how the different groups contribute to this fishery. The smallest vessels work pots or small otter trawls and are operated by people from a close-knit community with extensive family histories in the area. Larger, more powerful vessels use scallop dredges or beam trawls and these tend to be based in ports further afield. In addition, these vessels tend to be owned by management companies, with the skipper and crew as employees. Their main objective is to make as much money as possible for their owners and they may be

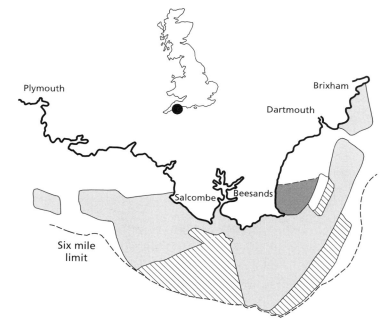

Fig. 6.6 A map of the English territorial sea between Plymouth and Brixham in Devon showing the agreed partitioning system of the fishing grounds between fishers that use fixed and mobile gear. The 6-mile territorial limit is shown by the dashed line. Areas reserved for potting are indicated by light stippling; potting or trawling exclusion zones are indicated by dark stippling; hatched areas are closed to trawling at some time of the year. All other areas are permanently open to trawling and dredging. After Hart (1998).

paid according to the value of the landed catch. This situation makes these vessels less likely to adhere to voluntary agreements such as area closures.

The voluntary agreement exists between several parties that are represented by South Devon and Channel Fishermen Ltd (potters), the Trawler Owners Association (towed gear fishers), and the Devon Sea Fisheries Committee (local enforcement agency). All fishing with towed fishing gear is prohibited in areas reserved for the setting of crab pots. Within each of these areas, each crab potter sets gear in the same area which is then regarded as 'territory'. Some areas are open to trawlers at certain times of the year, but not others. The area concerned is entirely inside the 6-nautical-mile exclusive fishing limits of the United Kingdom and as such only boats registered in the United Kingdom can use it. This partitioning of resources between interested parties has two advantages. Firstly, it reduces conflict between different gear users, and secondly, by excluding towed fishing gear from certain areas it promotes sustainable use of the seabed environment. The other notable feature of this management system is that it seems to work most of the time.

Why so few fishermen cheat the system? Hart (1998) used a game theory model that originated in the behavioural sciences to examine what motivates fishers to cheat or cooperate. **Game theory** deals with interactions between individuals whose reward (pay-off in behavioural jargon) depends on the action chosen by the other individual(s). This assumes that the individuals are able to choose from the same set of strategies and that the choice will be made such that it maximizes the **pay-off** (benefit) for an individual. The optimal outcome is that which provides an equal pay-off for all individuals (Maynard-Smith, 1982). Within game theory, the prisoner's dilemma model of conflict and cooperation captures most accurately this tension between the need to cooperate and the temptation to grasp today's catch before someone else does (Axelrod, 1984; Hart, 1998). Table 6.2 shows the reward for two options of cooperate or defect (i.e. grab today's catch before someone else). The ideal is for both to cooperate as each will gain a score of 3 each time the individuals interact. However, it is clearly beneficial to defect as an individual gains a score of 5 by defecting while the other cooperates. The expected outcome is thus that both defect and each gains a score of 1. This is known as the prisoner's dilemma paradox as the expected final

Table 6.2 The prisoner's dilemma pay-off matrix. Scores are given for both individuals in the competition for the resource. Subscript 1 relates to individual 1 and subscript 2 to individual 2. R_1 and R_2 denote the reward for mutual cooperation: S_1 and S_2 represent the sucker's (loser's) pay-off: T_1 and T_2 are the temptation to defect (i.e. not cooperate); P_1 and P_2 represent punishment for mutual defection. For this situation to become a dilemma, then $T > R > P \gg S$ and $R > (S + T)/2$ must be true.

		Individual 2	
		Cooperate	Defect
Individual 1	Cooperate	$R_1 = 3$ $R_2 = 3$	$S_1 = 0$ $T_2 = 5$
	Defect	$T_1 = 5$ $S_2 = 0$	$P_1 = 1$ $P_2 = 1$

position is to not cooperate, even when it is clearly beneficial to both participants. The depressing expectation that individuals should not cooperate even when it is clearly beneficial to do so only applies if individuals meet rarely (as would be the case in a long distance fishery). However, if individuals meet repeatedly (e.g. as for an inshore fishery) then the situation is the reverse. These costs and benefits are usually based on energetic calculations, but in this situation costs are related to the financial cost of replacing damaged gear, lost fishing time or physical violence. The benefits are measured in the increased value of improved catches and savings on gear maintenance. Whatever the true costs and benefits, as long as their rank orders meet the conditions in Table 6.2, these results will be general.

Axelrod (1984) introduced an additional variable ω ($0 \leq \omega \leq 1$) that provided a measure of the importance of the future relative to present actions. As ω approaches zero, the strategy that should be adopted is one of mutual defection. In other words, if the motivation of the fishers is to make as much money as possible today without any consideration for future earnings from the fishery, they should look after their own interests immediately. This is most likely to apply for the larger vessels whose crew work for an employer with little or no interest in the long term viability of the fishery. As the importance of the future increases (i.e. ω approaches 1), then a strategy of **tit-for-tat** (TFT) turns out to be the best strategy for cooperation (Box 6.1). In other words, when different groups of fishers both have future interests in the fishery it pays them to co-operate according to a set of rules similar to those predicted by TFT. This approach is similar to that of some economic models that seek to explain why users may compete for a resource (Chapter 11).

TFT is only likely to work as a strategy if long–term interactions are possible, such that individuals are likely to meet again, and on these occasions, they remember what happened at their last meeting (Axelrod, 1984). These objectives are met when the importance of the future (e.g. in this case the sustainability of the fishery) is more important relative to the present (e.g. the size of today's catch). Such a system is likely to occur in a small community of fishers. However, this system breaks down when fishers from outside the area, perhaps from a foreign country, are able to access the same resources. They have no interest in cooperation and have nothing to lose by 'breaking the rules'. While the Devon Management System works extremely well at a local scale, problems have arisen when large vessels from outside the area have violated the no-trawling zones. The only deterrent for these transgressors are large fines imposed by government enforcement agencies. Despite this threat of punishment, prosecutions are rare given limited enforcement resources, and often even large fines are small in comparison to the value of one catch.

Behaviour of competing fishers
An understanding of behaviour helps to explain how fishers respond to management measures. Rijnsdorp

et al. (2000a, b) considered how fishers would distribute themselves in space and time as they compete to catch fish. Since fish, in this case the plaice (*Pleuronectes platessa*) and sole (*Solea solea*) targeted by the North Sea beam-trawl fishery, are not spread uniformly throughout the sea, we expect the trawler 'predators' to distribute themselves in response to the abundance of their prey. In section 4.3 we considered the theoretical form of this distribution when fish compete for habitat or prey, the ideal free distribution. In this case, predators pick habitats with the greatest realized suitability. As the most desirable habitats become increasingly populated, so it is worth moving to others. Ultimately, all predators would be expected to use areas of similar realized suitability. Is this true of boats competing on fishing grounds?

The ideal free distribution would predict that fishers have revenues that are independent of vessel density because no vessel could move to a more suitable location. Rijnsdorp *et al.*'s results provided some evidence for this. In 65% of trips considered, revenue was independent of vessel density.

The existence of competition between vessels helps us understand how a fleet may respond to management measures. For example, if the number of vessels in the fishery were reduced then the number of competitive interactions would also fall and catchability would increase. This would have a different effect on actual fishing mortality than attempts to reduce effort using methods such as closed areas or seasons that may increase competitive interactions. Competitive interactions also affect the social stability of fishing communities and their willingness to behave in a cooperative manner. The patterns of fishing effort that emerge from competition between different vessels also affect the impact on seabed communities (Chapter 14). Hence, areas in which the largest vessels congregate will be subjected to a greater degree of physical disturbance than those frequented by the smaller vessels that fish lighter gear.

6.4.2 Fish wars

Although it may be hard to believe that fish would be worth fighting over—they were and still are. Fish are a highly valuable commodity and their historical importance as a source of protein and object of trade are described elegantly by Kurlansky (1997). For example,

---

Box 6.2

Conflict at sea.

From a newspaper report, March 9 1995:

• 12 : 50 h—Two Canadian fisheries department vessels and the Canadian Coast Guard ship *Sir Wilfred Grenfell* encounter the Spanish fishing boat *Estai*. They attempt to board the *Estai*. Spanish vessel cuts its own nets and speeds away.

• 13 : 40 h—Royal Canadian Mounted Police team attempts another boarding but is turned back by bad weather.

• 16 : 33 h—Four bursts of gunfire at *Estai*. It stops, and Spanish crew ordered belowdecks.

• 16 : 52 h—Department of Fisheries officials and RCMP board *Estai* and arrest captain. Ships head back toward St John's.

---

the United Kingdom has been involved in three **cod wars** with Iceland since 1950. Prior to 1950 UK fishers were permitted with other nationalities to fish without restriction in the waters around Iceland. Eventually, the Icelandic government realized the threat to their marine resources. In April 1950 Iceland annulled the 1901 Anglo-Danish Convention and extended its territorial limit to 4 nautical miles from its coastline. This was a radical move when we consider that it was commonly held that the sea belonged to everyone according to international law (section 1.2.1). This limit was further extended to 12 miles in 1958 in response to a 16% decline in groundfish catches. Later, the limit was extended to 50 miles in 1972 and then 200 miles in 1975. On each occasion, warships were dispatched by the British to protect their trawlers from the Icelandic coastguard vessels that tried to arrest the British trawlers or prevent them fishing by cutting their nets adrift or ramming the British boats. It wasn't long before all nations had declared similar exclusion zones around their coasts to protect their marine resources.

Fishing rights and territorial limits are guarded jealously by each nation. Of course, fish don't recognize national boundaries or demarcation lines. So while a shoal of fish may be protected in the national waters of one nation, there may be dozens of boats waiting to catch them in a neighbouring area. Some countries have taken action beyond their borders as happened on 9 March 1995 (Box 6.2). This dispute between Canada and Spain arose because fishers were competing for the same resources, Greenland halibut, *Reinhardtius hippoglossoides* (Pleuronectidae) off the Grand Banks. While the Spanish were fishing in international waters, the Canadian government took action against the Spanish vessels because they suspected that the Spaniards were fishing with nets that had illegal meshes.

6.4.3 Fishers in the political process

Although fishing generates a substantial proportion of the income of many of the poorer nations around the world it represents a relatively small component of Gross Domestic Product in most industrialized countries (Chapter 11). Nevertheless, fishing is intricately bound up with nationalistic feelings and in many ways is given a disproportionate level of importance in world politics. Lequesne (1999) has examined the influence of fishing communities on local and national politics. Commercially important fishing communities have tended to become concentrated in particular areas of each nation, e.g. Cornwall in England, Mallaig and Peterhead in Scotland, and Galicia in Spain. Fishing issues assume proportionally greater importance in these areas as the local community is closely bound to, or aware of the fishing industry and its problems. This awareness is no doubt a product of the relatedness of individuals in the area and also the political campaigning undertaken by the fishers and their representatives. As a result, fishers may often be supporters of regionalized political parties such as the Scottish National Party in Scotland and Herri-Batasuna in the Basque region of Spain. The political stance adopted by fishers in relation to local politics also affects their attitude towards wider political issues such as the Common Fisheries Policy that affects all fishing nations in Europe. Lequesne (1999) realized that the reasoning behind the campaign undertaken by Basque long-line fishers against drift-netters in the province of San Sebastian, Spain, was intimately connected with political affiliations. The industry leaders that represented the Basque long-liners in their campaign against French, English and Irish drift-netters were highly sympathetic to the Herri-Batasuna party. Their arguments against the use of the foreign fishers' techniques contained elements of Basque nationalist ideology regarding the protection of the local way of life from encroaching outside influences. Although the main issue of the conflict was the protection of the price of fresh albacore tuna, *Thunnus albacares* (Scombridae) on the Spanish market from cheap imported products, the actions of the Basque liners were certainly influenced by political ideology.

6.4.4 Traditional management systems

Despite many problems with fisheries management in the developed world, the approaches used are often recommended to non-industrialized nations. This is somewhat ironic as non-industrialized societies may already have effective management systems in place. Here, local customs and behaviour discourage the race to fish that has affected common property resources elsewhere (Ruddle *et al.*, 1992).

Traditional management systems are often based on use rights to a property. These rights are a claim to a resource and/or the benefits derived from it (Hall, 1999). Such rights operate at two levels. At the primary level, the exclusive use of local resources is enforced by the right of a community to prevent poaching. Such exclusive access to a resource is more likely to motivate its sustainable use. At the secondary level, management is directed at the allocation of the shared resource among the communal users. These systems are usually well enforced as they tend to be self-policed by fishers. Any activities or restrictions that apply to the use of the management area will have direct consequences for the owners. Consequently there is sense that the fate of the resource is under the control of the local community and not subject to human influences outside their control.

Many coastal fisheries in the Pacific Basin, Indo-Pacific, Caribbean, Latin America and Asia-Pacific are managed using traditional systems. These evolved through the need to conserve resources or through conflict. At the present time many of these systems have disintegrated, often as a result of population growth, the arrival of Christianity, changes to legal systems, urbanization, commercialization, technological change and economic pressures during the transition from primitive to market economies. Systems that once existed in places such as Hawaii and American Samoa have largely disappeared, but in Kiribati, Niue, Tokelau and Western Samoa, for example, they are still operational. Before further pressures force more of these systems to collapse, there is much to be gained from understanding them. Understanding the management systems does not just mean understanding the responses of fished species, but also the attitudes of the individual fisher. Knowledge of the latter is essential to comprehend management systems used by people who see the world differently (Johannes, 1978a).

Unlike much biologically based management, traditional systems focus on resolving gear use or allocation problems. Access control is enforced by fishers and by local moral and political authority. Many traditional systems include elements of 'contemporary' strategies such as closed areas and limited entry, with sanctions applied to fishers who violated them. Some of the sanctions applied were reviewed by Ruddle (1996). For example, in Indonesia, social, economic and supernatural sanctions would be used against poachers. These would involve corporal punishment such as caning, public shaming, confiscation of boats and gear, monetary fines and 'supernatural' events such as the disappearance of gear, sickness and death. Clearly, illegal fishing was unlikely to be rewarding! Similar sanctions are used in other parts of the Pacific, especially public shaming, the confiscation of gear, compensation payments and corporal punishment. However, the management systems are not as inflexible as the tough sanctions might suggest. For example, in Palau, fishing rights are controlled by chiefs who represent each village and these chiefs would arrange for fishers to gain access to each others' grounds (Johannes, 1978a).

Many traditional management systems broke down following western and colonial intervention. This was brought about by the change from primitive economies, in which fish and goods were bartered, to market economies where goods were bought and sold. Colonial powers also weakened much of the local and traditional authority and transferred it to central government. Colonial government sought to let villagers see the value of money for buying goods; if they wanted people to work for money they needed to encourage them to value it. In later years, fishers were encouraged to borrow money to pay for boats and to enter schemes that simplify export of valuable species such as sea cucumbers (*bêche de mer*) or live fish. These financial pressures and incentives have led to breakdowns in traditional management systems and a race to fish. It is not so much that all development was bad, since it brought other benefits such as improved health care and opportunity. Rather, traditional management methods were largely unappreciated by the West who still believed in the freedom of the seas.

6.4.5 Customary marine tenure

Local fishery management systems that regulate access to, and exploitation of fisheries are collectively known as **customary marine tenure** (CMT) systems: customary because the current approach to management, although dynamic, is linked to past experience, marine because the systems that are managed include marine areas such as reefs and lagoons and tenure because socio-cultural processes allow the fishers to control their territory and access to its resources (Johannes, 1978a). Customary marine tenure is the most widely used form of traditional management.

There are a range of CMT systems in the Pacific basin, and an appreciation of their role has sometimes been hampered by the western assumption that fish are a common property and an open-access resource. In reality, coastal fishing grounds and coastal land are often viewed as a single territory and are owned by kinship-

based groups (Fig. 6.7). These may be the clans or families that control the area. The groups did not identify strict boundaries between ownable land and unownable sea that apply in many western cultures. Indeed, territories such as the Fijian vanua, Hawaiian ahupua'a and Yap tabinau can include both land and sea.

With few available sources of animal protein, islanders of the Pacific Ocean have had to use marine resources sustainably or move from islands. Indeed, human migration between islands was often driven by resource shortages. Islanders had recognized the limitations of marine resources many thousands of years ago, unlike many westerners. Indeed, when T.H. Huxley made the frequently cited remark that the great sea fisheries were inexhaustible (section 1.2.2), many Pacific islanders had been aware of the consequences of fishery collapse for centuries and had invoked management measures that guarded against overexploitation. Now, while much of the western world appreciates the benefits of fisheries ownership and limited entry (17.3.3), the traditional systems that used these approaches can be stressed by western involvement in their societies.

There are various theories as to how CMT systems

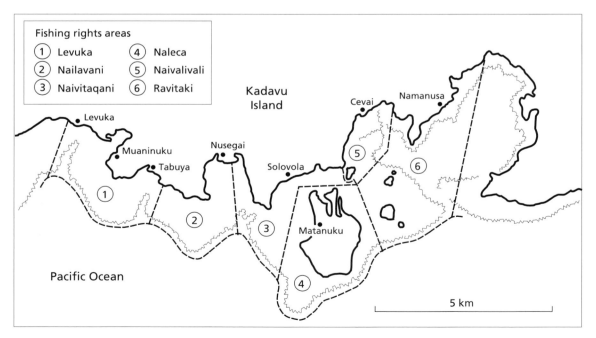

Fig. 6.7 Map of traditional fishing rights areas on the southern coast of Kadavu Island in Fiji. Fishing grounds are numbered and villages named.

evolved. In the simplest case, their evolution has been attributed to trial and error. Many oceanic islands have relatively limited resources and have been inhabited for 2000–3000 years. The large area of coastal waters in comparison to available land in the Indo-Pacific may have a strong influence on the likelihood of the evolution of a system such as CMT. If the resources of an island were not used in a sustainable manner then the inhabitants would suffer food shortages. Only islanders that fished in a sustainable fashion and managed their resources would survive. Thus, effective methods of conservation and management could have arisen by trial and error (Zann, 1989). In other cases, the islanders themselves had a strong conservation ethic and made deliberate attempts to conserve resources they saw as exhaustible (Johannes, 1978a).

While there appears to have been a strong and recognized conservation ethic in the western Pacific, this does not necessarily exist elsewhere. In Papua New Guinea and Indonesia, CMT is not universal and people give more attention to land rights than the sea (Polunin, 1983). When tenure did develop it was often the result of conflict over resources rather than the desire to conserve, and thus ownership existed for gain rather than constraint. Once methods for improving production from the land were developed, the land offered greater food security than the sea. However, in the areas that were fished and where more valuable species such as trochus shells *Trochus niloticus* were abundant, there was a strong history of dispute. This is probably because people took the trouble to defend what they saw as valuable (Polunin, 1984).

Many modern societies still perceive the sea as a common resource, in contrast to those in geographical regions in which resources are strictly controlled by kinship groups and communities. CMT systems have been weakened by western influence and population growth and it is questionable whether these CMT systems, without government support, are sufficiently adaptable to cope with the increasingly rapid effects of economic development. In Fiji, for example, the CMT system has collapsed around urban population centres while, with government support, it remains intact in many outlying islands.

6.4.6 Co-management

If the only aims of management are simple biological or economic criteria such as maximization of yields or profits, then regulation to achieve these aims may not work as expected when applied to real people. Indeed, many proposed management measures are just plain silly from a fisher's perspective. For management to be effective it is usually necessary to recognize the history and traditions of fishing communities. **Cooperative management,** or **co-management** formally recognizes the fishers' role, and fisheries departments provide scientific information and advice while local communities take responsibility for management. Co-management is particularly relevant to smaller community based fisheries in isolated locations since the resources do not exist to police them externally. For fisheries without a strong tradition of resource ownership and self-policing, co-management may simply allow fishers to behave as they wish!

Strategies that favoured the success of government-supported co-management on the Pacific island of Vanuatu were reported by Johannes (1994). These included publicizing the government's desire to collaborate and concentrating management efforts on villages where CMT and local authority were already strong and the community was cohesive. It was also important to ensure that national law supported communities in regulating the fishery but was not too prescriptive and left the final management decisions and enforcement to the villagers.

Some countries are making strong attempts to introduce co-management. In Fiji for example, we have seen that the traditional management system based on customary fishing rights areas has been pressured by the development of a stronger cash economy. The government now seeks to support villagers in maintaining this system. In Fiji, linkages between the community and government are facilitated. Thus, the government formally recognizes the traditional rights of communities to exploit reef resources (even though it retains ultimate ownership), allows communities to restrict access to these areas, formally records traditional management boundaries, and mediates in disputes. Moreover, for commercial fishers who want to fish in traditional areas, the government will only licence them after they have received permission from the owner and paid any charges that the owner may request (Adams, 1996).

The inshore fisheries of Japan also show how traditional practices can be incorporated into modern legislation (Ruddle, 1989). Japanese small-scale fishers have legalized access to, and ownership of, fishable resources in inshore waters. In Japan there is no conceptual difference between land and sea tenure and fisheries rights have legal status equivalent to land ownership.

In feudal Japan (1603–1867), common land (**iriaichi**) was shared by several villages within each fiefdom, the forerunner of today's prefectures. This extended to sea territories too which were known as iriai. Within the iriai, the use of a species, or type of fishing gear was reserved for one village. Regulations were also imposed to limit seaweed harvests during the spawning season when fish eggs were attached, the use of small mesh nets, and the use of bottom set gill nets that also caught non-target benthos. In some fiefs, licences would be issued for specific gears.

The feudal order collapsed throughout Japan with the Meiji restoration in 1868, and in 1876 the new federal government took ownership of all fisheries. A race to fish began, with many examples of disputes and over exploitation. Eventually, controversy and dispute led the government to abandon the 'revolutionary' common access system. However, larger fishing operations in cooperatives were allowed to carry on fishing and this created conflicts with those who had local rights. Only with the climate of economic stresses caused by World War II did fishers regain control of their local fishing grounds. Management based on fishing rights and licensing was put into statute again at the end of the war, and the system is still based on this approach today. Continuity of tradition has had many benefits and locals are empowered with control of their resources. The local management units, known as Fisheries Co-operative Associations (FCAs) have fishing exclusive rights and distribution of those rights is resolved by the membership.

It is wrong to suggest that the co-management systems we have described do not have their problems. But they have provided fishers with statutory rights to fishing areas and encouraged them to use resources in a sustainable and probably environmentally friendly manner.

Summary

- Fisheries management involves regulation of human activities and as such it crosses the boundaries of the social sciences.
- Social aspects of assessment and management have often been neglected in favour of purely biological investigation.
- Management is rarely effective without some level of community support.
- Theoretical predictions of fishers' behaviour are unable to account for the influence of factors such as religion, political affiliation and the pursuit of an activity for an end other than gaining food or money.

- Humans are able to adapt their behaviour to counteract any management measures that might be imposed upon them.
- Fishing is not just about catching fish and making money; rather it is bound up in the culture of coastal societies. Any factor that affects the viability of the fishery affects that society, the two are inextricably linked which is why fisheries students and scientists should never forget that humans should be considered as part of the marine ecosystem.

Further reading

For further details of how game theory can be used to predict how fishers interact in cooperative management systems, consult Hart (1998). For an analysis of the influence of fisheries interactions on political affiliations and vice versa see Lequesne (1999). Ruddle (1996) reviews traditional management systems in tropical reef fisheries.

7 Single-species stock assessment

7.1 Introduction

It is not easy to keep track of the rates of birth, death and reproduction of a large number of individuals. It is even more difficult to predict how a change in any one of these parameters, such as increased mortality, will affect the population in the future, because this will depend on numerous interactions within and among species, often against a backdrop of year-to-year variability in the environment. Indeed, fishery scientists have the additional problem of studying animals that are usually visible only when they have been brought over the side of a ship! Yet understanding the population biology of fished species is essential to meet one of the main objectives of fishery science, that of maximizing yields to fisheries while safeguarding the long-term viability of populations and ecosystems (Chapter 1).

The aim of this chapter is to provide a brief overview of single-species stock assessment. 'Brief' is certainly an understatement, given that many books are devoted primarily to this subject, including specialist volumes relevant to particular taxa or regions. For derivations of the methods introduced in this chapter we particularly recommend Hilborn and Walters (1992) and Quinn and Deriso (1999). Fisheries science is a quantitative subject, so students should be prepared to embrace the minimal mathematics included here. We have put much of this in separate boxes and point out key references for those who want more detail.

7.2 Balancing birth and death

Traditionally, the Holy Grail in fisheries science was to find the **maximum sustainable yield (MSY)**, the largest catches that can be taken over the long-term without causing the population to collapse. From a strictly biological point of view, this makes sense. However, this is by no means the only objective of fishery management,

and it may ignore the goals of the fishers themselves, who are often more concerned with employment and maximizing profit from the catch (Chapters 1, 6 and 11). Even if the objective is to maximize yield, precaution suggests managers should aim for yields below the theoretical MSY (section 7.3). In the meantime, MSY is a good starting point for understanding the biology of exploitation.

For a given level of fishing mortality to be sustainable, there must be a balance between the mortality, which reduces population biomass, and reproduction and growth, which increase it (Russell, 1931; Fig. 7.1). Mortality and reproduction are not entirely independent, but fluctuate within limits set by abiotic factors such as weather, and biotic factors such as competition and predation (Begon *et al.*, 1996a; Chapter 4). The balance that is struck through biotic factors is due to density dependence, the relationship between population density and per capita birth, growth or death (Fig. 7.2). Density dependence gives populations the resilience required to sustain elevated mortality from fisheries.

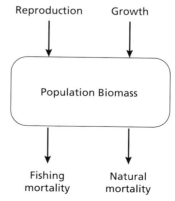

Fig. 7.1 Population biomass depends on growth, reproduction, natural mortality and fishing mortality.

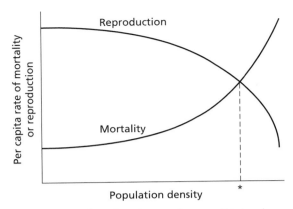

Fig. 7.2 Density dependence in per capita rates of birth and death. At the equilibrium (*) these processes balance one another.

7.3 **Surplus production models**

Surplus production models are used to search for the largest fishing mortality rates that can be offset by increased population growth, normally measured as changes in population biomass per unit time. They are a good starting point because they capture the basic logic of density dependence, and the simplest ones can be thought of as 'null models' that underlie the theory of sustainable exploitation for terrestrial as well as aquatic organisms (Milner-Gulland & Mace, 1998; Reynolds *et al.*, 2001). Surplus production models use data that have been aggregated to some extent across age classes. These models appear in the literature under several aliases, including **production models, stock production models, surplus yield models** or **biomass dynamic models**.

Suppose that a population grows in a classical logistic (sigmoid) fashion, beginning slowly at first and then reaching a maximum rate of increase, before slowing down again as it reaches maximum total biomass, B_{max} (Fig. 7.3a). The maximum biomass of the population is traditionally said to occur at the carrying capacity of the environment, although many fish populations fluctuate so wildly that it is easier to deal with this concept in theory than in practice (Chapter 4). The slow-down in population growth at high densities would be due to density-dependent processes such as competition for resources, cannibalism or the spread of disease (section 4.2.4). The impact of density dependence can be seen most clearly by plotting the rate of change of the total population biomass against total population

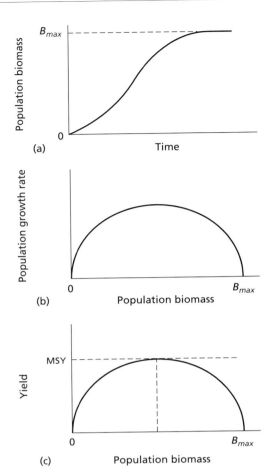

Fig. 7.3 (a) Logistic population growth. (b) Populations grow most quickly at intermediate sizes up to a maximum total biomass, B_{max}. (c) The maximum sustainable yield (MSY) in biomass occurs at a level of fishing mortality that places the population at an intermediate size.

biomass (Fig. 7.3b). The rate of population growth also shows the 'surplus' yield available to a fishery (Fig. 7.3c), and so the maximum sustainable yield (MSY) is found at the highest point on this curve.

7.3.1 **Stability**

We have now found MSY on a surplus yield curve, but in the real world MSY is a very small target indeed! Furthermore, it is a moving target, due to temporal changes in the productivity of the ocean and the fisheries that can be supported (Chapter 2). Before we discuss

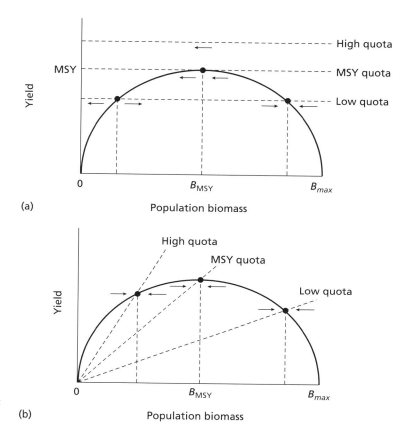

Fig. 7.4 Stability of surplus production models under various fishing quotas. Arrows indicate population trajectories. (a) Constant biomass caught. (b) Constant proportions of biomass caught.

ways of finding MSY using real data, we must consider what happens when we miss the target, because this is inevitable; our population estimates will never be perfect, nor will enforcement of quotas (catch controls, section 17.2.2) or other restrictions be sufficiently exact to score a direct hit on MSY. The yield curves in Fig. 7.4 correspond to those shown in Fig. 7.3 with arrows added to indicate stability of various catch rates.

First, consider the case where a constant biomass is caught (Fig. 7.4a). Suppose that this fixed quota had been set too high (above MSY—'high quota' in Fig. 7.4a). For all population sizes, yield would exceed the surplus production so the population would be driven to extinction. Now consider the case where MSY is estimated perfectly, and the quota is equal to MSY. The fate of the population depends on whether it is initially above or below B_{MSY}. If the population is larger than B_{MSY}, it will stabilize at B_{MSY}. This is due to density dependence (i.e. initial productivity will be less than

mortality), and as it is fished the population will decrease until production and mortality balance. However, if the population is initially smaller than B_{MSY}, the surplus production will always be less than the quota and the population will go extinct. What if the quota is too low ('low quota' in Fig. 7.4a)? If the population is larger than B_{MSY} a stable equilibrium will be reached, though the yield to the fishery will be less than at MSY because density dependence reduces productivity. If the population is smaller than B_{MSY}, the equilibrium is unstable and the population will either increase to the equilibrium point at the higher population size or crash. This consideration of stability conditions shows that one should *never* try to exploit populations at the MSY using constant catch rates: any reduction of the population below the theoretical point for maximum yield will crash the population. If a constant number of individuals is removed from the population, the MSY equilibrium is not stable.

The situation is better if we exploit at levels that are in proportion to the size of the population (Fig. 7.4b). For example, a fixed percentage of the population might be caught ($Y = pB$), where Y is the yield (biomass caught) and p is the proportion of population biomass, B. Here, perturbations of the population in either direction from the MSY point will be followed by a return to the equilibrium as long as the quota crosses the yield curve. For example, if the population is larger than B_{MSY}, the quota will exceed MSY and the population will be driven downwards to a stable equilibrium. Similarly, if the population biomass is less than B_{MSY}, the quota will be less than the surplus production that results from a release from density dependence in small populations. The population biomass will therefore increase to a stable equilibrium with the quota. But we are not out of the water yet!

7.3.2 Models of population growth

We derived the dome-shaped production model (Fig. 7.3) from a function that describes the growth rate of the population according to population size. The general form for continuous time is:

$$\frac{dB}{dt} = g(B) - Y \qquad (7.1)$$

where B is the exploitable population biomass at time t, $g(B)$ is the surplus production as a function of biomass, and Y is the yield to the fishery (in biomass). In words, the rate of change with time in the population biomass is equal to the surplus production minus the yield to the fishery. The equation we used in Fig. 7.3 for $g(B)$ will be familiar to population biologists as the classical logistic equation of population growth. It is usually called a **Schaefer curve** by fisheries biologists, after Schaefer (1954) who used it to develop a mathematical basis for fitting surplus production models. This equation expresses the change in the biomass of the population with time:

$$g(B) = rB\left[1 - \frac{B}{B_{max}}\right] \qquad (7.2)$$

where r is the intrinsic rate of population increase, i.e. the difference in biomass between per capita birth and death rates in the absence of density dependence, and B_{max} is the maximum biomass of individuals that the population can contain, i.e. the so-called carrying

capacity. In a fished population the yield, Y, is subtracted from the right-hand term (as in equation 7.1). At equilibrium, where density-dependence compensates for the additional mortality from exploitation, $dB/dt = 0$, and $g(B) = Y$.

The **Fox curve** is an alternative to the Schaefer (i.e. logistic) model of population growth (Fox, 1970). It is often used, because it may be more appropriate for biomass measurements than the logistic equation which is traditionally used in other contexts for numbers of individuals:

$$\frac{dB}{dt} = rB\left[1 - \frac{\log_e B}{\log_e B_{max}}\right] \qquad (7.3)$$

Here the inflection point corresponding to that in Fig. 7.3(a) occurs at less than half of the maximum theoretical population size, and so the maximum population growth rate and MSY are also to the left of the logistic cases shown in Figs 7.3(b) and 7.3(c).

A third alternative is the **Pella–Tomlinson model** (Pella & Tomlinson, 1969). This function has a parameter, m, added to the Schaefer logistic model, such that

$$\frac{dB}{dt} = rB - \frac{r}{B_{max}}B^m \qquad (7.4)$$

If $m = 2$ this equation is identical to Schaefer's original equation. When $m < 2$ the production model produces a maximum toward the left, and when $m > 2$ the maximum is toward the right. This Pella–Tomlinson approach allows for flexibility in the shape of production curves.

7.3.3 Fitting models to data

The choice of production curves is actually the least of our worries. More serious is the second phase of the procedure, namely how to fit these models to real data, to estimate MSY and the level of fishing effort at which it occurs. As we shall see, failure to do this properly has been implicated in the most dramatic stock collapse in the history of fishing.

The methods of fitting these models rely on the assumption that an index of abundance (such as commercial catch rates) can be related to true abundance, e.g.

$$I_t = qB_t \qquad (7.5)$$

where I_t is the index of relative abundance at time t (usually measured in years), B_t is the population biomass at time t, and q is the **catchability coefficient** (section 10.2.4). The latter term relates the **catch per unit effort** (CPUE) to population biomass:

$$CPUE = qB \qquad (7.6)$$

It is important to remember that catchability will change with improvements in technology and in response to changes in the distribution and behaviour of the population (sections 4.3 and 10.2.4).

Techniques for model fitting fall under two main categories, **equilibrium methods** and **non-equilibrium methods**. The latter can be further subdivided into process-error or observation-error methods.

Equilibrium methods

Equilibrium methods are often used to fit the Schaefer model, but they can be applied to the other models too. Schaefer (1954) developed a method for estimating MSY and the level of fishing effort at which MSY is achieved f_{MSY}, based on catch and effort data. The basic idea is to make the *very dangerous assumption* that each year's catch and effort data represent an equilibrium (or steady-state) situation, where the catch is equal to the surplus production at that level of fishing effort. Effort might be measured as the number of boats at sea per year, or number of traps baited per year or number of person-days spent spear fishing. CPUE is then regressed against effort over a series of years, producing a negative relationship (high fishing effort yields low CPUE). In part, this negative relationship may be driven by the fact that both the independent and the dependent variables contain fishing effort, providing some correlation even where none exists! This problem notwithstanding, with a little mathematical juggling, the parameters from this regression can be used to fit the familiar dome-shaped production curve.

The equilibrium method depends critically on the assumption that historical catch rates are in equilibrium with the population (Hilborn & Walters, 1992). This assumption is dangerous because temporal changes in CPUE will rarely be a sole reflection of density-dependent responses of the population to fishing mortality. Rather, CPUE reflects ongoing reductions in standing stock, because fishing effort often increases year by year as a fishery develops. Consider the case of the orange roughy, *Hosplostethus atlanticus* (Trachichthyidae) fishery off north-western New Zealand (Fig. 7.5). This fishery built up rapidly until declining catch rates precipitated a considerable reduction in the

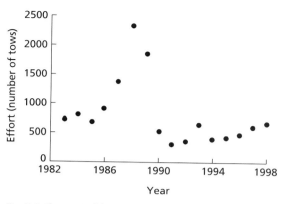

Fig. 7.5 Changes in fishing effort over time for an orange roughy fishery at the Challenger Plateau, New Zealand. The fishery first developed in 1981. After Field & Clark (1996, unpublished data).

total allowable catch in 1989–90 (Field & Clark, 1996). In this case, it would have been wrong to assume that the yield in any one year resulted from density-dependent responses bringing the population into equilibrium with high rates of fishing mortality, especially for a species like the roughy which does not reach maturity until it is over 20 years old. Instead, this fishery was mining the population, and the time series of CPUE data could not be used with any confidence. For this reason, the researchers stayed well clear of the equilibrium assumption and employed more sophisticated methods. Many fisheries take 'one-way trips' toward increased effort over time, and the result is that stocks are given credit for greater resiliency than they deserve.

Non-equilibrium methods

Process-error methods first transform the production curves into linear forms, and then use multiple regression to fit the models to data (Walters & Hilborn, 1976; Schnute, 1977). Catch and effort data are still used, but without the assumption that the population is in equilibrium. As with all statistical fitting techniques, the parameters in the fitted equation depend on the assumption made about the ways in which errors are distributed in the data. Process-error methods assume that catch and effort data have been measured without error, and all error is attributed to the functional relationship between population growth rate and population size (Hilborn & Walters, 1992; Polacheck *et al.*, 1993; Quinn & Deriso, 1999). That is, we assume error in equation 7.1 rather than 7.5. With this assumption, multiple linear regression is used to estimate the parameters of the production curve. In the case of a Schaefer curve, this means estimating r, B_{max} (equation 7.2) and the catchability coefficient, q, to relate catch per unit effort to population biomass. This method depends on having good variation (contrast) in the time series of catch and effort data. Otherwise, the estimates can go seriously astray.

Observation-error methods assume that the underlying production relationship is correct, and that all the error occurs in the relationship between true stock size and the index used to measure it (Pella & Tomlinson, 1969). Formally, this is the opposite of the process-error method above, because now the error is assumed to be in equation 7.5 rather than in equation 7.1. The time series for stock sizes is estimated by making an initial estimate of stock biomass. Then the model is used to predict stock sizes for the rest of the time period. One then compares observed and expected population sizes or catches, and uses statistical methods to adjust the parameter values to minimize the difference between the observed and expected values. For the Schaefer curve one thus estimates the same three parameters as with the regression methods, as well as the initial stock biomass. The latter might be the same as B_{max} if the time series extends back to the start of the fishery. This technique is reviewed by Hilborn and Walters (1992), Polacheck *et al.* (1993), and Quinn and Deriso (1999). Again, the quality of the estimates depends greatly on the quality of the data. We return to a direct comparison of the performance of the three methods for fitting models to data after we see one of them in action.

7.3.4 Surplus production models in action

Of the three methods discussed, the classical equilibrium method has the longest and most sordid history, principally because it presided over the infamous collapse of the Peruvian anchovy *Engraulis ringens*. This fish accounted for > 25% of global marine landings in 1970 (Fig. 1.4; section 4.2.2). It is instructive and sobering to see the disastrous consequences of the equilibrium assumption before we move on to a direct comparison of its performance with the other two methods.

The Peruvian anchovy population is found in the Peru Coastal Current, which runs close the shore of Peru and northern Chile. Anchovy are pelagic fishes, filtering chains of phytoplankton as well as zooplank-

ton, fish eggs and fish larvae from the water. Spawning peaks in September and October, with a smaller secondary peak in February and March (Laws, 1997). The young grow rapidly and recruit to the fishery when they are about 5 months old and 8–10 cm in length. They spawn at age 1 year and may live for 4 years. Most anchovy are caught by Peruvian purse-seiners, and converted to fishmeal to be sold to foreign countries for use in animal feeds. In the 1970s, the fishery accounted for about a quarter of Peru's foreign revenues.

Figure 7.6(a) shows the relationship between CPUE and effort in the anchovy fishery. The parameters from this relationship were used to fit the production curve in Fig. 7.6(b), using Schaefer's equilibrium method. The data include estimates of the biomass taken by seabirds, some 18% of the human catch. You may not be impressed with the extrapolation on the right side of the curve! Such is the nature of the parameter fitting technique, which allows extrapolation well into the unknown, for those brave enough to make the journey. The curve indicates a potential MSY of about 11 million tonnes. After subtracting an average of 1.5 million tonnes for seabirds, this leaves about 9.5 million tonnes for the fishery. It was reassuring to see that since the mid-1960s, the fishery's effort was about right for taking this yield. This was no accident since, in 1965, the Peruvian government brought in regulations that limited annual catches to 7.5 million tonnes. However, it proved difficult to enforce the regulations because the fishing fleets and fishmeal processing plants were greatly overcapitalized and, by 1970, the annual target of 7.5 million tonnes could have been processed in less than 40 days (Laws, 1997). Thus, there were considerable economic pressures for catches to exceed limits.

In January–February 1972 a research vessel survey recorded unusually low numbers of juvenile anchovy. At the same time, oceanographers recorded an intrusion of warm tropical water off the Peruvian coast. This soon developed into a full-blown El Niño event (section 2.3.2). The adult fish relocated to pockets of cool water and tended to move south. Adults were easy to catch in large numbers during March and April, but catches soon declined markedly. By July the earlier hints of low recruitment developed into loud alarms. Faced with this double blow of declining catches and failed recruitment, a management panel recommended a halt to the commercial fishery until they were sure that recruitment from the current cohort of adults proved successful. In

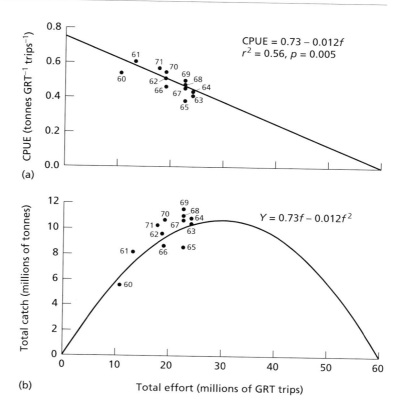

Fig. 7.6 Surplus production model for the Peruvian anchovy (*Engraulis ringens*). (a) Catch per unit effort (CPUE) vs. fishing effort, including both the commercial fishery and the catch taken by seabirds. (b) Schaefer production curve fitted from parameters derived from (a). GRT is gross registered tonnage, a measure of fishing vessel size. After Boerema and Gulland (1973).

fact, some fishing was allowed in the southern part of the fishery in November, because the adult stock seemed reasonably healthy there. But this was false hope. The stock collapsed and the fishery failed. The collapse put fishers out of work, left processing plants idle, and led to serious economic and social problems in Peru. Moreover, many of the seabirds that once fed on the anchovy starved and died. The anchovy stock did not recover quickly after this collapse, contrary to the hope for a stock with a high intrinsic rate of natural increase. Instead, it fluctuated at low abundance, suffering from the continuing effects of an overcapitalized industry and further moderate El Niños in 1976, 1982–83 and 1987. In the 1990s, however, the stock finally grew in size, and anchovy yields now exceed those of any other fished species (Fig. 1.4, Table 3.3).

A tournament of production models
Would the other two techniques for fitting production curves ('process-error' and 'observation-error' methods) have done any better? Polacheck *et al.* (1993) made a

direct comparison between all three methods by using them with three data sets from very different fisheries. Each method was used to fit the Schaefer form of the surplus production function (equation 7.2) to these data. Figure 7.7 shows that in two cases the equilibrium method was the most optimistic, with a process-error method being marginally more optimistic for one species, the New Zealand rock lobster *Jasus edwardsii*. A hint that the classical equilibrium method is too optimistic is illustrated by the data for the South Atlantic albacore *Thunnus alalunga* (Scombridae). This method predicted an MSY of just over 28 000 tonnes, which would be caught at fishing effort of just over 100 million hours (Fig. 7.7a). The fishery has been under this in most years, yet CPUE declined steadily over the period to only 30% of its initial value. The overfished state of this fishery is better predicted by the observation-error method, which proved to be the most conservative in all three cases. The highly over-optimistic predictions of the equilibrium method for Namibian hake (Fig. 7.7c) illustrate the problems

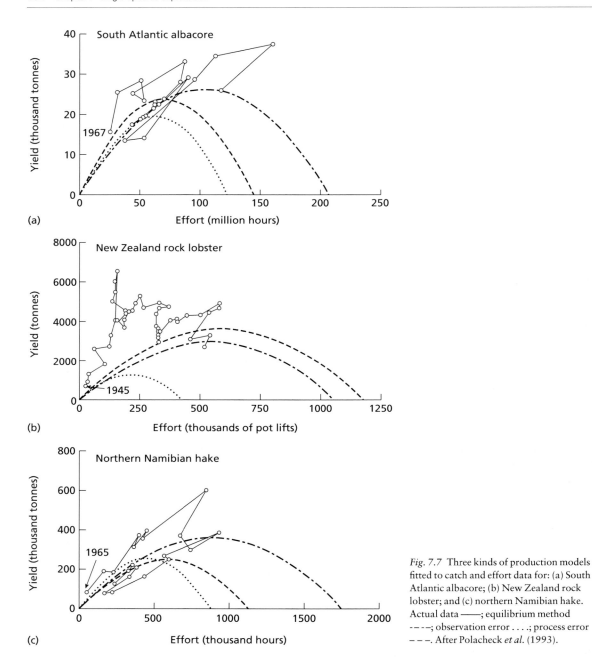

Fig. 7.7 Three kinds of production models fitted to catch and effort data for: (a) South Atlantic albacore; (b) New Zealand rock lobster; and (c) northern Namibian hake. Actual data ———; equilibrium method - - - -; observation error . . . .; process error – – –. After Polacheck *et al.* (1993).

mentioned in section 7.3.3 for the orange roughy (Fig. 7.5): in the early years of the hake fishery some of the 'surplus' production estimated by the equilibrium method included removal of some of the initial standing stock (Butterworth & Andrew, 1984).

Monte Carlo simulations were used by Polacheck *et al.* (1993) to compare the performance of the process-error and observation-error techniques. These added noise to the catch rates as predicted by the model to see how each technique handled it in terms of bias and precision. The observation-error method performed best, with the process-error method proving very imprecise.

The authors concluded that 'under no circumstances should agency staff, conference organizers, reviewers, managers or journal editors accept assessments or publications that are based on (equilibrium) or process-error estimators only'.

General lessons about production curves and MSY
Several lessons have been learned the hard way about the use of production curves and MSY, particularly from traditional methods that rely on the equilibrium assumption. First, fisheries are rarely in equilibrium. The build-up of the Peruvian anchovy fishery in the early 1960s is typical (Fig. 1.4). Such build-ups render catch and effort data much less informative about density-dependent changes in population growth than they appear. Second, everything we know about marine ecology suggests that stability is the exception rather than the rule (Chapters 2 and 4). In the case of the anchovy fishery, the El Niño had a catastrophic effect on the productivity of the anchovy stock. Third, catch and effort data are difficult to work with because CPUE is affected by advances in fishing technology and changes in the behaviour of the quarry, both of which are difficult to account for. Thus, the tendency of the anchovy to concentrate in pockets of cool water allowed fishers to maintain high catch rates despite sharp reductions in total stock size. Fourth, production curves suggest that the surest way to find the optimal fishing effort is to overfish the population, so that a predicted drop in yield at high effort is clearly discernible! In practice, overfishing often happens because fishers are attempting to earn a livelihood by competing for a common resource (Chapters 6 and 11), but this leads to a pervasive ratchet mechanism of resource exploitation, whereby it is much easier to allow effort to increase than to bring in regulations that decrease it. So, by the time we find out where yields drop with high fishing mortality, it will probably be too late to restrict the fishery. Finally, we must remember that surplus production models pool the various processes that determine population productivity. For some fisheries this may not be good enough. For example, in most marine fishes larger/older individuals will contribute disproportionately to reproduction, and they are often more valuable per kilogram. Yet their greater percentage contribution to the catch is ignored by surplus production models.

While early experiences such as the Peruvian anchovy collapse gave production models a bad name,

it would be wrong to paint them all with the same brush. The more sophisticated production models described are considerable improvements over equilibrium methods, and can outperform some of the more complex approaches presented later in this chapter (Punt, 1992). Furthermore, most methods of parameter fitting do not rely on complicated computer models and expensive fisheries data. Thus, they can provide useful guidance in fisheries where there are insufficient resources for time-consuming and costly research vessel surveys and analyses of age structure and growth rate. As we will see, the more auxiliary data that are available, the more sophisticated the model that can be used.

7.4 Delay–difference models

Delay–difference models, also known as **Deriso/Schnute** models, are surplus production models in the sense that data are aggregated over most age classes. But this technique goes further, by using not only population biomass data from previous years, but also incorporating information concerning instantaneous rates of natural mortality, body growth and recruitment (Deriso, 1980; Schnute, 1985; Fournier & Doonan, 1987). Thus, the approach is intermediate between the simple methods based on relative abundance, catch and effort data described above, and fully age-structured models such as the statistical catch-at-age and yield-per-recruit methods described in sections 7.6 and 7.7. Delay–difference models are so named because they allow for a time delay between spawning and recruitment, and they use difference equations, in which time changes in discrete steps (as opposed to differential equations, where time is continuous). The basic approach is to build a population model out of submodels that can describe survival, body growth and recruitment next year. Thus, the surviving biomass next year is predicted from the surviving biomass from last year, adjusted for body growth, plus next year's recruitment. Delay–difference models assume 'knife-edge' maturity and vulnerability to fishing (i.e. all individuals reach maturity at the same age and become equally vulnerable to fishing at the same age, sections 9.3.1 and 9.3.4), and that natural mortality is constant with age.

The simplest delay–difference model, which dates back to Allen (1963) and Clark (1976), is derived in Box 7.1.

The Allen–Clark model has only two components, survival and recruitment. Modern models add a third

Box 7.1

The Allen–Clark delay–difference model.

The following treatment, which follows Quinn and Deriso (1999), excludes stability conditions and generalizations to other models. First, we derive the equilibrium population size, and then add fishing. The adult population abundance next year, N_{t+1}, is the sum of the number of survivors this year, plus recruitment next year. Thus,

$$N_{t+1} = l_t N_t + D(N_{t+1-b}) \qquad (1)$$

where l_t is the annual survival against natural sources of mortality and D represents a stock-recruitment function, applied to adults backdated to the time of hatching b years ago. Note that this assumes that l_t is the same for all recruits. We can use whatever function seems appropriate to express recruitment according to population size, such as a 'Beverton–Holt' curve or a 'Ricker' curve (section 4.2.1).

The equilibrium population size, N_*, is given by

$$N_* = \frac{D(N_*)}{1 - l_*} \qquad (2)$$

Equation 2 meets the condition that there should be a balance between natural mortality $(1 - l)$, and recruitment: if mortality is high, recruitment must be high too, thereby offsetting these losses. This concept is developed further in section 7.8.

Now we can add fishing. Let:

$$S_t = N_t - C_t \qquad (3)$$

where S_t is the number of adults that escape from the fishery and C_t is the number of adults caught. Equation 1 becomes

$$N_{t+1} = l_t S_t + D(S_{t+1-b}) \qquad (4)$$

This means that the number of individuals next year will be the sum of adults that have escaped both fishing and natural mortality, plus recruitment. Note the assumption that fishing occurs in a single pulse at the start of the year and that there is only natural mortality for the remainder. We can use this equation to find the maximum sustainable yield with respect to optimal escapement from the fishery. Since $C_t = N_t - S_t$ (equation 3),

MSY = maximum (with respect to S) of $N_* - S$

= maximum (with respect to S) of $lS + D(S) - S$

At this point the population is not changing in size. Therefore, the first derivative of $[lS + D(S) - S]$ with respect to S will be equal to zero. This is solved to give

$$\frac{dD(S_m)}{dS} = 1 - l \qquad (5)$$

In other words, the MSY occurs at the value of optimal escapement, S_m, where the rate of change in recruitment is equal to the rate of natural mortality. Clark (1976) gives conditions for determining whether this equilibrium point is stable.

As an example, Clark (1976) used this model for Antarctic fin whales *Balaenoptera physalus*. This required a stock-recruitment function for equation 4. Clark used the following equation, with r representing the intrinsic rate of population increase and N_{max} as the maximum number of individuals that the population can contain:

$$D(S) = rS\left[1 - \frac{S}{N_{max}}\right] \qquad (6)$$

The parameters were $r = 0.12$, $N_{max} = 600\,000$, and $l = 0.96$. When these are used in equation 2, the result is solved to calculate the equilibrium population size in the absence of fishing:

$$N_* = N_{max}\left[1 - \frac{1-l}{r}\right] \qquad (7)$$

$$= 400\,000.$$

MSY is found by solving equation 5 to find the optimal escapement:

$$r\left[1 - \frac{2S_m}{N_{max}}\right] = 1 - l \qquad (8)$$

$S_m = 200\,000$ whales

While this is an elegant example of a simple delay–difference model, the predicted optimal escapement seems quite low, and depends critically on the parameters used for life histories and maximum population sizes, as well as the form of the stock-recruitment relationship.

component, body growth, which is applied to both those animals that had already been born, as well as new recruits. Thus, we now model next year's biomass, B_{t+1}, rather than just the number of individuals. A full derivation of this technique has been carried out by Schnute (1985), and the approach has been reviewed by Hilborn and Walters (1992) and Quinn and Deriso (1999). The method assumes that both fishing and natural mortality are constant for all individuals after recruitment. As before (Box 7.1), recruitment is modelled by whatever stock-recruitment relationship best fits the data (section 4.2.1).

The incorporation of fundamental life-history and recruitment information makes it possible to obtain good fits of the model to relative abundance data (e.g. CPUE or survey information). However, good fits may be achieved by several different combinations of growth, mortality and recruitment parameters. It is therefore important to pin down as many parameters as possible beforehand, using auxiliary information on life histories, other similar stocks, and so on. A good technique for doing this formally involves Bayesian inference (section 7.9.1).

7.4.1 Delay–difference models in action

A delay–difference model was used by Collie and Walters (1991) to calculate potential equilibrium yields for stocks of the yellowtail flounder *Limanda ferruginea* (Pleuronectidae), a commercially important flatfish that is exploited in coastal waters from southern New

England to the Grand Banks of Newfoundland. This species generally recruits to the fishery at 1–2 years of age in the southern part of their range, and 4–5 years in the north. Catches from several stocks of this species have followed a roller coaster pattern, rising rapidly during the late 1960s, dropping equally rapidly in the 1970s, and rising again in the early 1980s. These changes have roughly followed population abundance.

Collie and Walters (1991) assumed a natural survival value of 0.82. The authors set proportions of fish recruiting at each age according to findings from previous studies, and used commercial CPUE as their main index of population abundance, supported by research vessel surveys. They converted CPUE to population biomass using equation 7.6 and used weight at age data to determine growth parameters. Recruitment parameters were calculated by fitting a Ricker curve to spawner and recruit abundance data.

The resulting delay–difference model provided a good fit to the CPUE data, as shown for the Grand Banks stock (Fig. 7.8). However, the authors noted that the parameters had large confidence limits, and many different combinations of parameters could achieve this fit. Thus, equally good fits could be obtained if the population were small but productive or large but unproductive. The importance of this became clear when they looked at the predicted equilibrium yields for the Grand Banks fishery for different values of parameter *a*, the density-independent parameter from the Ricker spawner–recruit relationship (equation 4.2, section 4.2.1). Recall that higher values of *a* indicate that

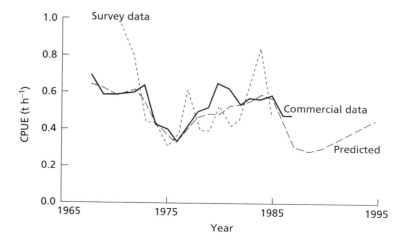

Fig. 7.8 Delay–difference model fit to catch per unit effort data of Grand Banks yellowtail flounder. After Collie & Walters (1991).

Table 7.1 Annual equilibrium yields (thousands of tonnes) from a delay–difference model for Grand Banks yellowtail flounder. The data show different combinations of potential exploitation rates and values of *a* from Ricker spawner–recruitment functions. The **bold** type shows the highest yields for each value of *a*. From Collie and Walters (1991).

Exploitation rate (%)	Ricker *a* value				
	0.0	0.3	0.6	0.9	1.2
20	14.7	13.5	12.7	13.4	16.8
30	**16.6**	16.8	16.8	18.5	23.7
40	13.6	**17.0**	18.9	22.1	29.6
50	2.4	12.0	17.6	**23.5**	33.8
60	0.0	0.0	9.9	20.1	**34.2**
70	0.0	0.0	0.0	5.4	24.8

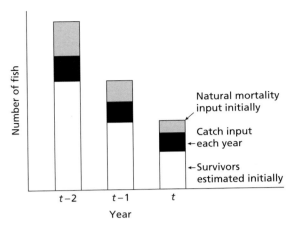

Fig. 7.9 Virtual population analysis for a fish with three age cohorts. Beginning with the current year, *t*, we can rebuild the historical stock sizes by adding the numbers that were caught by the fishery or that died from natural causes to the numbers of survivors.

the gradient of the spawner–recruit curve is steeper at the origin (more recruits per spawner) and that the curve will be more domed. As Table 7.1 shows, the more domed the spawner–recruit relationship, the higher the equilibrium yields will be.

This example thus illustrates that the benefits of the excellent fit that can be achieved by delay–difference models may come with a cost of considerable uncertainty about management recommendations if the parameters are uncertain. In particular, as with the other surplus production models presented above, time-series data must be handled cautiously when used to make inferences about population dynamics. Results such as those in Table 7.1 can show the model's sensitivity to parameter uncertainty, and point toward auxiliary information that should be collected to solve the problem. Many scientists now prefer to take advantage of advances in computing power to use fully age-structured models rather than the two age groups represented by delay–difference models.

7.5 Virtual population analysis

Virtual population analysis (VPA) uses commercial catch data to calculate stock sizes and mortality rates of age-based or length-based cohorts. VPA does not by itself indicate how many individuals can be caught to meet a given objective, nor does it predict the future. In fact, it explains the past. For, if we know the historical age structure of a population, we can then see the consequences of changes in mortality rates, based on

methods such as yield-per-recruit calculations (section 7.7). Our use of the term VPA is equivalent to **sequential population assessment** as reviewed along with other age-structured models by Megrey (1989).

If we know how many fish from a given cohort were caught one year, we have a minimum estimate of how many must have been alive the previous year. If we add natural mortality, we have the total mortality for that year and the total number that must have been alive the year before. One can then work backwards, year by year, deriving annual estimates of numbers of survivors and mortality rates. With a good picture of past and present population dynamics, one can then make the forecasts needed to assess management options. For fisheries targeting particularly short-lived species, these calculations can be done on a shorter time scale such as months rather than years, provided that the catch and age or length-cohort data are collected on that time scale.

The basis of VPA is illustrated in Fig. 7.9. For a given cohort (year class) this year, we calculate the number that must have been alive the previous year by adding the number caught by the fishery this year to the number estimated to have died of natural causes over the same period. First, we adopt a standard formulation in population biology to account for both natural and fishing mortality, which we call the **exponential decay equation** (Box 7.2):

Box 7.2

Equations of death.

Suppose we wish to follow the survival of a cohort (year class) though time. The number of individuals alive at any time t in the future will be a function of the number alive now, minus mortality. If we call the initial population size N_t, then the number alive one time unit later, N_{t+1}, will be given by the exponential decay equation:

$$N_{t+1} = N_t e^{-(F+M)} \qquad (1)$$

By convention, the symbol F is used to denote the instantaneous rate of fishing mortality. (Note the capital letter, to distinguish it from f for fishing effort.) M is the instantaneous rate of natural mortality and e is the base of the natural logarithm ($e = 2.71828$). Thus, the total mortality is given by $F + M$, which, again by convention in fisheries, is denoted Z.

An example of how the number of individuals alive in the future depends on instantaneous rates of mortality is shown in Fig. B7.2.1.

We can use equation 1 to calculate the rate of change in numbers alive with time. This is simply the first derivative of the equation:

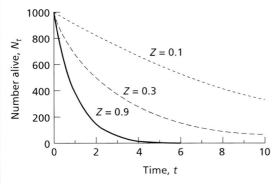

Fig. B7.2.1 Death of a cohort: number alive over time, for three levels of instantaneous total mortality (Z).

$$\frac{dN}{dt} = -FN - MN \qquad (2)$$

We can also use equation 1 to calculate the number of individuals caught by the fishery. First we calculate the number of fish that die. This is the difference between N_t and N_{t+1}, i.e.

$$\text{deaths} = N_t - N_t e^{-Z} = N_t(1 - e^{-Z}) \qquad (3)$$

The number of individuals caught over that time is the proportion of deaths due to fishing mortality:

$$C_t = \frac{F}{Z} N_t (1 - e^{-Z}). \qquad (4)$$

This is referred to as the **catch equation**, with $Z = F + M$.

For example, suppose we start with 1000 fish, and they die from fishing at the instantaneous rate of mortality $F = 0.5$, and from natural causes $M = 0.1$ per year. How many will be alive after 4 years? The exponential decay equation 1 can be generalized to

$$N_t = N_0 e^{-Zt} \qquad (5)$$

where N_0 is the number alive initially. Therefore

$$N_t = 1000 e^{-(0.6 \times 4)} = 91 \text{ fish.}$$

How many fish were caught by the fishery in the first year? From the catch equation 4 we calculate the proportion of mortality due to fishing:

$$C = \left[\frac{0.5}{0.6} \right] 1000 (1 - e^{-0.6})$$

$$C = 376 \text{ fish}$$

Similarly, 75 fish died of natural causes, producing a total first year's mortality of 451. This total could, of course, also have been calculated by putting $Z = 0.6$ into the exponential decay equation 1 and subtracting the result from 1000.

$$N_{t+1} = N_t e^{-(F+M)} \qquad (7.7)$$

Here N_{t+1} is the number of individuals alive at time $t + 1$, N_t is the number alive at time t, M is the instantaneous rate of natural mortality, and F is the instantaneous rate of fishing mortality. We will be seeing a lot

of this equation in this chapter, and we describe ways of estimating natural mortality in section 9.3.6.

We also calculate the numbers caught at time t, C_t, from fishing mortality and natural mortality at time t using the **catch equation**, which is also derived in Box 7.2.

$$C_t = \frac{F_t}{F_t + M_t} N_t (1 - e^{-(F_t + M_t)}) \qquad (7.8)$$

When catch is multiplied by the mean weight of individuals, and summed over age classes, we have the total yield in biomass to the fishery.

To do a VPA, we use the exponential decay and catch equations 7.7 and 7.8. Here we illustrate the concept by following one cohort backwards through time, and hence use only one subscript, t, although it is customary to add a second subscript to denote age, in order to keep track of multiple cohorts.

Step 1. Calculate the 'terminal abundance' (i.e. for the most recent year, t) of the oldest cohort from the catch equation 7.8, rearranged to give N_t. Use an estimate of terminal F_t, as well as M, which is assumed constant for all ages in all years.

$$N_t = \frac{C_t}{(F_t/Z_t)(1 - e^{-Z_t})} \qquad (7.9)$$

Step 2. Calculate the previous year's fishing mortality for the cohort by substituting equation 7.7 into the catch equation 7.8. The previous year is now t:

$$C_t = \frac{F_t}{Z_t} N_{t+1} e^{Z_t} (1 - e^{-Z_t}), \text{ hence}$$

$$C_t = \frac{F_t}{Z_t} N_{t+1} (e^{Z_t} - 1) \qquad (7.10)$$

We can substitute for $Z_t = F_t + M$ to get

$$C_t = \frac{F_t}{F_t + M} N_{t+1} (e^{F_t + M} - 1) \qquad (7.11)$$

Since we know the catch, C_t, the stock size from step 1, N_{t+1}, and M, F_t is the only unknown in equation 7.11. Unfortunately, equation 7.11 cannot be solved for F_t directly, but it can be solved numerically with a computer.

Step 3. For the younger ages, calculate N_t again, by inserting F_t (as calculated in Step 2) into an appropriate modification of equation 7.7. That is,

$$N_t = N_{t+1} e^{(F_t + M)} \qquad (7.12)$$

After using equation 7.9 to calculate the number of oldest individuals, we work back year by year, using equation 7.11 for fishing mortality and equation 7.12

for abundance, as shown schematically in Fig. 7.9. A worked example of a VPA is given in Box 7.3.

Although VPA estimates of annual fishing mortality and population size depend on the initial estimates for the most recent year, their proportional dependence on these becomes smaller as one works backwards through time. This gives greater confidence in estimates from earlier years, which can be fine-tuned as new population estimates are generated each year. As a rule of thumb, the method works best when F/Z is between 0.5 and 1.0.

For the most recent years, where estimates of F and population size cannot benefit from much hindsight, the use of 'tuning fleets' is important to infer population numbers at age (Pope & Shepherd, 1985). Thus, although VPA in theory does not require fishing effort data, in practice it is hard to get away without it. **Tuning fleets** may be research vessels making regular surveys, or commercial fleets where fishing activity has been well quantified over a number of years. For example, we might know the total number of days fishing per year for a fleet of beam trawlers using a certain mesh size in a certain region over a 10-year period. If we know how mesh size or engine power have changed, we can also correct the data to account for this. Then, we can plot historical population numbers, based on VPA, against effort by our tuning fleets in those years. The relationship can be used to predict very recent population numbers from recent tuning fleet catches and effort. If the tuning fleet predictions are far off the VPA predictions, a closer look at the data would be warranted. Some VPA software, such as **extended survivors analysis (XSA)** used routinely in European fisheries, have options for 'tuning' the VPA according to these fleets. Thus XSA can weight different fleets to account for the efficiency with which they target different age classes.

7.5.1 Age-based cohort analysis

A clever approximation of virtual population analyses was developed by Pope (1972). Within limits, this gives very similar results to VPA, without the need for iterative calculations of fishing mortality, F. Thus, numerical solutions are unnecessary. As with VPA, we use catch data and initial guesstimates of F and M to work backwards through time to reconstruct previous values of F and stock structure. However, Pope's ingenuity was to simplify the calculations by assuming that all

Box 7.3

Example of virtual population analysis.

Table B7.3.1

Age (t)	Catch numbers	Stock size	Fishing mortality (F)
1	70	487	0.172
2	90	335	0.349
3	80	194	0.600

Table B7.3.1 shows catches through time for a cohort of fish. We assume that $M = 0.2$ and is constant for all ages. Of course in reality M will vary, but the smaller it is relative to F, the less important this is. We begin with the oldest fish, aged 3, and set the terminal value of $F_3 = 0.6$.

Step 1. From equation 7.9, the stock size of age 3 fish that would have yielded a catch of 80 will be:

$$N_3 = \frac{C_3}{(F_3/Z_3)(1 - e^{-Z_3})}$$

$$= \frac{80}{(0.6/0.8)(1 - e^{-0.8})}$$

$$= 194 \text{ fish}$$

Step 2. From equation 7.11 we can calculate the value of F that must have occurred the previous year to yield the catch of 90 age 2 fish while leaving 194 survivors.

$$C_2 = \frac{F_2}{F_2 + M} N_3 (e^{F_2 + M} - 1)$$

$$90 = 194 \frac{F_2}{F_2 + 0.2} (e^{F_2 + 0.2} - 1)$$

$$F_2 = 0.349$$

(solved by iteration: a simple approach is to insert the right-hand side of the equation into a spreadsheet and have F in the equation refer to a column in which values of F vary up and down).

Step 3. We use this value of F with equation 7.12 to calculate the stock size for age 2 fish.

$$N_2 = N_3 \, e^{(F_2 + M)}$$

$$= 194 \, e^{0.349 + 0.2}$$

$$= 335 \text{ fish (depending on rounding)}$$

We repeat Steps 2 and 3 to work backwards through time.

---

fish are taken instantaneously halfway through the year, rather than continuously (Fig. 7.10). Thus, the number of fish alive at the moment just before fishing mortality ($N_{t+0.5}$) will be solely a function of natural mortality acting on the cohort since the start of the year. Then fishing reduces the stock all at once, producing the entire year's catch (C_t), followed again by natural mortality.

The number of fish alive just before fishing takes place ($N_{t+0.5}$) will be the number alive at the start of the year (N_t), reduced by half of the year's natural mortality, $M/2$:

$$N_{t+0.5} = N_t e^{-M/2} \qquad (7.13)$$

After the entire year's catch (C_t) is taken instantaneously, we are left with

$$N_t e^{-M/2} - C_t \qquad (7.14)$$

Now these fish suffer natural mortality for the rest of the year, leaving the number alive at the end of the year (N_{t+1}):

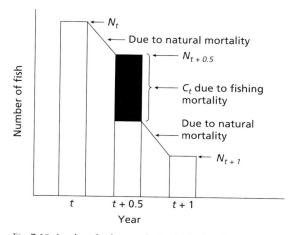

Fig. 7.10 Age-based cohort analysis. The logic is the same as for virtual population analysis (Fig. 7.9), but the year is broken into two parts, with the catch assumed to have occurred instantly in the middle of the year.

Box 7.4

Example of age-based cohort analysis.

Table B7.4.1 shows the same catches at age as for our VPA example (Box 7.3), with a comparison between the VPA results and cohort analyses, calculated below. As before, we take $M = 0.2$, and our initial estimate of F for the terminal age class (3-year-olds) as 0.6.

Step 1. The first step is identical to the VPA calculation, i.e. we calculate the number of age 3 fish, N_3, from the rearranged catch equation 7.9, to give:

$N_3 = 194$ fish

Step 2. The difference from VPA is that this is substituted into equation 7.16 as N_{t+1} to calculate the number of fish from that cohort that would have been alive the previous year:

$N_2 = (N_3 e^{M/2} + C_2)e^{M/2}$
$= (194e^{0.2/2} + 90)e^{0.2/2}$
$= 336$ fish

This is only one fish away from our calculation of 335 using VPA (Box 7.3).

Step 3. Now we calculate F on age 2 fish, using the exponential decay equation (equation 1 in Box 7.2):

$$N_{t+1} = N_t e^{-(F_t + M)}$$

This is rearranged to give:

$$F_t = \ln\left[\frac{N_t}{N_{t+1}}\right] - M$$

Thus, F for age 2 fish is given by:

$$F_2 = \ln\left[\frac{N_2}{N_3}\right] - M$$

$$= \ln\left[\frac{336}{194}\right] - 0.2$$

$$= 0.349$$

This is identical to the VPA result, or within about 0.002 of it, depending on rounding. Note that, unlike VPA, we have circumvented the iterative calculation of F. We continue working our way back through time repeating Steps 2 and 3 to build up the historical pattern of age-based stock sizes and fishing mortalities each year. When we perform Steps 2 and 3 again to calculate stock sizes and fishing mortality on age 1 fish, the results are also virtually identical to the VPA analyses shown in the table.

Table B7.4.1

Age (t)	Catch numbers	Stock size		Fishing mortality (F)	
		VPA	Cohort	VPA	Cohort
1	70	487	488	0.172	0.173
2	90	335	336	0.349	0.349
3	80	194	194	0.600	0.600

$$N_{t+1} = (N_t e^{-M/2} - C_t)e^{-M/2} \qquad (7.15)$$

Since we are trying to find the number of fish alive at the start of the year, we rearrange this equation to give the fundamental equation:

$$N_t = (N_{t+1} e^{M/2} + C_t)e^{M/2} \qquad (7.16)$$

This gives the number of fish alive at the start of the year based on the catch through the year and the num-

ber alive at the start of the following year. The catch is estimated in the usual way, from the fishery, and we can use whatever means are available to estimate M (section 9.3.6) and take an initial assumption about F, which generally proves fairly unimportant to the final result, as for the case of VPA. A worked example of age-based cohort analysis is presented in Box 7.4.

Pope (1972) showed that this method approximates VPA quite well for values of M as high as 0.3 and F as high as 1.3. If M and F are larger, then the method still works if the catch times are divided into smaller units than 1 year. One can also substitute $M/(1 - e^{-M})$ for $e^{-M/2}$, which gives an even better approximation to VPA, especially for larger values of M (MacCall, 1986b).

7.5.2 Length-based cohort analysis

Length-based cohort analysis was developed for species that cannot be aged. The principle is the same as for age-based cohort analysis, but animals are separated into length classes (Jones, 1981). The technique can be implemented by statistical packages such as FiSAT, developed by the Fisheries and Agriculture Organization (FAO) Fisheries Department and the International Center for Living Aquatic Resources Management (ICLARM).

First, length groups are converted to age groups based on the von Bertalanffy growth equation (sections 3.4.2 and 9.3.3). Thus,

$$t(L_1) = t_0 - \frac{1}{K}\log_e\left(1 - \frac{L_1}{L_\infty}\right) \tag{7.17}$$

where $t(L_1)$ is the age of individuals in length interval L_1. t_0, K and L_∞ are parameters of the von Bertalanffy growth equation (section 9.3.3). Thus, the time interval, Δt, between two successive age classes, $t(L_2) - t(L_1)$, becomes:

$$\Delta t = \frac{1}{K}\log_e\left[\frac{L_\infty - L_1}{L_\infty - L_2}\right] \tag{7.18}$$

The fundamental equation from age-based cohort analysis (equation 7.16) is modified to replace the time interval (assumed to be 1 year) with the converted age intervals. That is, the term $e^{M/2}$ is replaced with a term representing the fraction of a given length class that survive natural mortality from the time they are in L_1, until half of the time period has elapsed before they reach L_2:

$$N_{L_1} = [N_{L_2} T_{L_1,L_2}{}^{M/2K} + C_{L_1,L_2}]T_{L_1,L_2}{}^{M/2K} \tag{7.19}$$

where N_{L_1} is the number of individuals that survive to reach length L_1, corresponding to an age $t(L_1)$, C_{L_1,L_2} is the number of individuals between lengths L_1 and L_2 that are caught, and T_{L_1,L_2} represents the fraction from equation 7.18:

$$\frac{L_\infty - L_1}{L_\infty - L_2}.$$

The procedure for carrying out a length-based cohort analysis is similar to the age-based approach. First, we modify the age-based form of the catch equation (Box 7.2) for use with length classes, and apply it to the oldest length class:

$$C_{L_1,L_2} = \frac{F}{Z}N_{L_1}(1 - e^{-Z\Delta t}). \tag{7.20}$$

For the oldest age class we take $e^{-Z\Delta t} = 0$, because theoretically, the age corresponding to Δt being larger than the largest length class is infinite. We thus substitute zero for $e^{-Z\Delta t}$, plug the catch numbers for the largest length class into the left-hand side of equation 7.20, and make an initial guess of the terminal value of F/Z to calculate the N_{L_1} the number of individuals in the oldest length class. An example is shown in Box 7.5.

Once we have calculated the stock sizes corresponding to each length interval, we can calculate fishing mortality rates. The basic formula is:

$$F = M\frac{F/Z}{1 - F/Z} \tag{7.21}$$

where F and Z refer to length class L_1,L_2. F/Z is derived from catches and stock numbers, i.e.

$$\frac{F}{Z} = \frac{C_{L_1,L_2}}{N_{L_1} - N_{L_2}} \tag{7.22}$$

The final step is to convert numbers in each length class to actual numbers in the stock. However, we cannot do this until we know how long each individual spends in each time interval and sum across the intervals. The mean number of individuals in each age class per year is calculated as:

$$\overline{N}_{L_1,L_2} = \frac{N_{L_1} - N_{L_2}}{Z\Delta t} \tag{7.23}$$

Thus, the total is the sum across each length class, i, weighted by its time interval:

$$\sum_i(\overline{N}_{L_i,L_{i+1}})\Delta t. \tag{7.24}$$

As with age-based cohort analysis, this method requires an initial estimate of mortality (F/Z), but the

---

Box 7.5

Length-based cohort analysis for a hypothetical species.

Suppose the following data apply to a fished species: $L_\infty =$ 120 cm, $K = 0.3$, $M = 0.2$, and F/Z is initially guessed to be 0.8. Therefore $M/2K = 0.2/(2 \times 0.3) = 0.33$. The sizes of individuals in the last two length classes, the numbers caught, and their corresponding values of T are shown in Table B7.5.1.

Applying the length-based catch equation (equation 7.20) to the oldest age class:

$C_{100, \infty} = 25 = 0.8(N_{100})(1 - 0)$

$N_{100} = 31.2$

Working backwards to the next largest length class using the fundamental length-based cohort equation 7.19:

$N_{90} = (31.2 \times 1.5^{0.33} + 30)1.5^{0.33} = 75.1$

To calculate F from these data, we apply equations 7.22 and 7.21. Thus, for the length category 90–100 cm,

$F/Z = 30/(75.1 - 31.2) = 0.683$

$F = (0.2 \times 0.683)/(1 - 0.683) = 0.43$

Table B7.5.1

Length group (cm)	Number caught	T
L_1-L_2	C_{L_1, L_2}	$(L_\infty - L_1)/(L_\infty - L_2)$
90–100	30	$(120-90)/(120-100) = 1.50$
100–∞	25	–

---

calculations for older year classes do not rely heavily on the initial input. However, it is still important to estimate M and the growth parameters as well as possible.

7.6 Statistical catch-at-age methods

Statistical catch-at-age methods, also known as **stock synthesis**, or **integrated analysis**, are an alternative to VPA for estimating stock sizes. As their name suggests, these techniques are age based, which places them with age-based VPA and cohort analyses into the family of **age-structured stock assessment methods**. They are used routinely in many stock assessments. The basic idea is to develop a population dynamics model from first principles and then relate the model's predictions (e.g. of the annual catches) to the observed data. Statistical methods are used to find the best set of model parameters that will fit the observations. This is an extremely flexible approach. It also provides a way of overcoming a problem with VPA (section 7.5), namely that abundance estimates for the most recent year classes tend to be imprecise and require 'tuning'. Furthermore, most modern catch-at-age methods do not require estimates of M that are needed for VPA. One disadvantage of catch-at-age methods is that the

computational procedures are considerably more complex than those required for VPA. Thus, the estimation of parameters requires non-linear regression, which effectively precludes the use of computer spreadsheets. In the following, we collapse a great deal of detail into a snapshot of the method. The historical roots of the approach, which includes catch curve analyses and developments by Doubleday (1976), Paloheimo (1980), Fournier and Archibald (1982), and Deriso *et al.* (1985), are reviewed by Deriso *et al.* (1985), Megrey (1989), and Quinn and Deriso (1999). These references should be consulted by anyone wanting more than the brief overview we provide below.

A well-known early example of a statistical catch-at-age model is called **CAGEAN** (Catch-AGE Analysis; Deriso *et al.*, 1985). The model uses the same catch equation and exponential decay equation that we encountered in Box 7.2 and used for VPA. Rather than beginning with fishing mortality on the terminal (oldest) age class, CAGEAN estimates the annual recruitments and the numbers-at-age in the first year of the analysis. To simplify the estimation, it is assumed that there is **separable fishing mortality**. That is, fishing mortality can be described by the product of an age-dependent vulnerability to the fishery, and a year-specific factor.

The values for these parameters are obtained by fitting the model to the observed data. CAGEAN can use catch-at-age and fishing effort data, and place constraints on the extent to which recruitment deviates from a pre-specified spawner–recruit relationship. We define a function based on the weighted sum of squared deviations between the model's predictions and the observed data. Non-linear least-squares regression is used to minimize these deviations to find the best fit parameters. In contrast to VPA, CAGEAN does not assume that catches are known exactly. For catch-at-age data, the sum of squares term is given by:

$$SSQ(\text{catch}) = \sum_a \sum_t [\text{predicted } C_{a,t} - \text{observed } C_{a,t}]^2$$

$$(7.25)$$

where $C_{a,t}$ is the catch of animals age a during year t. The predicted catches are based on the catch equation and the exponential decay equation (Box 7.2).

The basic approach can be extended in many ways. For example, Smith and Punt (1998) assess a stock of gemfish *Rexea soladri* (Gempylidae, close relatives of the Trichiuridae) off eastern Australia for which catch-at-age data are missing from some years, although length-frequency data are available for most years. The underlying model is sex- as well as age-structured and most of the parameters are therefore sex specific, to capture the impact of sexual dimorphism.

The statistical approach used by statistical catch-at-age models to estimate parameters is an improvement over the *ad hoc* approach used by VPA. Furthermore, one need not guess at terminal fishing mortalities. Various authors (e.g. Deriso *et al.*, 1985) have developed methods for quantifying the uncertainty surrounding key model outputs such as the time series of spawner and recruit abundance. Incorporation of auxiliary information such as fishing effort or independent estimates of natural mortality are very helpful for estimating the parameters needed to fit models to catch data. Indeed, the flexibility of this technique for accommodating such additional information is a major asset. Proponents of VPA-related techniques such as cohort analysis counter that these methods are computationally simpler and more transparent, with a solid track record in many well-studied fisheries.

7.7 Yield-per-recruit models

At the beginning of this chapter we identified four factors influencing population biomass that must be in balance if we are to exploit sustainably: reproduction, body growth, fishing mortality and natural mortality (see Fig. 7.1). We also saw that surplus production models lump these together and ignore age structure. Models that keep the components separate fall under the 'dynamic pool' approach. With the information provided by analyses such as VPA and statistical catch-at-age analyses, we can use yield-per-recruit models to seek fishing mortality rates that achieve the best trade-off between the sizes of individuals caught, and the number of individuals available for capture. If fishing mortality rates are set too high, too many individuals will be taken before they have had a chance to grow. This is loosely termed 'growth overfishing'. If fishing mortality is too low, although the individuals will be large when captured, the total yield will be low. This logic can be seen if we follow the fate of a single cohort through time (Fig. 7.11). The optimal age at which to capture these fish is A.

The fundamental yield-per-recruit model assumes a steady state, i.e. that recruitment is constant, and hence the age structure of the population is the same as we would see if we followed a single cohort through time. Hence, yield is measured 'per recruit' (Beverton & Holt, 1957). We will therefore need to incorporate recruitment in a later section of this chapter. The model also assumes that fishing and natural mortality are constant from the moment that the fish become vulnerable to fishing gear. We can actually relax these assumptions when yield-per-recruit models are put into practice to see the effects of different ages at first capture and exploitation patterns, thereby informing decisions for management. Obviously, fishers will want to catch the largest total biomass rather than the largest yield per recruit. We keep track of total biomass later, by multiplying yield per recruit by the projected numbers of recruits, given various potential rates of fishing mortality.

The fundamental yield-per-recruit model gives the yield, Y (in biomass) to the fishery as:

$$Y = \sum_{t_c}^{t_{max}} F_t N_t W_t \qquad (7.26)$$

where t_c and t_{max} are the ages of first capture and maximum ages of cohorts, respectively, F is the instantaneous rate of fishing mortality, N is the number of individuals alive, and W is their mean weight. Strictly, the summation sign should be an integral, since age is a continuous variable. However, age is usually recorded

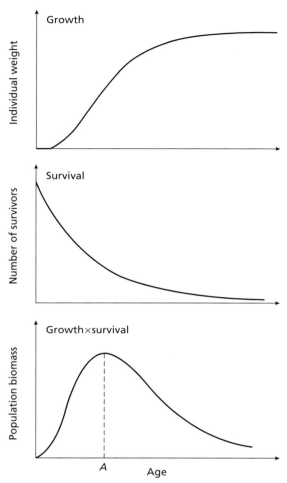

Fig. 7.11 The logic of yield-per-recruit models, based on the trade-off between growth and mortality of individuals. Here, the optimal age at which to catch the fish is at *A*.

in discrete categories (e.g. 1-year-olds, 2-year-olds, and so-on), so the weights of individuals caught in each age class can be simply added together.

Box 7.6 provides a worksheet for calculating yield per recruit for a single cohort of fish. This shows that if $F = 0.6$, about 1 kg of fish would be caught for every fish that initially recruited to the fishery. The total biomass of fish left behind in the sea would be about 2.4 kg. We usually want to find the value of F that meets some objective such as providing a high yield to the fishery, while maintaining the stock at a reasonable size. Thus, we repeat the calculations for a range of

fishing mortalities, and plot yield per recruit and population biomass per recruit vs. F (Fig. 7.12). The maximum yield per recruit occurs at $F_{max} = 0.23$. If current F were 0.9, then a reduction in fishing mortality to F_{max} would nearly double the yields and increase the stock size fivefold. The latter would give us a better safety margin against recruitment failure, as shown by the population biomass per recruit. For species with high M, yield-per-recruit curves are often much more flat-topped than in our example, which is typical for relatively slow-growing species such as cod. If curves are very flat-topped, practitioners are forced to rely on targets other than maximum values of these curves (section 7.10).

It would be unrealistic to expect M and F to be constant with age, since younger fish tend to suffer high natural mortality, but low fishing mortality until they are large enough to become vulnerable to fishing gears. Yield-per-recruit analyses can accommodate age-specific mortalities. Indeed, researchers often investigate the effects of changing mesh size by setting $F = 0$ for the youngest age classes of fish ('knife-edged selection'). This allows exploration of the joint effects of changing age at capture and mean F.

Long-term forecasts from yield-per-recruit models typically assume a stable age structure through time. This may often be unrealistic, given the erratic temporal changes in recruitment typical of most fish stocks (Chapter 4). Indeed, yield-per-recruit models by themselves ignore impacts of fishing mortality on recruitment: this requires an explicit link to a stock-recruitment curve (section 7.8). Moreover, errors in M may have important effects on the predictions when M is high relative to F. Temporal variation in M and F are also ignored, as is density dependence in growth and maturity, which may change with different levels of fishing mortality.

7.7.1 Yield-per-recruit models in action

Atlantic croaker *Micropogonias undulatus* are members of the family Sciaenidae, known as drums. Croakers are named after the noises they make during the spawning season. They are one of the most abundant inshore bottom fishes along the east coast of the United States, including the Gulf of Mexico. They are also one of the most important species for commercial and recreational fisheries. Croakers spend the autumn

Box 7.6

Example of yield-per-recruit calculations for a single cohort of fish.

Table B7.6.1 follows the fate of a single cohort of fish. The weights are of individuals at each age. The number alive is at the start of the year per 100 fish recruiting, based on an assumed natural mortality (M) of 0.2, and a hypothetical fishing mortality (F) of 0.6. The numbers

alive and catch numbers at the end of the year are based on the exponential decay equation and the catch equation, respectively (Box 7.2). The population biomass is calculated as (weight × number surviving) and catch weights (yields) are (weight × catch numbers).

These calculations are repeated for a range of potential fishing mortalities, yielding the data in Table B7.6.2, which are plotted in Fig. 7.12.

Table B7.6.1

Age (years)	Weight (kg)	Number alive	Population biomass (kg)	Catch numbers	Catch weight (kg)
1	0.6	100	60	41	25
2	0.9	45	40	19	17
3	2.1	20	42	8	17
4	4.1	9	37	4	15
5	6.3	4	26	2	11
6	8.4	2	15	1	6
7	10.0	1	8	0.3	3
8	11.2	0.4	4	0.2	2
9	12.6	0.2	2	0.1	1
10	13.5	0.1	1	0.0	0
Sum (kg)			237		98
Sum/Recruit (kg)			2.37		0.98

Table B7.6.2

Fishing mortality (F)	Yield per recruit (kg) (Y/R)	Biomass per recruit (kg) (B/R)
0.0	0	21.46
0.1	1.12	12.94
0.2	1.36	8.27
0.3	1.32	5.60
0.4	1.2	4.00
0.5	1.08	3.01
0.6	0.98	2.37
0.7	0.89	1.93
0.8	0.83	1.63
0.9	0.77	1.42
1.0	0.73	1.26

and winter in oceanic waters. They spawn from July until November along the US east coast, probably on the continental shelf edge. In spring, adults as well as larvae and juveniles move into estuaries, where they remain until early autumn. Most individuals reach maturity at 1 year of age.

Atlantic croaker are caught by haul-seines, pound nets and gill nets (Chittenden, 1991). Once they move

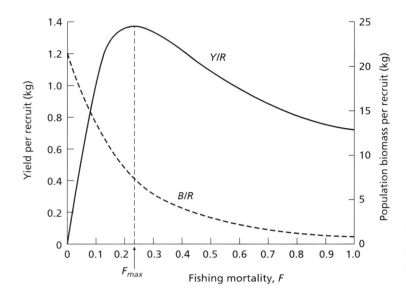

Fig. 7.12 Yield per recruit (*Y/R*) and population biomass per recruit (*B/R*) for a single cohort of fish, for various potential fishing mortalities, *F*. The results are for fish of ages 1–10, and natural mortality, *M* = 0.2 (see Box 7.6).

offshore they are caught by otter trawl and gill-net fisheries. Commercial landings have fluctuated drastically over the past 60 years, ranging from 20 000 tonnes from 1937 to 1940, to 29 000 tonnes in 1945, and plunging to less than 1000 tonnes from 1967 to 1971 before recovering to 13 000 tonnes in 1977 and 1978 (Barbieri *et al.*, 1997). There has been a decline since 1987, and recreational catches peaked in 1991 when approximately 21 million fish were caught.

Barbieri *et al.* (1997) used a yield-per-recruit model to investigate whether mortality was too high in two Atlantic croaker fisheries, and to make recommendations about the effects of reduced mortality on young fish through bycatch reduction devices and minimum size limits. Figure 7.13 shows the predicted effects of various values of *F* and mean age at first capture, *t*$_c$, on yield per recruit. Two fisheries are shown, each simulated twice—once with *M* = 0.20, and once with *M* = 0.35, to cover the probable range of the true value of *M*, which is uncertain. In lower Chesapeake Bay (Fig. 7.13a, b), analyses of landings showed that the current value of *t*$_c$ is 2 years of age. Surprisingly, it appears that a reduced age at first capture might actually increase yields, but this might be a risky strategy for a small benefit. The relationship with *F* is flat-topped for all ages of first capture. Total mortality *Z* (= *F* + *M*), is thought to be about 0.6. However, the breakdown between *F* and *M* is unclear. If *M* = 0.20 (Fig. 7.13a), *F*

would be 0.4, which is below but near the maximum. If *M* = 0.35 (Fig. 7.13b), *F* would be 0.25, and there is more room for higher yields with increasing *F*.

The situation is quite different for the North Carolina fishery (Fig. 7.13c, d). Here, *Z* is much higher, at 1.3. If *M* is low, reducing the age at first capture would be a bad idea. Indeed, the mean age at first capture, which is 1 year in this fishery, should be increased considerably. This is not so important if *M* is high (Fig. 7.13d). Note how much more peaked the yield-per-recruit surface is if *M* is low than if *M* is high. *F* is estimated to be either 1.1 if *M* is low, or 0.75 if *M* is high. In either case, this stock appears to be overexploited.

Since there was no indication of growth overfishing in Chesapeake Bay, Barbieri *et al.* (1997) suggested that that fishery should be regulated at the *status quo* until estimates of current mortality rates are improved. In contrast, the North Carolina fishery needed a higher mean age at first capture and lower fishing mortality. The authors noted that the differences may actually reflect the fact that the North Carolina data came from 1979 to 1981, whereas the Chesapeake Bay data came from 1988 to 1991. The early period coincided with the occurrence of unusually large fish. Whether the differences between the Chesapeake Bay and North Carolina fisheries reflect temporal or spatial patterns, this study shows that it is important to be careful about generalizing from one fishery to another.

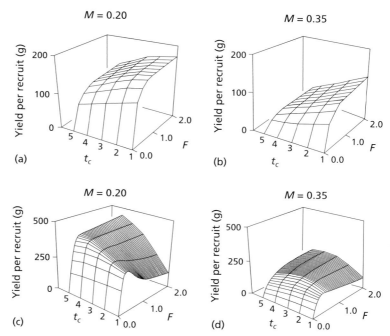

Fig. 7.13 Yield-per-recruit curves for Atlantic croaker *Micropogonias undulatus.* These are based on various combinations of mean age at first capture, t_c, and fishing mortality, F. (a) Lower Chesapeake Bay, assuming $M = 0.2$; (b) Lower Chesapeake Bay, assuming $M = 0.35$; (c) North Carolina, assuming $M = 0.2$; (d) North Carolina, assuming $M = 0.35$. After Barbieri *et al.* (1997).

7.8 Incorporating recruitment

We discussed recruitment briefly in section 7.4.1, but largely ignored it in our discussion of yield-per-recruit models. This is risky when recruitment is related to spawning stock size (Chapter 4). Thus, while yield-per-recruit models handle 'growth overfishing' very elegantly, they need to be integrated with recruitment if we are to avoid '**recruitment overfishing**', which is defined as a reduction in spawning stock biomass to the point where recruitment is impaired.

The simplest way to incorporate recruitment is to consider semelparous species with a life span of 1 year. These species, such as squid, spawn once and then die (section 3.4.3). Each year's fishery consists entirely of last year's recruitment, and the stocks are not buffered by multiple year classes, nor by small individuals that are temporarily safe from the fishery (e.g. Rosenberg *et al.*, 1990; Basson *et al.*, 1996). Here, spawner–recruit relationships must be used to find a target for **escapement**: the number of fish allowed to survive (Pauly, 1985; Beddington *et al.*, 1990). If recruitment is plotted against spawning stock size over a number of years, we can ask what level of recruitment would be needed to

maintain the population in the face of fishing mortality. The higher the mortality, the higher the recruitment has to be for the population to be in equilibrium.

7.8.1 Replacement lines

Replacement lines help us to understand how changes in F affect the recruitment rates of exploited stocks.

The theory of replacement lines was first developed by fishery scientists (Beverton & Holt, 1957), but has since been applied to other studies of population dynamics. As an example, we have chosen a Ricker spawner–recruit curve (Fig. 7.14). First, consider Fig. 7.14(a), with a diagonal line that has a slope of 1.0, showing the replacement level where recruitment balances spawning stock size. We can predict the trajectory of a population from any starting stock size by following from a point on the x-axis to the recruitment curve, and then taking this as the next generation to be plotted again on the x-axis, following up to the recruitment curve, and so-on. This is equivalent to reflecting the recruitment back to the spawning stock size in the following generation through the one-to-one replacement line. In Fig. 7.14(a), the population will approach a stable

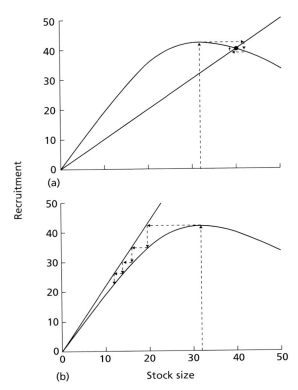

(a)

(b)

Recruitment

Stock size

Fig. 7.14 Population trajectories for a Ricker spawner–recruitment relationship. (a) The replacement line intersects the spawner–recruit curve, leading to a stable equilibrium. (b) The replacement line corresponding to higher fishing pressure exceeds the spawner–recruit curve, leading to a population collapse.

equilibrium, oscillating above and below it as it does so. In Fig. 7.14(b), a steeper replacement line has been drawn, to accommodate higher fishing mortality. Here, recruitment is not high enough to sustain the population, and it will crash.

Replacement lines can be used to predict the sustainability of fisheries if the slope of the replacement line is adjusted according to the level of fishing mortality (Shepherd, 1982). A doubling of mortality, for example, will result in a doubling of the slope of recruitment needed to replace the stock (if patterns of growth and maturation are constant). Yields can also be predicted, thereby building a useful bridge to methods such as yield-per-recruit models, which otherwise ignore impacts of fishing mortality, F, on future recruitment (section 7.7). Recall from section 4.2.1 (equation 4.3) the flex-

ible Shepherd spawner–recruit relationship that relates the abundance of spawners (S) to that of recruits (R). This can also be written as:

$$R = \frac{aS}{1 + (S/b)^c} \tag{7.27}$$

where a is the slope of the curve at the origin (maximum R/S at low stock sizes), b is the biomass at which recruitment is reduced to half the level it would have been under density independence only, and c controls the degree to which the spawner–recruit curve is asymptotic or dome shaped. If stock size S is measured as stock biomass B, we can substitute B for S and rearrange the equation to estimate B (Shepherd, 1982):

$$B = b(aB/R - 1)^{1/c} \tag{7.28}$$

equation 7.28 contains biomass per recruit (B/R), which we can calculate using yield-per-recruit models (section 7.4). We can also translate B/R into recruitment as $R = B/(B/R)$. Finally, if we know recruitment, we can calculate yield based on yield per recruit as $Y = R(Y/R)$. This concept is shown graphically in Fig. 7.15. Fig. 7.15(a) and (d) are the standard outputs from a yield-per-recruit model, showing the effect of different values of fishing mortality on population biomass per recruit and yield per recruit, respectively. Fig. 7.15(c) shows the relationship between recruitment and stock sizes (e.g. based on observations over a number of years). The replacement line has been added to Fig. 7.15(c) for a trial value of fishing mortality (F_{trial}) by giving it a slope that is the inverse of B/R for F_{trial}, using the translation step shown in Fig. 7.15(b). Equilibrium stock size is found where this line crosses the function relating recruits to population biomass. This is multiplied by yield per recruit to give the total yield (Fig. 7.15e).

7.8.2 Replacement lines in action

The concepts discussed above are illustrated for the North Sea cod *Gadus morhua* stock in Fig. 7.16. In this study, Cook *et al.* (1997) fitted a Shepherd spawner–recruit curve to data for 1963–94. The worrying result is that the spawner–recruit curve is below the replacement line. This is especially clear during the early years when stock sizes were high. Indeed, as predicted by the theory, this stock has been declining toward the origin. Recruitment exceeded the replacement line in only 8

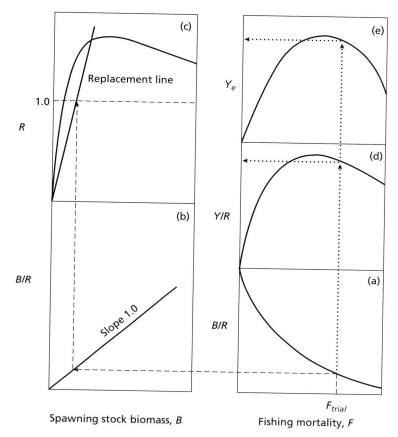

Fig. 7.15 Graphical links between yield-per-recruit calculations and a spawner–recruit relationship. A trial value of fishing mortality (F_{trial}) in (a) predicts spawning stock biomass per recruit (B/R), based on a yield-per-recruit calculation. This is converted to spawning stock biomass, B, in (b). A replacement line in (c) has slope $= 1/(B/R)$. This is used to convert B/R into equilibrium spawning stock biomass, at the intersection of the replacement line and the stock-recruitment relationship. Equilibrium spawning stock biomass is then multiplied by yield per recruit in (d) to give total equilibrium yield, Y_e, in (e). After Sissenwine and Shepherd (1987).

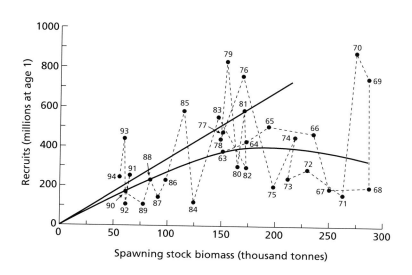

Fig. 7.16 Stock-recruitment curve and replacement line for North Sea cod. The numbers represent data for various years. After Cook *et al.* (1997).

of 31 years. Cook *et al.* (1997) also fitted Ricker and Beverton–Holt spawner–recruit curves to the same data. The values of F at which the population was predicted to crash were 0.91 (Shepherd), 1.05 (Ricker) and 1.13 (Beverton–Holt). These were perilously close to the contemporary estimate of $F = 0.91$ for that stock. We appear to be playing Russian roulette with the whims of recruitment. As we go to press, a major re-building programme is being proposed for this stock.

7.9 Confronting risk and uncertainty

There is uncertainty in most fisheries data. The values used for population size, commercial catches, natural mortality and fishing mortality are all estimates rather than true values. Even if we could estimate these perfectly, our projections into the future would still be uncertain. Although these facts are obvious to everyone, the structure of most decision-making processes still encourages the production of single 'best estimates' by fisheries biologists (Hilborn *et al.*, 1993a; Francis & Shotton, 1997). This is risky because it involves throwing away information.

Formal decision analyses allow us to face uncertainty directly. The process can be broken down into five parts (Hilborn, 1996; Punt & Hilborn, 1997).
1 Identify alternative hypotheses about the state of the fishery. For example, the population biomass may be 750 000 tonnes or 950 000 tonnes.
2 Determine the relative weight of evidence in support of each alternative. For example, the probability of 750 000 tonnes is 0.1 and the probability of 950 000 tonnes is 0.5.

3 Identify alternative management actions. For example, consider setting the quota to 100 000 tonnes or 150 000 tonnes.
4 Evaluate the distribution and expected value of each performance measure, given the management action and probability of each alternative hypothesis about the state of the stock. For example, if the biomass is 750 000 tonnes and you set the quota to 100 000 tonnes, you can expect the stock size after 5 years to be 50% of the virgin biomass.
5 Present these results to the decision-makers.

An example of this approach is a decision table for the New Zealand hoki *Macruronus novaezelandiae* (order Gadiformes) fishery in New Zealand (Table 7.2). This table focuses on alternative hypotheses for virgin stock size, but it could be adapted to whatever uncertain aspect of the fishery was of interest. The first row gives alternative potential virgin stock sizes, and beneath each of these (in parentheses) is the probability that it is true. We describe how the probabilities are calculated in the next section (7.9.1). Three potential management actions (in this case quotas) are considered in the left column. Each cell of the table shows the predicted outcome of each management action, in terms of the ratio of stock biomass after 5 years of exploitation to the hypothesized virgin biomass. These values are usually calculated using Monte Carlo simulations (reviewed by Punt & Hilborn, 1997). The expected values in the right-most column are the means for these cells, e.g. $0.66 = (0.51 \times 0.099) + (0.63 \times 0.465) + \dots (0.81 \times 0.003)$. If only the most probable virgin stock size (950 000 tonnes) had been considered, comparison with the mean expectations shows that this would have led to underestimates of the stock sizes after 5 years of

Table 7.2 A decision table to evaluate the consequences of a variety of alternative catch quotas, given various potential virgin biomasses, for the New Zealand hoki *Macruronus novaezelandiae*. Values in parentheses are the probabilities of each virgin biomass, cell entries are the ratio of the stock biomass after 5 years to the virgin biomass, and expectations are means of each of these ratios when multiplied by their probability. After Hilborn *et al.* (1994).

| Quota (10^3 t) | Alternative hypotheses (virgin biomass 10^3 t) | | | | | | Expectation |
	750 (0.099)	950 (0.465)	1150 (0.317)	1350 (0.096)	1550 (0.020)	1750 (0.003)	1047
100	0.51	0.63	0.70	0.75	0.78	0.81	0.66
150	0.26	0.45	0.56	0.63	0.69	0.72	0.49
200	0.22	0.26	0.42	0.52	0.59	0.64	0.34

exploitation. This is particularly true under the higher quotas, because, for example, the 31.7% chance that the virgin stock size was actually 1 150 000 tonnes would have been ignored.

7.9.1 Bayesian analysis

Two main approaches have been used to calculate probabilities of alternative hypotheses being correct: Bayesian analyses and resampling methods.

Bayesian analysis provides a way of making probabilistic inference by combining prior information with current information. It is based on Bayes' theorem, developed by the Reverend Thomas Bayes in 1763 (Box 7.7). The good reverend has reached sainthood in the eyes of many contemporary theorists.

Bayesian inference was used to establish the probabilities of the alternative hypotheses for virgin stock biomass of the New Zealand hoki (Table 7.2). We will not go into the mathematics here, but refer readers to McAllister *et al.* (1994), Walters and Ludwig (1994), Punt and Hilborn (1997), and Francis and Shotton (1997). Bayesian analysis contains two key elements: specification of prior distributions of each alternative hypothesis, and calculation of the goodness of fit of available data to each of these alternatives.

The **prior distribution** summarizes all information about a parameter, except for the data used in the likelihood calculations, in the form of a probability distribution. This information can include previous experience with other stocks, knowledge about the behaviour or life history of the animal, and so on. Specifying these 'priors' is not easy, and is sometimes criticized on the grounds that the final probabilities of each parameter may be warped by errors in the prior distribution. But this criticism can be levelled at any modelling technique to some extent. For example, non-Bayesian analyses of fish stocks often make a default assumption that the instantaneous rate of natural mortality, M, is 0.2. Effectively, this is also a specified 'prior', although it is not called that in traditional analyses. In practise, one should check how important such assumptions are in all analyses, including ones using Bayes' Theorem, by running sensitivity analyses on a range of parameter values. An advantage of Bayesian analyses is that specification of priors is a formal procedure that cannot be swept under the carpet, and it can incorporate many kinds of information simultaneously. The data that can be used in a Bayesian analysis are exactly the same as those used in non-Bayesian analysis (e.g. indices of population size, age-composition data, etc.).

Box 7.7

Bayes' Theorem.

Suppose we wish to calculate the probability that either of two hypotheses, H_1 and H_2, is correct, based on prior information and a current observation (R). In its simplest form for discrete parameters, Bayes' Theorem says that the probability of H_1 given the prior information and the current observation R is proportional to the product of the prior probability of H_1 and the probability of R, given H_1 (Edwards, 1992). In formal notation, the latter is written $P(R|H_1)$. Thus:

$$P(H_1|R) = k \times P(H_1) \times P(R|H_1) \qquad (1)$$

where k is a constant of proportionality, which is the same in the similar equation for H_2. It is given by

$$1/k = [P(H_1) \times P(R|H_1)] + [P(H_2) \times P(R|H_2)] \qquad (2)$$

For example, suppose you know from experience that 1/3 of lobsters in a population are males and 2/3 are females. Furthermore, 3/4 of males have large claws and 1/2 of females have large claws. You trap a lobster that has large claws. What is the probability that it is a male?

Let H_1 be that the lobster is a male, so that $P(H_1) = P(\text{male}) = 1/3$.

Let H_2 be that the lobster is a female, so that $P(H_2) = P(\text{female}) = 2/3$.

Let R be the fact that the lobster you have caught has large claws.

Based on the above prior information, $P(R|H_1) = 3/4$ and $P(R|H_2) = 1/2$. Therefore, $1/k = (1/3 \times 3/4) + (2/3 \times 1/2) = 7/12$. Hence, $k = 12/7$. Therefore, $P(H_1|R) = 12/7 \times 1/3 \times 3/4 = 3/7$. The probability that the large-clawed lobster is a male is 3/7.

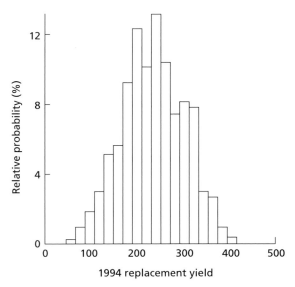

Fig. 7.17 Probabilities of various potential replacement yields for the Bering–Chukchi–Beaufort Seas stock of bowhead whales *Balaena mysticetus*. After Punt and Hilborn (1997).

The results of a Bayesian analysis are usually summarized by the **posterior distributions** for quantities of interest to management. The posterior probability of some variable is the probability that it is the true value after accounting for its prior probability and the information content of the available data.

An example of the use of Bayesian methods is provided by recent assessments of the Western Arctic population of bowhead whales *Balaena mysticetus*, conducted by the Scientific Committee of the International Whaling Commission (IWC). A key output from the assessment is the current replacement yield (i.e. the catch that will keep the population of animals aged 1 and older at its current size). This distribution is used by the IWC to set catch limits for this population. The assessments were based on an age- and sex-structured population dynamics model (Punt, 1999) and used data collected during visual and acoustic surveys off Point Barrow, Alaska. Various prior distributions were specified, including the pre-exploitation size of the population, the population size at which MSY is achieved, the age at maturity, and survival rates for adults and juveniles (IWC, 1995). The data are not particularly informative about current replacement (Fig. 7.17). The IWC based its catch limits on the lower fifth percentile of this

distribution to allow a high probability of some further recovery.

7.9.2 Resampling methods

An alternative approach to setting probabilities is to use resampling methods such as bootstrapping. These methods use existing data from the fishery to generate probabilities for alternative hypotheses. They are straightforward and often faster computationally than Bayesian analyses, although this advantage is becoming less important as computers get faster and Bayesian algorithms improve. It may also be comforting to avoid the formal setting of priors, although it is still important to admit the full extent of uncertainty in each parameter.

As an example of the resampling approach, we will consider a study of the orange roughy *Hoplostethus atlanticus* fishery on the Chatham Rise, east of New Zealand (Francis, 1992). The response that this study generated, including debate about whether Bayesian methods would be superior, shows that computational approaches to risk assessment are still experiencing growing pains (Hilborn *et al.*, 1993b; Francis, 1993). Orange roughy are deep-water fish, which aggregate on seamounts such as the Chatham Rise during spawning. Trawlers target spawning aggregations, and catch most fish below 750 m from mid-June to mid-August. As with the roughy fishery on the Challenger Plateau (Fig. 7.5), this fishery grew rapidly in the early 1980s, and well beyond the sustainable level for a fish that matures in its mid-20s.

Francis (1992) used a population model that combines yield-per-recruit analysis with life-history parameters and a Beverton–Holt spawner–recruit relationship to estimate maximum sustainable yield. The management objective of the New Zealand Ministry of Agriculture and Fisheries was to aim for 'maximum constant yield', defined in this case as two-thirds MSY. This required a reduction in the total allowable catch (TAC) toward 7500 tonnes. This was only a quarter of the existing TAC for 1989–90 of 28 637 tonnes. How quickly should these painful cuts be brought in? Socio-economic concerns needed to be weighed against the probability of a collapse. This is where risk analysis came in, and embraced the uncertainty in stock assessment.

Figure 7.18 shows the probability of fishery collapse within 5 years for alternative rates of TAC reduction. A collapse was defined as the stock being reduced to the

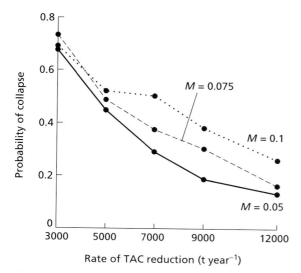

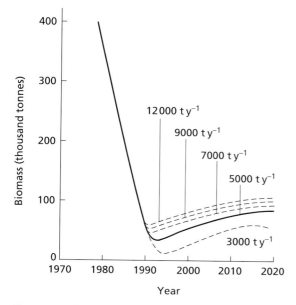

Fig. 7.18 Risk analysis for orange roughy *Hoplostethus atlanticus* on the Chatham Rise, New Zealand. This shows the probability of collapse within 5 years according to various rates of reduction of the total allowable catch toward a target of 7500 t y^{-1}. Each line represents a different assumption for the instantaneous rate of natural mortality, *M*. After Francis (1992).

Fig. 7.19 Traditional presentation of management advice for orange roughy *Hoplostethus atlanticus* on the Chatham Rise, New Zealand. Each projection of stock biomass from 1990 assumes an instantaneous rate of natural mortality, *M* of 0.05, for various rates of reduction of the TAC. After Francis (1992).

point where the TAC was not catchable with a fishing mortality rate of less than 1.0 per year. The solid line in Fig. 7.18 was generated with an estimate of natural mortality, *M* = 0.05, which is the best guess. If higher rates of *M* are assumed, there is greater urgency to reduce the catch because these suggest that total mortality must be higher.

Figure 7.19 shows the more traditional way of presenting management advice from a stock assessment. These forecasts of stock biomass for different rates of TAC reduction show different management options, but they do not incorporate uncertainty in natural mortality, survey indices or recruitment. This therefore does not show the dangerous situation of the fishery depicted in Fig. 7.18, i.e. a 30–50% chance of collapse even if the TAC is reduced at the rate of 7000 t y^{-1}, depending on the natural rate of mortality.

In summary, there is little excuse for ignoring the uncertainty that is inherent in stock assessments. We have the mathematical underpinnings (e.g. Bayes' Theorem), and the computational power. Perhaps the biggest remaining obstacle is a pervasive feeling that decision-makers can only handle recommendations based on single point values. This can be an expensive mindset: during the collapse of the northern cod *Gadus morhua* in the north-western Atlantic in the 1980s there was conflicting information about the state of the stock, depending on inferences from offshore catches, inshore catches or government surveys. Yet uncertainty was filtered during the decision process under the ethos that 'one message goes to the minister' (Harris, 1998). Unfortunately, the wrong message got through, contributing to the destruction of the fishery and the loss of 40 000 jobs. An alternative approach, advocated by Hilborn *et al.* (1993a), would be to allow decision-makers to have decision tables such as Table 7.2 and graphs like Figs 7.17 and 7.18. These could be presented to them by fishery scientists who could run on-the-spot scenarios in response to 'what if' questions asked by the decision-makers. These could include economic as well as biological alternatives. We return to methods of incorporating uncertainty in bioeconomic models in Chapter 11.

7.10 Biological reference points

Biological reference points are derived from models of populations, to serve as benchmarks in making man-

Table 7.3 Some examples of biological reference points. F = instantaneous fishing mortality; B = spawning stock biomass; R = recruitment.

Symbol	Definition
F_{MSY}	F giving maximum sustainable yield (also called F_m)
F_{MCY}	F giving maximum consistent yield, i.e. largest long-term yield without reducing population below a predetermined level, such as 0.2 virgin biomass, with a prespecified probability. May also be given relative to F_{MSY}, e.g. $(2/3)F_{MSY}$
F_{crash}	F that would drive the population to extinction
F_{max}	F where total yield or yield per recruit is highest
$F_{0.1}$	F where slope of yield per recruit vs. F is one-tenth of its value near the origin (Fig. 7.20)
F_{pa}	F set according to a specific precautionary approach
F_{lim}	F set as the highest that is acceptable by some specified criterion
F_{low}	F in an equilibrium population where recruitment per spawning stock biomass in 90% of years has been above the replacement level
F_{med}	F in an equilibrium population where recruitment per spawning stock biomass in half of the years has been above the replacement level
F_{high}	F in an equilibrium population where recruitment per spawning stock biomass in 10% of years has been above the replacement level
F_{LOSS}	F that would drive the stock to the lowest observed spawning stock size (LOSS)
$F_{x\%}$	F in an equilibrium population where recruitment per spawning stock biomass is x% of the corresponding unfished population
B_{MSY}	B corresponding the maximum sustainable yield
B_{LOSS}	The lowest B ever observed
B_{lim}	The B limit set as the lowest that is acceptable by some specified criterion
B_{pa}	The B set within a precautionary approach
$B_{x\% R}$	The B at which the average recruitment is x% of the maximum of the underlying spawner–recruitment relationship

agement recommendations. Fishing rates that would give the theoretical MSY were once considered a good target, but there has been a stampede away from this objective due to the difficulty of estimating MSY accurately (section 7.3.1; Punt & Smith, 2001). Indeed, the FAO's Code of Conduct and the UN Agreement on Straddling Fish Stocks and Highly Migratory Fish Stocks posit MSY as a limit, not a target. Fishing mortality should not exceed the theoretical point at which MSY would be achieved (F_{MSY}) and stock biomass should not drop below the MSY level (B_{MSY}).

A variety of biological reference points are listed in Table 7.3. Many of these are reviewed by Smith *et al.*

(1993) and Caddy (1998). Some apply specifically to particular kinds of models, and different ones may be adopted as either target reference points or as limit reference points (not to be exceeded). For example, production models show that F_{MCY} (maximum consistent yield) is a safer target than F_{MSY}, and F_{crash} is definitely a good thing to avoid (Fig. 7.20a).

The yield-per-recruit curve in Fig. 7.20(b) shows F_{max} and $F_{0.1}$, which are usually reported as limits and targets, respectively, when this modelling technique is used. $F_{0.1}$ is the value of F on the yield-per-recruit curve where the slope of yield is one-tenth of its initial slope at the origin (Gulland, 1983; Deriso, 1987). For each

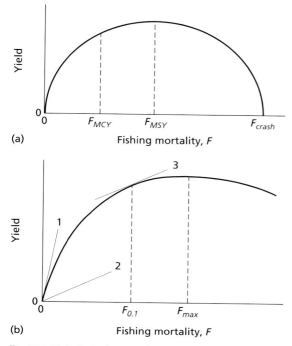

(a)

Fishing mortality, *F*

(b)

Fishing mortality, *F*

Fig. 7.20 Biological reference points. (a) Surplus production model; (b) yield-per-recruit model. $F_{0.1}$ is found by following the numbered steps indicated: (1) find slope at origin; (2) plot line with 10% of this slope; (3) find tangent to curve at this slope.

unit increase in F, the yield per recruit will increase by one-tenth of the amount at which it was first increasing when F was very low. $F_{0.1}$ will always be less than F_{max}, thereby maintaining the stock at a safer level. Indeed, $F_{0.1}$ also has the very useful practical advantage of allowing a more precise target than F_{max} when the yield-per-recruit curve has a wide flat top, as in the Atlantic croaker analyses for lower Chesapeake Bay (Fig. 7.13). For that case, $F_{0.1}$ was 0.27 and 0.64 assuming low and high values of M, respectively (Barbieri *et al.*, 1997). Remember that there is no theoretical underpinning for selecting the value of 0.1 as the percentage of initial F.

The study of the pros and cons of various biological reference points, including their integration with socioeconomic objectives, is a hot topic in fisheries. Reference points are intimately linked to considerations of risk, because they are the targets or limits considered as acceptable benchmarks to aim for or to avoid at all costs. We have come a long way from targeting MSY using equilibrium production models. But can we really get away with considering exploitation of one species at a time, and ignoring interactions among species? This is the subject of the next chapter.

Summary

- The aims of stock assessment are to describe the population biology of fished species and to find ways of maximizing yields to fisheries while safeguarding the long-term viability of populations.
- For a given level of fishing mortality to be sustainable, there must be a balance between the forces that reduce population biomass (natural and fishing mortality) and those that increase it (reproduction and growth).
- Many quantitative methods have been developed for single-species stock assessment. Fishery scientists who work on stock assessment have made key contributions to the wider understanding of the dynamics of exploited animal populations.
- Surplus production models aggregate production across age classes. Equilibrium surplus production methods have

received a bad name because they were based on faulty assumptions, but modern 'observation-error' methods are a considerable improvement and can perform well.
- Delay–difference models are a stepping stone between production models and fully age-structured models, requiring a minimum of data on body growth and recruitment.
- Virtual population analyses is used to describe age-specific stock structure. VPA does not by itself indicate how many individuals can be caught to meet a given objective, nor does it predict the future.
- Statistical catch-at-age analysis is an extremely flexible method for estimating stock size. Unlike VPA, estimates of natural mortality are not required, but one must be cautious about juggling numerous parameters at once.

Continued p. 158

Summary (*Continued*)

- Yield-per-recruit models are run using the information provided by analyses such as VPA or statistical catch-at-age analyses, to seek fishing mortality rates that achieve the best trade-off between the sizes of individuals caught, and the number of individuals available for capture.
- Yield-per-recruit models need to be integrated with recruitment if recruitment overfishing is to be avoided.

Recruitment overfishing is a reduction in spawning stock biomass to the point where recruitment is impaired.
- Advances in methods of stock assessment are being matched by new ways of converting this information into management advice. These include explicit incorporations of risk and uncertainty and the use of precautionary biological reference points.

Further reading

Many books are devoted to single-species stock assessment, and many fisheries laboratories do their assessments using software that they have developed or customized. The excellent books by Hilborn and Walters (1992) and Quinn and Deriso (1999) give detailed accounts of stock-assessment methods and contain many worked examples. They have to be consulted if you intend to do your own assessment! Sparre and Venema (1998) provide a clear description of stock assessment methods in tropical fisheries.

8 Multispecies assessment and ecosystem modelling

8.1 Introduction

Individual species do not live in a vacuum. They eat each other and may compete for food and space (Chapter 4). These **biological interactions** mean that the population dynamics of different species are inevitably linked. Furthermore, overlaps among species in body size and habitats means that it is impossible for even the most careful fisher to catch only one species at a time with most gears (Chapter 5). These **technical interactions**, i.e. fishing mortality on more than one stock by a single fishery, whether occurring intentionally or as a bycatch, present a considerable challenge to management regulations. Much of this ecological and practical common sense was left behind on purpose when fisheries science was developing in the 1950s and 1960s. The point of the single-species approaches described in the previous chapter was to strike a balance between abstraction, with intentional loss of reality, and the practicality of not asking for impossible estimates of parameters. But has the compromise gone too far? This would be ironic, since fisheries science was once at the forefront of the development of some key aspects of ecological theory.

In this chapter we focus on the two main aspects of multispecies theory: biological interactions among species, and technical interactions through fishing fleets catching more than one species at a time. We show how fishery scientists account for these interactions and we give examples of multispecies assessment in action. As we shall see, by including behavioural, ecological and technical interactions in models, we can reach some conclusions about fisheries management that are exactly the opposite of those based on single-species assessments. In the latter part of the chapter, we introduce models that have been used to describe the structure of marine ecosystems and predict how these ecosystems respond to fishing.

8.2 Multispecies surplus production

The simplest way to account for the multispecies nature of catches is to use a surplus production model (section 7.3) for aggregated yields of individual species. These multispecies yields can then be compared with the total effort in the fishery of interest (FAO, 1978). However, we must remember the biology of the component species. For example, consider three hypothetical species from a fishery, with individual yield curves plotted against fishing mortality (Fig. 8.1). Species A might

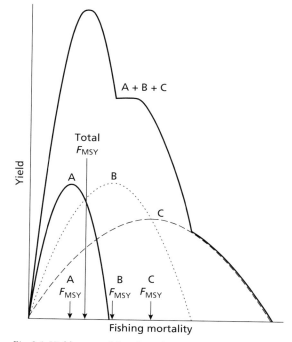

Fig. 8.1 Yield curves of three hypothetical species A, B, and C plotted against fishing mortality, *F*. Exploitation at the theoretical MSY for all three species combined would overexploit A and underexploit B and C. After Sparre & Venema (1998).

be a late-maturing species with a low intrinsic rate of natural increase such as a ray (order Rajiformes), species B might be an intermediate species such as a small grouper (Serranidae), and species C might be a small wrasse (Labridae). The aggregated yield curve reaches a maximum at a level of fishing that will be too high for the ray, and underexploit the wrasse. Indeed, exploitation at the theoretical maximum sustainable yield (MSY) level for the wrasse (C) would wipe out rays entirely. The loss of rays means that the effects of fishing are not reversible, unlike the case hoped for single-species production models. Moreover, note that the surplus production approach considers only technical interactions, i.e. capture of more than one species by a fishery. It ignores the fact that species may eat each other or compete with each other. Such biological interactions are considered later in this chapter.

8.2.1 Multispecies surplus production in action

We illustrate briefly a technique that we refer to as a **temporal multispecies production model**. This is a direct analogue of the equilibrium surplus production model for single species (section 7.3.3). This approach has been used most often in tropical fisheries, where the huge diversity of fishes is often reflected in fishers' catches, which may contain 40 or more species of economic importance. Figure 8.2 shows an example from the Hawaiian archipelago (Ralston & Polovina, 1982). Here, an equilibrium surplus production model (section 7.3.3) was based on catch and effort data for 13 bottom-dwelling deep-water species combined. Figure 8.2(a) shows catch per unit effort (CPUE) plotted vs. effort. This provides the parameters to fit to a Schaefer curve, shown in Fig. 8.2(b). This indicates a maximum sustainable yield of 106 tonnes at a fishing effort of 901 fisher-days. This was similar to the most recent estimate of actual yield of 96 tonnes.

The same strong reservations apply to the temporal multispecies production model as to the single-species analogue, namely that it assumes that the fish populations are in equilibrium with each level of fishing effort (section 7.3). In fact, in Hawaii, there was a consistent increase in effort throughout the 20 years of the study. An additional problem of the multispecies application of the method is that, unlike the single-species case, it is not clear that the total biomass of an unfished population

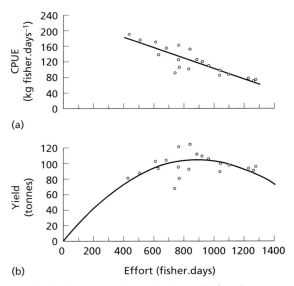

(a)

(b)

Effort (fisher.days)

Fig. 8.2 Multispecies surplus production model for a deep-sea handline fishery in Hawaii. (a) Catch per unit effort in various years; (b) Schaefer-type production model fitted using parameters from the regression line. After Ralston and Polovina (1982).

would be much higher than that of a fished population. This will depend on the predator–prey relationships within the assemblage, a point we will return to later in this chapter. The authors of the Hawaiian study minimized this problem by choosing species that did not eat one another.

Another approach is to use the same method, but rather than using a time series of data to estimate the necessary parameters for a production model, each data point in the analysis comes from a different site. This **spatial multispecies production model** treats different sites, such as reefs or small islands, as replicate fisheries and assumes that the productivity of their fish communities is the same. An example was a study of two Caribbean reef fisheries in south Jamaica and Belize (Koslow *et al.*, 1994). The authors collected catch and effort data from six areas in Jamaica and seven in Belize. Again, we have the nefarious assumption that fisheries are in equilibrium. In fact, this assumption proved to be violated for the Jamaican data set, and the authors are careful to point out this and other shortcomings in the data. Indeed, this study illustrates a typical 'warts and all' example from the real world.

Table 8.1 Data for a spatial multispecies production model in the Caribbean. After Koslow *et al.* (1994).

Country	Site	Area (km²)	Total yield, Y (t)	Total effort, f (thousands hook h)	Yield/area (kg km⁻²)	Effort/area (hook h km⁻²)
Belize	A	312	30	22	96	71
Belize	B	123	168	89	1366	724
Belize	C	43	24	24	558	558
Belize	D	231	18	15	78	65
Jamaica	A	46	55	286	1196	6217
Jamaica	B	252	344	2038	1365	8087
Jamaica	C	607	265	1303	437	2147
Jamaica	D	652	266	1448	408	2221
Jamaica	E	115	37	265	322	2304
Jamaica	F	135	31	257	230	1904

Table 8.1 shows total yield (Y) and effort (f) data for 10 sites. The authors actually surveyed landings from 13 sites initially, but decided that it would be safer to combine some sites when interviews with fishers in ports in Belize indicated some overlap in landings from multiple sites. The manner in which these sites were combined has caused some small discrepancies between the published figures and our own calculations presented here. The table includes all species landed, although the prime commercial species were from the families Lutjanidae (snappers), Serranidae (groupers) and Haemulidae (grunts). For area-based studies, it is important to correct for the extent of area fished. Thus, both catches and effort are expressed per square kilometre. Note that all of the data were also converted to a common currency of effort: hook hours per square kilometre.

A plot of CPUE against log-transformed effort showed a significant decline (Fig. 8.3a). Log transformation was used because the raw data were clearly non-linear, with CPUE dropping steeply at first, then levelling off at low values. Note that this decline was due to the contrast between the heavily exploited Jamaican sites and the weakly exploited Belizian sites. Indeed, had the authors not pooled the data from the two countries they would have seen no indication of any decline in catch rates with increasing effort in either country. Log effort explained 55% of the variation in CPUE. The regression equation is CPUE = 2.69 − 0.68 log(f). This gives us the parameters needed to fit a Schaefer curve that has the form $Y = af + bf^2$, where a is the intercept and b is the slope of the above regression

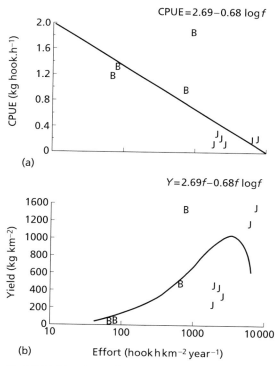

Fig. 8.3 Multispecies surplus production model for 10 fishing areas in Belize and Jamaica. (a) Catch per unit effort vs. effort on a log scale, showing individual sites (B = Belize, J = Jamaica) and regression line. (b) Yield vs. effort. The curved line is an equilibrium production model fitted to a Schaefer curve with parameters from the regression line in (a). After Koslow *et al.* (1994).

equation. For the log-transformed data used here, the yield curve is given by $Y = 2.69f - 0.68f \log f$. This production curve is shown in Fig. 8.3(b). According to this, the MSY would be about 1 t km^{-2} of productive habitat, defined as sea grass or coral areas. This would be caught with an annual fishing effort of approximately 3600 hook h km^{-2}.

Koslow *et al.* (1994) warn of several limitations, which carry general lessons for this technique. First, there was considerable heterogeneity among the fisheries in species caught and methods of capturing them, so we may be dealing with an apples and oranges comparison of productivity. Second, fishing effort has increased over time at the Jamaican sites, thereby violating the key assumption that populations are in equilibrium with fisheries. We have harped on about the problems with this assumption in our discussions of equilibrium production models in Chapter 7. Third, the necessity of pooling the data between the countries means we are effectively comparing just two sites, rather than many. This reduces the generality. The authors conclude that this method may provide a first approximation of the fishery potential of these countries, but it would be inappropriate to use it for management.

8.3 Multispecies yield per recruit

Another way of modelling technical interactions in multispecies fisheries is to use multispecies yield-per-recruit analyses. These are similar to single-species yield-per-recruit calculations (section 7.7), but the yields are summed among species (Murawski, 1984). This is done by calculating the fishing mortality that each species will face in the multispecies fishery. One can then model the effects of varying fishing mortality (e.g. by changing effort) and the proportion of individuals in each cohort from each species that will be susceptible (e.g. by changing mesh size). To forecast equilibrium yields to the fishery, one must take account of the different mean recruitment of each species. Thus, a relative recruitment multiplier is employed, based on survey (section 10.2.4) or VPA (section 7.5) recruitment estimates for each species (Murawski, 1984).

8.3.1 Multispecies yield per recruit in action

Murawski (1984) applied his method to a multispecies fishery on the Georges Bank region off the north-east United States and adjacent Canada. During the period 1977–79, four species provided 85% of the yields to a fishery dominated by otter trawls, in the following order: Atlantic cod *Gadus morhua* (Gadidae), haddock *Melanogrammus aeglefinus* (Gadidae), yellowtail flounder *Limanda ferruginea* (Pleuronectidae), and winter flounder *Pseudopleuronectes americanus* (Pleuronectidae). Stomach content analyses suggested that these species rarely ate one another and competition between them was considered minimal. This is important, because the yield-per-recruit analyses discussed here account only for technical interactions, and ignore biological interactions. The selectivity of nets for each species was known, including the proportion caught by length class in relation to mesh size (section 9.3.1). Information for standardizing mean recruitment among species ('**relative recruitment multipliers**') came from research vessel surveys and catch compositions in the commercial fishery. After some tweaking and simulating, the multipliers were as follows: cod (0.8), haddock (1.0), yellowtail flounder (5.0) and winter flounder (0.7).

The total yield per recruit from the four species is shown in Fig. 8.4. This shows that total yields can be increased by increasing mesh size. More interestingly, the analysis brings out the fact that the optimal total fishing effort for one species is not necessarily best for another. For example, at the largest mesh size (152 mm), the optimal annual effort for the entire fleet was about 9000 days fished (Fig. 8.4). But the optimal efforts for the individual species were scattered widely around this figure. This illustrates a concern raised for multispecies surplus production models (Fig. 8.1), namely that optimal effort for the combined species would overfish some (e.g. yellowtail flounder) and underfish others (e.g. haddock and winter flounder). Cynics might call this is a moot point, since at the time of the study the whole fishery was overfished at 19 000 days per year!

8.4 Multispecies virtual population analysis

We move now to biological interactions. There are several ways of capturing these formally; but we begin with **multispecies virtual population analysis** (MSVPA). As its name suggests, this is a logical extension of single-species virtual population analysis, which we

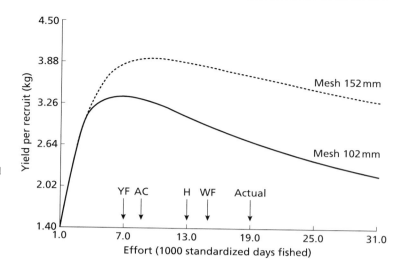

Fig. 8.4 Total yield per recruit vs. fishing effort for four species: Atlantic cod (AC), haddock (H), yellowtail flounder (YF), and winter flounder (WF). Two mesh sizes are shown. Each arrow corresponds to the maximum yield for individual species. The actual fishing effort for the period of the analysis (1979) is also indicated. After Murawski (1984).

will continue to denote VPA (section 7.5). As with VPA, MSVPA is essentially an accounting technique for calculating numbers of individuals from each age cohort and fishing mortality, F, based on catch statistics. The main difference from VPA is that natural mortality, M, is broken down into two components: $M2$ denotes predation due to those predators that are specifically included in the model, and $M1$ denotes residual mortality due to all additional factors that are not included explicitly in the model.

MSVPA is shown schematically in Fig. 8.5, which shows mortality of cohorts of two species, the first of which is eaten by the second. Thus, species 1 dies from predation ($M2$) by species 2, as well as all other natural causes lumped together ($M1$) and fishing mortality (F). The equivalent diagram for single-species VPA would show only one of these species, without $M2$. This general scheme can be enlarged to take account of many predators and predator–prey relationships in both directions. For example, the predator–prey relationships of species 1 and 2 could become reversed as members of species 1 grow large enough to prey upon small members of species 2. Cannibalism can also be included.

MSVPA was developed independently by Helgason and Gislason (1979) and Pope (1979), following a ground-breaking paper by Andersen and Ursin (1977). Reviews of MSVPA, including derivations of the equations used in the simulations, are provided by Sparre (1991) and Magnússon (1995). The essential feature of

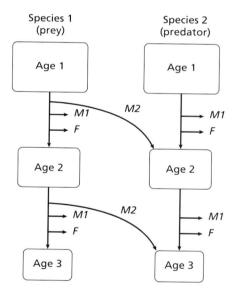

Fig. 8.5 Schematic overview of multispecies virtual population analysis. Arrows indicate losses of biomass to either the predator indicated ($M2$) or other predators ($M1$) or fisheries (F).

the method is that mortality due to predation is added to the two 'equations of death': the exponential decay equation, and the catch equation (see Box 7.2). This is easier to do in theory than in practice, since it requires detailed knowledge (or risky assumptions) about the food habits of predators. This includes total intake rates, and the manner in which feeding preferences may

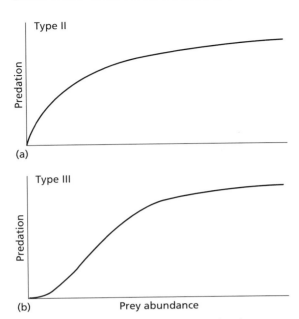

Fig. 8.6 Two potential behavioural responses of predators to abundance of prey. (a) Type II responses mean taking prey in proportion to their occurrence, whereas (b) Type III responses involve switching prey when the main prey becomes scarce. After Holling (1959).

change with the density of potential prey. For example, it is generally assumed that the quantity of a given prey consumed is proportional to its relative abundance in the 'community' of potential prey. This leads to a relationship between predation and prey abundance known from foraging theory as a **Type II functional response** (Fig. 8.6(a)). Alternatively, predators may switch prey species at low densities of prey (Type III, Fig. 8.6b), which would lead to overestimates of predation on rare prey species. Forecasts from MSVPA assume that total intake by predators remains constant from year to year for a given size class of predator, and that this is independent of prey abundance.

Once the feeding habits of key predators are estimated, they are summed across all predators that feed on a particular prey species. The resulting estimate of $M2$ as well as residual $M1$ is used for backward calculations in a similar manner to single-species VPA. Box 8.1 lists the main computational steps in algorithms for MSVPA.

8.4.1 Multispecies VPA in action

European fisheries biologists have invested tremendous effort in developing MSVPA, with particular focus on the North Sea (reviewed by Daan, 1987; Pope, 1991; Sparre, 1991; Magnússon, 1995). To obtain the predation data necessary, two massive sampling schemes were carried out in conjunction with the regular groundfish surveys coordinated by the International Council for the Exploration of the Sea (ICES), in 1981 and 1991. These two sampling schemes have become known as the 'Years of the Stomach', and with good reason. The first time the analyses were carried out, 54 000 fish stomachs were examined! These came from the five principal fish predators. The 1991 survey had expanded to include 14 fish species as well as grey seals

---

Box 8.1

Computational steps for MSVPA, based on Magnússon (1995), which gives the underlying equations.

1 Guess an initial value of $M2$ (mortality due to predators in the model) for all species and ages.

2 Solve the VPA as you would for a single species by iteration to get F and hence total mortality, Z (Section 7.5).

3 Calculate the average numbers of all prey at each age, based on Z.

4 Calculate the suitable prey biomass for each predator at each age, based on the biomass of the species in the model weighted by their suitability as determined

from the predation ecology of the predator, as well as a combined value for all other food available in the ecosystem that is not examined separately in the model.

5 Calculate new values of $M2$ for each prey species at each age based on their suitability to all predator species at each age, the total amount of food the predators eat, and the population sizes and biomasses of each prey at each age.

6 If the differences between the new values of $M2$ calculated above and the previous values of $M2$ in the iteration are too large (according to a prescribed error tolerance), then go back to Step 2 with a new $M2$ and start again. If the error is small enough, you've got it.

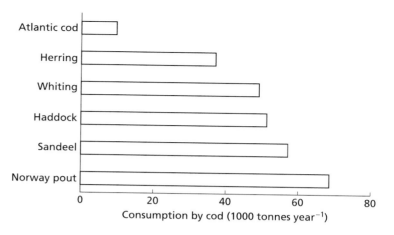

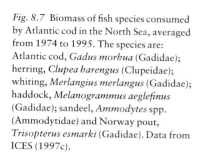

Fig. 8.7 Biomass of fish species consumed by Atlantic cod in the North Sea, averaged from 1974 to 1995. The species are: Atlantic cod, *Gadus morhua* (Gadidae); herring, *Clupea harengus* (Clupeidae); whiting, *Merlangius merlangus* (Gadidae); haddock, *Melanogrammus aeglefinus* (Gadidae); sandeel, *Ammodytes* spp. (Ammodytidae) and Norway pout, *Trisopterus esmarki* (Gadidae). Data from ICES (1997c).

and a category that included seabirds. Ten of the fish species had catch-at-age data available and were used in the main assessments (ICES, 1997c). Additional data on feeding habits have been used to refine model inputs.

By scaling up the stomach samples to the entire populations of predators and prey, it was possible to estimate the total biomass of each of 10 fish species consumed by predators for each year from 1974 to 1995. An example is shown for Atlantic cod *Gadus morhua*, based on averages over the time period (Fig. 8.7). This indicates some cannibalism (primarily on the youngest age classes). Researchers also looked for four species which proved not to comprise significant parts of the cod's diet: saithe, mackerel, plaice and sole. The mean total consumption rate of the 10 assessed species combined was an average of 271 000 t y^{-1}! In addition, cod consumed another 492 000 tonnes of other species whose population dynamics were not assessed in the model. By comparison, the mean yield of cod to human fishers was 188 000 tonnes.

Now we can turn the tables and consider the cod's position as a prey item in order to estimate $M2$ in comparison to F on cod (Fig. 8.8). This shows that mortality due to predators is very high on the youngest age classes, but falls to zero by the time cod are 7 years old. This is not surprising since few animals in the North Sea could eat such a large fish. When cod are between the ages of 1 and 2, humans become the most important predators, and fishing mortality remains consistently high beyond age 2.

Instantaneous rates of predation mortality are shown for each of six key fish species in Fig. 8.9. These

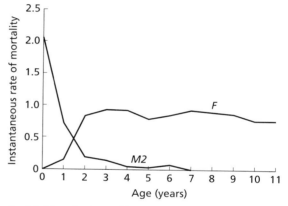

Fig. 8.8 Mortality due to fishing (F) and key predators ($M2$) on Atlantic cod in the North Sea. Data are averaged from 1987 to 1994. After ICES (1997c).

are for the 1991 Year of the Stomach survey; the overall patterns were similar in the analysis a decade earlier. Because predation is so high on young fish, each prey species has two panels, with the one on the left showing species in the 0 age group, i.e. less than 1 year old, and the right panel showing older age classes combined. Each bar within a panel is for a particular predator, as well as predatory seals, birds and 'other' predators, lumped together. This shows that predation on the 0-group cod is primarily due to grey gurnards, whiting and other cod. Cod return the favour to older age classes of whiting, while whiting turn out to be their own worst enemy when it comes to the youngest age class. Although we will see a great deal about seabirds eating sandeels in section 15.2.1, note that sandeels

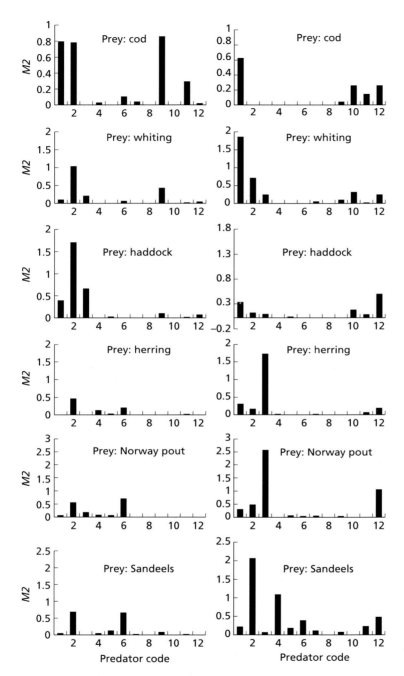

Fig. 8.9 Mortality due to key predators (*M2*) acting on six fish species in the North Sea. Panels on the left are for the 0 age group (i.e. less than 1 year old), and those on the right are for older age classes summed together. Predator codes: 1 = cod, *Gadus morhua*; 2 = whiting, *Merlangius merlangus*; 3 = saithe, *Pollachius virens* (Gadidae); 4 = North Sea mackerel, *Scomber scombrus* (Scombridae); 5 = haddock, *Melanogrammus aeglefinus*; 6 = western mackerel, *Scomber scombrus* (different stock of the same species as 4); 7 = starry ray, *Raja radiata* (Rajiidae); 8 = horse mackerel, *Trachurus trachurus* (Carangidae); 9 = grey gurnard, *Eutrigla gurnadus* (Triglidae); 10 = seals; 11 = birds; 12 = other. After ICES (1997c).

have far more to fear from whiting, mackerel and western mackerel.

This MSVPA is summarized for cod in Fig. 8.10. The first four panels show a standard description of the fishery, including a steep decline in the total stock size through the 1980s as well as a decline in the spawning stock biomass (a). These patterns are matched by recruitment (b) and changes in yields to the fishery (c).

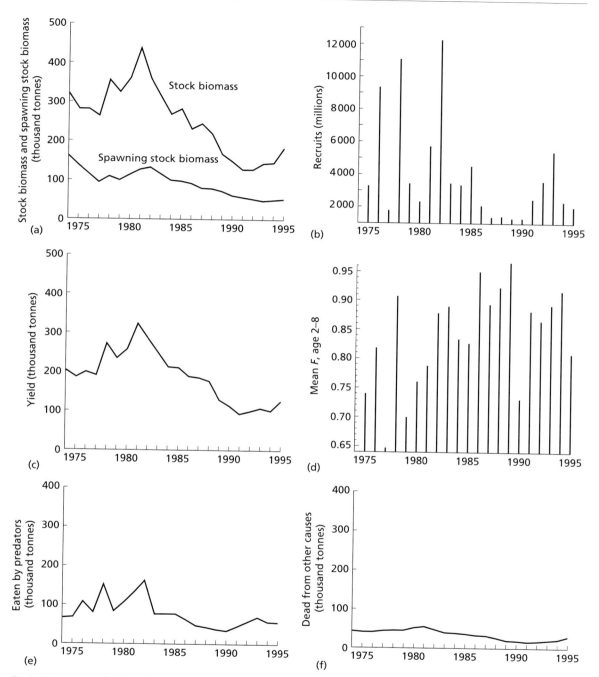

Fig. 8.10 Summary of MSVPA from 1974 to 1995 for Atlantic cod in the North Sea. (a) Stock biomass and spawning stock biomass; (b) recruitment; (c) yields to the fishery; (d) mean fishing mortality of fishes aged between 2 and 8 years; (e) biomass of cod eaten by those predators other than humans that were included in the analysis; (f) biomass of cod that were killed by predators that were not studied. After ICES (1997c).

They show less relationship to fishing mortality (d). The contribution of the MSVPA is shown in the bottom two panels. Predators eat roughly half as much cod as humans (e), while other animals, principally species that were not assessed in the 'Years of the Stomach', kill about half as many cod as the species that were assessed individually (f).

The North Sea study can be used to test a key assumption of MSVPA models, namely that suitability of individual species as prey remains constant according to their biomass as a proportion of potential food items (section 8.4). This study showed that although there was some evidence of predator switching, the similarities between the 1981 and 1991 samples were far stronger than their differences. This was also suggested by an earlier study of four years of stomach data, showing that predator preferences and prey vulnerability were fairly stable (Rice *et al.*, 1991). This stability is particularly true when M2s are aggregated among predators of a given prey species, which is important because prey populations will be affected by total mortality, not according to how mortality is distributed among predators. Indications of stability are a considerable source of relief, because otherwise forecasting would be impossible.

We have heard a lot about the North Sea in relation to MSVPA because the development of this technique arose from a concerted effort by European scientists (Daan & Sissenwine, 1991). But how typical is this region in terms of the majority of fish biomass going to other fish? A survey of five other regions of the world found the same result (Fig. 8.11). Note that humans play a larger role in the North Sea than in other areas, and that there is strong variation in the importance of other taxa. Clearly, we are not the only predators worth considering.

What is the future for MSVPA? After exploring this new technique with many alternative runs of the model, including changes in parameters, assumptions, and predators, the Multispecies Assessment Working Group concluded in their 235-page report 'Overall, our confidence in MSVPA continues to grow (as does our exhaustion with it)' (ICES, 1997c). Yet MSVPA has still not travelled beyond Europe very often, and even in its original North Sea home it has not replaced traditional single-species VPA, although it has clearly helped our understanding of the ecology of the North Sea, and improved our estimates of natural mortality (see below). The insatiable appetite of this technique for data is a major hindrance to its implementation, and until length-based versions are improved, we can forget about using

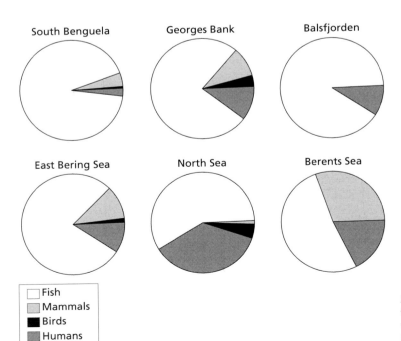

Fish
Mammals
Birds
Humans

Fig. 8.11 Relative loss of fish biomass to fish, mammals, birds and humans, in various marine ecosystems. After Bax (1991).

it in the tropics (Christensen, 1996). Note that MSVPA ignores the impacts of food supplies on growth of the predators. Thus, predation is a one-way street, with only impacts on prey being recognized. Furthermore, multi-species spawner–recruit relationships are not included. Despite these limitations, the findings so far have been quite promising for regions that can afford the surveys. For example, a user-friendly computer package is available for MSVPA in the Baltic, the so-called 4M package (Vinther *et al.*, 1998; ICES, 1999b). This includes tuning options, and is used regularly to update estimates of natural mortality. Indeed, it is expected that the approach will probably be used routinely for assessments of cod, herring and sprat *Sprattus sprattus* (Clupeidae).

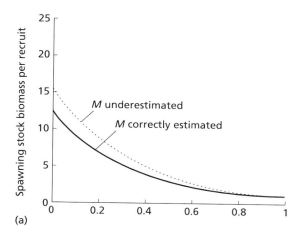

(a)

8.4.2 Applying MSVPA data to single-species models

The information gained from MSVPA, particularly concerning natural mortality and stock structure for young age classes, can be used to improve our estimates of fishery yields in single-species techniques. In the previous chapter, we saw the importance of shifts between natural and fishing mortality in yield per recruit predictions for the Atlantic croaker (Fig. 7.13). MSVPAs generally indicate that natural mortality rates on juveniles are much higher than traditionally assumed for such analyses. This means that the single-species approach must have underestimated the numbers of juvenile fish that were present initially and sustaining this mortality. Therefore, spawning stock biomass per recruit must have been overestimated, and yield per recruit must also have been overestimated (Fig. 8.12). These effects of underestimating *M* are smaller at very high levels of fishing mortality.

8.5 Predators, prey and competitors

So far, we have used multispecies interactions to improve estimates of natural mortality and stock structure, but we have not captured anything of the dynamics between species in relation to exploitation of one or more components of the ecosystem. This is where it gets really interesting.

8.5.1 Predator–prey dynamics

Suppose you have a predator–prey relationship between

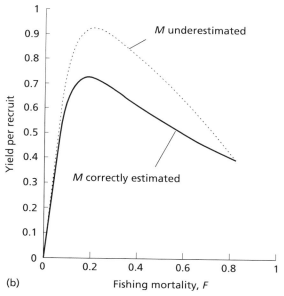

(b)

Fig. 8.12 The effects of underestimating natural mortality, *M*, as was typical of single-species assessments, compared to the true (higher) values revealed by multispecies assessments. Mistakenly low estimates of *M* mean recruitment must also have been underestimated. Therefore spawning stock biomass per recruit was overestimated (a), and so was yield per recruit (b). After ICES (1997c).

two species, both of which are the targets of fisheries. As an example, we might consider the fisheries for krill, *Euphausia superba* and baleen whales in the Southern Ocean, prior to the implementation of a whaling moratorium. One can plot the yield of one of these species, such as krill, as a function of fishing mortality (Fig. 8.13a).

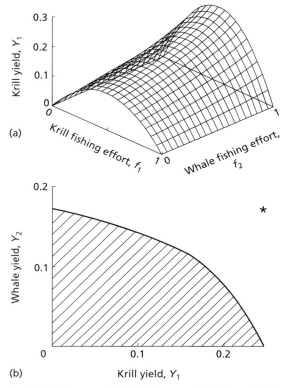

(a)

(b)

Fig. 8.13 Yield curves for (a) krill alone, and (b) krill combined with baleen whales. Yield is presented in arbitary units. After May *et al.* (1979).

This example, from May *et al.* (1979), scales fishing effort from 0 to 1, where a value of 1 corresponds to extinction of the stock. This is based on equilibrium yields under logistic models of population growth (section 7.3.2). Note that in this model of the predator–prey system, the yield of krill is maximized at 50% effort regardless of whaling effort, and that this yield is highest when whales are driven to extinction. One can also examine the yield of each species when plotted against the yield of the other (Fig. 8.13b). This shows that reductions in the krill fishery will increase yields of whales, and vice versa, with the shaded area under the curve indicating the zone where yields are sustainable. Whereas the whale yield is maximized when krill fishing stops, the krill yield is maximized when whales are exterminated. The asterix in the upper right corner shows the point at which both krill and whale yields would be maximized at their individual values. This is well outside the zone of sustainability. This illustrates

the general point that maximum yields of single species are incompatible with predator–prey systems.

Analyses such as these led to the 'krill surplus hypothesis' (section 15.3.2). This suggested that the depletion of whales by hunting would lead to the proliferation of krill, and the krill 'surplus' would provide more food for other species such as seals and penguins. In fact, the empirical evidence for such effects is weak (section 15.3.2) and May *et al.*'s (1979) prediction that models incorporating spatial structure would be more realistic has proved to be correct.

8.5.2 Competition, an unexpected result

Consider the schematic food web shown in Fig. 8.14 (May *et al.*, 1979). Predator species A eats both prey

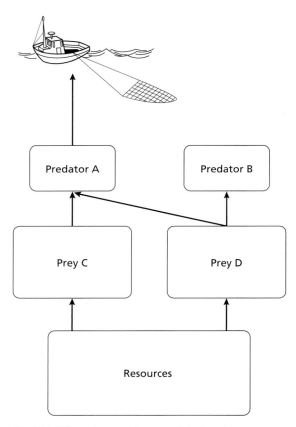

Fig. 8.14 Effects of competition on exploitation. Arrows indicate predator–prey relationships. Whereas predator A eats both C and D, predator B specializes on D. After May *et al.* (1979).

species C and D, whereas predator B eats only D. The two prey species feed on a common resource such as zooplankton. Because of the link to the common resource at the base of the food web, one might expect that if predator A were reduced in number by the fishery, this would be good for predator B. But it is unlikely that the two prey species will ever be exactly equal competitors. Suppose prey C is competitively superior to D, and its numbers are kept in check by predator A. Then if predator A becomes exploited heavily, C will increase and thereby deplete the resources needed to sustain D. As the population of D falls, so will predator B. Thus, exploitation of A may cause B to crash—a result opposite to the expectation if one ignores ecological interactions among their prey. In Chapter 15, we look in more detail at possible relationships between predators and their prey, in the context of interactions between marine mammals and fisheries.

One might expect that the chances of this sort of surprise arising would be diminished if there are many links in the food web, such that many other predators keep populations of non-target species in check. Indeed, as we see in Chapter 12, this is often the case in marine ecosystems which show less dramatic responses to changes in predator abundance than many freshwater lake and terrestrial ecosystems. But in those marine ecosystems where one or two species dominate the total biomass, such as the Barents Sea (section 12.4.4), it would be dangerous to take much comfort from this comparison.

8.5.3 Management implications

By this point we should be prepared for the dynamics between predators, prey and competitors to have some serious surprises in store for fishery management. One of the commonest ways of reducing fishing mortality is to increase the mesh sizes of nets (section 17.2.4). This allows smaller individuals to escape. Figure 8.15(a) shows a typical forecast of long-term changes in yields due to increased mesh sizes for fleets taking several species in the North Sea. Larger meshes are predicted to increase yields of the four large demersal (bottom-dwelling) species: Atlantic cod, whiting, saithe and haddock. But Fig. 8.15(b) shows that the multispecies predictions are just the opposite for three of these species! This is not because natural mortality rates are higher with the multispecies estimates; the single-

species estimates were calculated with the same mean natural mortalities as in the multispecies case. The key is in the predator–prey relationships, whereby larger mesh sizes may cause a decrease in yields of some species because predators escape the nets and prey on juvenile size classes (see also section 15.3.1).

8.6 Size spectra

Many of the biologists who were developing MSVPA in the 1980s also took a step back to look at the larger picture of whether patterns of exploitation could be deduced from large-scale changes in ecosystem composition. This began with a return to the basics of ecology. A fundamental feature of ecological food webs is conservation of mass through energy conversions through production, respiration, growth and predation (Chapter 2). These processes are all known to be functions of body size. Thus, aquatic trophic pyramids look like Fig. 8.16a, with the largest species tending to occur at the top. If we turn this pyramid on its side, we obtain the classical plot of numbers or biomass vs. weight (Fig. 8.16b). Under the law of biomass conservation, if we perturb the system by removing individuals of a particular size class, this should be reflected by a change in the slope of the relationship between abundance and body size. Suppose we reduce the numbers of fish at the top of the pyramid. This will give the pyramid a sharper peak, and hence a steeper slope in Fig. 8.16b. **Size spectra** therefore offer a way of looking at changes in the structure of exploited ecosystems, through comparisons of slopes of abundance vs. body size (section 12.4.3).

An example of this approach comes from a comparison of demersal fisheries in the North Sea with those on Georges Bank (Pope et al., 1988). In this case, the spectra were based on the sizes of fish caught in trawl surveys rather than the size composition of all individuals in the ecosystem. The result is shown in Fig. 8.17. The steeper lines for the North Sea suggested heavier exploitation. This can be seen more clearly by plotting the slopes for each area against time (Fig. 8.18). The slopes are more variable over time for Georges Bank. This variation matched changes in management, with steep slopes (more negative numbers) in the early years when there was strong and increasing international exploitation, leading to few fish surviving to large sizes. In 1970, catch quotas led to slightly shallower slopes,

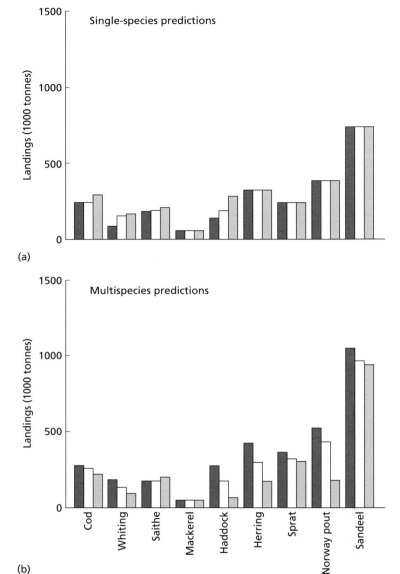

(a)

(b)

Fig. 8.15 Comparison of long-term forecasts of yield changes in the North Sea for (a) single-species predictions, and (b) multispecies predictions under different mesh sizes of 70 mm (dark shaded), 85 mm (unshaded) and 120 mm (light shaded). After Pope (1991).

although the pattern was erratic. A coastal state management scheme in 1977 led to shallower slopes than in previous years. Size spectra are increasingly employed to describe changes in fish community structure due to exploitation. We develop this topic in section 12.4.3.

Size spectra provide a nice integration between fisheries and community ecology, but what can we actually do with them to improve our management? A recent study by Duplisea and Bravington (1999) used models to explore the implications of various potential size-based exploitation regimes for maximizing yields, as well as testing for impacts on the community as revealed by stability and persistence of size spectra. One might expect that the best policy for maximizing yields would be to exploit lower trophic levels most heavily, so that biomass is not lost through trophic conversions as it goes up the food chain (section 2.5.2). In fact, the preliminary conclusion was that larger fishes

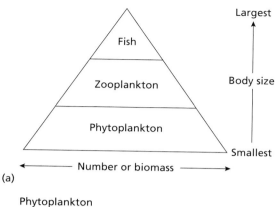

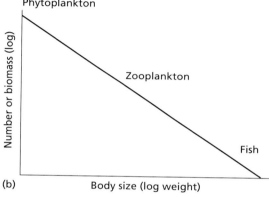

Fig. 8.16 (a) Trophic pyramid, where width is proportional to the number or biomass of organisms, and height is proportional to body size of individuals. (b) The same information displayed as a plot of number or biomass against body size.

should be taken preferentially, since this would reduce the predation on smaller fishes and lead to increases in their production. The results held for optimal yields in terms of both biomass and economics, although the economic benefits of targeting large species may be reduced if the costs of exploitation are taken into account. More work remains to be done to fill in the details of how size spectra may be used for management of fisheries and ecosystems.

8.7 Ecosystem models

Another way to tackle biological interactions in multi-species fisheries is to return to the basics of ecosystem functioning, by quantifying the flows of energy or biomass in the food web. If you understand what eats what, then you can see what happens when one com-

ponent such as humans increases its appetite. MSVPA is a good start in this direction, but we saw that it demands a massive sampling effort to deduce predation mortalities, and it inevitably leaves out many important species, particularly at the lower trophic levels. Is there a middle ground between the ecological principles that yield crude patterns of size spectra, and the sea of stomachs that feed into MSVPA? By returning to first principles of ecosystem biology, it should be possible to understand links between exploitation and other components of the ecosystem in a much more holistic way than afforded by either of these alternatives. There is considerable demand for this, given current concerns about the ecosystem effects of fishing (Chapters 1 and 17).

A major initiative towards incorporating trophic interactions into fisheries biology has culminated in a software package called '**Ecopath** with **Ecosim**'. The development of these methods followed from the work of Polovina (1984), who produced an ecosystem model for a coral reef in Hawaii. The current model is a joint production by scientists at the International Center for Living Aquatic Resources Management (ICLARM) in the Philippines, and at the Fisheries Centre at the University of British Columbia in Canada (Christensen & Pauly, 1992, 1993; Walters *et al.*, 1997; Pauly *et al.*, 2000). The software can be downloaded free of charge from www.ecopath.org. Ecopath describes ecosystems by balancing flows between trophic groups. Ecosim is a dynamic simulation model that uses this information to predict how changes in fisheries or climatic events will alter ecosystems.

The principle assumption underlying Ecopath is that if ecosystems are in equilibrium, then trophic flows will be balanced. For each of the living groups (phytoplankton, zooplankton, fish, etc.) in the system, this implies that input = output such that:

$$B_i \left[\frac{P}{B} \right]_i EE_i = Y_i + \sum_{j=1}^{n} B_j \left[\frac{Q}{B} \right]_j DC_{ji} \qquad (8.1)$$

In the first term, B_i is the biomass of i during the relevant time period, $(P/B)_i$ is the production/biomass ratio of i, and EE_i is the ecotrophic efficiency of i (the fraction of production consumed within, or caught from, the system). The second term, Y_i, is the yield in biomass of i to the fishery (where relevant). The last term is a summation of the effect of all predators, j, on i. B_j is the

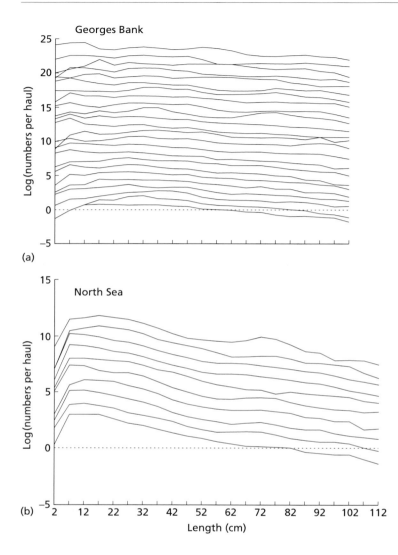

Fig. 8.17 Size spectra for (a) Georges Bank (1963–86), and (b) the North Sea (1977–86). The bottom lines refer to the earliest years, and each line above is a successive year with a value of 1 added to the *y*-axis. After Pope *et al.* (1988).

biomass of predator j, Q/B_j is the food consumption per unit biomass for predator j, and DC_{ij} is the fraction of i in the diet of j.

The whole ecosystem is modelled with a set of simultaneous linear equations. Each equation (as equation 8.1) refers to one group i in the ecosystem. Once the equations are solved (Pauly *et al.*, 2000), a network of quantified flows can be constructed from biomass, production ($P = B(P/B)$) and consumption ($Q = B(Q/B)$) estimates for each group. The data needed for Ecopath modelling are usually taken from published studies and new research, and the routines allow many missing parameters to be estimated. At the time of writing, over

100 models have been produced for marine and freshwater ecosystems. Many are reviewed by Christensen & Pauly (1993).

8.7.1 Ecosystem models in action

An example of an Ecopath analysis for the central South China Sea pelagic ecosystem is shown in Fig. 8.19 (Pauly & Christensen, 1993; Walters *et al.*, 1997). These diagrams have a number of interesting uses, such as suggesting the impacts of predation by fish and marine mammals relative to fishing. Estimates of fractional trophic level for different groups are an output

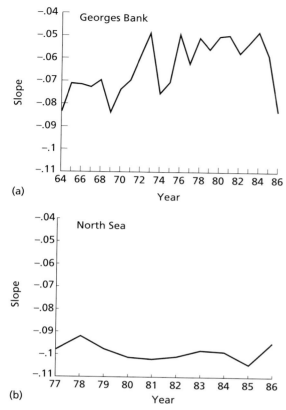

(a)

(b)

Fig. 8.18 Slopes from the size spectra relationships between numbers and length of fish in Fig. 8.17 plotted against time for (a) Georges Bank, and (b) the North Sea. Steeper slopes are indicated by more negative numbers. After Pope *et al.* (1988).

of Ecopath models (these are shown on the *y*-axis in Fig. 8.19). These estimates can be used to assign trophic levels to species caught in fisheries, and to examine long-term changes in the trophic structure of global catches (section 12.4.2; Pauly *et al.*, 1998). Estimates of trophic levels, transfer efficiencies and landings were also used to estimate the proportion of primary production required to sustain existing global fisheries (section 2.5.4; Pauly & Christensen, 1995).

While Ecopath provided a static description of ecosystem structure, it did not help managers who wanted to predict the effects of different fishery management actions, such as targeting forage fish, on the structure and functioning of marine ecosystems. Ecosim was developed to address this problem. In Ecosim, the linear equations that describe fluxes (equation 8.1) have been replaced with coupled differential equations that can be used for dynamic simulation and the analysis of changing equilibria. It also uses a functional relationship to predict changes in the production to biomass ratio of primary producers according to their biomass, and replaces static consumption, Q, with relationships that allow consumption to change in response to the biomass of predators and prey.

Preliminary trials with this new approach have provided some fascinating glimpses of the potential ecosystem impacts of fishing. For example, Fig. 8.20 shows a 30-year prediction of a simulated increase in fishing mortality for sharks in the Gulf of Mexico, followed by a return to low levels (Walters *et al.*, 1997). Figure 8.20(a–f) show the biomass of the food web in descending order from top predators to primary producers. The simulation assumes tight 'top-down' control in this ecosystem (Chapter 12). The predictions of large oscillations in tuna, billfish and demersal fish are not meant to be taken too literally at this preliminary stage of the development, but visual inspection of the original Ecopath flow diagram that underlies this simulation (Browder, 1993) gives little hint that the impacts would be so large. The impacts are predicted to be smaller if the ecosystem is characterized by greater 'bottom-up' control.

Ecosim predictions can be sensitive to violations of the equilibrium assumption, although temporary changes on a seasonal cycle are not a problem if the time scale is lengthened accordingly. Another problem is that Ecosim tends to predict overly high fishing mortalities for maximum yields to fisheries involving mammals and sharks, because these animals cannot translate higher food availability into increased fecundity as easily as other animals. Future developments of the software may allow users to limit the efficiency of such animals to convert high food levels into biomass, or set upper limits to their recruitment (Walters *et al.*, 1997).

Ecopath and Ecosim did not account for the spatial distribution of predators and prey in the marine ecosystem, and yet the degree of spatial overlap between predator and prey populations has a critical effect on consumption rates. Ecospace is a relatively new model, that incorporates spatial dynamics. Users can specify transfer rates between specified areas, changes in predation risk and feeding rate in different habitats and spatial patterns in fishing effort. Ecospace may help to evaluate the effects of area closures (no-take zones, section 17.5.3) on the marine ecosystem.

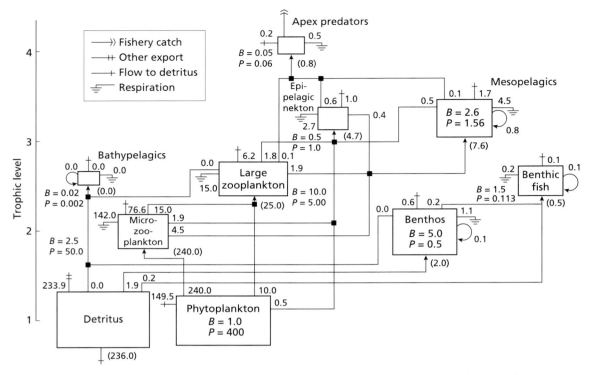

Fig. 8.19 Flow diagram of the central South China Sea pelagic ecosystem in the 1980s. Arrows indicate the flow of production (P; t km^{-2} y^{-1}), and the size of boxes is approximately proportional to log-transformed biomass (B; t km^{-2}). Thus, production moves up the trophic levels, and is lost to respiration, detritus, other exports, and fisheries for the apex predators. After Walters *et al.* (1997).

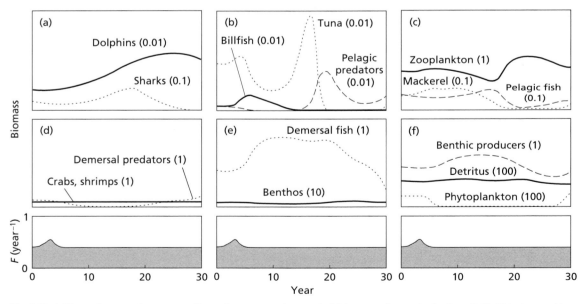

Fig. 8.20 A 30-year forecast of ecosystem effects of a temporary increase in fishing mortality on sharks, for a Gulf of Mexico trophic model. The bottom panels show the blip in fishing mortality, while a–f show simulated changes in biomass as one moves down the food web. The biomasses of all groups have been altered to fit onto the same relative scale. The true biomasses can be found by multiplying them by the values in parentheses. This simulation assumes strong 'top-down' predation effects, and is intended to show one *potential* outcome from a preliminary analysis. After Walters *et al.* (1997), based on data in Browder (1993).

Summary

• There are good reasons to switch from single-species to multispecies assessments, because the former generally ignore the fact that we rarely catch only one species at a time (technical interactions), as well as trophic interactions among species (biological interactions).

• For technical interactions, the simplest approach is to aggregate catches from all species and use surplus production models, but species do not share common dynamics, and parameter fitting is risky if equilibrium assumptions are made.

• Multispecies yield-per-recruit analyses are more sophisticated, but require additional information including age structure.

• Multispecies virtual population analysis is a logical extension of single-species VPA, incorporating predator–prey relationships into estimates of natural mortality. Although the method shows promise, it requires a massive amount of information about predator behaviour.

• Examination of size spectra can provide a broader ecosystem perspective on the effects of fishing.

• Ecosystem models such as Ecopath and Ecosim are nicely grounded in the study of community ecology, and although early results of simulations are fascinating, it is too early to say how useful they will be for fisheries assessment and management.

Further reading

Early collections of papers on multispecies methods were edited by Mercer (1982) and Daan and Sissenwine (1991). Multispecies virtual population analysis was reviewed more recently by Magnússon (1995). A key ICES Working Group report (1997c) provides compre-hensive exploration of North Sea multispecies assessments, and this has been updated by ICES (1999b). A series of technical papers on models for the Barents Sea has been edited by Rødseth (1998). Ecopath and related software are reviewed by Pauly *et al.* (2000) and are available at www.ecopath.org/.

9 Getting the data: stock identity and dynamics

9.1 Introduction

Stocks are treated as the basic management units in fisheries because birth, growth and death within the stock have greater effects on dynamics than immigration or emigration. In this chapter we describe methods for identifying stocks and estimating parameters that describe their dynamics. **Parameters** are quantitative descriptors of the stock and are assumed to remain constant from the time when they were estimated until the time when they are used for prediction.

The models used to describe how fished species, fisheries or ecosystems behave are simplified abstractions (Chapters 7 and 8). Model parameters are key inputs and describe average properties across many individuals. If the parameter estimates are inaccurate then model outputs will be wrong. The largest bias in parameter estimation often results from unrepresentative sampling of the stock. Representative sampling is difficult because sampling methods are highly selective and because stocks migrate, change distribution with abundance and form size-segregated groups (section 4.3).

The methods used to estimate parameters have developed rapidly in recent years. Many methods that were developed in the middle of the last (!) century have been superseded or modified, and most assessment scientists now pay more attention to error and uncertainty in data and model specification. Most fisheries laboratories use in-house or other sophisticated software for analytical work. The details of these routines are beyond the scope of this book, but Hilborn and Walters (1992) and Quinn and Deriso (1999) give excellent overviews. The aim of this chapter is to demonstrate how stocks can be identified and to describe some basic methods of parameter estimation that serve as an introduction to more advanced methods.

9.2 Stock identification

9.2.1 The stock concept

Fished species rarely reproduce randomly with conspecifics throughout their geographical range, but form a series of **stocks** that are reproductively isolated in space or time. As a result, stocks respond more or less independently to the effects of fishing, and their own birth and death rates have a more consequential effect on their dynamics than those of adjacent stocks. The assessment models described in Chapter 7 were based on estimates of birth, growth and death in closed populations. They did not account for emigration and immigration. Without knowledge of stock structure they cannot be parameterized or applied correctly.

Whether stocks differ genetically depends on their isolation, and a lack of genetic differentiation does not imply that the stock is not a useful management unit. For example, a transfer rate of 5% of individuals each year between two stocks may lead to genetic homogeneity, but for practical management purposes the stocks' population dynamics will be largely independent. The differences between fishery managers with their short-term perspective and evolutionary ecologists with their longer-term perspective have led to a range of stock definitions applicable to different situations (Gauldie, 1991; Carvalho & Hauser, 1994). The stock of interest to a fishery manager with a short-term perspective is sometimes referred to as a **harvest stock**.

9.2.2 Methods for stock identification

Approaches to stock identification range from simple studies of distribution and abundance to those based on genetics and life histories. In this section we describe methods used to identify stocks and their relative merits

and disadvantages. An ideal method of stock identification would provide an indication of the spatial and
temporal integrity of a stock and allow individuals of a
fished species to be assigned to a given stock. This is
because individuals from several stocks may mix in the
same area at the same stage of their life history, and
parameter estimates for each stock will be biased unless
samples can be divided before they are counted, aged or
measured. In practice, such division is usually impossible, and sampling must be conducted with great care
to ensure that only one stock is present in the sampling area at the time of sampling. Methods of stock
identification are reviewed by Ihssen *et al.* (1981) and
Pawson and Jennings (1996).

Distribution and abundance
Methods such as visual census, acoustic, long-line,
pot and trawl surveys provide descriptions of the distribution and abundance of fished species (Chapter 10).
From commercial fisheries, additional information
comes from seasonal and spatial records of catches.
Different spawning grounds, such as the distinct areas
used by herring, *Clupea harengus* in northern Europe
(Fig. 9.1), may be used by different stocks and thus eggs

or larvae can be followed in space and time to link
spawning and nursery areas. Stock identification is
rarely the main aim of distribution and abundance
studies, and many surveys are too infrequent or too
geographically imprecise to provide good evidence of
stock separation. They also provide no help with identifying individuals in areas where stocks mix.

Natural marks and tags
Fished species may encounter parasites or chemicals in
specific geographical regions, and when they move they
carry the evidence of their former habitat with them.
Natural tags offer a number of advantages when
compared with artificial tags and marks. They identify
a greater proportion of the population than when
individuals have to be captured, and may be found on
species that are too delicate to tag artificially. Parasites
can provide preliminary information that helps with
the design of more expensive artificial tagging studies
and may allow fish to be assigned to different stocks in
areas where they mix (MacKenzie, 1982).

The main problem with parasite studies concerns
data interpretation, because the taxonomy and life histories of many parasites are poorly understood. The
parasite's prevalence, longevity and capacity to infect
fished species at various stages of their life history must
be known. Changes in the abundance of parasites may
prevent the successful planning and implementation of
long-term studies. The characteristics of the best parasite tags are listed in Table 9.1.

Species which spend part of their life history in
chemically contaminated water may absorb or ingest
these chemicals. These may remain in body tissues and,
when the fish is caught in later life, indicate its origins.
For example, American eels *Anguilla rostrata* (Anguillidae)
in the St Lawrence River, Canada, picked up heavy
metals, Mirex pesticide and various organochlorines
that allowed them to be identified in catches elsewhere

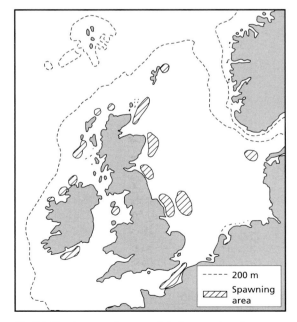

Fig. 9.1 Herring spawning grounds in the seas of northern
Europe. After Cushing (1988a).

Table 9.1 Criteria for selecting parasitic tags.
From Lester (1990).

1. Prevalence of parasite should be stable from year to year
2. Parasite should be easy to detect and identify
3. Parasite should not have a debilitating effect on the host
4. Parasite should have a direct life cycle (not be transported via
intermediate host)
5. Parasite should not affect the behaviour of study species

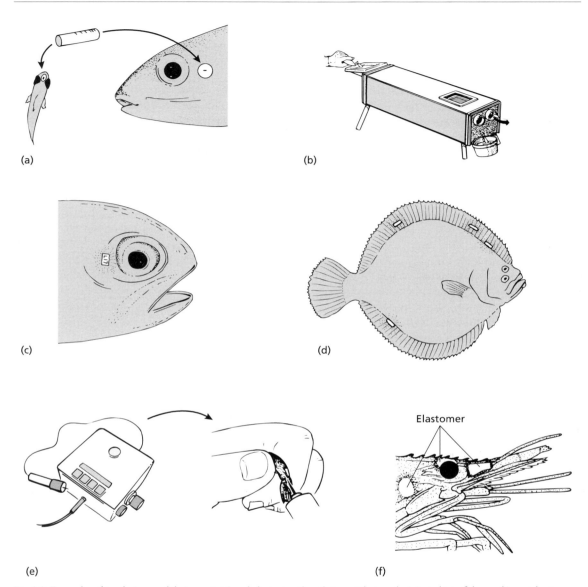

Fig. 9.2 Examples of tag designs and their use. (a) A coded wire tag (length 1 mm) that can be injected into fishes and invertebrates, and is detected by passing recaptures through a detector (b); (c) a visible implant tag injected close to the eye and viewed under ultraviolet light; (d) visible implant elastomer injected into a juvenile flatfish; (e) injecting the tail of a shrimp with an elastomer through a fine needle, and (f) close-up of a shrimp after marking the base of the rostrum and side of the head with an elastomer. After North-west Marine Technology International Ltd, Shaw Island, USA.

(Moreau & Barbeau, 1982; Dutil *et al.*, 1985; Caston-guay *et al.*, 1989). If chemical contamination is to be used to track stocks, then defined chemical sources must exist in areas they use. Thankfully, few such sources exist in the open sea, although levels of radioac-tive contamination in fishes caught around the British Isles have been used to identify their origins (Pawson & Jennings, 1996). Naturally occurring chemicals or trace elements may also be absorbed, particularly in the otoliths and bones of fishes (see below).

Artificial marks and tags

Most information on stock identification and migration has been derived from artificial tagging. Tagging also provides the fisheries scientist with information on growth rates and behaviour. Tagging is rather different from the ringing of birds or marking of terrestrial mammals since individuals are usually killed at recapture. As such, patterns of migration are inferred from the tag and recapture locations of many individuals rather than from sightings of the same individuals at many sites on many occasions. The main exceptions are tagged reef fishes that can be observed underwater, crabs and lobsters that are caught and returned when undersized, species tagged with tags that store environmental data that can be used to reconstruct tracks, and large or rare species of great conservation value, such as the common skate *Raja batis*, which may be returned alive to the water after capture.

Because marking and tagging studies often have multiple aims we will describe them in general terms and not solely in relation to stock identification. The simplest methods of **marking** do not identify individuals, and 'batches' are marked with similar fin clips, shell notches or dyes. However, modern **tagging** methods allow the rapid tagging of individuals with unique tags and so batch marking methods are used less frequently. The appropriate tag depends on the aims of the study, the size and species to be tagged, the expected time between release and recapture and whether the tag is likely to be recovered by fishers or fishery scientists. Some commonly used tags are illustrated in Figs 9.2 and 9.3. These include serially numbered or coded tags attached to external parts, coded wire microtags which are injected into tissues and accompanied by a general identification mark such as a fin clip, simple internal unnumbered tags which may be recovered by magnets in fishmeal plants and dye, latex or chemical marks (Parker *et al.*, 1990; Nielsen, 1992).

In recent years, **data storage tags** (DST) have been used to collect information on the depth, light level or temperature that a tagged individual experiences (Fig. 9.4). These computerized tags record data at intervals of minutes to hours and can usually store several thousand readings. When the animal is recaptured, the tag can be removed and the data downloaded. The information on physical parameters may allow movements to be reconstructed, since day length will vary with latitude, depth with location, and temperature

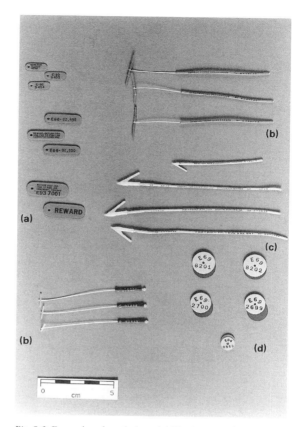

Fig. 9.3 Examples of tag designs. (a) Howitt's tags that are attached with thread beside the dorsal fin on round fish; (b) Floy or anchor tags that are anchored to the fish (usually just beside the dorsal fin) when the 'T' bar is injected; (c) dart tags that are anchored to the fish by the dart; and (d) Petersen discs that are used to tag flatfishes by passing a wire through the body and securing a disc on each side. Photographs S. Davies.

with season and location. In addition, depth records allow the examination of movements on a smaller scale, such as the vertical migrations of fishes using selective tidal stream transport (section 3.5.2). As electronic technology evolves, DSTs have become smaller and carry more information. Some of the latest designs can be programmed to be released from the fish automatically after a set time period. They float to the surface and download their location and stored data to a satellite. This technology is only suited to larger fish such as tunas at present, but these tags are likely to be miniaturized in the the future.

Capture and tagging cause some trauma, and the

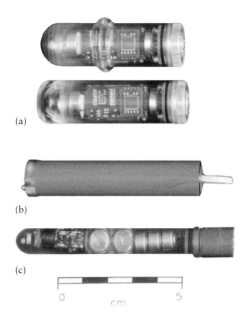

(a)

(b)

(c)

0   cm   5

Fig. 9.4 Examples of data storage tags and transponders.
(a) CEFAS/LOTEK data storage tags; (b) VEMCO temperature
and depth recording tag; (c) 300 kHz depth sensing telemetry
transponder. Photographs S. Davies.

Fig. 9.5 Posters encouraging fishers to return tags. Photograph
S. Davies.

condition of tagged individuals at the time of release
determines the probability of the tag being returned,
as does the size and colour of the tag and the method
of capture and handling. Commercial fishers may not
notice tags or not be keen to return them, so many reward
and publicity schemes have been devised (Fig. 9.5).
Because the cost of capturing, releasing and tagging
may be very high, increasingly large rewards have been
used to encourage returns. In the Atlantic Ocean, fishers
are offered a reward of $US1000 for each tag they
return from bluefin tuna *Thunnus thynnus* (also known
as tunny). In a plaice *Pleuronectes platessa* tagging programme in the North Sea where valuable data storage
tags are returned, all fishers who return a tag are
entered into a draw for a prize of $US1600. If scientists
need the fish as well, they usually pay full market value
as well as the tag reward.

Tagged individuals are usually measured and
weighed at the time of release, so if recaptured and
remeasured, their growth rate may be calculated. One
problem though, is that tagged and untagged fish may
have different growth rates. Tagging can also be used to
estimate abundance and mortality rates (section 10.2.6)

and to identify individuals used in age validation studies (section 9.3.2). There are several methods that allow
stock structure and migration rates to be determined
from tagging data. These are reviewed by Seber (1982)
and Quinn and Deriso (1999).

Crustaceans such as shrimps, crabs or lobsters are
difficult to tag because they lose their exoskeleton
during moulting and any tag attached externally will
be lost. Shrimps may be injected with elastomers that
contain fluorescent pigments, often at the base of the
rostrum or side of the head, and these marks are
retained during moulting (see Fig. 9.2). These elastomers are also used for marking very small fish when
injected just beneath the skin. Coded wire microtags
are made of stainless steel wire, usually 1 mm in length
and 0.25 mm diameter. The wire is marked with a
unique code of notches and magnetized. The microtag

can be injected into body tissue beneath the exoskeleton, at sites such as the base of legs or claws. Tags are detected by passing recaptures through a magnetic detector. When animals with tags are identified, the tags are removed by dissection and the codes read under a binocular microscope (Bailey & Dufour, 1987).

Shellfish are often marked by making notches in the shell. These marks are incorporated as the shell grows and may be identifiable for many years. Notches provide time markers that can be used to validate the periodicity of growth ring formation. For shorter-term studies, shellfish are also marked using numbered tags glued to the shell.

Mersitics and morphometrics

Taxonomists have long been aware that counts of elements such as vertebrae, fin rays or scales vary among individuals. In many cases, such counts are used to identify species. There is also intraspecific variation in counts and they can be used to identify individuals from different stocks. For example, Newfoundland cod *Gadus morhua* stocks can be distinguished using vertebral counts (Templeman, 1981). There is considerable variation in meristic characters between individuals and, while mean values may differ between stocks, the method cannot be used to definitively assign an individual to a stock. Statistical stock discrimination tools, however, do allow us to estimate the probability that an individual comes from a given stock.

The shape of individuals can be described by ratios between the sizes of body parts. Examples are head depth and body length or, more recently, the outline of a fish's shape as decomposed into a series of linear measurements, ratios or other mathematical descriptors which are amenable to statistical analysis. When such morphometric characters are compared in many individuals, distinct patterns can emerge that indicate differences between stocks. As with meristic characters, the variation between individuals is large and morphometric measures are unlikely to help with assigning individuals to a given stock.

Despite a range of difficulties, meristic and morphometric characters are widely used by taxonomists and provide the basis for many initial stock separations. Morphometric and meristic differences may result from genetic or environmental factors. The relative roles of these factors are rarely known, but dominant effects of the environment may not be a problem in themselves since consistent environmental differences in, for example, the temperatures in nursery habitats used by two stocks, may lead to consistent differences in vertebral counts. Problems arise when the environment is changing continually and the role of genetic factors is not known. In these circumstances it would be better if genetic and meristic studies were conducted in parallel. The relative roles of such factors are nicely demonstrated by comparing the studies of Creech (1992) and Leslie and Grant (1990). Creech found that almost all the meristic variation between two European smelt *Atherina* (order Atheriniformes) stocks which had more recently been treated as a single species was due to genetic factors while Leslie and Grant showed that environmental factors determined differences among anglerfish *Lophius* (Lophiidae) stocks in southern Africa.

Calcified structures and scales

Characteristics of fish ear bones (otoliths, section 9.3.2) such as weight, dimension, shape, optical density and chemical composition may be stock or species specific. The mean values of such characteristics often differ when samples of fish from different stocks are compared, but the measurements rarely allow individuals to be assigned to specific stocks. The chemical composition of otoliths may also differ among stocks. Thus, migratory and non-migratory stocks of New Zealand common smelt *Retropinna retropinna* (order Osmeriformes) were identified using oxygen and carbon isotopic analyses of the otolith, because the calcium carbonate deposited in the otolith was in isotopic equilibrium with that in the different environments they inhabit (Nelson *et al.*, 1989).

Scales may differ between stocks in terms of circuli spacing, extent of damage or chemical composition. The shape and development of calcified structures may be affected by phenotypic or genotypic factors. Calcified structures are useful for stock identification because they are easy to collect from fish markets or during monitoring programmes. Scales can be collected without sacrificing the fish.

Genetics

Differences in morphology and life history within species have been noted for many years, and this led to the expectation that there would be genetic differences among stocks (Jamieson, 1974). Heritable discrete protein and nucleic acid variability provides a universal and frequently abundant array of markers for the examination of stock identity and genetic tools are increasingly used to investigate stock structure. The techniques used include allozyme electrophoresis, mitochondrial and nuclear DNA analysis. Genetic studies have provided particularly useful information on evolutionary relatedness and genetic variation for conservation purposes. If genetic differences exist between stocks then genetic approaches may allow individuals to be assigned to those stocks (Carvalho & Hauser, 1994).

When genetic analyses indicate that more than one reproductively isolated unit is present, the result helps to identify management units. However, problems arise when a genetic study fails to provide evidence of stock separation, since this does not prove that there are no components in the fishery which respond more or less independently to fishing pressure. Tagging studies have shown that some species form a number of stock units which have sufficient integrity to be managed independently and yet genetic differences cannot be detected. The transfer of a few individuals between these stocks may be all that is needed to maintain genetic homogeneity. Although genetic studies have provided important biological and evolutionary descriptions of population structure, they may not be valuable to stock managers working in the short term (Waples, 1998).

Life-history parameters
Individual traits such as growth rate or maximum size can be described parametrically (section 3.4.2), and provide a basis for discriminating stocks. Different stocks have characteristic life histories, although these are often phenotypic responses to the environment (section 3.4.2). Studies of differences in life-history parameters are most useful for identifying tentative stock units that can be confirmed using methods such as tagging. This is because data on life-history parameters such as growth rate or age at maturity are relatively easy and inexpensive to collect (Chapter 10), while tagging is expensive and time consuming.

9.3 Stock dynamics

9.3.1 Sampling

A proportion of a stock must be sampled to collect biological data. Calculated parameter estimates will be biased if the samples used are unrepresentative of the stock or if the results cannot be corrected to account for sampling bias. Sampling bias can occur when fishing or surveying the stock (Chapter 10) and when selecting individuals from the catch.

In collecting biological data there is sometimes a tendency to use a method because 'it has always been used' rather than because it is best. This applies, in particular, to sampling fish from commercial catches, when very little is known about the locations of fishing, the selectivity of the fishing gears and the sorting procedures used by the fishers (e.g. high grading and discarding, section 13.2). If a fisheries scientist is given a few boxes of fish from a market and told to do a stock assessment, the scientist's efforts are most unlikely to provide useful information.

Catch sampling also requires a basic training in the biology of the target species. The methods we describe for estimating age, growth or mortality require individuals of the same species from the same stock. This may seem obvious, but knowledge of species identification, for example, cannot be taken for granted. We have seen people measuring emperors (Lethrinidae) from two species with the intention of using the data for a single-species length-based stock assessment.

Catch sampling
Catches are usually too large to be examined in their entirety and must be sampled to provide individuals for study. There are several approaches to catch sampling based on the ease with which various types of biological data can be collected. Thus, it may be easy to measure all the fish in a catch, because this is a fast process, whereas it is unlikely that otoliths can be collected from every fish for age determination. As such, one might age and measure enough fish to estimate the length-frequency distribution within each age class and use this information to assign ages to fish that can only be measured (Doubleday & Rivard, 1983).

Methods of sampling catches for ageing include taking a fixed number of individuals in each length class or taking individuals in proportion to their abundance in each length class. The first approach can result in too much effort being allocated to the less abundant groups because individuals in the smallest and largest length classes will be scarce. The second approach may mean that unrealistically few individuals are aged in the smallest and largest size classes. Another option is a gradual stepped increase in the number of individuals aged in each length category to allow for increasing uncertainty about age at length with increasing size. For example, while fast-growing fish with a maximum size of 100 cm would all be expected to be age 1 in length class 10–15 cm, several ages may be present in the length class 70–75 cm. In practice, many fishery scientists have avoided methods that are too prescriptive when working at sea and tend to aim for a minimum level of sampling in all size categories with increased replication in larger and more abundant size classes (Pope, 1988; Cotter, 1998).

Age is more difficult to measure than length and hence **age–length keys** (ALK) are used to assign ages to large numbers of fish. These keys give the proportion of individuals, by age, in each length class. The length distribution of the whole catch is then used to estimate the age distribution of the whole catch by assigning ages to individuals in each length class according to their frequency distribution in the ALK (Table 9.2). Length is not a direct surrogate for age. Rather, the ALK gives the probability that an individual fish of given length has a given age. Since larger fish show a much wider range of body sizes for their age than smaller ones, this probability may be low for larger individuals.

The ALK may be specific to fishing stations, strata, regions or the entire survey area. Station-based ALKs will be unaffected by spatial variability in length at age across the region, but regional keys may be more precise because they aggregate otoliths from many catches. However, regional keys may yield biased estimates of numbers at age if length at age varies across the region. Clearly, length frequency distributions at the smallest sampling scale should be compared first and aggregated only if there appear to be no spatial differences in length at age.

When catches are sampled to determine mean age and size at maturity then levels of replication in age and size groups close to the estimated size at maturity should be increased, with a few larger and smaller individuals taken to confirm that they are all mature and immature, respectively. When catches are sampled for fecundity studies then fish will have to be taken from a range of size and age categories, but smaller immature individuals need not be sampled. When measuring diversity in fish communities, the whole catch must be sorted since any sampling will reduce the probability of encountering rare species. Once the whole catch has been sorted then sampling methods can be used to assess the age and size structure of individual species.

Gear selectivity

All gears used to sample fished species are selective to some degree. For example, the size of entrances to a pot or creel will determine the size of fish and crustacea that can enter, and the size of a hook will determine the lower and upper sizes of fish that can be hooked. We need to know about selectivity to understand biases in sampling programmes and to correct for their effects. For example, if the smallest size classes of fish caught in a trawl were aged to give an estimate of size at age in the stock, it is likely that size at age in the stock would be overestimated. This is because fish that were small for their age would be more likely to escape through the meshes of the trawl and not appear in the sample.

An understanding of selectivity is important for other reasons too. Mesh size can be used to control the size of catches and to change mortality rates for different size classes (section 17.2.4). To match the size of the mesh to the levels of escapement, the selectivity of the net must be known. In addition, knowledge of net selectivity may be used to correct length-frequency distributions of samples, by adjusting for differential catchability at size.

In towed and encircling nets, fish below a certain size range will pass through the mesh while all larger fish are caught. With meshing nets like gill or trammel nets, fish below a certain size will pass through without being caught but fish above a certain size will also escape as they are too big to be snagged. In some cases, however, the larger fish may be caught because they become tangled. Some fishers deliberately set the nets quite loosely to encourage this.

Mesh size is usually measured by forcing a tapered and calibrated guage into the mesh and seeing how far it will penetrate at a given pressure (Fig. 9.6). Enforcement agencies will measure a number of meshes to check consistency in mesh size. The orientation of the netting in relation to the direction of force on a towed gear has an important bearing on the shape of the holes in the net and hence the size of fish they retain. Diamond meshes tend to close when force is applied (Fig. 9.6b) while square meshes remain open (Fig. 9.6a, section 5.5). As such, selectivity is not based solely on mesh size and would have to be expressed for diamond and square mesh independently. Mesh selectivity of

Table 9.2 An age-length key (ALK) for the amberjack *Seriola dumerili* (Carangidae) giving the number of individuals by age in each length class (a) and the percentage of individuals in each length class at a given age (b). The lengths are the lower limits of each 5 cm length class. Data from Manooch and Potts (1997).

(a)

Age (years)	1	2	3	4	5	6	7	8	9	10	11	12	13	14	15	16	17
Length (cm)																	
50	1	1															
55		1															
60			1														
65	3		3														
70		1	1	2													
75			1														
80			2	1	1												
85			4	1	1												
90			2	6	1	2											
95				13	6	3	2										
100					10	3	5	6	2								
105					1	7		4	6								
110						4	4	5	5	2							
115						1	3	8		3	1	1					
120								4	3	8	3	2	1				
125								1	2	6	5	3					
130										1	4	6	3	1		1	
135											1		2				
140												1		2	2		
145												1		1	1		
155										1							1
175														1			
N	1	5	2	13	34	24	17	30	18	21	14	12	8	5	2	1	1

(b)

Age (years)	1	2	3	4	5	6	7	8	9	10	11	12	13	14	15	16	17
Length (cm)																	
50	50	50															
55		100															
60			100														
65		50		50													
70			25	25	50												
75				100													
80				50	25	25											
85				67	17	17											
90				18	55	9	18										
95					54	25	13	8									
100					38	12	19	23	8								
105					6	39		22	33								
110						20	20	25	25	10							
115						6	18	47		18	6	6					
120								19	14	38	14	10	5				
125								6	12	35	29	18					
130										6	25	38	19	6		6	
135											33		67				
140												20		40	40		
145												50		50	50		
155										50							50
175														100			
N	1	5	2	13	34	24	17	30	18	21	14	12	8	5	2	1	1

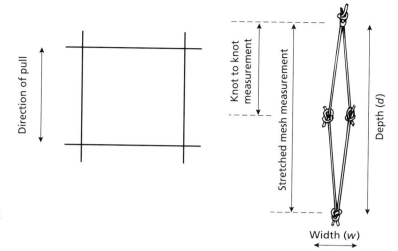

Fig. 9.6 (a) Square and (b) diamond mesh. The ratio between *w* and *d* is a function of the hanging ratio. Mesh is usually measured by forcing a guage through the stretched mesh.

trawls may be determined by placing fine mesh covers over the cod-end to catch escapees. The size distributions of catches in the cod-end and cover are compared to assess selectivity. Another way of assessing selectivity is to compare catches when alternating between two trawls of different mesh size.

The change in the number of fish retained by a cod-end as fish size increases is widely described using a logistic curve:

$$P = 1/(1 + e^{-r(L-L_c)}) \quad (9.1)$$

Where P is the proportion of the total catch of length L caught in the cod-end, r is a constant and L_c is the mean length at which 50% of fish are retained in the cod-end (mean length at first capture). P is calculated as the number of fish of length L in the cod-end divided by number of fish of length L in cod-end and cover. In the case of an experiment with two different trawls, P will be the proportion of fish of a given length caught in the larger mesh net. Equation 9.1 can be rearranged in linear form:

$$\log_e((1-P)/P) = rL_c - rL \quad (9.2)$$

Thus, a plot of $\log_e((1-P)/P)$ against length (L) will have a slope of $-r$ and the mean length at capture L_c will be given by a/r, where a is the y intercept. When L_c and r have been calculated they can be fed into equation 9.1 to produce the selection curve (Sparre & Venema, 1998).

Figure 9.7 shows selection curve for winter flounder *Pseudopleuronectes americanus* when using an otter

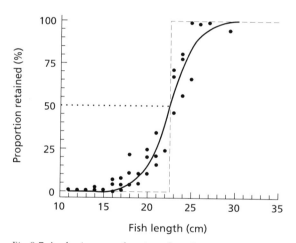

Fig. 9.7 A selection curve for winter flounder in a 102-mm diamond mesh cod-end. The dashed line shows the assumed shape of this curve when there is knife-edge selection. After Simpson (1989).

trawl with 102 mm mesh fished in Long Island Sound, off Connecticut, USA. The curve was derived by comparing catches in the 102-mm cod-end with those in a 51-mm cod-end (Simpson, 1989). We have also used Fig. 9.7 to demonstrate the concept of knife-edge selection. Although the logistic curve fits the real data rather well, it is often convenient to assume knife-edge selection at the length when 50% of fish entering the cod-end would be retained. In this case, the actual catch of fish smaller than the length at knife-edge selection would be larger than expected (i.e. > 0) and the catch of fish larger than the length at knife-edge selection would be smaller than that expected. Nevertheless, the concept of knife-edge selection is often used to make simple comparisons between gears.

The selectivity of passive gears such as gill nets is largely determined by mesh size and hanging ratio. **Hanging ratio** is the ratio between the length of the fishing net when made up, divided by the length of the original sheet of netting. The mean ratio between width and depth of meshes is a function of the hanging ratio (Fig. 9.6). Selectivity curves for gill nets of any given mesh size and hanging ratio are characterized by increasing retention with fish length followed by a decrease in retention as length increases further. This can often be modelled with a normal curve. In other cases, loosely hung gill nets entangle large fish that would not be snagged by the gills, and the selectivity curves become very complex.

The derivation of selection curves for gill nets is more complex than for trawls, because we rarely know the size distribution of the entire stock that is fished, and our assessments have to be based on comparisons between the length-frequency distributions taken by different nets. If we assume that the logs of the numbers of fish caught per length group by two different mesh sizes (C_1 and C_2) are linearly related to fish length (L, expressed as midpoint of length class):

$$\log_e(C_1/C_2) = a + bL \tag{9.3}$$

Since this takes the form of a linear relationship, a plot of $\log_e(C_1/C_2)$ against L produces estimates of a and b. Holt (1963) showed that a and b could be used to estimate selection factors (SF) and to produce selection curves.

$$SF = -2a/b(m_1 + m_2) \tag{9.4}$$

where m_1 and m_2 are the mesh sizes of the nets that resulted in catches C_1 and C_2, respectively. The selection factor is used to calculate the length class caught most effectively by each net (L_{m_1} and L_{m_2}). For mesh sizes m_1 and m_2 these are:

$$L_{m_1} = SF \times m_1 \tag{9.5}$$

$$L_{m_2} = SF \times m_2 \tag{9.6}$$

The common variance (s^2) that describes the spread of the selection curve is:

$$s^2 = -2a(m_2 - m_1)/b^2(m_2 + m_1) = SF(m_2 - m_1)/b \tag{9.7}$$

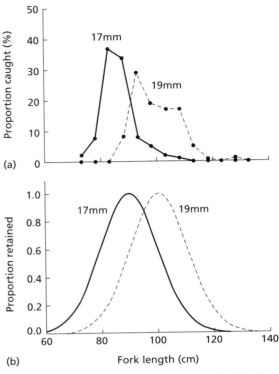

(a)

(b)

Fig. 9.8 (a) Retention of sea bream *Diplodus annularis* in gill nets of 17 mm and 19 mm mesh, and (b) the calculated selection curves. Data from Petrakis and Stergiou (1995).

The probability (P) of capture for a fish of a given size with each mesh is calculated as:

$$P(L) = e^{-(L-L_{m_1})^2/2s^2} \tag{9.8}$$

$$P(L) = e^{-(L-L_{m_2})^2/2s^2} \tag{9.9}$$

P can be calculated for a range of lengths to derive the selection curves for the two mesh sizes. The location of these curves will be defined by L_{m_1} and L_{m_2}, and they will have the same variance. The Holt (1963) method can also be expanded to allow for comparisons between many mesh sizes.

As an example of this method we can consider a study of gill-net selectivity for the sea bream *Diplodus annularis* (Sparidae) in South Euboikos Gulf, Greece. This fish is an important species in many Mediterranean artisanal fisheries. Petrakis and Stergiou (1995) compared several gill nets with a 0.6 hanging ratio, including those of 17 mm and 19 mm mesh. When data shown in Fig. 9.8(a) were entered into equation 9.3, they produced values of $a = -18.956$ and $b = 0.200$.

Given that $m_1 = 17$ and $m_2 = 19$ it can be seen from equation 9.4 that the selection factor was 5.27 and that the lengths caught most effectively by the two nets would be 89.6 and 100.1 mm, respectively. The variance from equation 9.7 is 52.7, and this was entered into equations 9.8 and 9.9 to produce the selection curves shown in Fig. 9.8(b).

9.3.2 Length, weight and age

Fished species have to be sized and aged for most stock assessments. The length of fish can be measured in various ways (Fig. 9.9), and total length (TL) and fork length (FL) are widely used for measuring adult fish. Fork length (FL) is length from the snout to the median caudal fin rays. FL is often used for species with forked tails since it is not biased by damage to the ends of the tail. Standard length (SL), from the tip of the upper jaw to the posterior end of the hypural bone (or to end of caudal peduncle) is usually used by taxonomists. Length is often measured on board with a 'snout stop'. The snout of the fish is held against the stop while the length is recorded at the tail. TL measurements based on compressing the caudal lobes dorso-ventrally can be longer than TL measured when the fin is in a 'natural' position, so TL must be recorded using a standard procedure. Because several methods of measuring length are used,

conversion tables have been published to allow species-specific conversion between TL, FL and SL. For larval fish, notochord length is usually measured because tissue behind the notochord is easily damaged and lost during capture. Larvae can be measured under a binocular microscope with a micrometer or miniature measuring board.

Length measurements are generally made to the nearest centimetre or millimetre, or to the centimetre or millimetre below. If measured to the unit below then 0.5 units are subsequently added to calculated mean lengths. Many electronic measuring devices have now been developed for use on large-scale surveys where thousands of fish are measured. For example, length may be recorded using an optical device that reads a length-specific bar code on the measuring board. The same optical pen can also be used to record species, site and station codes from other bar codes and obviate the need to key or write data in wet laboratories, harbours or fish markets.

The length of bivalve mollusc shells is recorded as the greatest distance along a line approximately parallel with the hinge axis (Fig. 9.10). For gastropod molluscs, length (height) is usually measured along the axis of greatest distance from the base of the shell to the tip of the shell. Decapod crustaceans are usually measured using carapace length or total length from the tip of the rostrum to the median uropod. Crabs are often measured using carapace width (Fig. 9.10). Echinoderms such as sea cucumbers create many measurement problems because they lack skeletal material and are composed largely of water. As a result they can expand and contract freely. Length is often measured when fully contracted, although this may be hard to define and is still subject to bias.

Larval and juvenile fish may shrink substantially after death. Shrinkage can be exacerbated by mechanical damage in nets. If larvae cannot be measured while alive (rarely a possibility for larvae caught in plankton nets) shrinkage will bias estimates of growth rate (Fig. 9.11). Larval fish also shrink in fixative and preservatives. There are various published conversion factors available that allow estimation of live length from preserved length (Pepin *et al.*, 1998).

It is more difficult and time consuming to record weight than length, especially on a bouncing boat, and so weight–length relationships can be calculated for a sample of individuals and used to estimate the weights of other individuals. The whole fish is usually weighed,

(a)

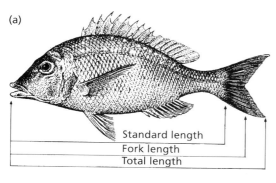

Standard length
Fork length
Total length

(b)

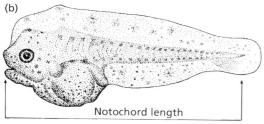

Notochord length

Fig. 9.9 Measuring the length of fish.

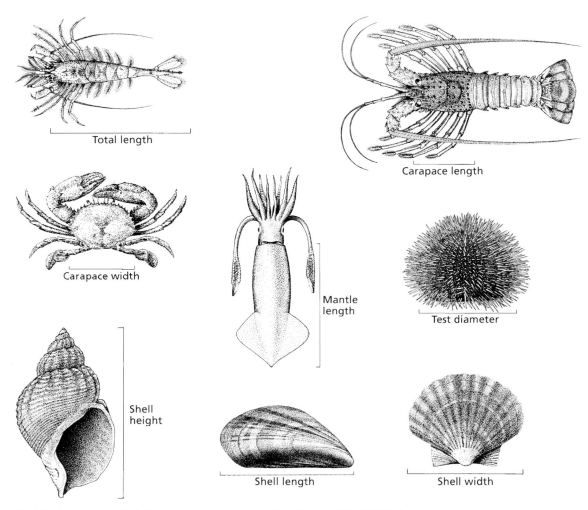

Fig. 9.10 Measuring the body size of crustacea and molluscs. Drawings copyright S.R. Jennings.

but eviscerated or gonad free (somatic) weight may be recorded. Conversion factors are calculated between gutted and whole weight so that the whole weight of fishes taken from commercial catch samples can be estimated. Clearly, these are rather crude conversions since the weight of material removed by gutting is very variable and depends on the maturity stage and recent feeding history of the fish.

The relationship between length and weight can provide information on the condition of fish. The weight of a fish increases in relation to the increase in volume and so the relationship between length (L) and weight (W) can be described using a power function of the form:

$$W = aL^b \tag{9.10}$$

If length and weight are transformed to logs then the relationship becomes:

$$\log_e W = \log_e a + b \log_e L \tag{9.11}$$

In this form, a and b can be determined from a plot of $\log_e W(y)$ against $\log_e L(x)$ where $\log_e a$ is the intercept and b is the slope of the fitted linear relationship. An example of a weight–length relationship for the red snapper *Lutjanus bohar* (Lutjanidae) based on samples collected by line fishing in Seychelles is shown in Fig. 9.12. It is $\log_e L = -11.11 + 3.045 \log_e W$. Thus, $a = e^{-11.11} = 1.496 \times 10^{-5}$ and $b = 3.045$.

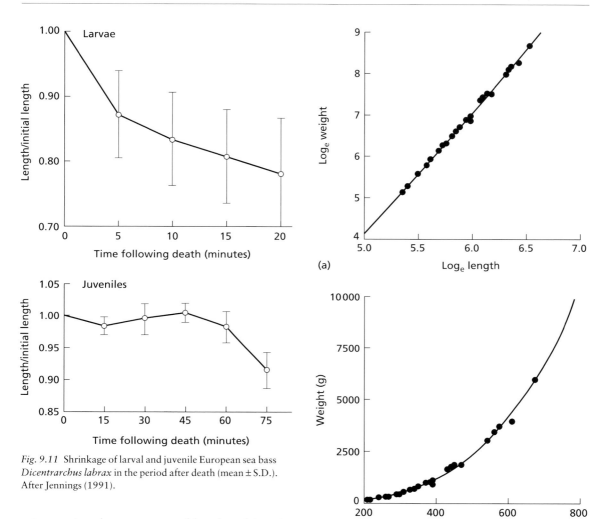

Fig. 9.11 Shrinkage of larval and juvenile European sea bass *Dicentrarchus labrax* in the period after death (mean ± S.D.). After Jennings (1991).

In practice, the parameters of length–weight relationships are fitted using models that account for the error structure of the data. Models with multiplicative error structure are appropriate when variability in growth increases as a function of the independent variable, and models with additive error structure are appropriate when variability in growth is constant as a function of the independent variable. Equation 9.11 can be modified for multiplicative error structure by the addition of an error term. Models with additive error structure are fitted with non-linear least squares (Quinn & Deriso, 1999).

The value of *a* in equation 9.11 has widely been used as an index of fish condition, such that a fish with a higher value of *a* is heavier for its length. This is calculated as:

Fig. 9.12 (a) Calculation of the length–weight relationship for the snapper *Lutjanus bohar* and (b) the fitted relationship.

$$a = W/L^b \qquad (9.12)$$

A number of indices, known as **condition factors**, are based on this relationship. Their use is not recommended since they cannot be used for comparisons among populations, if length distributions or values of '*b*' differ (Cone, 1989). Rather, in order to see if one sample of fish are heavier for their length than another, it is best to calculate weight–length relationships for each sample and to use these to calculate predicted weights (and confidence limits) at specified lengths.

Alternatively, to look at seasonal changes in condition, length–weight relationships are determined for each of a series of samples and used to calculate predicted weights at length.

Age

Growth is expressed in relation to time, so it is useful if target species can be aged. Fortunately, many fished species carry records of time in hard parts such as shells, otoliths, scales or vertebrae. When ageing with hard parts is not possible, changes in size over a known time interval may be used to estimate growth. This approach is often used for crustaceans, where hard parts are shed during moulting. The change in size of individuals between tagging and recapture also provides data that allows growth to be estimated.

Environmental stimuli cause variations in the growth rates of bony structures. Such variation occurs on time scales from days to years. Growth may be slower in some seasons due to reductions in temperature or production. The otoliths or scales in bony fish reflects these growth patterns and are usually used for ageing. Ageing has become a field of study in its own right and many books and symposia discuss it in detail (Bagenal, 1974; Summerfelt & Hall, 1987; Pentilla & Dery, 1988; Anon, 1992; Secor *et al.*, 1995; Fossum *et al.*, 2000).

Scales have long been favoured for ageing freshwater fish because they can be collected without killing the fish. This is rarely a concern for marine fishes, although scales do, for example, allow fish to be aged at the time of tagging and release, and can be taken from fish whose market value would be affected if the otoliths were removed. On scales, fine ridges called **circuli** are used for ageing. Circuli start to appear in spring as growth accelerates and cease being laid down when growth slows in winter. There are two types of scales, cycloid and ctenoid (Fig. 9.13). On **cycloid scales**, circuli extend all around, but on **ctenoid scales** the exposed part of the scale rarely has clear circuli and may be covered in spines. In this case, the buried (anterior) part of the scale is used for ageing. Scales are often lost by fish and new ones regenerate. Regenerated scales cannot be used for ageing and are easily distinguished because the central area appears unstructured and lacks circuli (Fig. 9.14). In older fish so many scales are regenerated that 20 or more may have to be collected to find one for ageing. The favoured area for collecting scales that are easy to read and least likely to be

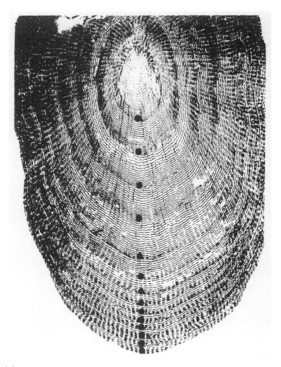

(a)

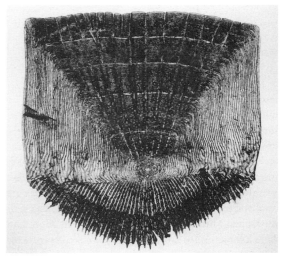

(b)

Fig. 9.13 (a) Cycloid and (b) ctenoid scales. The cycloid scale is from a 14-year-old haddock *Melanogrammus aeglefinus* from Georges Bank (Pentilla & Dery, 1988) (annuli are marked) and the ctenoid scale is from a 5-year-old European sea bass *Dicentrarchus labrax* (Eaton, 1996). Photographs copyright US National Marine Fisheries Service (a) and UK Crown Copyright (b).

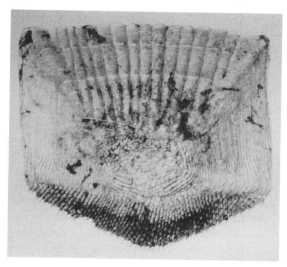

Fig. 9.14 A regenerated scale from a European sea bass *Dicentrarchus labrax* (Eaton, 1996). Photograph copyright UK Crown Copyright.

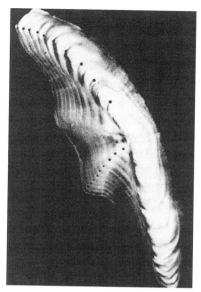

Fig. 9.15 A sectioned American plaice *Hippoglossoides platessoides* otolith showing annual growth increments (Dery, 1988). Photograph copyright US National Marine Fisheries Service.

regenerated is often from the flank, near to the tip of the pectoral fin, although this varies from species to species. Scales are read directly under a microscope or projecting viewer, although in some cases they are used to make impressions on acetate sheets and the impressions of circuli are counted. One limitation of scales is that they can be difficult to read in older fish.

Otoliths lie over sensory tissues in the ear and stimulate hair cells when they move or vibrate. This allows fish to detect sound, gravity and acceleration. There are three pairs of otoliths in the ear, but only the largest, the **sagittae**, are usually used for ageing. Otoliths are also found in elasmobranchs but they consist of gelatinous rather than bony material and cannot be used for ageing. Unlike scales, otoliths grow through deposition rather than ossification. Since they cannot be lost, and since there is no evidence for their resorption during periods of poor growth or stress, they are ideally suited for ageing. This is particularly true for marine fish, where killing the fish is usually acceptable.

Otoliths are dissected from the head of the fish, wiped clean and stored dry until processed. In some cases, annual growth increments can be viewed directly under a microscope with transmitted light. In others, the otolith is split and charred with a flame (**break-and-burn**) for direct viewing under a binocular microscope or mounted in resin, sectioned and viewed on a slide

(Fig. 9.15). Methods of preparation are described by Pentilla and Dery (1988). The bands seen on an otolith are known as **hyaline zones** (translucent) and **opaque zones**. The opaque zone is denser (richer in calcium carbonate). One opaque and one hyaline zone are deposited each year. The timing of zone formation varies among species, age classes and according to geographical location, but opaque zones are usually formed during periods of relatively rapid growth and hyaline zones during periods of slower growth (e.g. Williams & Bedford, 1974; Beckman & Wilson, 1995). In the past it was thought that tropical fishes could not be aged from otoliths as the tropical environment was aseasonal. This is wrong, and with appropriate preparation the otoliths of many tropical species have been read (Fowler, 1995). Before otoliths are used for ageing, it is important to validate whether hyaline and opaque zones are deposited annually. Validation should mean that the error associated with ageing has been quantified (e.g. 90% of age estimates are correct with an error of ± 1 years) (Francis *et al.*, 1992a). Methods of validation are discussed below.

Otoliths also provide a record of daily growth and, since the 1970s, they have been used for daily ageing

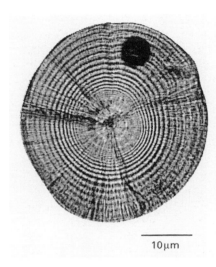

10μm

Fig. 9.16 Otolith of a juvenile European sea bass *Dicentrarchus labrax* showing daily growth increments. Photograph copyright W. Reynolds.

(Panella, 1971; Secor *et al.*, 1995). Ageing fish by the day has allowed larval life spans and growth rates to be estimated, hatching dates of larval and juvenile fish to be back-calculated and age cohorts in larval fish assemblages to be identified (Brothers *et al.*, 1983; Victor, 1986, 1991; Wellington & Victor, 1989). Daily ageing is fastest and easiest with fishes up to a few months old, but with appropriate preparation an age in days can be determined for older fish. Daily ages can be estimated from examination of the continuous and discontinuous zones that are deposited on the otolith as it grows (Fig. 9.16). These zones are a few micrometres in thickness. **Continuous zones** (also known as incremental) contain dense clusters of aragonite (calcium carbonate) crystals in a fibrous protein matrix. In the **discontinuous zones** the growth of aragonite crystals is interrupted. One continuous and one discontinuous zone is usually deposited each day, although zones may be deposited at faster or slower rates and daily ageing techniques should always be validated. Validation can be achieved by rearing larvae of known age or by marking the otolith with a chemical such as tetracycline (see below). **Daily growth zones**, widely termed **daily growth rings** or **daily growth increments**, in larvae may be observed at magnifications of 400× and more beneath a standard microscope. In older fish, the larger otolith needs to be ground to a thin section before examination. This can

be done by mounting the otolith in a drop of resin on a slide and lapping with very fine abrasive paper and an aluminium oxide block. If heat-sensitive resin is used then it is very easy to warm the resin to release the otolith, and to reverse it for grinding on the other surface. Methods of preparing otoliths for daily ageing are reviewed by Stevenson and Campana (1992).

Other hard parts such as spines, opercular bones and vertebrae can also be used for ageing. Vertebrae and spines are favoured for ageing elasmobranchs. Vertebrae have been used to age species such as the shortfin mako *Isurus oxyrinchus* (Lamnidae) and lemon shark *Negaprion brevirostris* (Carcharhinidae) (Gruber & Stour, 1983; Pratt & Casey, 1983). The centra are removed from the vertebral column and cleaned in alkali to remove connective tissue and notochord. They are then stained and cut or ground for examination (Stevens, 1975; Gruber & Stour, 1983). The spines found in front of spurdog (spiny dogfish) *Squalus acanthias* dorsal fins can be sectioned and used for ageing (Holden & Meadows, 1962).

Ageing techniques can be validated using tetracycline antibiotic. This combines with bone calcium during calcification and produces a mark that fluoresces under ultraviolet light (Weber & Ridgeway, 1962). Large fish can be tagged and injected with buffered tetracycline solution before release. On recapture, growth rings that have formed after the deposition of the fluorescent mark can be counted and compared with the time at liberty. Small fish and larvae, that would be injured by injection, will take up tetracycline if immersed in a buffered tetracycline solution. Tetracycline has been used for validating daily and annual growth increments on scales, otoliths, bones and spines.

Squids are aged using the **statolith**, but octopods cannot be aged. The statolith, the organ responsible for detection of acceleration, is composed of aragonite crystals in an organic matrix. Growth rings may be laid down in the statolith on a daily basis, and can be seen when sectioned. Sectioning techniques are described in detail by Rodhouse and Hatfield (1990). Many species, excepting the cephalopods, have shells with a calcium carbonate matrix secreted by specialized cells in the mantle. This shell provides protection, but is useful to fishery biologists since it is laid down in layers on daily, seasonal and annual cycles.

Sea cucumbers, shrimps and crabs cannot be aged as they lack hard parts or shed them during growth.

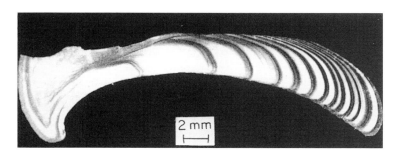

Fig. 9.17 Growth rings on a sectioned shell of the surf clam *Spisula solidissima* (Mactricidae) (Ropes & Shepherd, 1988). Photograph copyright US National Marine Fisheries Service.

Growth estimates for these species are usually obtained by tagging or length frequency analysis (section 9.3.3). Bivalves and gastropod molluscs may be aged from seasonal growth patterns in the shell matrix. Shells of bivalves are usually cleaned, bleached and sectioned from the base of shell across the valve to the margin (Ropes & O'Brien, 1979; Ropes & Shepherd, 1988). The section is mounted in resin for examination (Fig. 9.17).

9.3.3 Growth

For species that can be aged, growth is determined directly from size at age data, or back-calculated from scale readings. If the species cannot be aged, then growth may be estimated from length frequency distributions by treating length as a non-linear measure of biological time.

Back-calculation of growth using scales
Back-calculation is a technique that uses fish length and fish scale measurements at the time of capture to infer length at times in the past (Francis, 1990). Back-calculation requires that a relationship between fish length (L) and scale radius (S) can be established. If fish length at capture is denoted by L_c, and scale radius at capture by S_c, then L_i and S_i can be taken as the corresponding measurements at the time when the ith annulus was formed. Back-calculation formulae allow the calculation of L_i from L_c, S_c and S_i and are reviewed by Francis (1990).

As an example, we will consider the formula of Whitney and Carlander (1956) that assumes a constant proportional deviation of the length or scale measurements of an individual from the mean length or scale measurement of the population.

$$f(L_i) = (S_i/S_c)f(L_c) \tag{9.13}$$

where f is the function describing the relationship between body length and scale radius, such that $f(L)$ is the mean scale radius for fish of length L. If the body length to scale radius relationship is linear, then:

$$L_i = -(a/b) + (L_c + a/b)(S_i/S_c) \tag{9.14}$$

where a and b are the parameters from a linear regression of the form $y = a + bx$.

There are many potential errors associated with back-calculation, and the methods used should be validated. Francis (1990) recommends any validation must show:
1 that the radius of a scale mark is the same as the radius of the scale at the time the mark was formed;
2 that the supposed time of formation of the mark is correct;
3 that the formula used accurately relates scale radius and body size for each fish.
Back calculation of fish length from annuli is easily automated using digitizing tablets and simple software, so that scale measurements are immediately converted to fish lengths (Pickett & Pawson, 1994).

Growth models
The fitting of growth models to age and size data is almost a field in its own right, but choice of model is generally less important than obtaining representative samples. For many stock assessments it is acceptable to calculate mean body size at age in a tabular form and enter this into analyses directly. Those tables should be updated annually to allow for growth changes. Such an approach is used with yield per recruit analysis (section 7.7).

The only growth model we describe here is the **von Bertalanffy growth equation** (**VBGE**) because it provides a good description of growth in most species and because the parameters have been widely compiled and have considerable utility in life-history studies (section 3.4.2). It is given as:

$$L_t = L_\infty(1 - e^{-K(t-t_0)}) \qquad (9.15)$$

Where L_t is the length at age, L_∞ is the asymptotic length (the length at which growth rate is theoretically zero), K is the Brody growth coefficient (rate of growth towards asymptote) and t_0 is the time when length would have been zero on the modelled growth trajectory. t_0 can be seen as a parameter that shifts a growth trajectory defined by K and L_∞ along the x-axis.

If this equation is combined with the length–weight relationship $W = aL^b$ (equation 9.10) and the asymptotic weight is defined as W_∞, it can be rearranged to give the von Bertalanffy growth curve for weight:

$$W_t = W_\infty(1 - e^{-K(t-t_0)})^b \qquad (9.16)$$

Where W_t is weight at age and b is the parameter from the length–weight relationship (equation 9.10).

Methods used to fit the von Bertalanffy growth equation should account for the error structure in the data and indicate the precision of parameter estimates. Many computer programs allow von Bertalanffy growth parameters to be fitted using models with additive or multiplicative error structure and these are described in Quinn and Deriso (1999).

Growth parameters can also be derived from tagging data if size is measured at release and recapture and if the interval between release and recapture is known (Francis, 1988; Quinn & Deriso, 1999). The following approach is taken from Quinn and Deriso (1999). Suppose that fish i was of length $L_{1i} = L_{(t_{1i})}$ at age t_{1i} when tagged and of length $L_{2i} = L_{(t_{2i})}$ at age t_{2i} when recaptured. The elapsed time between tagging and recapture can be denoted as $\Delta t_i = t_{2i} - t_{1i}$ and growth as $\Delta L_i = L_{2i} - L_{1i}$. If we assume that the VBGE describes growth (equation 9.15) and that an additive error structure applies to growth then:

$$\Delta L_i = (L_\infty - L_{1i})(1 - e^{-K\Delta t_i}) + \varepsilon_i \qquad (9.17)$$

Where ε_i is a random error term with mean 0 and variance σ^2. Parameters L_∞ and K can be estimated by non-linear least squares. The value of t_0 is obtained by solving the VBGE when length at age is known.

With seasonal samples taken from fluctuating environments, growth can vary with season. Various modi-

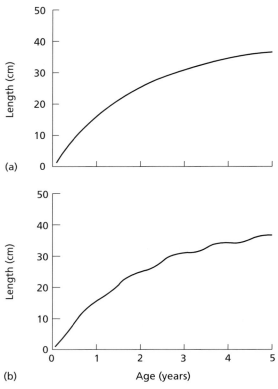

(a)

(b)

Fig. 9.18 (a) A von Bertalannfy growth curve ($C = 0$), and (b) its seasonally adjusted equivalent when $C = 0.2$.

fications of the VBGE have been produced to allow for this. One seasonally modified growth model is:

$$L_t = L_\infty(1 - e^{-K(t-t_0)-(CK/2\pi)\sin(2\pi(t-t_s))}) \qquad (9.18)$$

This is simply a VBGE with the added term $(CK/2\pi)\sin(2\pi(t - t_s))$ to give a seasonal oscillation (Pauly, 1982). The constant C describes the amplitude of seasonal oscillations and t_s the phase shift of these oscillations. An example of the seasonal growth curve is given in Fig. 9.18.

Growth parameters from length-frequency data

The problems with ageing some fished species have encouraged some biologists to use length as a non-linear measure of biological time (Rosenberg & Beddington, 1988; Sparre & Venema, 1998). The aim of estimating growth from length frequencies is to separate a complex length-frequency distribution for the whole population into cohorts and to assign ages to them (Fig. 9.19). We used this approach for length-

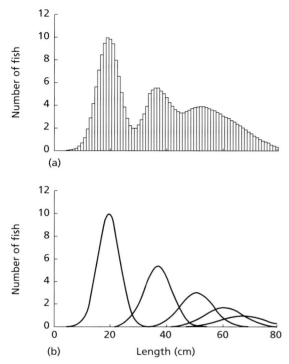

Fig. 9.19 (a) A composite length-frequency distribution, and (b) component cohort length-frequency distributions.

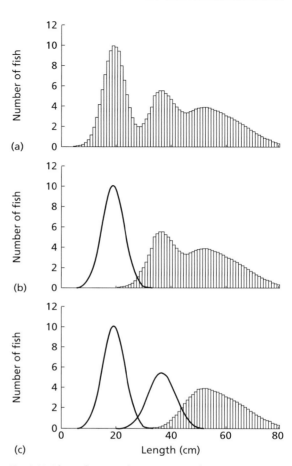

Fig. 9.20 Bhattacharya analysis. Starting with a composite length-frequency distribution (a), the first cohort is identified and separated (b). This gives a new composite distribution (b) from which the next cohort is identified and separated (c).

based cohort analysis (section 7.5.2). The existence of cohorts is very clear in some life-history data for short-lived species but very unclear for longer-lived species where the variation in length at age increases with age.

The practical validity of length-based methods is often questioned, and it is often true, as emphasized by MacDonald (1969) and Hilborn and Walters (1992), that if the cohorts are not clear to the eye then the outputs of the analyses we describe may not be valid. Indeed, Rosenberg and Beddington (1987) made the general observation that the difference between the mean size of two cohorts must be greater than twice the smaller of the standard deviations of their lengths if they are to be separated. In practice, this often means that only the youngest two or three cohorts can be distinguished (Fig. 9.19). For relatively short-lived species that cannot be aged, length-based methods have provided useful growth data that could not be obtained in other ways. Most length-based analyses are computationally complex, but software is available to implement them (Sparre & Venema, 1998).

Some methods of identifying cohorts make assumptions about the form of the length-frequency distribution within each cohort, usually that it is normal. Bhattacharya (1967) developed a method that can be used to split a composite distribution into a series of normal distributions representing cohorts. **Bhattacharya analysis** relies on identifying the length-frequency distribution of the youngest cohort and subtracting the numbers of fish in each length group within the first cohort from the total length-frequency distribution. The approach is then repeated with the next cohort, and continued until no more normal distributions can be identified (Fig. 9.20). At this stage, ages are assigned to each cohort to get mean length at age. This analysis can be done using the FAO-ICLARM Stock Assessment Tools (FiSAT) package (Gayanilo *et al.*, 1994). Another approach to length-frequency analysis is modal progression analysis where changes in the size of individuals within a cohort are followed in successive samples. Clearly, the same fishing gear with the same selectivity must be used to take these samples and the same population must be fished.

Other approaches to length-frequency analysis do not make any distributional assumptions. These include Shepherd's length composition analysis (Shepherd, 1987) in which a test function is calculated from a specified set of growth parameters to predict the form of the composite length-frequency distribution. Comparisons between the actual and predicted distribution provide the basis for selecting the most appropriate growth parameters. A more widely used method is **Electronic Length Frequency Analysis** (ELEFAN) (Pauly & David, 1981). This 'smoothes' the entire length-frequency distribution by calculating a running average frequency and identifying length classes for which actual frequencies exceed this average (peaks). The peaks are treated as modal lengths in each cohort and the program tests possible growth curves to see which provides the best fit. The most recent version of ELEFAN is available on the FiSAT package.

The ELEFAN approach is also applicable to modal progression analysis where the growth of cohorts is followed through time. An example of growth of the sea scallop *Argopecten purpuratus* (Pectinidae) in Chile is shown in Fig. 9.21 (Stotz & González, 1997). Scallop samples were collected at intervals of 1–8 months and ELEFAN was used to fit the growth curves shown. The cohort that recruited to the samples in May 1993 (Fig. 9.21a) had the fitted growth curve shown in Fig. 9.21(b), and encouragingly, this curve was similar to one produced by measuring the growth of tagged individuals. ELEFAN can also be used to fit seasonal growth curves using equation 9.18 (Pauly, 1982).

An alternative method of length-frequency analysis is MULTIFAN (Fournier *et al.*, 1990). This approach uses a likelihood-based estimation procedure and is more objective than the widely used ELEFAN. Other methods of deriving growth from length-frequency data are described by Rosenberg and Beddington (1988), Gulland and Rosenberg (1992) and Sparre and Venema (1998). Biases in some methods are considered by Basson *et al.* (1988).

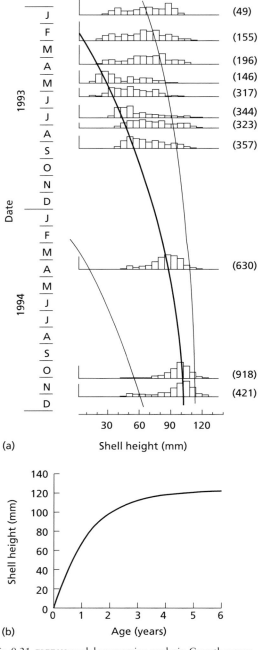

Fig. 9.21 ELEFAN modal progression analysis. Growth curves have been fitted to sequential length-frequency data for Chilean sea scallop *Argopecten purpuratus* (a). By calculating changes in shell height over time we can see that there is good correspondence between this growth estimate and that obtained from tagging data (b). After Stotz and González (1997).

Table 9.3 Examples of maturity stages.

Stage	Description	Characteristics
1a	Immature	Thin transparent ovaries with no blood vessels
1b	Late immature	Longer than 1a, thickening
2	Spent, recovering	Flattened from larger size, no visible eggs, large empty lumen
3	Early ripening	Small developing eggs. Ovary not full, blood vessels apparent
4	Late ripening	Ovary full with visible but not hydrated eggs
5	Ripe	Hydrated eggs present but not released through oviduct
6	Running	Eggs released through oviduct with pressure on ovary
7	Spent	Flat deflated ovary, eggs in resorption, slime in ovary

9.3.4 Maturity

It is rare for all individuals in a fished stock to mature at the same age, the major exceptions being some squids, salmon and small, fast growing and short lived fishes such as anchovies. In most other species, females attain first maturity at a range of ages. To estimate spawning stock biomass it is necessary to know the proportion of mature females in each age group.

Mature fish are identified by examination of the gonads. Gonadal development is 'staged' to allow seasonal patterns in the reproductive cycle to be identified by showing when the fish begin to mature, spawn and recover. A typical series of maturity stages for a fish is given in Table 9.3.

The length or age at which 50% of females attain maturity (L_{mat} and T_{mat}, respectively) is usually estimated by fitting a logistic curve to the relationship between proportion mature (P) and length or age:

$$P = 1/(1 + e^{-r(L-L_{mat})}) \qquad (9.19)$$

$$P = 1/(1 + e^{-r(T-T_{mat})}) \qquad (9.20)$$

where r is a constant. These equations can also be expressed in linear form:

$$\log_e((1-P)/P) = rL_{mat} - rL \qquad (9.21)$$

$$\log_e((1-P)/P) = rT_{mat} - rT \qquad (9.22)$$

So, a linear fit of $\log_e((1-P)/P)$ against length (L) or age (T) (of form $y = a + bx$) can be used to estimate L_{mat} and T_{mat} because $r = -b$ and L_{mat} or $T_{mat} = a/r$. Of course, biased samples taken with selective gear will give inaccurate L_{mat}. An example of L_{mat} for the American lobster *Homarus americanus* (infraorder Astacura) from Cape Cod Bay on the coast of Massachusetts is given in Fig. 9.22 (Estrella & McKiernan,

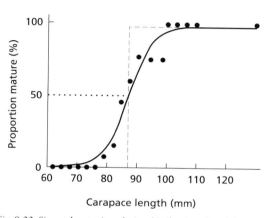

Fig. 9.22 Size and maturity relationship for American lobster *Homarus americanus* in Cape Cod Bay. The dashed line shows the assumed shape of this curve when there is knife-edge maturity. After Estrella and McKiernan (1989).

1989). In this case maturity was assessed from development of the cement gland which produces the 'glue' used to attach eggs to the pleopods during incubation. Maturity, like net selection, may be treated as knife-edge for descriptive or analytical purposes. Thus, if the length at 50% of maturity for the lobster is 87 mm, the equivalent knife-edge assumption would be that all lobsters mature at this size.

9.3.5 Fecundity

Fecundity estimates are needed to estimate stock abundance using the annual egg production method (section 10.2.7) and to describe life histories (section 3.4.3). In **determinate spawners**, the eggs due to be spawned are all present as developing oocytes in the ovary prior to the spawning season and fecundity is estimated by counting yolked eggs. In **indeterminate spawners**, fecundity estimation is much more complex, because the eggs to be spawned are not all present in the ovary at the start of the spawning season. Fecundity estimates do not necessarily equate with the numbers of eggs spawned since some yolked eggs may be resorbed in the ovary (section 10.2.7).

In indeterminate spawners, counts of yolked eggs do not indicate annual fecundity since the fish continuously mature new batches of eggs throughout a protracted spawning season. This is reflected in the continuous size distribution of oocytes in the ovary (because batches are maturing one after another). Conversely, the size distribution of oocytes in determinate spawners has two distinct modes corresponding to oocytes that are either very small and undeveloped or at a similar stage of later development (Fig. 9.23) Some fishes may actually spawn several times during the season, but are called determinate because the number of advanced oocytes at the beginning of the season is equal to annual fecundity (less atresia – the resorbtion rather than spawning of fully developed eggs, see section 10.2.7).

Most marine teleosts have very high fecundity (section 3.4.3) and it is not practical to count all eggs in the ovary. Therefore, subsamples of the ovary are taken and related either to ovarian weight (gravimetric methods) or to the total volume of an aqueous suspension of all oocytes in the ovary (volumetric method). In the former case, the subsample is simply cut from the ovary. In the latter, the ovary is stored in Gilson's fluid (a very toxic cocktail of acid, alcohol and mercuric chloride (Simpson, 1951) that breaks down the connective tissue and frees oocytes). The oocytes are then suspended in water. Full details of the methods are given by LeClus (1977), Bagenal and Braum (1978), Hunter *et al*. (1985) and Anon (1997).

The **batch fecundity** (number of eggs produced in a batch) of indeterminate spawners is often estimated by counting **hydrated oocytes** (those about to be spawned and identifed by their larger size and translucence) using a gravimetric method. The fish and ovary are weighed accurately (since batch fecundity has to be expressed in relation to body weight). Small samples are cut from areas on both ovaries, weighed then added to glycerin on a slide and the oocytes loosened by careful probing. Hydrated oocytes are counted under the microscope. Numbers are multiplied up by ovary weight

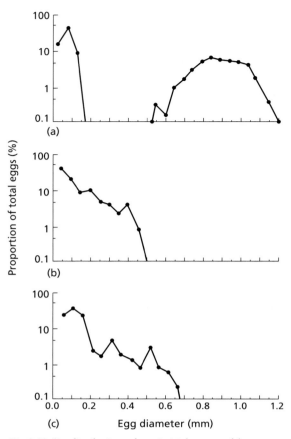

Fig. 9.23 Size distributions of eggs in (a) the ovary of the determinate spawning Atlantic herring *Clupea harengus*, (b) the indeterminate spawning Atlantic pilchard *Sardina pilchardus* and (c) northern anchovy *Engraulis mordax*. After Hunter *et al*. (1985).

to give batch fecundity for the individual fish (Hunter *et al*., 1985; Anon, 1997). Sometimes this method is not possible because the fish are sampled when hydrated oocytes are not present. In this case the so-called size-frequency method can be used. With this method the number of oocytes in the largest size category is determined and assumed to represent batch fecundity. The sample is treated as before but a micrometer is used to measure the size of all oocytes on the slide. Levels of replication needed to get acceptable confidence limits around the batch fecundity estimate can be high. For northern anchovy *Engraulis mordax*, 50 fish were needed to get a coefficient of variation (CV) of less than 10% for the sampled population (Hunter *et al*., 1985).

Batch fecundity of a population can be calculated from the mean weight of females using a relationship between batch fecundity and female weight. Batch fecundity varies from year to year. To get annual fecundity (the product of batch size and number of batches produced) it is necessary to know how many times indeterminate spawners spawn in a season (Hunter & Goldberg, 1980; Hunter & Macewicz, 1985). Spawning frequency is estimated from histological sections of the ovary. The sections are examined for the presence of hydrated oocytes and postovulatory follicles. Hydrated oocytes are surrounded by a follicle of thin cell layers and connective tissue. When the oocyte is released at spawning, the postovulatory follicle retains its integrity for a day or more, before being resorbed. Degeneration is staged by examining fish of known spawning history in the laboratory. Degeneration may be more rapid for species living at higher temperatures. If the persistence of degenerative stages is known, then the proportion of fish in a sample that have spawned in a given period may be calculated. Similarly, if the duration that oocytes are hydrated before spawning is known, then frequency of spawners in a sample can be determined from the number with hydrated oocytes.

The fecundity of determinate spawners must be estimated before spawning begins. Spawning fish can be identified by the presence of postovulatory follicles or hydrated oocytes and must be excluded from the sample. Fecundity can be estimated using gravimetric, volumetric or stereometric methods. The gravimetric method follows that described for indeterminate spawners, except that all oocytes are counted. When using volumetric methods, particle counters can be used to automatically count and size larger oocytes (200–2500 μm). These are put into aqueous suspension after the ovary has been digested with Gilson's fluid. Smaller oocytes are hard to count with this method as they tend to be confused with debris (Witthames & Greer-Walker, 1987).

A stereological method can also be used to estimate fecundity. This is based on the **Delesse principle** that the fractional volume of a component in a tissue is proportional to its fractional cross-sectional area. The ovaries must first be sectioned and stained. The proportional area, and hence volume, occupied by oocytes is determined by recording the proportion of points on a Weibel grid (a grid with equally spaced marks) that

overlie oocytes. The number of oocytes per unit volume is raised by the total volume of the ovaries to give the number of oocytes present. The method has a number of advantages. Toxic Gilson's fluid is not used, and both atretic oocytes and postovulatory follicles can be identified and counted (Emerson *et al.*, 1990).

9.3.6 Mortality

In general, there are fewer larger and older individuals in a population than smaller and younger ones because the larger and older individuals have been exposed to the risk of fishing or natural mortality for longer. Total mortality (Z) is the sum of fishing (F) and natural (M) mortality. We have already described methods such as VPA that can be used to calculate rates of fishing mortality (section 7.5). In this section we look at other ways of calculating Z, F and M, beginning with catch curves.

Age-based catch curves
Recall that as a cohort gets older the decrease in number of survivors at time t (N_t) can be described by the exponential decay equation (Box 7.2). This assumes that total mortality is constant beyond some reference age.

$$N_t = N_0 e^{-Zt} \tag{9.23}$$

We can take the natural logarithm of this equation to get:

$$\log_e N_t = \log_e (N_0) - Zt \tag{9.24}$$

Although abundance can be determined using the fishery independent methods in Chapter 10, we often have to rely on indices of abundance from commercial catches or survey results. So, to make the above relationship useful we can assume that catch rates are proportional to abundance and convert them to catches at age, C_a (or catch per unit effort):

$$C_a = N v \tag{9.25}$$

where v = vulnerability to the fishery expressed as the proportion of the population that are caught. Substituting C/v for N in equation 9.24 gives us:

$$\log_e \left[\frac{C_a}{v} \right] = \log_e (N_0) - Za$$

$$\log_e(C_a) - \log_e(v) = \log_e(N_0) - Za$$

$$\log_e(C_a) = \log_e(N_0 v) - Za \tag{9.26}$$

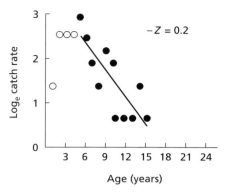

Fig. 9.24 An age-based catch curve for the surgeonfish *Acanthurus nigrofuscus* (Acanthuridae). The line is not fitted to the data represented by open circles because these age groups are not fully recruited to the fishery. After Hart and Russ (1996).

Now this is something we can use, because a plot of $\log_e(C_a)$ vs. a will have a slope of $-Z$, the instantaneous rate of total mortality (Fig. 9.24). This is an age-based **catch curve**. The standard error (SE) of the estimate of Z is equal to the SE of the slope.

We should be wary of some implicit assumptions in this derivation. The main danger with catch-curve analysis occurs when the data used in the analysis are taken from a stock where recruitment is declining. This will still produce a linear catch curve, but the value of Z will be underestimated. If trends in recruitment for a stock are not known, then catch curve analysis should not be applied to data from a single year.

Catch-curve analysis also assumes that all age classes have equal natural mortality rates, are equally vulnerable to the fishery, and that recruitment is constant from year to year. However, if there are many age classes, variations in recruitment should be apparent as variance around the fitted line and may not affect the slope. It is preferable to fit the catch curve to abundance data from successive samples that track the abundance of individual cohorts through time. In this case, we assume that catchability is constant from year to year. This is unlikely to be the case because the habitats used by fish may change with age and younger groups may escape through the net. For these reasons the line should only be fitted to age groups that are fully recruited to the fishery. Catch curves can equally be fitted to catch per unit effort data.

Other methods of estimating Z and their advantages

and disadvantages are discussed by Murphy (1997) and Quinn and Deriso (1999). There are also length-based methods for estimating Z that use length rather than age as measure of biological time. There are the same problems with identifying cohorts that we outlined in section 9.3.3. Length-based methods are described in Sparre and Venema (1998) and Quinn and Deriso (1999), and are available in packages such as FiSAT.

Natural mortality (M)

It is not that difficult to estimate Z but it is very hard to partition Z into F and M. If a stock is unexploited, then M will be equivalent to Z if the mortality imposed by sampling is negligible. However, such circumstances are rare and the investigation of a stock is almost always driven by exploitation. In addition, estimates of M made before exploitation may not be meaningful once the stock is exploited due to phenotypic changes in the life history.

If a stock has been fished at a range of intensities over a period of time, and if fishing effort and estimates of Z are available at each of those times then M may be estimated. However, this approach assumes an equilibrium, and must be applied with appropriate caution (Chapter 7).

Since $Z = F + M$ (Box 7.2) and F will be the product of catchability q and fishing effort f, it follows that

$$Z = M + qf \qquad (9.27)$$

This takes the form of a linear relationship. Thus, if Z is plotted against f then M will be the value of Z when $f = 0$. Once again, if the values of Z are biased (see above) the value of M will not be valid. An example of an estimate for M for the smooth-tailed trevally *Selaroides leptolepis* (Carangidae) from the Gulf of Thailand is given by Pauly (1982) (Fig. 9.25). In this case $M = 1.9$, a high value as would be expected for a fast-growing tropical fish. This approach is useful for getting a first estimate of M, but must be verified by calculating M in other ways.

In section 3.4.2 we saw the close links between life-history parameters. There have been many attempts to derive general equations for M based on relationships between M and other aspects of the life history, such as maximum size or growth rate, that can be more easily measured (Beverton & Holt, 1959; Beverton, 1963; Pauly, 1980; Hoenig, 1983; Gunderson & Dygert, 1988). Pauly (1980) derived the following equation for

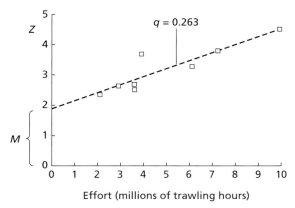

Fig. 9.25 Determination of M from a plot of Z against f for the smooth-tailed trevally *Selaroides leptolepis*. After Pauly (1982).

natural mortality using multiple regression analysis. It was based on a collation of life-history parameters for many stocks where M, L_∞, K and the mean temperature where they lived ($T°C$) were known:

$$\log_e M = -0.0152 - 0.279\log_e L_\infty + 0.6543\log_e K$$
$$+ 0.4634\log_e T \qquad (9.28)$$

There are some problems with this approach in that the data used to produce the plots are not independent, although these can be resolved to some extent by deriving equations within taxonomic groups. Both Pauly (1980) and Hoenig (1983) have done this. There is also fairly wide scatter around the fitted line, but in many cases this remains the best approach for getting an initial estimate of M.

Mark-and-recapture data can also be used to estimate M because the tagged fish will have to die or emigrate. Seber (1982) and Quinn and Deriso (1999) suggest a number of approaches.

Even in the most intensively studied fisheries, fishery scientists are often uneasy about their estimates of M.

The problems that ICES faced with estimating M for north-east Atlantic groundfish stocks such as cod and haddock led John Pope, the originator of cohort analysis, to provide the ingenious derivation shown in Box 9.1.

9.4 The impact of errors

Most of this chapter has been concerned with the estimation of parameters that describe the mean attributes of a population. It should be clear from our comments that there will always be errors associated with such parameters. These could arise because of measurement errors in the data or because samples from the population were not representative. If the assumptions on which an equation was based were not met, then model error would result. The Bayesian approaches discussed in Chapter 7 and 11 are rapidly becoming *de rigeur* when dealing with uncertainty in stock assessment and fisheries management.

The effect of errors that arise during parameter estimation can be assessed using various forms of sensitivity analysis. Pope and Gray (1983) looked at the CV in a North Sea cod total allowable catch (TAC) for various errors in fishing effort, catch at age and recruitment data. For example, to set a TAC that aimed to give $F = 0.5$, and with input CVs of 15%, 10% and 30% for fishing effort, catch at age and recruitment, respectively, the CV in the TAC was 33%. Tabulating such errors for a range of options helps to make cost benefit decisions about where sampling effort should be targeted. For example, halving the CV of the recruitment data would have reduced the CV for the TAC by 7% while if it were possible to reduce it to zero, it would only reduce the CV for the TAC by a further 4%. Given the effort needed to get low CV for recruitment data, this was unlikely to be the best approach to reducing the CV associated with the TAC.

Box 9.1

Pope's derivation of natural mortality rate.

The evolution of $M = .2$

$M = ?$ $? \rightarrow 7 \rightarrow ,7 \rightarrow ,2 \rightarrow .2 \rightarrow .2$ **EUREKA !**

Summary

• Stocks respond more or less independently to the effects of fishing because their dynamics are governed by birth, growth and death within the stock rather than immigration and emigration.

• If stocks are reproductively isolated for long periods they may differ genetically. Tagging often provides the most useful information on stock identity and migration.

• Stocks are sampled to collect the biological data needed to parameterize models. Considerable care is needed to ensure that samples are not biased.

• Fished species are aged and measured to describe growth. Otoliths, scales and other hard parts can be used for ageing. Species with no hard tissues and species that moult during growth cannot be aged in this way and growth is estimated from tagging data or length frequency analysis.

• Maturity and fecundity data are key inputs to spawning stock biomass calculations.

• Estimates of natural and fishing mortality are required for stock assessment. There are several approaches for making these estimates and they all make some assumptions that may not be reasonable.

• There are errors associated with parameter estimates because samples are not representative or the assumptions of an estimation method are not met. These are a source of uncertainty and should be considered explicitly. We should determine how input errors will affect the output of assessment models.

Further reading

Various contemporary tagging methods and their applications are described by Nielsen (1992). Methods for annual and daily ageing of many fished species are described by Pentilla and Dery (1988) and Secor *et al.* (1995) respectively. Methods of parameter estimation are discussed by Hilborn and Walters (1992) and Quinn and Deriso (1999), two books that are essential reading if you are actually going to do a stock assessment.

10 Getting the data: abundance, catch and effort

10.1 Introduction

This chapter describes methods for estimating the abundance of fished species. It deals with direct methods of estimating abundance rather than the analytical approaches of Chapter 7. Abundance estimates are needed to follow changes in stock size and to calculate parameters such as mortality and recruitment rates.

Relative abundance (as catch per unit effort) and fishing effort were used to predict yield before the failings of equilibrium-based surplus production models were widely recognized (Chapter 7). Nowadays, other abundance estimates are needed to calculate the size of a fishable resource, recruitment to the fished stock (pre-recruit surveys), year-class strength, biomass per unit area and population parameters (Chapter 9). Direct abundance estimates are also used for validation. Thus, virtual population analysis (VPA) abundance estimates for the western stock of mackerel *Scomber scombrus* in the north-east Atlantic are validated using egg production methods.

For fisheries that target sedentary species such as urchins, abalone or sea cucumbers in shallow water, direct counts by divers are possible and can provide accurate estimates of true abundance. Other fished species, however, live at many depths in association with many habitats, change their range and distribution with abundance, and are only seen when caught. For these species, getting good abundance estimates is a continuing challenge. Even the validity of estimates made by the largest and best resourced laboratories are still routinely questioned.

In this chapter we emphasize the problems and potential errors associated with abundance estimation and the difficulties of designing effective surveys. As in the preceding chapter on data collection, we emphasize how poor input data may mean that the most impressive analytical techniques give unreliable outputs. The analyst must be aware of the constraints under which abundance data are collected and the biases that exist. We describe all the methods in common use: visual census, acoustic survey, trawl survey, depletion, mark recapture and egg production. Some methods, such as egg production and acoustics, are largely independent of fishery data. Egg production methods are described in detail since they provide a context for introducing methods of plankton sampling and staging egg development. We end the chapter by asking what can be learned about abundance using fisheries data.

10.2 Abundance

10.2.1 Survey design

Accuracy and precision
Before we consider the mechanics of abundance estimation we will look at general features of survey design. The aim of surveys is usually to get the best estimate of total stock abundance or the abundance of part of a stock such as an age group. 'Best' can mean a number of things, but usually means an estimate of the highest precision, highest accuracy or both. **Precision** is the closeness of repeated measurements of the same quantity to each other and **accuracy** is the nearness of a measurement to its actual value (Zar, 1996). In many cases we focus on precision because various biases make accurate measures impossible. The precision needed is generally better than ± 20% rather than an order of magnitude, but costs of improving precision escalate rapidly because the relationship between error and sample size is not linear (Pope, 1988). In general, the reduction of error is proportional to the square root of sample size, meaning that a fourfold increase in sample size is necessary to reduce the error by half.

Surveys never record every individual in the stock, so we need to decide how the stock can best be sampled with available resources. For example, should one sacrifice accuracy to improve precision or sacrifice temporal replication to improve spatial replication? There are no firm and consistent answers to such problems. Rather, the decision will depend on the aims of the survey and the statistical power needed to detect change in abundance.

Sources of error around abundance estimates can be divided into bias and variance. Bias is systematic error that cannot be reduced by additional replication. It is often necessary to accept that bias exists, but to try to ensure it is consistent from site to site or year to year by employing the same sampling technique in the same way. Consistently biased abundance estimates would, for example, adequately describe temporal trends in relative recruitment rates. Variance is not systematic and is reduced by improved survey design and increased replication. A third source of error is often overlooked, those mistakes due to human error. For scientists collecting data in field conditions, procedures to check and control for human errors are essential (Pope, 1988).

Stratification

Some stratification is essential in most surveys. Most areas where fished species are found are very heterogeneous and contain a range of depths and habitat types. As such, it is unrealistic to calculate variance in abundance across the whole area because this will be determined by biogeographic effects. The alternative is to sample within predefined strata and combine abundance estimates. **Stratified sampling** is preferable whenever the variance among observations in a stratum is less than the variance in a random sample of the same number of observations taken from the whole survey area (Hilborn & Walters, 1992). **Random sampling** within strata gives an unbiased estimate of the mean, but systematic sampling can give a more precise estimate. For trawl surveys, systematic sampling is often favoured (Fig. 10.1) because it increases the rate of replication per unit time (by reducing vessel track between a given number of points), and decreases the probability of net loss or damage on rough ground (Pope, 1988; Hilborn & Walters, 1992). Systematic samples may be biased if they follow gradients in stock abundance, but a well-chosen systematic grid design will cut across abundance gradients. Systematic grid systems are also good for mapping boundaries, contouring abundance patterns and for making temporal

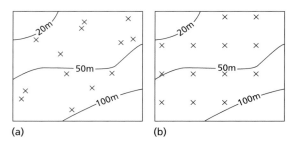

(a)   (b)

Fig. 10.1 (a) A stratified random, and (b) systematic survey with stratification areas based on depth contours.

comparisons (Hilborn & Walters, 1992). In other circumstances, unbiased estimates of the mean may be needed and stratified random sampling is used. This may apply to visual census work on reefs for example.

Biogeographic and multivariate analyses help to define suitable strata, and in most cases they will be based on unambiguous physical features such as depth or substrate type. Once strata have been defined they should be kept the same from season to season or year to year. Too many 'long-term' data sets cannot be used for looking at long-term changes in community structure due to changes in sampling areas, sampling methodology and level of replication. But this rationale is also unwisely invoked to avoid making necessary improvements to survey design and promoting 'institutional lethargy'.

Many fished species show density dependent habitat use, increasing in range with increases in abundance (section 4.3). As such, surveys should be designed to cover the potential range of a stock during the periods of greatest abundance. Since scientific study is usually preceded by exploitation, most stocks are below their peak abundance when study begins and current distributions may not indicate those in future years. Some methods of abundance estimation, such as acoustic and egg production methods, are rather more flexible than trawl surveys and can be modified on an annual basis to account for changes in distribution.

10.2.2 Visual census methods

In relatively clear and shallow waters divers can make direct records of size and abundance using visual census (Fig. 10.2). During the census, target species are counted in set areas or over set time periods. Visual census

Fig. 10.2 A diver estimating reef fish abundance by the underwater visual census method. Photograph copyright S. Jennings.

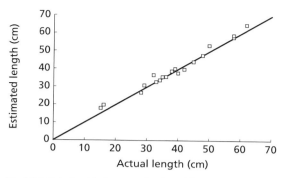

Fig. 10.3 Relationship between the sizes of fish estimated underwater and their actual lengths. After Carlos and Samoilys (1992).

works most effectively for non-cryptic, diurnally active species that tend not to avoid divers. Species in the census can include many reef fishes, sea urchins, abalone and sea cucumbers. A major advantage of visual census methods is that habitat data can be collected at the same time and that the divers gain an understanding of the fished ecosystem that is often missed by those who only see target species dead and dying on deck. Decompression requirements limit working time for divers and visual census work is usually conducted at depths of 20 m or less. This is not necessarily a problem since many important species on shallow reefs are found and fished at these depths. Area-specific abundance estimates are possible if reef area can be measured using aerial photographs or remote sensing (Green *et al.*, 1996). In deeper water visual census counts have been made from submersibles, such as those of snappers, groupers and carangids off Johnston Atoll in the central Pacific (Ralston *et al.*, 1986). Visual census methods are reviewed by Harmelin-Vivien *et al.* (1985).

The main **visual census methods** are transects, point counts and timed point counts. These have been developed from the many census techniques of terrestrial ecologists (Seber, 1982; Buckland *et al.*, 1993). Lengths of target species can be estimated and converted to weight and/or age using length–weight relationships and age–length keys, respectively (section 9.3.2). Length estimates made by trained divers can be remarkably accurate (Fig. 10.3). When target species are very abundant then size classes and abundance categories (often on a log scale) are used. However, these are best

avoided if possible since data in this form cannot be used to calculate biomass.

Transect methods involve counting individuals along either side of a tape measure that is laid on the seabed (Fig. 10.4a). An area 50×5 m is used as the basic sampling unit for many reef species, but smaller areas are used for smaller fish or cryptic recruits and larger areas are used for scarce resource species like clams. Habitat data can be recorded at the same time using the **line intersect method** where the distance that the tape overlays each type of habitat (coral, sand, rubble, rock, etc.) is recorded and expressed as percentage of total distance covered. If there are net movements of fish then the direction in which transects are surveyed may bias counts (Watson *et al.*, 1995). Large areas of seabed are best covered with **Manta tows** where a snorkel diver is towed over fixed distances while holding onto a board towed behind a boat. The boat stops at intervals to let the diver record data. Manta towing is favoured for counting large visible species like giant clams that are relatively scarce and found in shallow water.

Point counts involve counting individuals in circular areas (Fig. 10.4b, c), usually with a radius of 5–7.5 m. Fish within a set distance of the seabed are recorded. The distance depends on the underwater visibility and the habitat of target species. Instantaneous counts are used to estimate absolute abundance. Timed counts measure relative abundance because counts will depend on the frequency of movement in and out of the count area. During instantaneous counts the abundance and size of target species is estimated by counting each individual in the census area and estimating its length to

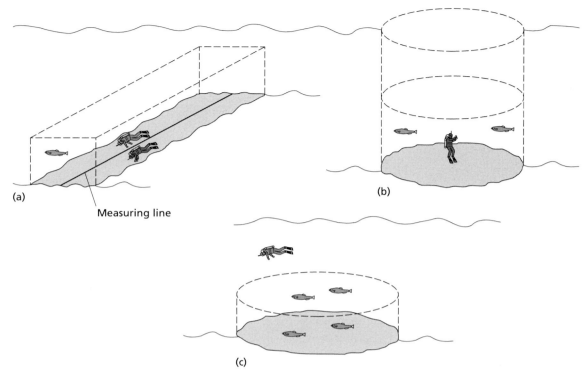

(a)

Measuring line

(b)

(c)

Fig. 10.4 Visual census methods. (a) Transect; (b) timed point count; and (c) 'instantaneous' point count. The area covered by the census is indicated by the dotted lines.

the nearest 1 cm. The count for each species group is completed as quickly as possible, usually beginning with the most wary species. In reality 'instantaneous' counts will take several seconds for each species group and provide a series of snapshots. When a count for one species is complete, all further movements of that species into or out of the census area are disregarded. The time required to complete a census is not standardized since this depends on the number and diversity of species present and the complexity of habitat to be searched.

Since point count techniques do not require tape measures they are particularly useful in strong currents and in areas where marine mammals such as sealions interfere with tapes. Locations of count boundaries are usually estimated, and this can be done very accurately with practice (Bohnsack & Bannerot, 1986). Point counts offer several advantages over transect counts since their location is easy to randomize and short count durations increase the potential for replication in a given survey period. Push–pull effects due to divers

attracting or repelling fish are reduced by the short counts, and net movements of fish in the census area do not create a bias.

Visual census abundance estimates can be biased in several ways. Many species are quite cryptic and may be overlooked if the diver's swimming speed is too high (Lincoln-Smith, 1988). Moreover, the ability of an observer to record fish falls rapidly within a few metres of the observation point and depends on underwater visibility. Some species are wary or disturbed by the diver's activities (Samoilys & Carlos, 1992), and estimates of fish abundance may fall when consecutive visual surveys are conducted in the same area (Harmelin-Vivien *et al.*, 1985). The detectability of fish in transect surveys may vary with human disturbance and this could bias comparisons between fished and unfished areas. Thus, area-specific lethrinid catches on Fijian reefs were higher than biomass estimates from visual census (Jennings & Polunin, 1995b) probably because these fish avoided divers (Kulbicki, 1998). Clearly, it is important to know the biology and behaviour of the target species before deciding whether to use visual census.

10.2.3 Acoustic methods

Sound waves travel much greater distances through water than light waves, and their passage is largely unaffected by turbidity. As such, they help us 'see' through water. We have already described how acoustic techniques are used by fishers to detect shoals of fish and to monitor the geometry and catching efficiency of their gears (Chapter 5), but fishery scientists also use acoustic methods to estimate the abundance and size composition of fish populations. **Acoustic methods** allow fish to be detected throughout the water column (excluding the few centimetres closest to the surface and seabed) and are particularly useful for estimating the abundance and distribution of pelagic fishes that could not otherwise be sampled across large depth ranges.

Sound pulses are produced, and their echoes received, by a transducer that is towed from, or mounted in, the hull of the survey vessel. When a sound pulse meets an object with a density different from the water, sound waves are scattered in all directions (Fig. 10.5). Since they spread spherically, their intensity decreases rapidly with distance. The intensity of echoes received from

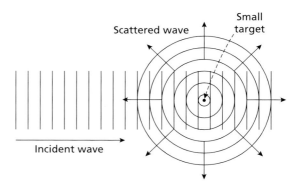

Fig. 10.5 Scattering of sound waves by an acoustic target such as a fish's swim bladder. After MacLennan and Simmonds (1992).

objects at different distances is magnified by a gain function to compensate for their attenuation.

Early devices recorded individual echoes, but a key development was the **echo-integrator** that allowed the strength of many echo returns to be summed over depth and distance (Fig. 10.6). If the **target strength** (acoustic reflectivity) of the fish is known (from calibration experiments), then the output from the integrator can be converted to estimates of fish density (MacLennan & Simmonds, 1992; Misund, 1997). Target strength is

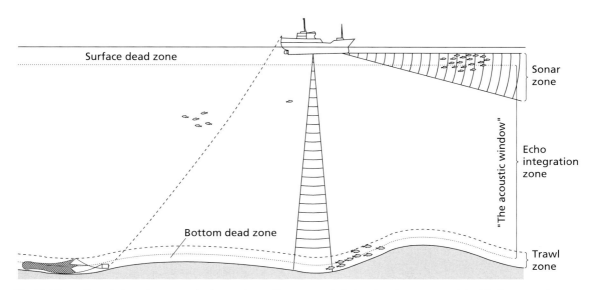

Fig. 10.6 Recording windows for sonar, echo-integration and bottom trawls during abundance surveys. After MacLennan and Simmonds (1992).

difficult to measure and depends on size, species and behaviour, but knowledge of target strength is critical to getting accurate biomass estimates. For fishes with swimbladders, the swimbladder is the main source of scattering. Mathematical theory can predict the target strengths of regular objects, but for fishes, target strength is usually estimated by empirical measurement, often using live fish in cages or *in situ* studies (Misund, 1997). The echo integrator is set to produce output within a specified depth range over the distance travelled. Output from echo integrators can be recorded and processed later to identify schools of fish, species of fish and to adjust target strength estimates by species. Most species cannot be identified without experimental fishing, and acoustic methods are usually used for large-scale surveys of pelagic species that can be identified from school structure and location. The results can be corroborated with other estimates of abundance (Hampton, 1996).

Many of our general comments on survey design (section 10.2.1) apply to acoustic surveys (MacLennan & Simmonds, 1992). The cruise track should cover the whole area occupied by the target stock and the area must be surveyed in the shortest possible time to get a spatial picture of distribution without temporal bias. This is best achieved by avoiding periods when fish are migrating or scattered over large areas. Another option is adaptive survey, where scouting tracks are followed by intensive surveying in areas where fish are located. This approach is useful for identifying and mapping aggregations. Some examples of common survey designs are shown in Fig. 10.7. Fish recorded on traces are sampled by fishing and the samples must be representative of the size and species structure of the shoal if valid target strength estimates are to be applied. Since surveys often involve several ships it is also important to calibrate between them. This is often done by comparing outputs as vessels take turns to lead around a set grid.

The process by which stock abundance and structure are determined from acoustic surveys and sampling of fish is shown in Fig. 10.8. Some famously inaccurate abundance estimates were made using acoustic techniques because the wrong species were identified or echoes misinterpreted. Shoals of fish may create acoustic shadows and fishes furthest from the transducer are poorly detected. This leads to underestimates of abundance when fish density is high. Correction factors have been derived to compensate for 'shadowing' (Toresen,

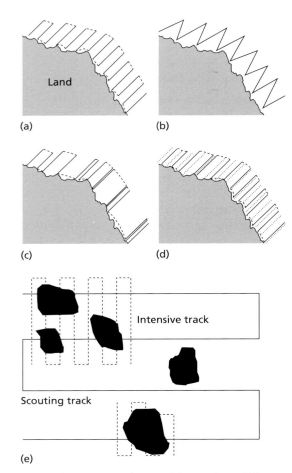

Fig. 10.7 Acoustic survey designs. (a) Systematic parallel transects; (b) systematic angled transects; (c) random transects; (d) random transects within strata (strata defined by dotted lines); and (e) outline survey where aggregations are identified on a scouting track and then surveyed intensively. After MacLennan and Simmonds (1992).

1991). As with trawl surveys, fish may avoid the survey vessel in shallow water leading to low abundance estimates. Despite these biases, acoustic technology is improving all the time, and we expect acoustic methods will be increasingly used to estimate the biomass of large pelagic stocks.

10.2.4 Trawl surveys

Commercial catch and effort data are unlikely to provide good estimates of relative abundance. This is

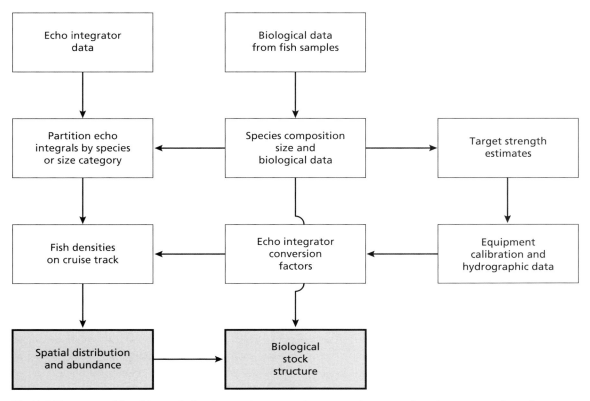

Fig. 10.8 The process of describing stock abundance and structure using an acoustic survey and supplementary trawl sampling. After MacLennan and Simmonds (1992).

because fishers target 'hot spots' where abundance remains high regardless of changes in overall stock size, and improvements in gear technology mean that effective fishing effort changes over time. Problems with commercial data have encouraged fishery scientists to obtain their own data using standard fishing techniques at standard locations. Most of these surveys are conducted with trawl nets, but some use fixed nets, longlines or pots. For many non-target species trawl surveys provide the only abundance estimates, so both precision and accuracy of the data can be important.

Annual **bottom trawl surveys** are widely used to measure variation in the size of commercially important stocks, to measure rates of recruitment and to sample fish for biological studies (Walsh, 1996). Nets with smaller meshes than those used commercially can be used to sample fishes before they recruit to the fishery. Some surveys are conducted through the cooperation of several countries. Surveys of the North Sea in Europe,

for example, are coordinated by the International Council for the Exploration of the Sea (ICES). Contributions from various western European countries in a number of surveys provide data that are pooled and analysed at regular meetings of multinational working groups of scientists.

Fish of different sizes and species have very different catchability, and catchability depends on gear design. Trawls never sample the whole community and apparent differences between fish communities may be due to selectivity if gear is not standardized. Many international surveys are plagued by problems of gear standardization, because the 'standard' trawl is rigged and fished differently by different countries.

The aim of a trawl survey is usually to get indices of abundance that are proportional to true abundance. The entire area inhabited by a fish stock should therefore be surveyed and the catchability (q) of the gear needs to be known. If catch (C) and true abundance (N) are related and f is fishing effort, then:

$$C = qfN \tag{10.1}$$

q is effectively the proportion of the stock caught for a given effort. In scientific surveys, attempts are made to keep f and q constant by using consistent tow durations, towing speeds and gear types. Thus catch per unit effort (CPUE) in scientific surveys can be defined as:

$$CPUE = qD \tag{10.2}$$

Where survey catchability (q) is the proportionality constant between CPUE and true abundance, and D is the density of fish in the path of the trawl. q may be decomposed into **availability** (q_a) and **catching efficiency** (q_e). Availability is the proportion of the stock in the survey area. Catching efficiency is the ratio of the number of fish caught and retained by the net to the number of fish in the trawl path. The area of the trawl path is the product of tow distance and wing spread for an otter trawl or tow distance and beam width for a beam trawl. From this it follows that:

$$D = CPUE/(q_a q_e) \tag{10.3}$$

which means that stock size (B) is given by:

$$B = DA \tag{10.4}$$

where A is the area used by the stock. If the catch efficiency, q_e, is constant but unknown, then B will be proportional to actual stock size. The biases in trawl surveys result from changes in q_a and q_e. q_a changes in response to the environment and to stock abundance (through density-dependent habitat use, section 4.3) while q_e changes in response to trawl efficiency. Neither is usually constant over time although it is often incorrectly assumed that size and species selection, trawl performance, and area swept by the trawl are known and constant.

Many biases affect abundance estimates from trawl surveys. These relate to changes in the efficiency of the net with depth, time of day, tow duration and vessel noise. In fact, there are so many biases that catch rates are never consistently proportional to abundance, and trawl surveys provide the best of the bad options for assessing abundance! We now consider some of these biases in more detail.

The geometry of otter trawls changes with depth. The swept area increases and the height of the trawl headline decreases in deeper water (Fig. 10.9). This may increase catches of bottom-associated species such

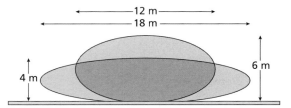

Fig. 10.9 Changes in trawl geometry at depths of 50 m and 450 m. The shaded area shows the shape of the mouth of the net, which is wider when fished at the greater depth. After Godø and Engås (1989).

as flatfishes in deep water and decrease those of species living above the seabed (Godø & Engås, 1989). Other factors such as current, towing speed and bottom type also affect trawl geometry. A varying swept area model can be used during data analysis to account for changes in trawl geometry. Changes in trawl geometry can be measured with net-mounted sensors that transmit positional data to the vessel. Beam trawls have fixed geometry but can only be fished on relatively smooth mud, sand, shell and gravel grounds, and are not suited to catching species living above the seabed.

Tow duration may affect relative catch rates since short tows may not allow time for larger fish to tire and fall back into the net. Thus, catches from short tows might be dominated by smaller fishes. However, a practical study by Godø et al. (1990) suggested that towing time had little effect on the size distribution of catch when increased from 5 min to 2 h. In this case, it would be cost effective to reduce tow duration in order to increase levels of replication. The amount of light that penetrates the water affects fish behaviour and catch rates (Walsh, 1991). Engås and Ona (1990), for example, studied fish behaviour by day and night in the mouth of a trawl using a high-frequency scanning sonar. At night, fish entered the middle of the trawl, close to the bobbins, and did not appear to escape over the headline. By day, fish entered more irregularly and escaped over the headline.

Noise made by fishing vessels affects the efficiency and selectivity of pelagic and bottom trawls. Most noise comes from propeller cavitation. In shallow water the noise created by vessels may lead to avoidance behaviour. Ona and Godø (1990) showed that haddock were disturbed to depths of 200 m, but not at 200–500 m. Vessel avoidance will affect trawl selectivity when a range of species and size classes, with different swimming capacities and behaviour, are encountered (Fréon, 1993).

10.2.5 Depletion methods

In the short term, the rate of reduction in abundance of a fished population is determined by catch rate and population size. **Depletion methods** use the relationship between abundance and catch rate to predict abundance. The methods we describe are valid if the fished population is closed, if the period of fishing is short relative to the time for population growth, and if catchability is proportional to abundance. These criteria limit the applicability of depletion methods, but they have been applied to small isolated stocks of reef fishes around atolls, and fish populations linked with single well defined geographical features such as seamounts or estuaries. We consider a basic depletion method (Leslie & Davis, 1939).

When a closed population is fished, the population at time t (N_t) will be equal to the population at $t = 1$ (N_1) less the cumulative catch prior to time t (K_t).

$$N_t = N_1 - K_t \tag{10.5}$$

If it is assumed that catch rate is proportional to effort, then

$$\text{CPUE}_t = qN_t \tag{10.6}$$

where q = catchability. Substituting 10.5 into 10.6:

$$\text{CPUE}_t = qN_1 - qK_t \tag{10.7}$$

Since this takes the form of a linear regression, a plot of CPUE_t (y) against K_t (x), will have a y-intercept (a) of qN_1. The slope of the line will be $-q$ and the x-intercept (given by qN_1/q) is N_1 (Ricker, 1975). Note that there is no reason why abundance cannot be measured independently of the fishing process that determines cumulative catch. Thus, in equation 10.7, any other measure of abundance could be substituted for CPUE_t. This could be an estimate of N from an egg production survey or a visual census biomass estimate for example.

As an example, we have applied the Leslie and Davis method to data collected by Smith and Dalzell (1993) (Table 10.1). They repeatedly spearfished an area of reef on Woleai Atoll in the Federated States of Micronesia over a 4-day period to estimate the biomass of surgeonfishes present. When these data were plotted

Table 10.1 Catch and effort during a depletion experiment at Woleai Atoll, Federated States of Micronesia where surgeonfishes (Acanthuridae) were depleted by spearfishing (Smith & Dalzell, 1993).

Day	Effort (person h)	Catch (kg)	CPUE (kg h^{-1})	Cumulative catch (kg)
1	30.75	18.44	0.60	18.44
2	23.92	12.73	0.53	31.17
3	24.50	8.61	0.35	39.78
4	25.08	8.04	0.32	47.82

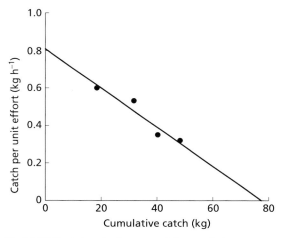

Fig. 10.10 Depletion plot for surgeonfishes caught by spear fishing at Woleai Atoll. Data from Smith and Dalzell (1993).

following equation 10.7, the fitted regression (Fig. 10.10) has parameter values $a = 0.81$ and $b = -0.0104$. Thus, $N_1 = 0.81/0.0104 = 77.88$ kg. The reef area was 20 400 m^2, so the estimated initial biomass of acanthruids is 3.82 g m^{-2}. If this estimate is to be used for assessment purposes, then the associated confidence limits would be calculated too. The formulae for these are given by Ricker (1975).

Our example from Woleai Atoll demonstrates some of the biases associated with depletion methods. The catches reported are family rather than species specific and different species within the family have different catchability. As such, catchability is not consistently proportional to abundance. The situation is not necessarily better with single species either, because different stages of the life history have different catchability.

Moreover, catchability is likely to fall as depletion proceeds because the most vulnerable individuals (easiest to catch) are depleted first. If catchability falls as depletion proceeds then N_1 will be an underestimate of true population abundance. Another concern is that some individuals are virtually uncatchable. For example, older, larger and 'experienced' fish may have learnt to avoid spear fishers. Biases associated with depletion methods and some approaches to assessing their validity are described in Ricker (1975) and Hilborn and Walters (1992).

The assumption that the population is closed and the need to conduct fishing over short time intervals limits the value of depletion methods for large stocks over large sea areas. Depletion methods have also been developed for use with open populations and these require estimates of recruitment, immigration and emigration (Seber, 1982; Hilborn & Walters, 1992). As more and more of the processes that describe stock dynamics are included in the methods, they become increasingly like the dynamic models described in Chapter 7.

10.2.6 Mark–recapture methods

Mark–recapture methods have been used most successfully to estimate the abundance of terrestrial animals and of freshwater fishes in enclosed lakes, but they can be used for marine fishes in confined areas such as estuaries or crustacean populations on rocky reefs. Since tags are used to mark the target species, it is possible to obtain information on additional aspects of their biology such as growth and movement. The methods assume that tagged fish mix randomly with untagged fish before capture and have the same catchability. In this case, the proportion of tagged fish (T) in a population of size N is equivalent to the proportion of tagged fish recaptured (R) in a catch (C).

$$T/N = R/C \tag{10.8}$$

so that abundance is estimated by:

$$N = TC/R \tag{10.9}$$

This mark–recapture method also assumes that tagging does not alter rates of mortality and that there is no recruitment, immigration or emigration between the times of marking and recapture. These assumptions mean the methods have limited applicability for stocks of adult fish that migrate over large areas. Methods for

calculating variance associated with population estimates and developments of mark–recapture methods are described by Seber (1982), Ricker (1975) and Youngs & Robson (1978). Some of these methods allow for multiple release of tagged fish into the stock and for immigration and emigration. A huge literature pertains to mark–recapture methods; consult Schwarz and Sibert (1999) for modern approaches.

A good example of the use of mark–recapture methods is the study of Pawson and Eaton (1999). They used a mark–recapture method to estimate the number of young European sea bass *Dicentrarchus labrax* in the Medway Estuary, England, and to estimate the proportion killed on the intake filter screens of a large power station that abstracted cooling water from the estuary. Mortality on the screens was about 15% of the population size of young-of-the-year fish. Mark–recapture was an ideal method in this situation because young-of-the-year bass were fully recruited to the nursery and tended not to immigrate or emigrate.

10.2.7 Egg production methods

The abundance of eggs and larvae may be used to estimate the abundance of spawners (Saville, 1964). **Egg production methods** provide fishery-independent estimates of biomass and can be used for fishes which spawn pelagically or demersally on defined spawning grounds. We describe them in some detail since they are used to estimate the biomass of some of the largest fished stocks and show how information on fish life histories, egg distributions and biological parameters is synthesized for assessment purposes.

There are two methods for estimating stock biomass (B) from egg surveys; the **annual egg production method** (AEPM) and the **daily egg production method** (DEPM).

The total egg production (P) of a fish stock is:

$$P = BRF \tag{10.10}$$

where B = biomass of spawning stock, F = fecundity (as number of eggs spawned per unit weight of female) and R = sex ratio or female fraction (abundance of egg-producing females as a proportion of the abundance of females and males) (Parker, 1980). Thus:

$$B = P/(FR) \tag{10.11}$$

The AEPM and DEPM methods differ in the way they integrate egg production (P) and fecundity (F) with respect to time. The choice of method depends on whether the study species are determinate or indeterminate spawners. In determinate spawners, potential annual fecundity is determined before the onset of spawning and the eggs may be spawned all at once, or in batches. In indeterminate spawners the potential annual fecundity is determined after the onset of spawning, and unyolked oocytes continue to be matured and spawned during the spawning season. The eggs are always spawned in batches (Hunter & Lo, 1993; section 9.3.5).

Annual egg production method

The AEPM (Fig. 10.11a) is applied to determinate pelagic spawners, such as the Atlantic mackerel *Scomber scombrus* and the horse mackerel *Trachurus trachurus* (Carangidae) (Lockwood, 1988). In these species, total fecundity, the total number of eggs spawned in a year, is equivalent to the potential total fecundity prior to spawning less losses due to atresia. When using the AEPM, P (equation 10.11) is calculated as total annual egg production by the stock (the area under the curve in Fig. 10.11a) and F is the mean annual fecundity per unit weight of female fish. Total fecundity (potential annual fecundity) is determined from a sample of fish collected before the spawning season. Egg production in the sea is estimated from a series of plankton surveys that cover the whole spawning area of the stock throughout the spawning season. The survey data are used to produce an egg production curve and integration of the area beneath this curve gives total annual egg production. Mean weight of fish (W) and the sex ratio (R) are estimated from trawl samples. To collect valid samples the stock identity of spawning fish must be clear and the sample representative (Picquelle, 1985). The plankton surveys are the most expensive and challenging part of this work since spawning areas

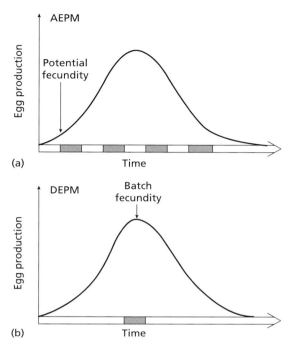

Fig. 10.11 Annual egg production (AEPM) and daily egg production (DEPM) methods of estimating biomass from egg production. With the AEPM (a), egg production is integrated over the whole spawning season using the results of several surveys (shaded) and potential fecundity measured before spawning is assumed to equal realized fecundity. With the DEPM (b), there is one egg survey close to the middle of the spawning season, and batch fecundity and spawning fraction are estimated from a trawl survey conducted at the same time. After Anon (1997) and Priede and Watson (1993).

may extend over thousands of square kilometres, the distribution of eggs may be very patchy and spawning seasons may last several months. An example of the use of the AEPM to estimate the biomass of two species of fish in the Irish Sea is given in Table 10.2.

Table 10.2 Estimation of plaice *Pleuronectes platessa* and sole *Solea solea* biomass in the Irish Sea using the annual egg production method. Data for 1995 from Anon (1997).

Parameter	Units	Plaice	Sole
Annual egg production (P)	10^9 eggs	1 850	2 814
Mean fecundity of female	eggs	69 012	152 532
Mean weight of female (W)	g	324	342
Mean fecundity (F)	eggs/g	213	446
Sex ratio (R)	weight females/ weight females plus males	0.466	0.612
Spawning stock biomass (B)	10^3 t	18.64	10.31

Parameter	Units	1980	1982	1984
Relative batch fecundity (F)	no. eggs batch^{-1} g female^{-1}	444.44	575.94	456.32
Spawning fraction (f)	g spawning females g mature female^{-1}	0.142	0.120	0.160
Adjusted relative batch fecundity F′ (= Ff)	no. eggs batch^{-1} g female^{-1}	63.11	69.11	73.01
Sex ratio (R)	g females/ g females plus males	0.478	0.472	0.582
Daily specific fecundity (F′R)	no. eggs g^{-1} d^{-1}	30.17	32.62	42.49
Daily egg production (P)	10^{12} eggs d^{-1}	26.34	13.51	12.98
Spawning stock biomass (B)	10^3 t	873	414	305

Table 10.3 Estimation of northern anchovy biomass using the DEPM. Data from Picquelle and Stauffer (1985).

Daily egg production method

The DEPM (Fig. 10.11b) is used for indeterminate spawners such as sardines and anchovies, although it can be applied to determinate spawners that release eggs in batches (Anon, 1997). With the DEPM, all parameters can be estimated in a single survey and it provides a rapid fishery-independent estimate of biomass. The DEPM requires that the entire stock of spawning and non-spawning adults are accessible to sampling gear, and that one plankton survey covers the whole spawning area around the time of maximum egg production. Practical experience shows that it takes 4–5 months to get a biomass estimate for the northern anchovy *Engraulis mordax* stock off the Californian coast.

Each female of an indeterminate spawning species produces several batches of eggs so equation 10.11 has to be modified. To relate egg production to female biomass, relative fecundity (F) is adjusted to allow for spawning frequency:

$$F' = Ff \qquad (10.12)$$

where f is the fraction of females spawning in a given time interval. It can be estimated if the females that have spawned or will spawn can be identified, if the longevity of the character used to identify spawning is known and if the spawning frequency remains constant over the sampling interval. Subject to these conditions, the spawning fraction (f) is the proportion of females displaying the spawning characteristic divided by the length of time for which it is detectable. Equation 10.11 can now be written as:

$$B = P/(RF') \qquad (10.13)$$

P can be estimated over any time interval provided that f is determined over the same interval. Thus f = 1 for a species that spawns once each year (Parker, 1980). Table 10.3 shows the derivation of spawning stock biomass estimates for the northern anchovy stock off California. Clearly, there will be errors associated with all the parameter estimates and the final biomass estimate. A formal procedure for carrying these errors through the calculation has yet to be developed.

Egg mortality and atresia

Both the AEPM and the DEPM methods are rather more complex than described, since fecundity and egg production values have to be corrected before they are entered into equations 10.14 and 10.16. Fecundity estimates are corrected to allow for **atresia** (the resorption rather than spawning of fully developed eggs) and egg production estimates are corrected for egg mortality.

Atresia can reduce potential fecundity (the stock of eggs in the ovary before spawning: F_{pot}) by more than 10%. Because some of the eggs will be atretic (F_{atr}), realized fecundity (F_{real}; eggs released into the water) is less than F_{pot}. Thus $F_{real} = F_{pot} - F_{atr}$. In the western stock of Atlantic mackerel in 1995, for example, $F_{real} = 1302$ eggs g^{-1}, $F_{pot} = 1473$ eggs g^{-1} and $F_{atr} = 171$ eggs g^{-1}. Thus, realized fecundity was 88% of potential fecundity (Anon, 1997).

Rates of atresia are determined by examining the ovary after spawning and estimating the relative numbers of post ovulatory follicles (follicles after oocytes have been released) and atretic oocytes (eggs that are never released). To ensure the postovulatory follicles and atretic oocytes can be reliably distinguished, oocyte atresia has been staged. Staging of oocyte atresia is also used to separate immature females from postspawning females when estimating size at maturity and spawning frequency.

Many of the eggs spawned will be eaten before they are caught in plankton nets (section 4.2.2). We have to allow for this when calculating the number of eggs present at the moment of spawning, $P(0)$. Production (P) of eggs at stage (I) is modelled as:

$$P(I) = P(0)e^{(-Z.\mathrm{age}(I))} \quad\quad (10.14)$$

The mortality rate (Z) is estimated as the slope of a linear relationship between $\log_e P(I)$ and the mean age of eggs. An estimate of $P(0)$ is obtained from equation 10.14. Since egg mortality can be very high, estimates of mortality rate, which may depend on temperature and other factors, can have an important effect on estimates of egg production. Methods for calculating variance are described by Picquelle and Stauffer (1985).

Egg production

The AEPM and DEPM both require estimates of egg production in a specified time interval (usually 1 day). Since eggs remain in the plankton for several days before hatching, plankton samples can contain eggs of different ages. To calculate egg production per unit time we need to know how old the sampled eggs are.

Eggs are aged by defining a series of clearly identifiable developmental stages and describing development rates in relation to stage and temperature (Lockwood *et al.*, 1981; Lo, 1985; Lockwood, 1988) (Fig. 10.12). For example, when estimating Atlantic mackerel *Scomber scombrus* egg production it is necessary to identify stage 1 (1A or 1B) eggs and to know how long they have been at that stage. The relationship between temperature (T, °C) and age (A, h) of stage 1 eggs is $\log_e A = -1.61 \log_e T + 7.76$. So, estimated production (P) of stage 1 eggs (m^{-2} d^{-1}) = ($24 \times$ number of eggs m^{-2})/exp $(-1.61 \log_e T + 7.76)$.

To estimate egg abundance by stage, eggs are sampled at stations across the entire spawning area of the target species. An egg survey aims to cover this area in the shortest possible period to reduce temporal bias. This is often achieved by many research vessels,

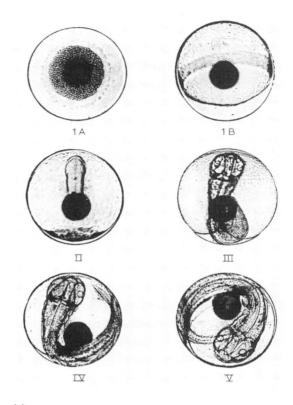

(a)

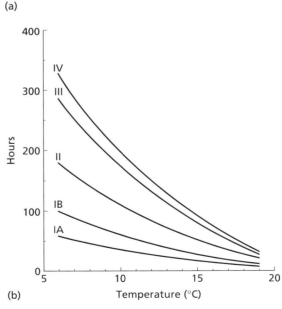

(b)

Fig. 10.12 (a) Staged development of mackerel eggs, and (b) the times to reach stages 1A to IV at different temperatures. After Lockwood (1988).

(a)

(b)

Fig. 10.13 (a) A high-speed plankton net used for mackerel egg surveys. Note the internal flowmeter in the nose-cone of the net and the external flow meter attached to the frame below. (b) The cable is used to tow the net and to transmit information on depth and flow rates. Photographs by S. Milligan, copyright British Crown.

sometimes from different nations, working at the same time. Since the survey is never truly instantaneous, the data are assumed to give a single estimate of egg abundance on the middle day of the survey.

Many nets have been designed for quantitative plankton sampling. Those used to catch eggs (Fig. 10.13) are also used for surveys of fish larvae and other types of plankton. They are usually equipped with internal and external flowmeters. The ratio between flowmeter readings provides an index of net clogging and allows the volume of water filtered to be calculated. Nets are usually fished on a double oblique tow: a V-shaped profile from the surface to 2 m above the bottom and back that is intended to filter the same volume of water

at every depth. Mesh size of the plankton net is adjusted to egg size and must be sufficiently small to catch all eggs but sufficiently large to avoid excessive clogging with smaller planktonic organisms. A mesh size of 150–250 µm is used for many species, sometimes with an outer protective sheath of stronger 1–2 mm netting.

Most plankton nets have a small nose cone to limit the rate of water inflow and to ensure that water passes through the filtering mesh at an acceptable rate. If the flow rate is too high, eggs may be extruded or the net may clog too quickly. The small nose cone means that the net can be fished at speeds of several knots, which is convenient for the research vessel and makes controlled double oblique tows possible. Most nets carry temperature sensors as temperature data are needed to calculate the developmental rates of eggs. Net and survey design are described in detail by Smith *et al.* (1985) and Smith and Hewitt (1985).

After each tow, the net is 'washed down' with a low-pressure hose to move all the captured plankton into a collection jar attached to the end of the net. The plankton are usually fixed and preserved in buffered formaldehyde solution, although alternatives are being developed because formalin is carcinogenic. Eggs are sorted from other plankton in the laboratory and eggs of target species are grouped by species and developmental stages.

The total number of eggs per unit area of sea surface (G) of stage (I) at station (j) is:

$$G(I)_j = (N(I)_j \text{depth}_j)/\text{filtered volume}_j \qquad (10.15)$$

G is converted to daily egg production (DEP) estimates using stage durations (D) in days and integrated water column temperature.

$$\text{DEP}_j = G(I)_j/D(I)_j \qquad (10.16)$$

DEP (as eggs m^{-2} by station) is used to estimate DEP in the entire sea area where spawning takes place. This can be done by multiplying the area of various strata by egg production at stations in those strata. More recently, gridding and contouring procedures available on software such as SURFER have been used to produce an egg production 'surface' and to calculate total production in the survey area (ICES, 1997a). Alternatively, egg production can be modelled using generalized additive models (GAM) (Borchers *et al.*, 1997; Augustin *et al.*, 1998). Needless to say, the survey grid should cover the entire area where spawning takes place.

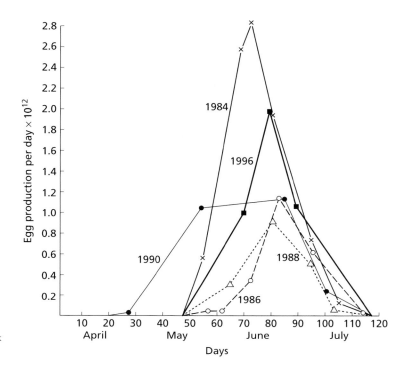

Fig. 10.14 Egg production curves for the western mackerel *Scomber scombrus* stock from 1984 to 1996. After ICES (1997a).

Estimates of DEP for the spawning area are integrated over the spawning season to give total egg production. Because it is not feasible to sample continuously over the spawning season, total production estimates are usually calculated by plotting the production of stage I eggs against the mid-date of each survey. The area beneath this seasonal production curve gives total egg production. The start date of the spawning season can be estimated from the age of the oldest eggs or larvae on the first survey and the end date of the spawning season from the last time that stage I eggs are observed. An example of an egg production curve is given in Fig. 10.14.

For demersal spawning North Sea herring *Clupea harengus*, surveys of larval abundance have been used to provide indices of stock biomass since 1946. Larval density is calculated from plankton samples on an area specific basis and used to calculate larval production estimates. Numbers of newly hatched larvae are back-calculated after allowing for size at capture, growth rate and mortality. Spawning biomass is calculated from egg production with knowledge of fecundity and the proportion of eggs surviving to hatch (Heath, 1993). For the Pacific herring, demersal egg production

is estimated by scuba divers and used to calculate spawning stock biomass (Schweigert, 1993).

10.3 The fishery

So far, we have looked at methods of abundance estimation that do not rely on fishery data. However, the scale and efficiency of commercial fishing activity far exceeds that of research surveys and the fishery can provide catch, landings and effort data from larger areas and over longer time scales. In areas where resources for research surveys are not available, fishery data may provide the only information on relative abundance. Catch data are also needed to estimate the numbers of fish removed from a stock. Catch and effort data are often collected together using the same recording system but, in large commercial fisheries, landings may be recorded without records of fishing effort. So-called catch-per-unit-effort (CPUE) data from commercial fisheries are usually **landings-per-unit-effort** (LPUE) because the weights recorded in ports exclude fish that are caught and then discarded at sea (Box 1.1; Chapter 13). Such landings and LPUE data will be biased by variable discarding. Because discard rates may exceed landings in some fisheries, it is vital to measure them, but this is still not done on a systematic basis. In general, in world fisheries we need vastly better data collection systems and to account for discards. The solution of placing independent observers aboard fishing vessels has been used in many US fisheries, but there are many fisheries in which the vessels are too small to carry observers and/or the costs of the programme would be too high.

It is usually impossible to record the catch and effort of every fisher or vessel that participates in a fishery. Thus, data are recorded for a proportion of fishers or vessels and extrapolated to the whole fishery. Fishery information can usefully be stratified. This ensures that data are collected for similar types of fisher or vessel and extrapolated to each stratum (Pope, 1988; Shepherd, 1988).

Catch, landings and effort data can be recorded by observers working with the fishers, in fishers' log books or data input systems, and by port sampling (Fig. 10.15; Box 10.1). Experienced observers working with fishers probably collect the most accurate data, but this relies on good cooperation if data collection is not to interfere with fishing activities. Log books, in which fishers

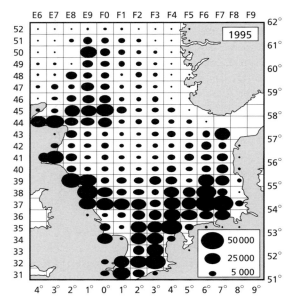

Fig. 10.15 The spatial distribution of international otter and beam trawling effort (as hours year^{-1}) in the North Sea during 1995. Fishing effort (as hours fished) is declared at the port of landing, and assigned to ICES statistical rectangles of 0.5° latitude by 1° longitude (shown by grid lines). After Jennings et al. (1999).

Box 10.1
Measurement of fishing effort.

Estimates of fishing effort are needed to examine spatial and temporal changes in fishing patterns and to calculate landings-per-unit-effort (LPUE) or catch-per-unit-effort (CPUE) for assessment purposes. Simple units of fishing effort such as days at sea by vessel type are recorded in

Table B10.1.1 Alternative measures of fishing effort in a beam trawl fishery

Measure	Requirement
Number of vessels	Occasional port monitoring
Days at sea	Regular port monitoring
Hours fishing	Logbooks/observers
Area swept by trawl	Satellite monitoring/observers

many fisheries because more detailed data are too difficult or too expensive to collect. Since the efficiency of fishers will almost always improve with time, simple units of fishing effort do not represent the same catching power from year to year. In modern trawl fisheries, for example, catch efficiency per hour will often increase by 5% or more each year.

Table B10.1.1 gives examples of different measures of fishing effort that may be used in a beam trawl fishery. As the complexity of the effort measures increases they become more useful for calculating CPUE or LPUE. However, as complexity increases, the effort data will also cost more to collect. In practice, the data collected reflect a compromise between information content and cost. Thus days at sea may be collected for the whole fleet while swept area data may be collected for a few 'representative' vessels and extrapolated to estimate swept area for the fleet.

record details of fishing effort, fishing locations and catches, provide useful information but require constant attention that may not be compatible with fishing activities. The most successful log-book schemes usually collect relatively simple data. If log-book schemes are not monitored and if there is little or no contact between the scientist and fisher, they are likely to fail. To overcome this, some schemes are compulsory, but this may encourage fishers to misreport catches. Log-book data collected in fisheries where catches are regulated through quotas and restrictions on fishing grounds are unlikely to be reliable (unless the system is very well policed) because there is a big incentive to cheat.

On-board data-collection systems are now used on some larger vessels, where fishers record data prior to landing the catch. Such systems are in use in some western Australian fisheries for example, and the fishers are encouraged to report accurate data because there are spot checks on landing vessels to ensure the data input corresponds with landings. If there are discrepancies then the skipper will be fined. Satellite tracking devices are also used on an increasing number of large vessels to monitor their movements. These may be combined with a catch and effort input system to provide total real time information. Port sampling is still used to collect landings and effort data. Landings are easy to monitor if there are a few key ports, but monitoring becomes increasingly difficult when large fleets land in many small harbours.

Many methods of fishery assessment rely on the idea that catch rates are proportional to stock abundance. Unfortunately, the analysis of data from commercial fisheries and independent surveys shows that LPUE from commercial fisheries is rarely proportional to abundance. Rather, the behaviour of fishes and fishers tends to mean that LPUE only falls rapidly when stocks approach collapse. The poor correlation between LPUE and abundance is well known to regulators who have difficulty in persuading successful fishers that stocks are dangerously low. Fishers maintain high catch rates because there is density-dependent habitat use and the fishing hot spots continue to be populated by fish moving there from other parts of their range. In other cases, LPUE will decrease faster than abundance (Hilborn & Walters, 1992). Thus, for some reef fishes and invertebrates, patches of high concentration are rapidly depleted and LPUE falls relatively fast as the distribution of target species becomes more homogeneous. If the relationship between LPUE and relative abundance is known it can be used as a non-linear abundance index.

Given the problems we have already identified with estimating LPUE from 'standardized' trawl surveys, it should be clear that the quality of LPUE data from fisheries is even lower. In real fisheries, gears are rarely standardized, the efficiency of fishers and gears increases with time and fishers never fish randomly. Thus, fishers enter the fishery at times when available technology differs and have different amounts to invest in vessels and gear, and different ways of maximizing their benefits from the fishery (Hilborn & Walters, 1992). There is one further but rather unusual problem in some fisheries, that of fishers not trying to maximize catch rates because fishing has social functions (section 6.3.1).

Summary

- Direct estimates of abundance help to predict the size of a fishable resource, rates of recruitment or year-class strength and to validate abundance estimates based on analytical methods.
- Errors around abundance estimates are due to bias and variance. Bias is systematic error. Variance is not systematic and can be reduced by improved survey design and increased replication.

- Visual census methods are best used for estimating abundance of non-cryptic, diurnally active species that do not avoid divers and live in shallow and clear water.
- Acoustic methods are good for estimating abundance of shoaling pelagic species found across a range of depths.
- Trawl surveys are used to estimate the abundance of groundfish resources. Surveys based on stratified systematic grids are widely used and may give biased

Continued p. 222

Summary *(Continued)*

estimates of mean abundance. However, they are good for mapping boundaries, contouring abundance patterns and making temporal comparisons.

• Depletion and mark–recapture methods are most useful for estimating abundance of fished species in confined habitats such as reefs and estuaries.

• Egg production methods are used to estimate abundance of large stocks that spawn pelagic eggs. The annual egg production method (AEPM) and daily egg production method (DEPM) differ in the way they integrate egg production and fecundity with respect to time.

• The scale and efficiency of commercial fishing activity means that catch and effort data can be collected from large areas over long time scales. The quality of such data are affected by misreporting and differences in fisher skill or behaviour.

Further reading

Seber (1982) and Buckland *et al.* (1993) describe methods for estimating animal abundance. The coverage of Seber (1982) is particularly comprehensive. MacLennan and Simmonds (1992) and Misund (1997) review acoustic methods, and Hunter and Lo (1993) describe egg production methods. Gunderson (1993) provides wide-ranging coverage of methods used to survey fish resources. Chapters in Gulland (1988) and Hilborn and Walters (1992) discuss survey design and collecting data from fisheries.

11 Bioeconomics

11.1 Introduction

Globally, fishing provides employment for millions of fishers and for workers in associated industries such as boat building, net making and retailing. Fishers buy boats and fishing gear, sell catches, spend income, invest profits and often receive subsidies. Given that fishing is the focus of so much economic activity, it is surprising that the roles of economic factors in driving fisheries exploitation are often ignored. To manage fisheries effectively, we need to know how economic factors affect them.

One might expect fishing to be a profitable business. Apart from the costs of boats and gear, access to fishing grounds is often free and fishers reap harvests that grow without being sown. However, in global terms, the fishing industry is highly inefficient and the costs of fishing, supported by government subsidy, have exceeded direct income by more than $US50 billion each year. Economic analyses help us to understand why resources are used as they are, why fisheries are economically inefficient, and how fisheries could be better managed. Moreover, in conjunction with biological data, they can provide a basis for choosing between management options. Bioeconomic analyses, for example, can help to determine optimal fleet sizes, configuration and employment, whether catch limits should be fixed or variable, and how taxation or licence fees would influence fishing effort.

The aim of this chapter is to review the economic significance of fisheries throughout the world and consider the economic reasons why fishers often exploit stocks beyond their biological capacity. We then describe some bioeconomic models and the ways in which they can help to inform management decisions.

11.2 The value of fisheries

11.2.1 Trade in fished species

Most fished stocks are exploited for economic gain. Even when fishers rely on fishing effort to get their own food, they usually sell catches as well as eat them (Fig. 11.1). The total value of fish trade between nations exceeds $US50 billion each year and the trade within nations is worth much more (FAO, 1999). In 1997, 95% of international trade involved only 20 countries. Norway, China, the United States, Denmark, Thailand and Canada were the main exporters, each exporting fish and fish products worth more than $US2000 million. Japan was the main importer, receiving $US15 540 million worth of fish and fish products in 1997. More than $US3000 million worth of imports were also received by the United States, France and Spain (FAO, 1999).

In most large and economically developed nations, fisheries make a relatively minor contribution to national economic activity. Thus, the contribution of fisheries to **Gross Domestic Product** (GDP), the total income of a country before costs, is less than 1% in Europe and the United States. National figures can be misleading, however, because the income from fisheries will often be an important driver of economic activity within coastal communities.

In small islands and developing countries with extensive coastlines, fishing can be a key contributor to national economic activity. In Iceland, where there are major fisheries for cod, herring and capelin, fishing contributes directly to 15% of GDP and to 35–50% of GDP via economic linkages and multiplier effects (OECD, 1997). In the Pacific Island state of Kiribati where there are many artisanal fisheries for reef fishes and tuna, fishing contributes directly to 54% of GDP (Dalzell *et al.*, 1996).

Fig. 11.1 Fishers selling their catches at a market in the Fijian capital of Suva. Photograph copyright S. Jennings.

11.2.2 Catch values and employment

In 1997, the total first-sale value of fish catches around the world was $US93 329 million of which marine species accounted for $US74 601 million (Table 11.1). The most valuable groups of marine fishes were the redfishes, basses and congers at $US13 687 million and the cods, hakes and haddocks at $US8423 million. Crustaceans were a high-value and low-weight catch accounting in total for 25.1% of value and 6.2% of weight, while fish landed for conversion to fish meal and oils (reduction) were a low value and high weight catch accounting for 3.3% of value but 28.6% of weight. Price differs between species because it reflects the supply of fish available and the demand for them. The maximum price of individual species can reach over $US100 kg^{-1} on resale. Giant grouper *Epinephelus lanceolatus* (Serranidae), favoured by the live fish trade, have sold for over $US10 000 each in Hong Kong. Big-eye tuna *Thunnus obesus* (Scombridae), northern bluefin tuna *Thunnus thynnus* and southern bluefin tuna *Thunnus maccoyii* destined for Japanese sashimi markets also fetch over $US10 000 each.

The relative value of fisheries and the proportion of people employed in fishing depends on many factors, but in general, they are highest in developing coastal and island states. In some of these countries, fishing may be the only source of food and income for many coastal dwellers. The Organization for Economic Co-operation and Development (OECD) collate data on the value of fisheries and fishery-related employment in member states (Table 11.2). In these developed countries, direct employment in fishing is rather low, only exceeding 1% of the workforce in Norway and Korea. This contrasts with employment levels of 10% or more in some developing countries. In the Pacific Island state of Fiji, for example, around 30% of the rural population fish at least once each week. On the largest island of Viti Levu, 37% of males, 48% of females and 5% of children are fishers (Rawlinson *et al.*, 1994).

In many developed nations, the majority of fish landed are sold, and thus their value is a good indicator of the economic significance of fisheries. In rural economies, landings may be eaten by fishers and their families, and thus the social and economic significance of landings may be overlooked by policy makers who base assessments on landed value. To overcome this problem, Dalzell *et al.* (1996) and others have calculated the replacement value for subsistence landings and included them in economic assessments. In the Pacific Islands, the replacement value of subsistence landings consistently exceeds that of landings that are sold (Table 11.3). This emphasizes the significance of fisheries to rural economies, even if they are not always trading in fish.

Although recreational fisheries are not the focus of this book, their value may exceed that of commercial fisheries for species such as European sea bass *Dicentrarchus labrax* in Europe (Dunn *et al.*, 1994). In

Table 11.1 Weight, value and prices of fished species landed during 1997. From FAO (1999).

Group	Weight (1000 t)	% of total weight	Price $US t^{-1}	Value ($US million)	% of total value
Fish					
Flounders, halibuts, soles	994	1.2	2 950	2 932	3.9
Cods, hakes, haddocks	8 022	9.6	1 050	8 423	11.3
Redfishes, basses, congers	6 003	7.2	2 280	13 687	18.3
Jacks, mullets, sauries	4 845	5.8	680	3 295	4.4
Herrings, sardines, anchovies	11 674	13.9	240	2 802	3.8
Tunas, bonitos, billfishes	4 851	5.8	1 510	7 325	9.8
Mackerels, snoeks, cutlassfishes	5 263	6.3	315	1 658	2.2
Sharks, rays, chimaeras	790	0.9	840	664	0.9
Fish for reduction	23 986	28.6	103	2 471	3.3
Other fishes	5 328	6.4	510	2 717	3.6
Crustaceans					
Sea spiders, crabs	1 183	1.4	2 900	3 431	4.6
Lobsters, spiny rock lobsters	231	0.3	11 800	2 726	3.7
Shrimps, prawns	2 535	3.0	3 800	9 633	12.9
Other crustaceans	1 258	1.5	2 300	2 893	3.9
Molluscs					
Abalone, winkles, conchs	106	0.1	6 000	636	0.9
Oysters	194	0.2	2 950	572	0.8
Mussels	224	0.3	420	94	0.1
Scallops, pectens	477	0.6	2 200	1 049	1.4
Clams, cockles, arkshells	831	1.0	880	731	1.0
Squid, cuttlefish, octopus	3 321	4.0	1 800	5 978	8.0
Other molluscs	1 648	2.0	450	742	0.9
Echinoderms					
All species	109	0.1	1 300	142	0.2
Totals	83 873			74 601	

some cases, this has led to calls for the exclusion of commercial fishers from fisheries.

11.3 Bioeconomic models

Economic analyses of fisheries are, like the biological analyses we have already described, based on models that abstract aspects of the system and attempt to describe or predict system behaviour. Bioeconomic models help us to understand why fisheries develop as they do and to predict their behaviour under different management regimes. The more advanced models deal explicitly with uncertainty in the parameter estimates.

Before we consider the ways in which economic models can help to explain patterns of exploitation in fisheries, we need to introduce some terminology. **Revenue** is the price of a product multiplied by the

amount that is sold. For fishers the product is usually fish! **Costs** are the amounts that need to be spent to produce revenue. They are commonly divided into **variable costs** (**short-term costs**), such as fuel for the boat, that can change over periods of a few days, and **long-term costs** or **fixed costs** that do not depend on whether a fisher is actually fishing. Variable costs are likely to be proportional to fishing effort while fixed costs are not. Typical fixed costs would include loan repayments or insurance on boats that still have to be paid when the boat is not at sea. **Profit** is the difference between revenue and cost.

Producers sell products that are bought by **consumers**. It is assumed that producers try to maximize their profits, and they do this by deciding what to sell, how much to sell and when to sell it. Prices reflect the desirability of the product for the consumer. Thus live Maori

Country	Value of Landings ($US million)	Employment (number of people)	Employment (% workforce)
Australia	1 200	–	–
Canada	964	140 000	0.94
EU countries			
Belgium	100	624	0.01
Denmark	438	5 299	0.21
Finland	30	2 948	0.12
France	962	27 598	0.11
Germany	171	4 979	0.01
Greece	5	40 164	0.98
Ireland	189	3 400	0.29
Italy	1 415	45 000	0.20
Netherlands	518	2 834	0.04
Portugal	367	30 937	0.66
Spain	2 080	79 369	0.67
Sweden	136	3 500	0.08
United Kingdom	682	20 751	0.08
Iceland	720	–	–
Japan	21 000	325 000	0.49
Republic of Korea	–	405 000	1.99
Mexico	–	–	–
New Zealand	459	10 000	0.57
Norway	1 034	23 000	1.07
Poland	–	11 500	0.07
Turkey	–	–	–
United States	3 800	300 000	0.23

Table 11.2 Value of landings and employment in fishing (not in related service industries) within OECD member countries. Data for 1993–95. From OECD (1997).

Country	Fish production (1000 $US y^{-1})		
	Subsistence	Commercial	Total
Federated States of Micronesia	11.24	1.48	12.72
Fiji	45.77	18.34	64.11
French Polynesia	14.47	14.37	28.84
Kiribati	13.37	4.77	18.14
Papua New Guinea	41.17	22.10	63.27
Solomon Islands	8.41	4.34	12.75
Western Samoa	5.07	0.32	5.39

Table 11.3 The value of fish production in some Pacific Island nations. The value of subsistence landings is their replacement value. Data for 1989–94. From Dalzell *et al.* (1996).

wrasse *Cheilinus undulatus* (Labridae) sold to restaurants in Hong Kong fetch $US90 kg^{-1} while sandeels *Ammodytes* spp. caught in the North Sea industrial fishery are worth $US0.1 kg^{-1}. If a product is expensive, then producers try to increase supply and make more profit. However, increasing the supply usually incurs greater costs and, because price is usually inversely related to supply (the law of supply and demand), this may cause prices to fall. When costs exceed revenue, profit cannot be made and it is no longer worth fishing.

11.3.1 Descriptive bioeconomics

Gordon model
Gordon (1954) made one of the first attempts to produce an economic analysis of a fishery when he tried to explain why Canadian fishers had such low incomes.

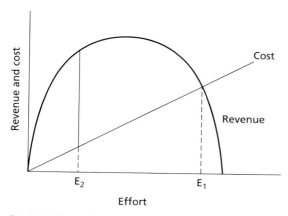

Fig. 11.2 The Gordon (1954) model showing the relationship between fishing effort, revenue and cost. In an unregulated fishery, fishing continues until revenue = cost (E_1) while the greatest profits (difference between revenue and cost) are made at E_2.

He based his model on the classical surplus production curve with all its inherent limitations (Chapter 7), but his analysis was very important because it suggested why an open-access fishery will be overfished and provide poor economic returns for the fishers.

If we assume that yield is proportional to revenue and that the cost of fishing is proportional to fishing effort, then the yield curve described in section 7.3 simply becomes a revenue curve and cost is linearly related to effort (Fig. 11.2). An unregulated fishery would be expected to expand until revenue = cost (point E_1 in Fig. 11.2). This is because fishers would lose money once fishing costs exceeded revenue. The fishery would be most profitable (highest difference between revenue and cost) at E_2. The model suggests that if the costs of entering the fishery and catching fish were low then the fishery would develop well beyond its biological limits and the stock would become depleted. Moreover, the fishery would become economically inefficient because too many fishers would be chasing too few fish.

Gordon thus provided an economic explanation for the low incomes of fishers in an open-access fishery: an open-access fishery would be expected to expand to a greater size than that which gives the highest yield and profitability. However, Gordon's model was not good for prediction. It was a static model, like the surplus production models described in Chapter 7, and suf-fered from similar weaknesses. Moreover, it did not demonstrate that there are costs associated with moving from E_1 where revenue equals cost to the point E_2 where profits are maximized. These costs exist because stock recovery takes time and revenue falls when effort is first reduced. When reduced effort causes immediate financial loss rather than gain then there is little incentive for fishers to reduce effort. This is especially true in an open access fishery where individual fishers have no control over the whole resource.

Tragedy of the commons

The idea that open-access fisheries will be exploited beyond their biological limits is an example of the tragedy of the commons as formalized by Hardin (1968). Here, competition between people who exploit a common resource depletes the resource beyond its biological limits. Many fisheries are common resources and fishers compete to exploit them. Even with the introduction of Exclusive Economic Zones (EEZ), that gave many countries sole fishing rights within 200 nautical miles of the coast, fisheries often remain open access to fishers within countries (Burke, 1994; section 1.2.1). In many developing nations, the sea is seen as the source of food and livelihood of last resort, from which no-one can be turned away (McManus, 1996). While access to marine resources is essentially free, the action of any fisher does not have a major effect on the dynamics of an exploited stock. There is little to be gained by one fisher trying to conserve fish because fish left in the water will simply be caught by someone else! Thus, it can be argued that a lack of access control, coupled with the common perception that everyone has a right to fish without cost, are the main reasons for the over-exploitation of marine resources. In section 17.3.3 we consider management methods that give fishers property rights to their resource and increase the probability that they fish sustainably. As we saw in section 6.4.4, a few such systems have already evolved in some small island states, particularly in the Pacific Ocean.

Non-malleable capital

The capital invested in fishing fleets is said to be **non-malleable** because fishing vessels can rarely be used for other things and can only be sold at considerable capital loss. As such, fishers are unwilling to reduce effort once they have invested in fishing capacity. The non-malleability of capital means that shutting down parts of a fishing industry while a resource rebuilds can have catastrophic economic consequences.

There are many examples of fishers' unwillingness to reduce effort once they have invested non-malleable capital in fishing boats. Munro (1992) gives a good example from the northern cod fishery in Newfoundland and parts of Nova Scotia. In 1977, Canada implemented Extended Fisheries Jurisdiction which meant that the cod stocks, that were once fished by many countries, were brought under Canadian control. The Canadian government decided that fishing mortality had to be reduced to allow the stock to rebuild. The initial reductions in fleet capacity were easy to achieve since foreign vessels had been ejected from the fishery. As the resource grew, the Canadian industry planned to expand, and began to invest in fishing and processing equipment. Such equipment was not useful for other purposes and thus the capital invested was non-malleable. Unfortunately, in 1987 fishery scientists realized that the cod stocks were not rebuilding at the rate predicted, and that the original estimates for stock recovery rates had been too optimistic. They called for drastic reductions in cod catches, but there was a strong adverse reaction from an industry that had invested non-malleable capital on the basis that cod were predicted to be abundant in future years.

Discounting

Even when fishers do not compete for a resource, the decision whether to catch species now or leave them in the sea will depend on their future value. If the value of a fish stock 5 years in the future is perceived to be less than the money that could be made by catching the fish now, selling them and investing the money in a bank for 5 years, then there is an economic incentive to fish.

Discount rates are used to measure the rate at which the perceived value of a resource, such as a fished stock, falls over time. Discount rates reflect the cost of return on alternative investments. Thus, if you 'invest' some money in fish by leaving them in the sea, you require that its value should grow at least as fast as the money you would get from selling the fish you caught and investing the money.

The present value (*PV*) of income *V*, occurring *t* years into the future is:

$$PV(V_t) = \frac{V_t}{(1+\delta)^t} \qquad (11.1)$$

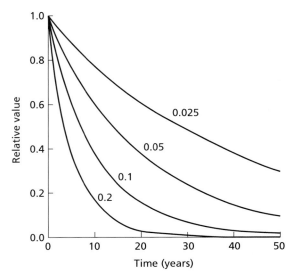

Fig. 11.3 The decline in perceived value of a unit of income at different discount rates.

where δ is the discount rate. The decline in perceived value of a unit of income at different discount rates is shown in Fig. 11.3. A 10–20% (0.1–0.2) discount rate rapidly reduces the perceived value of fish caught and sold in the future.

High discount rates are used by fishers because they reflect risk; fishers are uncertain about reaping the benefits from fishes left in the sea. Fishers' rates are, for example, typically higher than the difference between government interest rates and inflation which guide civil projects such as bridge construction. If fishers had secure rights to fish they might use lower discount rates, but the rates would probably still be higher than the return on other investments because there is always some uncertainty about the growth of fish in the sea.

If fishers use high discount rates they will want to catch fish as soon as possible, particularly if the costs of fishing do not increase rapidly as the stock is depleted. Investment in fish in the sea is less attractive than converting fish to money for investment, because the fish grow at a lower rate than invested capital. The use of high discounting rates explains why species such as whales, with very low growth rates, were 'mined' rather than fished sustainably. In fact, many stocks have growth rates that are lower than fishers' discount rates. The only reasons why these stocks are not routinely 'mined' to depletion are: (i) that the costs of fishing may rise at low levels of abundance and (ii) that some regulations to prevent 'mining' are enforced.

In those fisheries where no attempt is made to control 'mining' then history suggests that mining will occur. Thus, the sea cucumber (*bêche de mer*), trochus and clam fisheries in some Pacific islands have been characterized by short periods of boom-and-bust exploitation, with the fishers removing all the accessible animals and only returning many years or decades later, when stocks have recovered to economically viable levels. For example, a fishery for sea cucumber developed rapidly in the Galápagos Islands during 1992. Fishing was largely unregulated and very intensive because foreign buyers were paying high prices for the sea cucumbers. The economic boom far exceeded anything that had been witnessed in the islands before but, within 5 years, stocks of sea cucumber were so depleted that fishing was no longer profitable (section 1.3.1).

Given that fished species are almost always more valuable if caught today than if left in the sea, why do we bother to conserve fish at all? Is it not better to simply catch and sell fish as quickly as possible and then to invest the money? In purely economic terms this may be true, but we have reached a point at which economic analyses do not address all our concerns about fisheries management. In reality, society as a whole tends to favour the conservation of resources because they feel it is right to conserve them. The reasons include moral and ethical responsibilities to care for life on earth, a desire to preserve fishing communities and lifestyles, and to leave marine ecosystems in an acceptably 'natural' state (Kunin & Lawton, 1996). Clearly, quantifying these benefits is a very complex task, and outside the scope of this book, but the desire to conserve resources for future generations remains an overriding aim of fisheries management.

Clark (1985) has emphasized that a recognized discounting policy is not used when making fishery management decisions. As such, arguments over acceptable levels of catch between the regulators and the fishing industry are effectively an argument over discount rates. Regulators usually want to reduce current catches while fishers want to increase them. The regulator has a long-term view and thus works with a low discount rate while the fishers want to maximize immediate benefits from the fishery and use a much higher discount rate. To the fisher, anything left in water is virtually

		User 1	
		Conserve	Deplete
User 2	Conserve	$(Y*/2\delta, Y*/2\delta)$	$(0, B*)$
	Deplete	$(B*, 0)$	$(B*/2, B*/2)$

Fig. 11.4 Payoff matrix for the fishing game. The game is described in the text. After Clark (1985).

worthless. If 0% discount rates were used when planning fisheries development they would encourage the long-term rational exploitation of natural resources. Discounting rates also have critical effects on stock rebuilding initiatives. Thus, if regulators demand that current catches are reduced to boost catches in future, the fishers who used high discount rates would only be expected to support the policy if they expected future catch increases to be very high (Hilborn & Walters, 1992).

Clark's fishing game

Clark (1985) provides a bioeconomic explanation for overfishing in the form of a simple game (Fig. 11.4). This considers whether fishers, acting independently, would be motivated to limit their catches in order to prevent overfishing. He considers two users, which could be individual fishers, boats, fishing companies or nations, with access to a common resource. For simplicity, costs are ignored. Each user has one of two options: to conserve or deplete the resource. We further assume that no restrictions are imposed on fishing activities, that users have equal financial opportunities, and that total profit for both users would be maximized at sustainable biomass ($B*$) and sustainable yield ($Y*$). δ is the discount rate

The game begins with biomass $B*$. When $B* > Y*/2\delta$ then depletion is the best strategy for each user regardless of what the other does. So both deplete and obtain a yield of $B*/2$. When $B* < Y*/2\delta$ then a user loses income by the deplete strategy if the other decides to conserve. As such, there is an incentive to cooperate. Both conserve only because it is in their own interest. However, although the depleter loses revenue, if the other user conserves, the conserver will lose more. As a result, if the users do not trust each other it is likely that they will compete and both deplete the resource for fear of making nothing. So begins a race to fish.

In a more realistic situation with more users the payoff for conserving is $Y*/\delta N$ where N is number of users. If anyone can flout the regulations then they can get $B*$—a much greater share than if they participate in the conservation strategy. In these circumstances, the game is likely to degenerate such that the best strategy is to deplete regardless of what the others do. Unless all users can agree to conserve, and trust each other, they will inevitably deplete the resource. The solution deplete–deplete is seen as inferior to the cooperative solution conserve–conserve, but competition seems to force the inferior solution.

We can use a simple model such as this to look at the benefits of management strategies. An example would be restricted entry where the number of users is reduced. In fact, unless the number of users is reduced to one and the user is a monopolist, the results described above will always apply. Access restriction will not encourage conservation because there is still no incentive to conserve. Indeed, under access restriction, future benefits come to users in proportion to their fishing power. This is why real-world access restrictions have to be accompanied by catch or effort control (section 17.2).

Clark's game helps us to understand the basis for overfishing, but it is a 'one shot' game. Fishing in real oceans is a game that continues indefinitely and, in these circumstances, a few users with similar catching power may well cooperate (Hannesson, 1995a, b, 1997).

11.3.2 Optimal fishing strategies

For a single cohort

Most fisheries target fish of many ages from many cohorts, and these cohorts differ in abundance at the time they recruit to the fishery. Bioeconomic models can predict when these cohorts should be fished in order to optimize the value of the catch, and the ways in which factors such as discounting affect the time at which unregulated fishers will exploit them. There is much interest in economic optimization of fishing strategies, but we need to remember that the economically optimal strategy is not necessarily the most desirable strategy for policy-makers of fishers because other social, political and biological constraints may have more bearing on the management process.

As an introduction to bioeconomic models, we describe the approach of Clark *et al.* (1973) and Clark (1985) to determining the optimal fishing strategy for a single cohort. We will look at a single cohort as it passes through a fishery and see how the discount rate affects the time at which it should be exploited in order to give the greatest economic yield.

We have seen in Box 7.2 that the change in the number (N) of fish alive in a cohort with time can be expressed as a function of natural mortality (M) and fishing mortality (F). Since fishing mortality is a function of catchability (q) and fishing effort (f), equation 2 in Box 7.2 can be written:

$$\frac{dN}{dt} = -(M + qf)N \qquad (11.2)$$

The present value (PV) of profits from fishing such a cohort will be given by the expression (Clark *et al.*, 1973):

$$PV = \int_0^\infty e^{-\delta t} f_t (pqN_tW_t - c)\, dt \qquad (11.3)$$

where δ is the discount rate, f_t is fishing effort at time t, p is price, q is catchability, N_t is numbers at time t, W_t is mean individual weight at time t and c is fishing costs. For simplicity, Clark *et al.* assumed that the cohort recruited to the fishery at $t = 0$, that price was fixed and that fishing costs were proportional to effort.

The term $f_t (pqN_tW_t - c)$ expresses the profit to be made from fishing at time t, while the term $e^{-\delta t}$ reduces the present value of this profit as t or the discount rate rises. Note that the Clark *et al.* (1973) equation is simply a modification of the fundamental yield-per-recruit model (equation 7.26) where the summation sign has been replaced by an integral because time is a continuous variable.

Clark *et al.* (1973) solved equation 11.3 to determine the time at which the cohort should be fished to maximize PV. The solution was sensitive to the discount rate (Fig. 11.5). As the discount rate falls, so the catches are shifted towards the age at which the cohort reaches its maximum biomass (Fig. 11.5a). At high discount rates, growth overfishing, or fishing before the cohort reaches its maximum biomass, is increasingly likely to occur because catches are shifted towards the beginning of the cohort's life span (Fig. 11.5b). Total catch will thus fall as the discount rate increases.

The time at which the cohort should be fished does

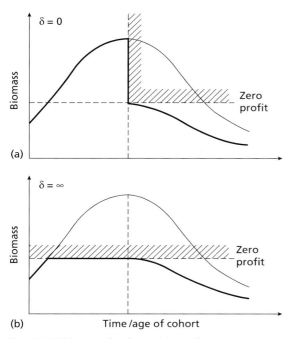

Fig. 11.5 Fishing a single cohort at (a) zero discount rate and (b) infinite discount rate. The fine lines show the natural biomass of the unexploited cohort and the bold lines show the biomass after exploitation. The 'fishing zone' is shaded. We assume that fishing effort is so high that the stock is instantaneously reduced by fishing. Zero profit occurs when $B = c/pq$. After Clark *et al.* (1973).

not depend on the history of fishing the cohort, but on the existing value of the cohort and its potential for future growth in biomass. Once the potential for future growth (and hence value) is less than the existing value then the economically optimal policy will be to catch the fish and sell them as quickly as possible.

The single-cohort model suggests that in an unregulated open-access fishery, fish will be caught as soon as there is a profit associated with catching them. High discount rates reduce the optimal age at first capture. As we have seen, fishers tend to work with very high discount rates but managers do not.

Multiple cohorts, species and fisheries

The models of Clark can be extended to more typical fisheries involving simultaneous harvesting of several cohorts or even other species. The optimal equilibrium, where one exists, is still associated with the marginal

total productivity equal to the discount rate, but is modified by a marginal stock effect when net rewards are a function of stock size(s). However, the basic models are mathematically and biologically simple. As biological and economic realism increases, and as the models become more non-linear, more complex optimal paths appear, equilibria may not be optimal, and the ability of the mathematics to identify a true optimum becomes more limited.

Optimal results are often criticized as being unrealistic, being of a 'bang–bang' nature—going from where we are now immediately to the optimal solution. But this is unfair. The bang–bang solutions only occur for the simplest of models (often used heuristically), and many optimal solutions are sensibly smooth. Indeed if we put a cost on changing fishing effort from year to year, the solution is likely to be smooth. Horwood and Whittle (1986) and Horwood (1990) show how solutions can be found to more non-linear models, and they provide approximate solutions to multispecies and multifleet problems.

When the future is uncertain

Fluctuations in the environment and recruitment mean that we have little ability to predict long-term changes in the abundance of exploited populations (Chapter 4). As such, the future is uncertain and the investment of capital in boats and equipment is risky. The basic bioeconomic model of Gordon (section 11.3.1) gave valuable insights into the tragedy of the commons and what managers should do in principle, but ignored uncertainty and the biological processes underlying the production curve. For this reason it did not really help us to give quantitative advice that would increase economic efficiency. Predictive models have to deal with uncertainty if they are to provide realistic outputs. The usual approach to incorporating uncertainty is to take a static model and to replace uncertain variables with random variables having specified probability distributions.

Given that fished stocks do vary in abundance over time, one of the most pressing questions that bioeconomic analyses can address is the optimum level of investment in a fishery that targets a fluctuating resource. Stock fluctuations are inconvenient for industry. When stock sizes are low, fishing may be unprofitable and fleet capacity is not used. When stock sizes are high, fishers may not be able to catch all the fish that biological analyses suggest are available and, even if

Table 11.4 Parameters used in the Hannesson (1993) model to predict optimal investment in a fishing fleet when potential catches vary from year to year.

Parameter	Meaning
R	Revenue (net of operating costs)
ER	Expected revenue (net of operating costs)
Q	Quota (total allowable catch)
δ	Discount rate
m	Maintenance and depreciation
p	Total value from selling catch
c_y	Costs per unit caught
c	Total operating costs of making catch
K	Fleet capacity represented by amount of money invested in fleet
K_{opt}	Optimal fleet capacity (investment) to maximize ER
Q^*	Limit to fleets catching capacity

they can, processors or markets may not be able to deal with all these fish. Bioeconomic models can describe the uncertainty in stock fluctuations and use this as a basis for optimizing investment.

We will now look at a model developed by Hannesson (1993). This shows how optimum fleet capacity might be estimated in a fishery where catch quotas are essentially random. We consider an application of this model to the Norwegian capelin *Mallotus villosus* fishery. Capelin are fished just before spawning by a modern purse seine fleet. The fishery targets only one or two mature age classes, and like many salmonid fishes, the capelin die after spawning. The aim of management has been to ensure that a minimum biomass escapes each year to contribute to subsequent recruitment. Given the great variability in year-class strength this means that the catch quota (Q) set each year is also very variable. We have adapted and simplified the model for the purposes of this example. So please do not use it for fishery management!

To formulate the model, Hannesson needed: (i) a probability density function for the quotas; (ii) a model for determining optimum fleet capacity; (iii) a description of the relationship between catching capacity and quota; and (iv) a description of the relationship between revenue, price and cost for a range of quotas. These relationships were synthesized to develop an overall model for estimating optimum capacity. The notation needed for these analyses is summarized in Table 11.4.

Note that economists routinely use notation that is used for other purposes by stock assessment scientists.

To predict the probability that any given quota will be set in the future, the frequency distribution of past quotas was described using a probability distribution. In a fishery based on one or two age classes, quotas reflect recruitment variability, and there is reasonable evidence to suggest that errors around a spawner–recruit relationship are log-normally distributed (section 4.2.1). In reality, the probability distribution will vary with time and the quotas set from year to year are not independent of one another. However, for the purposes of the model the time dependence and autocorrelation were ignored and the probability distribution of Q was treated as time invariant.

Hannesson assumed that fishing effort was proportional to the number of boats in the fishery, that vessels were identical and that the cost of a boat was constant and independent of the number built. Because we are interested in determining fleet size, and fleet size is equivalent to optimal investment, fleet size is represented by the amount of money invested in the fleet (K).

When the fleet is operational, the profit per year will be $R - mK$, where R is revenue net of operating costs and m is the maintenance and depreciation cost of capital invested in the fleet. Maximization of the present value of profits can then be expressed as:

$$\text{maximize} \sum_{t=0}^{\infty} [ER(K) - mK]/(1 + \delta)^t - K \qquad (11.4)$$

where ER is the expected revenue net of operating costs and δ is the discount rate. Note that the upper and lower limits of t have been set as 0 and ∞ for analytical purposes.

Hannesson shows that equation 11.4 can be solved and multiplied by δ to give:

$$\max ER(K) = (\delta + m)K \qquad (11.5)$$

In written terms we are now maximizing expected annual revenue from the fishery net of all costs. $(\delta + m)K$ is the annual capital cost which consists of the alternative cost of capital (δK) and maintenance and depreciation (m). The optimal solution of equation 11.5 is obtained by differentiation with respect to K:

$$ER'(K) = \delta + m \qquad (11.6)$$

where ER' is the first derivative of revenue net of operating costs. The term $ER'(K)$ gives the annual expected revenue from investing in an additional unit of fish production (i.e. boats!) and $\delta + m$ gives the annual cost of a unit of capital. Thus, it pays to invest in boats until the expected annual revenue from an additional boat is equal to the annual cost of an additional boat.

We will skip much of the mathematics that is concerned with deriving an expression for $ER(K)$, but this was described in full by Hannesson (1993). In brief, expected revenue is calculated as the product of all possible catch values (from 0 to ∞) and the probabilities that they occur (0–1: as defined by the log-normal distribution function for Q) summed over all possible catch values. Since catch is treated as a continuous variable the solution is calculated by integration. In words, we can express the solution as:

$ER(K)$ = (Revenue when quota is less than fleet capacity, i.e. actual catch equals quota) + (Revenue when quota is greater than fleet capacity, i.e. actual catch equals fleet capacity)

Because we want to know the size of the fleet at which the maximum revenue is expected, it is necessary to calculate the mean increase in catch value when fishing capacity is increased by a small amount. This is given by the first derivative of ER which, if price and catch rate are constant in relation to stock size, can be simplified to give:

$$ER'(K) = [1 - F(k_1 K_{opt})](pk_1 - c) \qquad (11.7)$$

where k_1 is the amount caught per unit capital invested and $F(k_1 K_{opt})$ is the probability that the quota will be less than the catch capacity of the fleet (F is used here to denote a function). If we now refer to equation 11.6, the equation we previously derived for expected revenue, this can be substituted into equation 11.7 and the equation rearranged to give:

$$[1 - F(k_1 K_{opt})] = (\delta + m)/(pk_1 - c) \qquad (11.8)$$

where K_{opt} is the optimal catching capacity (or investment) to maximize the expected revenue. We can now apply this equation to estimate K_{opt} in the Norwegian Barents Sea capelin fishery, and see how changes in various parameters affect K_{opt}.

First we have to develop a function to describe variation in the quotas. The data we use to do this are those presented by Hannesson for quotas (as catches) in the period 1970–85 (Table 11.5).

We have seen that the errors around a mean spawner–

Table 11.5 Capelin quotas (Q) 1970–85. Data from Hannesson (1993).

Year	Q (1000 t)	$\log_e Q$
1970	1131	7.031
1971	1393	7.239
1972	1592	7.373
1973	1336	7.197
1974	1149	7.047
1975	1417	7.256
1976	2545	7.842
1977	2940	7.986
1978	1894	7.546
1979	1783	7.486
1980	1648	7.407
1981	2006	7.604
1982	1760	7.473
1983	2304	7.742
1984	1461	7.287
1985	851	6.746

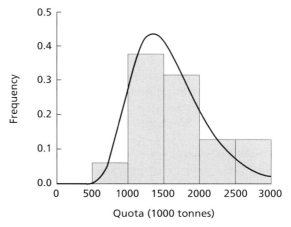

Fig. 11.6 Frequency distribution of quotas in the Norwegian capelin fishery and a fitted log-normal probability density function. After Hannesson (1993).

recruit relationship are often log-normally distributed (section 4.2.1), and since the fishable biomass of capelin is based on only one or two year classes then a log-normal distribution may be appropriate for describing the variation in Q. This is shown when a log-normal probability density function is plotted with the frequency distribution of quota values (Fig. 11.6). To describe the probability density function for Q, we determine the mean (μ) and standard deviation (σ) of the

Table 11.6 Inputs used to assess the optimum investment in the Norwegian capelin fleet.

Input	Value	Units
Price of capelin	80	$ t^{-1}
Operating cost	40	$ t^{-1}
Vessel cost	7×10^6	$
Amount caught per unit capital (k_1)	0.00357	t $$^{-1}$
Maintenance and depreciation (m)	0.05	y^{-1}
Discount rate (δ)	0.05 (& variable)	y^{-1}

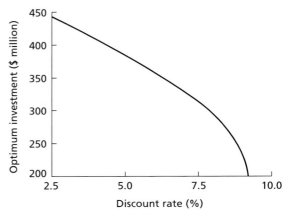

Fig. 11.7 Relationship between discount rate and optimum investment in the Norwegian capelin fishery.

$\log_e$-transformed values of Q. These are 7.392 and 0.318, respectively. We have plotted the log-normal probability density function onto the frequency distribution of quota values in Fig. 11.6.

Having described the probability density function for Q we can add the other parameters to the model. Values for these parameters are given in Table 11.6.

Equation 11.8 is used to estimate K_{opt}. We set both the discount rate and maintenance and depreciation at 5% (0.05) such that $(\delta + m) = 0.10$. Maintenance and depreciation consists of the expected costs of maintaining the fishing vessel and capital equipment plus the depreciation rate. The depreciation rate is the inverse of the lifetime of the fishing vessel and capital equipment. Thus, a vessel and capital equipment with an expected lifetime of 25 years would have a depreciation rate of 0.04.

The term $(pk_1 - c)$ is the net value of the landed quota, where p and c represent the total value realized from selling the catch and the total cost of making that catch, respectively. k_1 is the amount of capelin caught by one unit of capital invested. If the price of capelin is $80 t^{-1} and the operating costs per unit caught (c_y) are $40 t^{-1}, then the net price is also $40 t^{-1}. A large purse seiner can catch some 25 000 tonnes of capelin each year and costs around $7 million. With this information we can calculate k_1, the amount caught per unit capital invested as $25 \times 10^3/7 \times 10^6 = 0.00357$ t $$^{-1}$. Thus the term $(pK_1 - c)$ is equal to $0.00357 \times 40 = 0.1428$ and the right-hand side of equation 11.8 is 0.700.

On the left-hand side of equation 11.8, $F(k_1 K_{opt})$ is the probability that the quota will be less than the catch capacity of the fleet. Since Hannesson has shown that $k_1 K_{opt} = Q^*$, where Q^* is the limit to fleet capacity, the right-hand side can be expressed as $1 - F(Q^*)$. Since we

have assumed that quotas are log-normally distributed, the probability that the quota will be less than Q^* can be determined from statistical tables. For the input values given in Table 11.6 we get an optimal Q of 1372 000 tonnes. The optimum investment (K_{opt}) to achieve this quota can be obtained by dividing the quota by k_1, the catch of capelin produced for every $ invested. This is $384 million. Note that the optimal investment is less than that needed to catch the full quota every year, so some quota would be left uncaught.

The model can be used to show how changes in discount rates affect optimum investment. As discount rates increase from 2.5 to 7.5% (maintenance and depreciation costs were kept at 5%), there is a rapid fall in optimal investment (Fig. 11.7). When discount rates exceed 10%, any investment is unlikely to be worthwhile, and the fishery may have to be subsidized if past quota variation reflects future variation.

Changes in quota variability also affect optimal investment strategy (Fig. 11.8). Not surprisingly, with no variation in quota, the optimal investment provides enough boats to catch the whole mean (but constant) quota each year. As variance around the mean increases however, the optimum investment falls.

Equation 11.8 suggests that it is never optimal to invest in a fleet that can always take the quota unless the cost of capital is zero. Higher costs lower the amount of fish caught per unit of money invested and thus optimum capacity falls. These conclusions are similar to those of Charles (1983). He developed a general model that related optimal investment in fleet capacity to

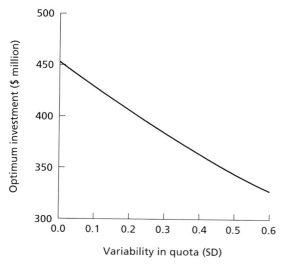

Fig. 11.8 Relationship between predicted variability in the quota and optimum investment in the Norwegian capelin fishery.

uncertainty. His model showed that it would only be worth investing more in fleet capacity when the future was uncertain if vessel capital was very malleable and the intrinsic rate of population increase for the resource was high. In the case of the Norwegian capelin fishery, like most other fisheries, vessel capital is non-malleable and it is economically desirable to invest less under uncertainty.

The choice of optimal capacity can be explained by considering two opposing effects: the **downside risk** of suffering idle excess capacity in bad years and the **upside risk** of lacking sufficient capacity to take full advantage of the resource in good years. Depreciation rates play a key role in the response. If there is no depreciation, then capital is infinitely long lived and the recurrent cost of capital is low. Thus, the downside risk of an increased investment is relatively small. As the rate of depreciation increases, so annual costs rise and the effective life of a unit of invested capital will fall. This causes the upside benefits of an extra unit of capital to be lower, because there are likely to be fewer years in which the fishery could take advantage of higher capacity. In general terms, the lower the depreciation rate the more likely it is that there will be higher investment under uncertainty.

The models we have considered are for single-species fisheries. If fishers can target several stocks, then they may be able to use their fishing capacity elsewhere during bad years. This reduces downside risk and would be expected to increase the optimal investment. Indeed, most fisheries are opportunistic to some extent, often targeting different species in different seasons.

In presenting this example of a model for the Norwegian capelin fishery we have based the output on catch quotas that were set from 1970 to 1985. In fact, Hannesson (1993) also presents data for the years 1986–90. In 1986 catches fell dramatically, and from 1987 to 1990 catches were zero. This demonstrates an important point, that past fluctuations in fish stocks are not necessarily a good guide to those that will occur in the future and may not be consistent with a fixed probability density function. There is always uncertainty about the future, e.g. Box 11.1.

When the future and present are uncertain
Horwood and Whittle (1986) provide another means of finding the optimal effort when there are both random fluctuations in recruitment, and when the size of the stock at age is only known with only some degree of error. Solutions can be found that take the form of a control law that regulates the fishing by each fleet, so that an optimal return can be obtained in the face of recruitment and measurement uncertainty. The actual values are updated as stock sizes change.

11.3.3 Bayesian methods

We have seen that one of the easiest, and still most widely used, approaches to dealing with uncertainty is to take a static model and to replace variables with random variables having specified probability distributions. In adopting this approach, we assume that a probability distribution obtained from models of historical data can reasonably be applied to estimating the probability of events in the future. This is true for card games, coin tossing and in other cases where the range of outcomes is limited and there is complete knowledge of the probabilities. However, this does not strictly apply to fisheries problems where distributions cannot be correctly determined from theory, and the empirical data used to compile distributions are estimated rather than known values (Clark, 1985).

Our capelin model in the previous section was a rather good example of this. The log-normal distribu-

---

Box 11.1

Optimal harvesting with risk of extinction .

Fossil and historical evidence shows that stocks eventually become extinct, and yet the risk of extinction has scarcely been considered in most fisheries and bioeconomic models. How should we harvest when extinction is possible? Lande *et al.* (1994) incorporated risk of extinction in a stochastic model to predict strategies that maximise the expected present value (PV) of cumulative harvest before extinction. They predicted that a stock should not be harvested unless it was above its equilibrium population size in the absence of harvesting.

This was an unexpected and unsettling result because it entails a complete halt to fishing except when random processes push the population above the so-called carrying capacity. This is a far cry from the classical goals of maintaining lower populations so that productivity is increased (Chapter 7). Whittle and Horwood (1995) confirmed this result, but reconciled it with classical theory by also considering a different objective: maximising rate of return per unit time (before extinction). These papers are important because they force managers to consider the likelihood of extinction, rather than assuming that stocks can automatically bounce back from low levels.

---

tion provided a good description of the probability of different quotas being set from 1970 to 1985 (see Fig. 11.6), but in the next 3 years the quota was reduced to nil, a pattern that was not in accordance with that suggested by the log-normal distribution. This should not surprise us when we look at the long-term dynamics of other stocks (Chapter 4), for many stocks undergo periodic cycles of collapse and recovery as their environment changes.

We need to realize that probabilities are uncertain in many areas of fishery science. Moreover, an approach to dealing with uncertainty that is based on replacing model parameters with probability distributions may not help with making management decisions. Confidence limits around parameters may indicate expected uncertainty but they do not provide management guidance. Thus, fishery scientists tend to say that quotas should be reduced in the face of uncertainty while the fishing industry takes the opposite view! (Clark, 1985).

An alternative approach for making decisions in the presence of uncertainty is decision analysis. Here, uncertainty is recognized and accepted at the outset using a prior distribution (Bayesian analysis, section 7.9.1). This distribution is subjective, based on the scientist's experience and any existing data. Since scientists have studied many fisheries, it is usually possible to make an educated guess about the form of the prior distribution. This may be as simple as setting minimum and maximum values for a variable and assuming a uniform distribution between them. Often, we can do much better.

As an example of an application of decision analysis we can consider how we might estimate optimal investment in a developing fishery (Clark, 1985). In most developing fisheries, little is known of the abundance or productivity of the newly exploited stock. Usually, fishery development will be driven by fishers and investors who see an opportunity to make profit, and overcapacity quickly develops. By the time scientific assessment has taken place, and regulations are imposed, the fishing industry suffers economically because they have invested non-malleable capital. Clearly, it would be better to impose regulation at the outset, to keep the stock biologically productive and the fishing industry economically efficient. How can we determine whether investment should be limited at the outset, and to what extent, when the future is uncertain?

Let us assume that, in the developing fishery, year-to-year fluctuations in recruitment are correlated with year-to-year fluctuations in production. If recruitment has been observed for a few years, or we have observed the dynamics of a similar stock elsewhere, we can get a prior distribution. This could, for example, be a log-normal distribution. This prior distribution can then be used to estimate the optimal investment. After a few years, we have more recruitment data, so we know more about the levels of uncertainty, and update the distribution to recalculate the optimal investment. The approach is conceptually simple, and Clark (1985) shows how it can be addressed mathematically. The outputs from his model show that uncertainty about long-term investment prospects calls for a conservative initial investment, since upward adjustments can be made later. This makes intuitive sense, since if the initial investment were too high, subsequent downward adjustment would cause economic hardship.

11.4 Economic vs. social management objectives

11.4.1 Subsidies

We have already shown that marine fisheries, when viewed on a global scale, are not profitable. Unprofitable industries usually collapse, but a large fishing industry persists because it subsidized by direct and indirect payments from governments. The main objectives of governments are to maintain employment levels and to ensure the fishers receive a reasonable income.

Global economic analyses of fisheries are notoriously difficult because a large proportion of fishing activity is not reported to the Food and Agriculture Organization of the United Nations (FAO) that collate these data. However, in 1993 the FAO (1993a) produced an analysis of costs and revenues for the global fishing fleet using data from 1989. The vessels they included were mostly those greater than 100 GRT (gross registered tonnage) in size, and so the reported values should be regarded as minimum estimates. Some of the results are shown in Table 11.7.

It is clear from Table 11.7 that the costs of operating the fleet exceeded revenue by $US54.1 billion in 1989. The fishing industry was overcapitalized and economically inefficient for the reasons we outlined in general terms in section 11.3.1. Notably, access to many fishing grounds was not restricted, and the best way to catch more fish was to invest in more fishing power. In addition, when prices were high or fish were abundant, more vessels were built, but these could not be used outside the fishery when prices are poor and fish were scarce. The investment of such non-malleable capital in vessels means that vessels often had to continue fishing, even when it is unprofitable, simply to cover some of their fixed costs.

There is no doubt that subsidies maintain an economically inefficient industry and increase the probability that fished stocks will be exploited beyond their biological limits. However, they do maintain employment in the fisheries sector and prevent the collapse of

Table 11.7 Operating costs and revenue for the global fishing fleet in 1989. From FAO (1993a).

Costs	$US billion	%
Cost of capital	31.9	25.7
Maintenance	30.2	24.3
Labour	22.6	18.2
Gear and supplies	18.5	14.9
Fuel	13.7	11.0
Insurance	7.2	5.8
Total	124.1	
Less revenue	70.0	
Deficit	54.1	

fishing communities. The fishing industry is subsidized in various ways. These can include artificial price control for catches, subsidies for fuel or gear purchase, low-cost loans or grants for boat and capital equipment purchase and the provision of ports and marketing facilities. In 1989, the Japanese fishing industry was subsidized by $US19 billion, and high levels of subsidy were also provided for Russian and East European fleets. The European Community paid the fishing industry $US0.6 billion and individual governments within the EC provided further subsidies to their own fleets. As we will see in Chapter 17, many of the current attempts to improve fisheries management are based on plans for reducing subsidy and improving economic efficiency.

11.4.2 The case for economic efficiency

Economic efficiency is just one possible aim of fishery management (section 1.5). In many cases, one large vessel fishing in areas where effort is strictly controlled to maximize biological yields may be more economically efficient than many small boats in an open-access fishery. Economic efficiency, as in many industries, may improve with increased mechanization and fewer employees. Since fishing has traditionally employed many people in coastal communities it is sensible to ask whether economic efficiency is a more or less desirable goal of management than high employment.

Hannesson (1996) provides an interesting perspective on this problem. He suggests that what is traditional and picturesque is not necessarily very productive and that turning fisheries into living museums will not be a solution to global overcapacity and inefficiency in the industry. Many who observe fisheries may feel that small is beautiful and would prefer to see many poor fishers in small boats trolling for tuna rather than modern and sophisticated purse seiners searching for entire shoals with helicopters. Managers may also prefer fisheries based on small boats with limited catching power because it is easier to implement effort controls. But Hannesson considers that economically inefficient fisheries, at least in developed countries, can have undesirable economic effects. This is because society as a whole has to bear the costs of reduced efficiency and productivity in the fisheries sector, and thus resources that could be better used elsewhere (to run hospitals and schools, for example) are needed by fishers.

In other industries, technical improvement leads, in the longer term, to the transfer of the workforce to other parts of the economy where they can be more productive. Witness the development of telecommunications, computing and service industries in the developed world as the extraction of natural resources and manufacturing were increasingly automated. In general, preserving a museum culture within the fishing industry will mean that its inefficiency has to be subsidized and that people working in other industries will have to pay for this through increased taxation or loss of services.

These arguments may be persuasive in the developed world, although many fishers will obviously prefer to carry on fishing rather than move to cities and work in telecommunications or computing. However, in the developing world, many countries have weak economies and fishers have few, if any, alternate sources of food, income and protein. These people cannot leave the fishery without financial support. Unfortunately, such support is rarely available.

Summary

- Most stocks are fished for economic gain and fishing provides employment for millions of fishers and workers in associated industries.
- In most large and economically developed nations, fisheries make a minor contribution to national economic activity. However, fisheries may support large sectors of the economy in small island and developing nations and their contribution to Gross Domestic Product can exceed 10%.
- On a global scale, marine fisheries are heavily subsidized and economically inefficient.
- Bioeconomic models help to explain why fisheries are overexploited and predict their response to different management measures.
- An unregulated fishery expands until revenue equals the cost of fishing. If costs are low, then the stock is likely to be fished beyond its biological limits. Fishers compete in unregulated fisheries because anything left in the sea could be caught by someone else.
- If the future value of fish is perceived to be less than the money that could be made by fishing now, selling the catch and investing the money, then there is an economic incentive to overfish.
- Uncertainty can be incorporated into static bioeconomic models by replacing uncertain variables with random variables that have specified probability distributions. This assumes that probability distributions from historical data can be used to estimate probabilities in the future.
- An alternative approach for making decisions in the presence of uncertainty is decision analysis, where uncertainty is recognized and accepted at the outset.

Further reading

Hannesson (1993) provides a clearly written and accessible introduction to the bioeconomic analysis of fisheries, and gives many examples of bioeconomic models in action. Clark (1985) is a more advanced treatment and is packed with interesting examples and ideas.

12 Fishing effects on populations and communities

12.1 Introduction

The aim of this chapter is to show how and why fishing affects fish populations and how these changes impact multispecies communities. We begin by describing how behaviour and life histories determine vulnerability to fishing and how fishing affects reproduction and genetic selection. This complements our previous discussion of fishing effects on the population dynamics of target species (Chapter 7). We then extend our analysis to the community level and see how changes in the abundance of individual species affect community structure. Throughout the chapter we emphasize the effects of fishing on non-target species. Non-target species are playing an increasingly important role in fisheries management decisions because they may be endangered by fishing and contribute to the diversity of marine communities.

12.2 Vulnerability to fishing

Not all species are equally vulnerable to fishing. Apart from the simple fact that a gear with a large mesh will only catch larger individuals (section 9.3.1), vulnerability is determined by behaviour and life history. Behaviour determines susceptibility to fishing gears and hence the fishing mortality. Life history determines how populations respond to a given level of fishing mortality.

12.2.1 Behaviour

Behavioural interactions lie at the heart of all predator–prey relationships including those between humans and target species. The ways in which fished species shoal, swim, feed and migrate can affect their vulnerability to fishing (Fernö & Olsen, 1994; Pitcher, 1986, 1995; Vincent & Sadovy, 1998; Reynolds & Jennings, 1999). In this section, we see how behaviour affects

vulnerability to mortality and the potential for population recovery.

Fish may shoal because this increases foraging success and their ability to avoid predators (Pitcher & Parrish, 1993; Mackinson *et al.*, 1997) (Fig. 12.1). However, fishers can take advantage of shoaling behaviour because whole schools of pelagic fish such as mackerel *Scomber* spp., herring *Clupea* spp. or anchovy *Engraulis* spp. can be surrounded with seine nets (section 5.3.2) and the aggregation of schools in hot spots means that fishing will be profitable even if total population size is low (section 4.3). For example, large catches of Peruvian anchovy *Engraulis ringens* were still taken in the early 1970s as the stock collapsed because the remaining fish formed shoals. If fish do not form shoals and are not particularly valuable then they may be safe from the attentions of fishers. Thus, the small pelagic fishes of the deep scattering layer (family Myctophidae) are one of the largest oceanic fish resources, but are not fished commercially because they are so widely dispersed that catching them would be prohibitively expensive (section 2.5.3). In traditional reef fisheries, fishers have long observed the spawning behaviour of target species and fished within spawning aggregations (Johannes, 1980).

The ability of fished species to avoid nets depends on swimming speeds and escape responses (section 5.3.1). Since sustained swimming speed is proportional to body length, smaller individuals cannot stay ahead of an approaching trawl for so long and are more likely to be caught (Wardle, 1986; Videler, 1993). However, even if fish manage to swim ahead of a trawl net and avoid capture, the subsequent stress may still be fatal. Species show different responses to an advancing trawl. As haddock *Melanogrammus aeglefinus* drop back into the net they tend to swim upward while cod *Gadus morhua* stay low (Main & Sangster, 1982a). Net-makers have taken advantage of these responses by adding

Fig. 12.1 Shoaling two-spot snapper *Lutjanus biguttatus*. Photograph copyright S. Jennings.

horizontal separator panels inside the net. These divert cod to the codend but let other species escape through panels of larger mesh on the top of the trawl. Trawls of this type are known as **separator trawls**. Such bycatch reduction measures could increase the survival of non-target species in mixed demersal fisheries. Separator panels have also been used to reduce bycatches of small gadoid fishes in Norway lobster *Nephrops norvegicus* trawls because the target *Nephrops* tend to stay within 70 cm of the seabed, while most young fish rise upward (Main & Sangster, 1982b). Understanding differences in the behaviour of shrimps and bycatch species have provided a basis for designing nets that reduce turtle mortality. **Turtle exclusion devices** (like **trawl efficiency devices**, known as TEDs) provide a large opening at the top of the net through which the turtles can escape (section 5.5). However, relatively few shrimps are lost through the opening because they tend to stay low in the net. TEDs also reduce fish bycatches (Rulifson *et al.*, 1992; Robins-Troeger, 1994).

Fishers must understand feeding behaviour to catch fish with baited hooks and traps. Subtle differences in the type of bait, its presentation, the depth of fishing and time of fishing all have marked effects on catches. An understanding of foraging behaviour helps to identify which individuals are vulnerable to capture and helps fishers to improve gear design. For example, red

king crabs *Paralithodes camtschaticus* make only a limited number of approaches to traps, and search for an access route to the bait over a very small area. Existing trap designs have small entrances, and the restricted searching behaviour of the crabs means that the probability of their entering is relatively low (Zhou & Shirley, 1997). Larger entrances that are easier to find would improve the traps, provided these did not allow the crabs to escape again. Dominance behaviour by feeding crustaceans can also affect which individuals are caught. Thus, the presence of lobsters *Homarus* spp. may deter crabs *Cancer* spp. from entering (Richards *et al.*, 1983; Addison, 1995). This may reduce crab mortality on fishing grounds where both species are present.

The migrations of fish are well known to fishers who often catch fish at bottlenecks in migration routes (section 3.5.2). Thus, when bluefin tuna (known also as tunny) *Thunnus thynnus* migrated past headlands or between islands in the Mediterranean they were often caught in seasonal fisheries. Many of these fisheries no longer exist because the tuna have been overfished. Similarly, when cod travel along migration highways they are easily targeted by trawlers (section 3.5.2). Ontogenetic changes in migration will change vulnerability to exploitation. Hence, European sea bass *Dicentrarchus labrax* make short, tidally related migrations in confined estuarine nursery areas until maturity and

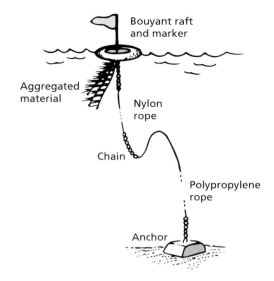

Fig. 12.2 A fish aggregation device. After King (1995).

then start to migrate over hundreds of kilometres (Pawson *et al.*, 1987). The tidal migrations in nurseries make young bass vulnerable to pollution, barrage construction, habitat reclamation and fishing. An understanding of behaviour can also help to mitigate these negative effects. Thus, sluices, fish bypasses, bubble screens and low-frequency sound emissions may deter fish from passing into turbines or cooling water intakes of power stations (Davies, 1988).

The habitats chosen by fished species will affect their vulnerability to fishing. Flatfishes that live on smooth, sandy sediments are readily accessible to many fishing gears, while conger *Conger conger* (Congridae) or moray eels (Muraenidae) living in deep gulleys on rocky reefs are not. Many pelagic fishes are known to aggregate around floating objects, and fishers use these as foci for their trolling or netting activities. Fishers now construct floating structures known as **fish aggregation devices** (FADs) to attract many pelagic fish species and provide a focus for recreational and commercial fishing. These usually consist of a floating raft of rope, matting and other material which is anchored to the seabed (Fig. 12.2). Artificial reefs also attract fish and can make them more accessible to fishers (Pickering & Whitmarsh, 1997). These are usually made of tyres, concrete blocks or scrap iron, and are sometimes treated as a cheap waste-disposal option!

12.2.2 Life histories

Fished species have evolved a range of life histories which meet particular ecological demands (section 3.4). Some of these combinations, although appropriate in an unfished environment, leave populations vulnerable to the additional mortality imposed by fishing. Others allow populations to sustain fishing mortality rates that exceed natural mortality by a factor of three or more. As a result, some fished stocks have collapsed following intensive fishing while others have proliferated (Jennings & Kaiser, 1998; Roberts & Hawkins, 1999; Hawkins *et al.*, 2000).

Theoretical analyses suggest that large, slow-growing and late-maturing species suffer greater population declines for a given mortality rate, because these attributes are associated with lower intrinsic rates of population increase (Adams, 1980; Beddington & Cooke, 1983; Roff, 1984; Kirkwood *et al.*, 1994; Trippel, 1995; Pope *et al.*, 2000). Empirical tests of these relationships are more difficult, since it is not simply a matter of relating trends in abundance (after accounting for differences in fishing mortality) to parameters that describe the life history. The problem is that each stock, or species, cannot be regarded as an independent data point when stocks have not evolved independently. For example, slow-growing species such as skates and rays (*Rajidae* spp.) have often decreased in abundance following exploitation (Brander, 1981; Walker & Heessen, 1996; Casey & Myers, 1998; Dulvy *et al.*, 2000). This response has been attributed to their advanced ages at maturity and low fecundity. However, because members of the genus *Raja* are closely related, they share other characteristics such as broad body shape and the laying of egg cases on the seabed which could also be responsible for their susceptibility. Therefore, if we were to compare eight species of ray with eight species in the herring family (Clupeidae), we may get a significant difference in vulnerability to fishing pressure, but we have no idea which of the many differences in life histories between these families led to differences in vulnerability. On the other hand, comparisons between pairs of species in different genera, or between stocks within species, would help eliminate the effect of other variables that they have in common and therefore yield evolutionarily independent data (Felsenstein, 1985; Harvey & Pagel, 1991). This phylogenetic comparative approach can therefore be used to test the hypothesis that taxa with late maturity, slow growth, large body size and low potential rates of population increase will decline more quickly under exploitation than related stocks or species with 'faster' life histories (Reynolds *et al.*, 2001).

Life histories are described using a variety of parameters. Growth rate and maximum size are usually expressed using the parameters K and L_∞ of the von Bertalanffy growth equation (sections 3.4.2, 9.3.3) and age and size at maturity are the age or size at which 50% of the stock have attained maturity (section 9.3.4).

The phylogenetic comparative approach has been used to relate population trends in 18 exploited stocks of the north-east Atlantic to their life histories. After accounting for differences in fishing mortality, stocks that decreased in abundance more than their closest relatives had lower ages at maturity and greater maximum body size. A good example of the role of life histories was provided by the two cod stocks that were examined. The North Sea cod decreased in abundance faster than the Irish Sea cod from 1975 to 1994, even though it was subject to lower fishing mortality. However, the Irish Sea cod matured at age 2.5 years rather than 3.8 years and had an asymptotic size of 99 cm rather than 123 cm. These life histories may have allowed the Irish Sea stock to sustain population size despite more intensive fishing (Jennings *et al.*, 1998).

For such well-studied stocks, the assessment of vulnerability could be made using conventional population analyses (Chapter 7). However, the value of relationships between easily measured life-history parameters (section 9.3) and vulnerability is that they can be used to assess vulnerability when we have limited knowledge of a species' biology. This is often true for non-target species and those exploited by small-scale fisheries in developing countries. Thus, for tropical reef species an easily measured parameter such as observed maximum size may be used to make preliminary estimates of the effects of fishing on grouper (subfamily Epinephelinae in family Serranidae), snapper (Lutjanidae) and parrotfish (Scaridae) species. The analysis showed that those species that decreased in abundance relative to their nearest relative had a greater maximum size. The quality of prediction was good for the intensively fished groupers and snappers but poor for the lightly fished parrotfishes (Fig. 12.3). This reflects the relative importance of fishing as opposed

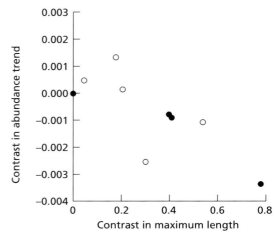

Fig. 12.3 Relationships between contrasts in trends in abundance and contrasts in maximum size for reef fishes. Trends in abundance were calculated as the slope of the relationship between log biomass and fishing intensity across 10 fishing grounds. Maximum length is the maximum observed length of the species and is closely correlated with asymptotic length (L_∞). Contrasts control for phylogenetic relationships and were calculated by subtracting the maximum length of the smaller species in each related pair from the larger one. Abundance contrasts were calculated in the same sequence. Open circles show contrasts between parrotfish species, and shaded circles show contrasts between grouper and snapper species. The plot shows that species that have a larger maximum size than their nearest relative are more vulnerable to fishing (decline in abundance more). After Jennings *et al.* (1999b).

to habitat and recruitment in structuring their populations (Jennings *et al.*, 1999b).

In general, empirical analyses of vulnerability show that species with short life spans and rapid population growth, which mature early and channel a large proportion of their resources into reproduction, may be fished sustainably at relatively young ages and higher levels of mortality. Fisheries based on slower growing species, which mature later and at a larger size, are likely to be vulnerable to intensive exploitation despite naturally more stable population sizes which, in the unexploited state, are buffered against occasional recruitment failure by numerous age classes.

12.3 Intraspecific effects

12.3.1 Age and size structure

Fishing is size selective because the meshes of nets or

traps allow smaller individuals to escape. As a result, fishing tends to change the size and age structure of a population, with mean body size and mean age decreasing as fishing mortality increases (section 9.3.6). This change affects other aspects of the life history. For example, egg production of the population may fall because smaller individuals are less fecund (section 3.4.3).

Life-history parameters of fishes are remarkably plastic and change in response to the environment and food supply. Decreases in population density following fishing may lead to increased growth rates and production (Chapter 7). We assume that increases in growth are often due to density-dependent effects, because lower population abundance should lead to increased food availability for the individual. However, density dependence is not easy to demonstrate when food supply and temperature are variable (section 2.3.1). Moreover, fishing activities such as trawling, while leading to decreases in the abundance of target species, may release additional food from the seabed or shift invertebrate communities to those dominated by faster-growing and more productive species (sections 14.3, 14.6). Despite these complications, density-dependent increases in the growth of juvenile and adult fishes do occur when fishing reduces their abundance (Rijnsdorp & Leeuwen 1996). These growth changes are accompanied by other changes in life histories which may compensate fitness when the life span is artificially curtailed by fishing. For example, Norwegian spring spawning herring matured at age 3 in the 1970s, 2–3 years earlier than in the 1950s and 1960s when they were more abundant (Toresen, 1986, 1990).

12.3.2 Reproduction

Size-selective fishing can change the sex ratios of fished populations and curtail reproductive life span. Tropical families contain many hermaphrodite species, and most individuals function as both sexes, either simultaneously or sequentially (section 3.4.1). In most fished hermaphrodites, sex change occurs at a critical size and is not a socially induced facultative response. As such, the larger sex will not be replaced quickly. Porgys (Sparidae) and groupers (Epinephelinae) are protogynous hermaphrodites (female first) and fishing has biased their sex ratios by taking out the larger males. Thus, porgy populations in a South African fishery were female-biased in comparison with those from an adjacent marine protected area (Buxton, 1993), while the proportion of male fish in gag grouper *Mycteroperca microlepis* (Serranidae) spawning aggregations in the Gulf of Mexico fell from 17% to 2% during a 10-year period of intensive fishing (Sadovy, 1996). Similarly, two grouper species on lightly exploited offshore Jamaican banks had male/female sex ratios of 1 : 0.72 and 1 : 0.85 as opposed to ratios of 1 : 5.6 and 1 : 6.0 at a heavily fished inshore site (Thompson & Munro, 1983). If the loss of males leads to reduced fertilization of females, depensation may occur at low stock sizes and lead to recruitment failure (section 4.2.3). These effects have scarcely been examined.

Changes in sex ratios are also observed in exploited invertebrates (Smith & Jamieson, 1991). Size limits in the trap fishery for Dungeness crab *Cancer magister* in British Colombia effectively bar females from the catch because they rarely grow to a size in excess of the limit. Consequently, in heavily exploited areas, males greater than the size limit are rare and the sex ratio favours adult females. Based on an assessment of male mating activity and the sizes of mating pairs, Smith and Jamieson (1991) proposed that mature females would struggle to find a sexual partner in intensively exploited fisheries and that the mating of fewer larger females would adversely affect egg production. This would lead to depensation in the spawner–recruit relationship and reduce the probability that an overexploited stock would recover.

Mating systems are also affected by fishing. During the rapid expansion of the tilefish *Lopholatilus chamaeleonticeps* (family Malacanthidae, order Perciformes) fishery in the mid Atlantic Bight from 1978 to 1982, the population size fell by 50% or more and males spawned 2–2.5 years earlier and 10 cm smaller (Grimes *et al.*, 1988). These authors suggested that the change in population density may have allowed smaller and younger males to claim mating territories on the spawning grounds. Fishing pressure also modified the social conditions in a Mediterranean wrasse *Coris julis* (Labridae) population and this induced earlier sex change (Harmelin *et al.*, 1995).

Shifts in the size and age distributions of fish populations have profound effects on reproductive output. The relative fecundity (number of eggs per unit of body mass) of fishes increases with body size, and thus a population of a given biomass has greater potential fecundity when composed of larger rather than smaller individuals (section 3.4.3). In addition, when reproductive life span is artificially curtailed by fishing, potential reproductive output will not be realized. Reproductive output at a given size or age may compensate through either genetic shifts in the population or through phenotypic plasticity. There is evidence for such changes in spiny lobster *Panulirus marginatus* (infraorder Palinura) where there was a 16% increase in size-specific fecundity after exploitation (De Martini *et al.*, 1992) and in North Sea plaice *Pleuronectes platessa*, where younger fishes of a given length had a higher absolute fecundity (Horwood *et al.*, 1986). When intensive fishing reduced the abundance of the east Tasmanian orange roughy *Hoplostethus atlanticus* stock by 50%, mean individual fecundity rose by 20%. The compensatory increase in individual fecundity, coupled with an apparent increase in the proportion of females spawning, limited the decline in egg production to 15% (Koslow *et al.*, 1995). Similarly, changes in growth, maturation and fecundity have compensated for about 25% of the losses in total egg production due to increased exploitation of North Sea plaice, cod and sole (Rijnsdorp *et al.*, 1991).

12.3.3 Genetic structure

Because fishing is selective with respect to a number of heritable life-history traits, exploited populations will evolve in response to harvesting. There are good precedents for such effects because life-history traits such as age and size at maturity, growth rate and reproductive output have a genetic basis, and selective predation on fishes by other fishes is seen as a major cause of evolutionary change (Hixon, 1991; Policansky, 1993). In experimental systems, selective culling has led to genetic change. When large freshwater tilapia *Oreochromis mossambicus* were selectively removed from the population, the growth rates of males decreased and the difference was heritable (Silliman, 1975). Similarly, selective culling experiments with *Daphnia magna* populations showed that the removal of larger individuals selected for slower growth (Edley & Law, 1988; Law & Grey, 1989). A classic series of studies showed rapid evolution of life history traits in guppies, *Poecilia reticulata*, from Trinidad. Guppies living in streams that contain large predators capable of eating adults were shown to have greater reproductive effort at an earlier age than guppies from low-predation streams. When researchers transferred guppies between high and low predation streams, they found that the fish evolved rapidly in the directions expected from their new predation regimes (Reznick *et al.*, 1990).

It has often been difficult to detect heritable responses to exploitation because they are masked by phenotypic effects (Stokes *et al.*, 1993). In North Sea plaice *Pleuronectes platessa*, age and size at maturity have fallen during intensive exploitation. Part of this fall is due to increases in growth and part is due to genetic effects (Rijnsdorp, 1993). For the North Sea cod *Gadus morhua*, Rowell (1993) calculated optimal age at maturity when reproductive output was assessed under different mortality schedules. The model predicted that mean age at maturity would fall with increasing mortality if age at maturity had a genetic component. Changes predicted by the model were consistent with those recorded in data from 1893, 1923 and the 1980s.

The evolutionary effects of exploitation can be investigated with quantitative genetics. Using a model that incorporated quantitative genetics into a model of population dynamics, Law and Rowell (1993) examined the effects of exploitation on body length in North Sea cod and suggested a small selection response of around 1 cm at age 1 year after 40 years of exploitation. Their results indicate that evolutionary change will be slow in comparison with phenotypic responses. This may explain why the evolutionary effects of exploitation have received remarkably little attention from fishery scientists who often base their management recommendations on the short-term forecasts provided by population models (Chapter 7). Since humans are likely to be managing marine fisheries for many years we should take the genetic effects of fishing much more seriously (Law & Stokes, 2000).

Fishing caused reductions in the genetic diversity of an orange roughy population well before the stock would be considered endangered by those concerned with fish population dynamics (Smith *et al.*, 1991). Reductions in intraspecific diversity have caused considerable concern in salmonid fisheries but are not well known for strictly marine species (Ryman *et al.*, 1995). While fisheries managers have treated stocks as operational rather than genetic units (section 9.2.1), there is a continued need to identify stocks that have defined genetic characteristics for the purposes of recording losses of intraspecific diversity and deciding how to protect it.

12.4 Community effects

12.4.1 Diversity

Extinctions

Extinction is the permanent loss of a species, while **extirpation** refers to local losses of stocks or sub-populations. Fishing has extirpated many fished species on scales of tens to hundreds of square kilometres. These include giant clams *Tridacna gigas* (Tridacnidae) on many of the Pacific islands and the Indo-Pacific bumphead parrotfish *Bolbometopon murica-tum* (Scaridae) that sleeps in reef crevices and is highly accessible to spear fishers. Fishing has also led to the extirpation of species such as giant sea bass *Stereolepis gigas* (Acropomatidae, close relatives of the Serranidae) in parts of California and skate *Dipturus batis* in the Irish Sea (Brander, 1981; Dulvy *et al.*, 2000). Extirpation is likely to cause reductions in genetic diversity. The white abalone *Haliotis sorenseni* (Haliotidae) of northern Mexico and southern California has a rela-tively restricted range and may be close to extinction as a result of fishing. The Banggai cardinalfish *Pterapogon kauderni* (Apogonidae) is collected for the aquarium trade and the young live behind the spines of sea urchins in shallow Indonesian lagoons. This species may be approaching extinction in the wild even though it now thrives in captivity (Tegner *et al.*, 1996; Roberts & Hawkins, 1999).

The factors that one would expect to make species vulnerable to extirpation by fishing are limited geo-graphical distributions, dependence on specific habitats slow life histories, and accessibility to fishers. For these species a series of extirpations may lead to extinction. Some marine species have remarkably small ranges. An assessment of 1677 coral reef fishes revealed that 9.2% were restricted to areas of less than 50 000 km^2 and much of that area was deep ocean rather than suit-able reef habitat (Roberts & Hawkins, 1999; Hawkins *et al.*, 2000).

For the major target species, the probability of their being fished to biological extinction is very low because economic extinction will occur first. Provided that the species is not increasingly valuable when scarce, and that there is no depensation in the spawner–recruit relationship (section 4.2.3), recovery should occur. However, this does not help if a species is vulnerable as bycatch (Chapter 13). Thus several skates in the Atlantic can only tolerate low levels of fishing mortality due to their low fecundity, large body size and slow life histo-ries (section 12.2.2). However, they suffer high mortal-ity because they are caught by trawlers chasing primary target species such as cod and haddock (Walker & Heesen, 1996; ICES, 1997b; Casey & Myers, 1998; Dulvy *et al.*, 2000). In the case of the North Sea skates and rays, four of the larger species are currently fished at levels that may be expected to extirpate them, though the patchiness of fishing efforts may mean that *de facto* refuges help to maintain their populations. These species are all vulnerable to groundfish trawls well before they attain maturity (Table 12.1). In general terms, however, the probability that fished species will become extinct as a result of fishing is low in compar-ison with the numerous and well-documented effects of hunting on terrestrial birds and mammals. For many species with restricted ranges and great habitat spe-cificity, it is likely that environmental changes, such as the effects of human activities on habitat and climate, will cause more problems than fishing.

Diversity

Even in intensively fished areas like the continental shelves of the north-east Atlantic, large-scale patterns in fish diversity are primarily governed by biogeo-graphic factors (Rogers *et al.*, 1999a, b). These patterns are modified by the direct and indirect effects of fishing.

Common name	Species	L_∞	L_{mat}	T_{mat}	Fecundity	$Z_{r=0}$	Z
Common skate	*Dipturus batis*	237	160	11	40	0.38	*
Thornback ray	*Raja clavata*	118	86	10	140	0.52	0.60
Spotted ray	*Raja montagui*	79	62	8	60	0.54	0.72
Cuckoo ray	*Leucoraja naevus*	75	56	8	90	0.58	0.69
Starry ray	*Raja radiata*	71	39	5	38	0.87	0.79

Table 12.1 Life-history characteristics of five skates and rays (Rajidae) caught in the North Sea, the predicted levels of mortality they can withstand in the long term (the level at which the intrinsic rate of increase, *r*, is reduced to 0, $Z_{r=0}$) and current estimates of fishing mortality (*Z*). The common skate is already too scarce to permit a realistic estimate of *Z*. From Walker and Hislop (1998).

Table 12.2 Some indices of species diversity. From Jennings and Reynolds (2000).

Index	Measures	Formula	Notes
Species richness (S)	Number of species	S	
Margalef (D)	Number of species for given number of individuals	$(S - 1)/\ln N$	1
Menhinick's (D)	Number of species for given number of individuals	$S/\sqrt{N}$	
Shannon Wiener (H')	Richness and equitability	$-\Sigma p_i \ln p_i$	2
Eveness (for H')	Eveness	H'/H_{max} or $H'/\ln S$	
Brillouin (HB)	Richness and equitability	$(\ln N! - \Sigma \ln n_i)/N$	3
Pielou Eveness (for HB)	Eveness	HB/HB_{max}	
Simpsons (D)	Dominance	$\Sigma (n_i(n_I - 1)/N(N - 1))$	4
Hill N_0	Number of species	S	5
Hill N_1	Number of 'abundant' species	$\exp H'$	
Hill N_2	Number of 'very abundant' species	$1/D$	6
Taxonomic diversity (Δ)	Species diversity with taxonomic separation	$[\Sigma\Sigma_{i<j} \omega_{ij} x_i x_j]/[n(n - 1)/2]$	7
Taxonomic distinctness (Δ*)	Taxonomic distinctness without species diversity	$[\Sigma\Sigma_{i<j} \omega_{ij} x_i x_j]/[\Sigma\Sigma_{i<j} x_i x_j]$	7
Taxonomic distinctness (Δ⁺)	Taxonomic distinctness for presence/absences data	$[\Sigma\Sigma_{i<j} \omega_{ij}]/[s(s - 1)/2]$	7

Notes: (1) Where S = number of species and N = number of individuals. (2) Where p_i is the proportion of individuals of the ith species. Shannon–Weiner index assumes random sampling from an infinitely large population and that all species present are represented in the sample. In reality the true value of p_i is unknown and is estimated from $p_i = n_i/N$. (3) Alternative to Shannon–Wiener used to describe a known collection. The form appropriate to an infinitely large community would be Σp_i^2 where p_i is the proportion of individuals in the ith species. n is the number of individuals in the ith species. (4) Gives probability that any two species drawn at random from an infinitely large community belong to different species. Since D increases with decreasing diversity this index is often expressed as $1 - D$ or $1/D$. (5) Hill proposed a family of diversity measures ranging from those that emphasize uncommon species (richness) to those that emphasize dominance. N_0, N_1 and N_2 cover most aspects of diversity and are usually reported together. Note relationships between Hill numbers and the Shannon–Wiener and Simpson indices. (6) Where D is Simpson's Index. (7) x_i is abundance of ith of s species observed, $n = \Sigma_i x_i$ is total number of individuals in a sample, and ω_{ij} weights the path length linking species i and j through the taxonomy.

Diversity can be measured in many ways (Table 12.2). They range from counts of the total number of species recorded (species richness) to statistics that indicate both richness and the way in which the total number of individuals is divided among the total number of species (equitability). When using equitability indices, a community of high eveness and low dominance is considered more diverse than one with the same number of species but low eveness and high dominance (where few species account for most of the total abundance). The Hill (1973) indices can be used to describe aspects of diversity ranging from species richness (N_0) through eveness (N_1, effectively the number of abundant species) to dominance (N_2, effectively the number of very abundant species).

Taxonomic-based diversity indices were introduced by Warwick and Clarke (1995) and Clarke and Warwick (1998). These incorporate the idea that a community containing three distantly related species, say a cod, herring and skate, is more diverse than one containing three species of anchovy (Fig. 12.4). The indices take account of the relatedness of species in the community and their abundance. Taxonomic diversity (Δ) is based on Simpson diversity, with an additional component of taxonomic separation. It is the mean path length along a taxonomic hierarchy between

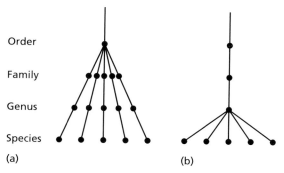

Fig. 12.4 Two examples of taxonomic hierarchies for species in samples. Mean taxonomic path lengths between individuals in (a) are greater than in (b). Thus sample (a) would be regarded as more diverse. After Clarke and Warwick (1998).

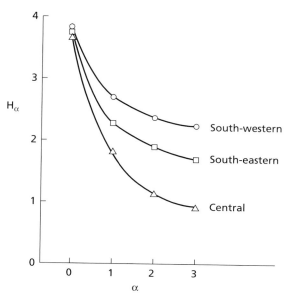

Fig. 12.5 Rényi diversity profiles for fish communities in three regions of the North Sea. The south-western North Sea is the most diverse because the profiles do not cross at all values of α and diversity in the south-western area is always highest. After Rogers et al. (1999b).

any two randomly chosen individuals (Fig. 12.4). Δ^* is similar, but with a reduced role of species abundance, such that it measures the mean path length between any two randomly chosen individuals, conditional on them being from different species. The simplest variant on this approach ignores abundance information altogether, such that Δ^+ measures taxonomic path lengths in presence/absence data. The taxonomic indices are independent of sample size. When the path length between individuals is determined from a taxonomy, rather than based on genetic distance, it is not a true measure of relatedness. These indicies were recently used to compare diversity of fishes in the north-east Atlantic (Rogers et al., 1999a).

Because different indices measure different things it has proved difficult to rank fished systems in terms of their diversity. Diversity ordering procedures overcome this by presenting diversity indices for a range of α values where α determines the relative weighting towards species richness or dominance. The Réyni (1961) index (H_α) has been used in studies of fishing effects on diversity:

$$H_\alpha = (\log_e \Sigma p_{i\alpha})/(1 - \alpha) \qquad (12.1)$$

Where p_i is the proportional abundance of the ith species. Once H_α has been calculated for a range of α values, H_α is plotted against α. If the trajectories for two communities cross they are not comparable, but when one consistently falls below the other that community is said to be less diverse (Fig. 12.5).

While diversity indices offer some statistical convenience they provide little information on the overall distribution of individuals between species. An alternative

is to look at relationships between species and abundance using graphical techniques. Changes in the relationships over time or comparisons among fished and unfished areas indicate the effects of fishing or other stresses on community structure. The most widely used are dominance and k-dominance curves. Dominance curves are the abundance of each species, after ranking by abundance, expressed as a percentage of the total abundance of all species and plotted against log rank (Fig. 12.6a). k-dominance curves are the cumulative ranked abundance against log species rank, such that the most elevated curves show the lowest diversity (Fig. 12.6b). The cumulative abundance scale is often log or modified log transformed.

Trawl surveys of the North Sea demersal fish community provide evidence for changes in diversity due to fishing. Between 1929 and 1953 and 1980–93, Hill's N_1 and N_2 for the whole demersal fish community declined, although there was no change in the non-target community. k-dominance curves showed that the community was dominated by fewer species in recent years (the curves were steeper and more elevated, Fig. 12.7) (Greenstreet & Hall, 1996). Species diversity declined in the areas where fishing intensity was

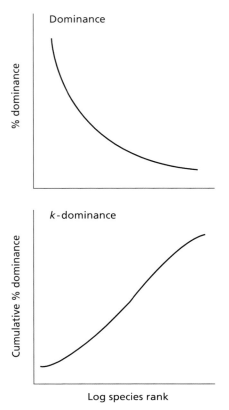

Fig. 12.6 Dominance and k-dominance curves.

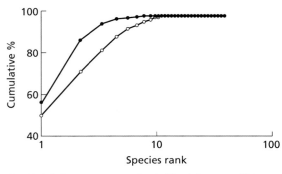

Fig. 12.7 k-dominance curves for the North Sea groundfish community off East Shetland from 1929 to 1953 (○) and 1980 to 1993 (●). After Greenstreet and Hall (1996).

highest, but fishing effects were largely confined to decreases in abundance of species which have slow life histories and are particularly vulnerable to exploitation (Jennings *et al.*, 1999a). This suggests that changes in the abundance of vulnerable indicator species, rather than broad measures of community diversity, provide a better measure of fishing effects in the North Sea.

Spatial variations in the diversity of fish faunas around the British Isles have been compared using the Réyni's diversity index family (Rogers *et al.*, 1999b) (Fig. 12.5). The central North Sea fauna is least diverse, partly as a result of the uniform nature of the seabed. To the west, in the English Channel, the more heterogeneous substrate supports a more diverse fauna of smaller fish. k-dominance plots also showed that the fish communities were less diverse in the North Sea than in the English Channel and Irish Sea. These broad patterns in diversity were the result of biogeographic factors, seabed structure and regional hydrography rather than fishing effects (Fig. 12.8).

On tropical reefs, the species richness of many target species is correlated with fishing intensity, but broad patterns in diversity are also determined by biogeographic factors and habitat structure. Indeed, the greatest impacts of fishing on diversity have been indirect, when habitat destructive techniques such as dynamite fishing reduce complex substrate to rubble (section 5.4) and when the indirect effects of fishing prevent coral growth and favour bioerosion (section 14.7).

On Fijian reefs, the species richness of target groupers (subfamily Epinephelinae in family Serranidae) and snappers (Lutjanidae) is significantly correlated with fishing intensity (Fig. 12.9), as is the species richness of snappers and emperors on Seychelles reefs (Jennings *et al.*, 1995; Jennings & Polunin, 1997). Comparisons between fished and unfished (marine reserve) areas on coral reefs have consistently shown that species richness is higher in the unfished areas (Roberts & Polunin, 1991). The studies on tropical reefs differ from those on temperate shelves because they are based on small areas and the target fishes are relatively site attached and non-migratory. Moreover, the whole North Sea has been intensively fished since the 1700s (Anon, 1885), whereas some small areas of reef have little or no history of fishing. In the North Sea, widespread changes in diversity may have taken place before scientific investigations began.

Local reductions in species richness are not necessarily associated with clear decreases in yield. Does this mean that some species are redundant? If fish production is the only ecosystem function of concern, then the loss of one or two grouper species on a tropical reef, where 15 or more species are relatively abundant, is unlikely to lead to detectable changes in total yield. However,

Species richness

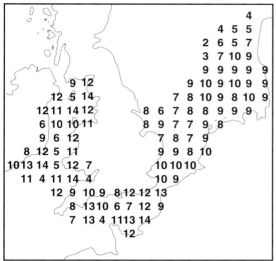

Shannon diversity

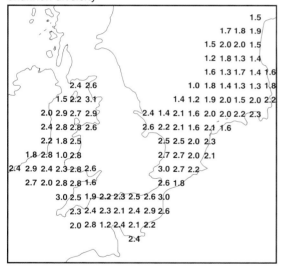

Fig. 12.8 Species richness of flatfish, and Shannon diversity in the seas around the British Isles. See Table 12.2 for a definition of these diversity indices. After Rogers *et al.* (1998).

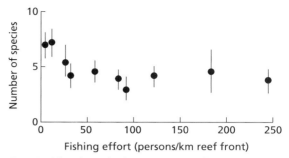

Fig. 12.9 The relationship between species richness (±95 C.L.) and fishing intensity for groupers (Epinephelinae) on Fijian reefs. After Jennings and Polunin (1997).

Even if species richness does not play an important role in maintaining ecosystem processes at present, it may ensure that there are species to fulfil new roles when conditions change. Moreover, since the recruitment of related species rarely varies in unison, diversity ensures that some species are abundant at any given time. It is notable that fishing effects have been most dramatic in ecosystems where few species fulfil key functional roles. Most 'long-term' ecological studies have been in progress for a few decades and do not allow us to predict the role of species in future years (Jennings & Kaiser, 1998).

Diversity loss and ecosystem stability

If fishing leads to extirpation and changes in diversity, how will this affect ecosystem stability? The links between diversity and ecosystem stability are an active field of research for terrestrial ecologists but virtually unstudied in the marine environment. In general terms, the extent to which a system is viewed as stable depends on the measure of stability, and the spatial, temporal, taxonomic and abundance scales considered. Thus, an ecosystem which is highly fragile over a few years may be highly stable over centuries, and an ecosystem which is fragile at small spatial scales may be stable when viewed at larger scales. In fisheries terms, one measure of stability is the total production of fish protein on temporal scales of years to decades and spatial scales of tens to hundreds of kilometres. Individual species show little stability in production on these scales while total fish production may be remarkably stable (Chapters 2 and 3). However, as species are removed, we approach a threshold where further removals lead to a shift in ecosystem structure or function. The threshold may be reached rapidly if the species interact strongly. Thus the

while the extirpated species may be 'redundant' at the present time, the value of ecological functions will change and there should be greater emphasis on functional similarity. This allows the level of similarity to be quantified (at a given place or time) whereas redundancy implies a non-constructive 'valuable or worthless' 'all or none' perspective (Collins & Benning, 1996).

removal of a few species of urchin-eating fishes by fishers has led to the proliferation of bioeroding urchins on Kenyan reefs (section 14.7).

12.4.2 Community structure

Most fisheries are relatively unselective and many species experience high levels of mortality as bycatch even if they are not targeted by the fishery (Chapter 13). The susceptibility of late-maturing and larger fishes to fishing suggests that small and early-maturing species would increase in relative abundance in an intensively exploited multispecies community (section 12.2.2). However, while the life histories of smaller species may enable them to sustain higher instantaneous mortality rates than larger species, they may also suffer lower fishing mortality simply because they are less desirable and escape through meshes in nets and traps.

Fish communities pass through a series of structural changes as they are increasingly heavily fished. At first, larger individuals of all target species decrease in abundance, and larger species form a smaller proportion of the total abundance (e.g. Haedrich & Barnes, 1997). Ultimately the whole community is dominated by smaller individuals and smaller species. These trends are quite predictable, both between areas subject to different fishing intensities and through time in a single area. Since fish communities are relatively difficult to study, changes in community structure are often interpreted from changes in catches. These usually show the same patterns, with catches of larger species and individuals dominating the fishery at first and decreasing as fishing intensity increases.

There are many examples of changes in community structure and catches on tropical reefs. In Jamaica, the largest fish caught with traps (the main fishing method), including groupers, snappers and parrotfishes, virtually disappeared from catches at all heavily fished study sites in a 20-year period, while smaller, less desirable species could still be caught (Koslow et al., 1998). On reefs around Apo and Sumilon Island in the Philippines, which were opened and closed to fishing as marine reserve management succeeded and failed, large species with slow life histories declined in abundance most rapidly when fished, and increased in abundance most slowly when protected (Russ & Alcala, 1988a, b). On Seychelles and Fijian reefs that were subject to high fishing intensities, larger species with slower life histories were less abundant than on lightly fished and unfished reefs (Jennings et al., 1996, 1999b; Jennings & Polunin, 1996b).

Similar patterns are involved in temperate waters. In the North Sea, there have been marked changes in the relative abundance of species since trawl surveys began in 1925 (Greenstreet & Hall, 1996; Greenstreet et al., 1999a). During this period, fishing effort increased steadily and the mean growth rate across all species and all individuals increased, while mean maximum size, age at maturity and size at maturity decreased (Fig. 12.10) (Jennings et al., 1999a). These effects could be explained in terms of the differential effects of fishing on species with contrasting life histories. A phylogenetic analysis showed that species that decreased in abundance relative to their nearest evolutionary relatives matured later at a greater size, grew more slowly towards a greater maximum size and had lower rates of potential population increase. This applied to both target and non-target species since North Sea demersal fishes are fished by trawls and most species are affected by fishing even if they are subsequently discarded as bycatch. The decline in relative abundance of species with slow life histories is the result of two factors: they can sustain lower mortality rates and they are often more desirable and heavily fished. It is usually impossible to separate the relative roles of these factors, even though we know life histories are important. This is because it is so difficult to estimate fishing mortality for non-target species (section 9.3.6).

Changes in relative abundance that result from the combined effects of mortality and life history are also reflected in trends in trophic structure. In general, species with faster life histories feed at lower trophic levels and have higher production to biomass ratios. On intensively fished reefs, fish biomass is dominated by herbivores, while invertebrate feeders and piscivores dominate on lightly fished reefs (Jennings & Polunin, 1996b). Studies of the relationships between fishing intensity and fish community structure in the tropics, where virtually unfished and yet accessible sites are often located, have indicated the rapidity with which the composition of target fish communities can change in response to fishing (Russ & Alcala, 1989, 1996a; Jennings & Polunin, 1996b, 1997). Low levels of fishing intensity, such as those which occur when a marine reserve is occasionally fished, are sufficient to cause dramatic changes in the trophic structure of the fish community (Fig. 12.11).

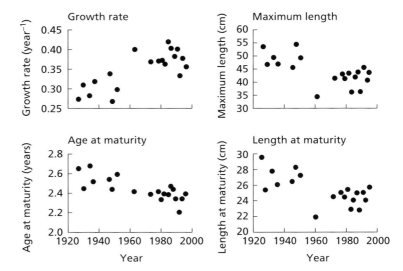

Fig. 12.10 Temporal trends in mean growth rate; maximum length; age at maturity; and length at maturity of individuals in the North Sea demersal fish community. After Jennings et al. (1999a).

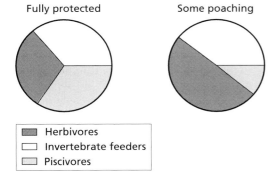

Fig. 12.11 Trophic structure (by biomass) of fish communities in Seychelles marine reserves that are either fully protected, or subject to a small amount of fishing and poaching. After Jennings and Kaiser (1998).

Changes in trophic structure observed in individual fisheries are also reflected in global landings statistics. Thus, the mean trophic level of marine landings fell between 1950 and 1993 as fishers 'fished down the food chain' (Pauly et al., 1998). In this analysis, the trophic levels of landings were calculated for Food and Agriculture Organization (FAO) species groups using Ecopath (section 8.7). The trophic level of global landings rose briefly in the early 1970s when the Peruvian anchovy fishery collapsed, but the general trend was downward (Fig. 12.12a). In heavily fished areas such as the north-east Atlantic and Mediterranean (Fig.

12.12b, c), there was a more consistent downward trend as fishers strived to maintain yields. While these data show clear trends in the trophic level of landings, we cannot necessarily infer that equally large changes in trophic level have occurred in fish communities in the sea. This is because fishers may change the gear they use or the fishing grounds they exploit as catching larger species becomes less economically viable. This effect may accentuate the succession from high to low trophic groups and large to small sizes.

12.4.3 Size structure

The size structure of the biota in a marine ecosystem follows regular patterns, and the relationship between size (as classes) and total normalized biomass (biomass in a size class interval divided by the interval width) can be predicted by models of energy flow from prey to predators. Fishing is likely to cause the size distribution of biota within an ecosystem to differ from that predicted, since it will selectively remove larger fishes and this may, in turn, affect their predators or prey (section 8.6).

Size spectra have been used to look at the effects of fishing on demersal fish communities such as those sampled by groundfish surveys. Within these communities, aggregate numbers (log) of fish (all species) combined by length class decline linearly over the range of length classes which are fully sampled (Fig. 12.13). The

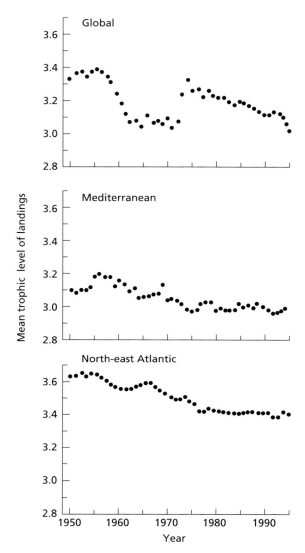

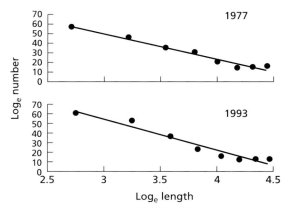

Fig. 12.13 Examples of size spectra for the North Sea demersal fish community in 1977 and 1993. The log$_e$ numbers of fish per log$_e$ 10 cm size class are shown. Note the steeper gradient and higher intercept of the fitted relationship in 1993. After Rice and Gislason (1996).

12.4.4 Competition and trophic interactions

Predation is a key structuring process in aquatic ecosystems, but empirical evidence suggests that most predator–prey relationships are not tightly coupled and that the removal or proliferation of one species which eats another does not have a dramatic and predictable impact on ecological processes.

Ecosystems controlled by predators are said to show **top-down control** while those controlled by the physical environment show **bottom-up control**. In many marine systems, bottom-up effects (Chapters 2 and 3) overwhelm top-down effects, and the removal of predators by fishing does not result in the proliferation of their prey. These observations may not be predicted by simplistic models of predator–prey interactions that take no account of prey switching, ontogenetic shifts in diet, cannibalism or the diversity of species in marine ecosystems. The main exceptions occur in communities of low diversity where one or two species perform keystone roles.

Fig. 12.12 The mean trophic levels of global fish landings; Mediterranean landings (FAO Area 37); and landings from the north-east Atlantic (FAO area 27). FAO areas are shown in Fig. B1.2.1. After Pauly *et al.* (1998).

The Barents Sea ecosystem
In some temperate ecosystems, only a few planktivorous species link zooplankton production to fish production, and there is evidence for tight predator–prey coupling. In the Norwegian-Barents Sea ecosystem, for example, fish biomass is dominated by Atlantic herring *Clupea harengus*, capelin *Mallotus villosus* and cod *Gadus*

slope and intercept of this relationship may indicate changes in the exploitation regime (e.g. Pope *et al.*, 1988; Murawski & Idoine, 1992) and, in a recent North Sea study, appeared to be linear functions of fishing intensity (Rice & Gislason, 1996; Fig. 12.14). These patterns were largely attributed to the selective removal of larger fishes by fishers rather than proliferation of prey.

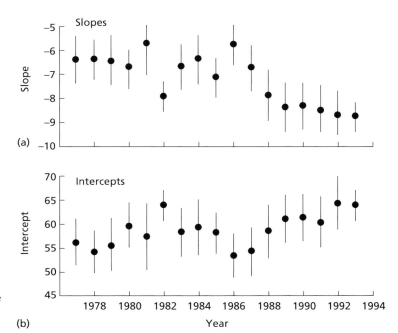

Fig. 12.14 (a) Slopes and (b) intercepts of linear regressions fitted to size spectra (Fig. 12.13) for the North Sea demersal fish community from 1977 to 1993. Ninety-five per cent confidence limits are shown. Data from Rice and Gislason (1996).

morhua. The diets of herring include capelin larvae and those of capelin include herring larvae. Cod eat herring and capelin. Following intensive fishing and falling sea temperatures in the 1960s there was a dramatic fall in herring biomass and reduced predation on larval and juvenile capelin. Reduced predation, coupled with lower sea temperatures, led to better capelin recruitment. In the early 1980s, however, sea temperature rose again and cod recruitment increased. The warmer conditions did not favour capelin and their recruitment failed in 1984 and 1985. This poor recruitment, coupled with the effects of intensive fishing, led to the collapse of the capelin stock (Fig. 12.15). Because herring biomass was already very low, cod had few alternative prey and started to cannibalize their young. Cod weight-at-age (by year class) decreased by 20–70% from 1984 to 1988 and individual food consumption fell by 40–70% (Hamre, 1988; Mehl & Sunnanå, 1991; Blindheim & Skjoldal, 1993). Fortunately, in recent years, the capelin stock has recovered.

In temperate waters, studies of predator–prey relationships are complicated by the direct effects of fishing since many prey species are taken as bycatch or may be the targets of directed reduction (industrial) fisheries (Chapter 13). Thus, models used to investigate

predator–prey relationships need to incorporate fishing mortality and rapidly become very complex (Pope, 1979; Sparre, 1991; see Chapter 8). In the North Sea, Pope and Macer (1996) investigated whether predation effects accounted for trends in the recruitment of cod and whiting *Merlangius merlangus*. These species primarily prey upon their own young, haddock *Melanogrammus aeglefinus*, Norway pout *Trisopterus esmarki* and sandeel *Ammodytes* spp. Using a model that assumed fixed feeding behaviour by each species, they estimated predation rates on cod and whiting from 1921 to 1992. It appeared that predation effects did not account for trends in cod or whiting recruitment and that the direct effects of fishing had the overwhelming impact on population history.

Species replacement in pelagic fish communities
Some of the most dramatic shifts in fish community structure have occurred in the highly productive fisheries for clupeoids in upwelling ecosystems (Schwartzlose *et al.*, 1999). Are these the result of the depletion of one species by fishing opening a 'window of opportunity' for another? When the fishery for Californian sardine *Sardinops sagax* collapsed (section 2.1) the biomass of northern anchovy *Engraulis mordax*, another

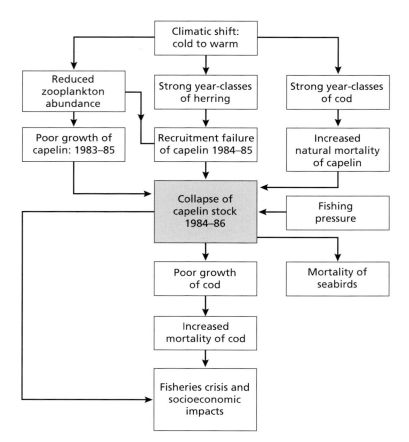

Fig. 12.15 Relationships between ecological events in the Barents Sea ecosystem. After Blindheim and Skjodal (1993).

fast-growing planktivorous species, began to increase (MacCall, 1986a, 1990). The northern anchovy may have been in competition with the sardine because they had similar feeding habits and life histories. The collapse of the sardine stocks was purported to have released additional food for the anchovy and led to reduced predation on anchovy larvae. However, although an interaction did exist, the environment appeared to play the key role in governing abundance. Thus, changes in the abundance of the two species as determined from scale deposition in anaerobic sediments (section 4.1) showed that there were several periods during the last 2000 years when both species were very abundant. In general, it appears that intensive fishing of the sardine did not lead to the proliferation of the anchovy stocks, since all species did well during periods of high productivity and low fishing effort. While fishing has acted in synergy with environmental factors to magnify and accelerate the collapse of pelagic stocks,

the collapses often occurred at times of marked environmental change, and the replacing species may have proliferated in any case.

Comparing reef and lake ecosystems
On coral reefs, where the functional and species diversity of fishes is relatively high, the indirect effects of fishing on the abundance of prey fish appears to be minor. On the scale of kilometres to tens of kilometres, significant decreases in the abundance of piscivorous target species following fishing (often an order of magnitude or more) were not associated with corresponding increases in the abundance of their prey (Bohnsack, 1982; Russ, 1985; Jennings & Polunin, 1997; Russ & Alcala, 1998b). On reefs, as in many marine ecosystems, the indirect effects of fishing on fishes are smaller and less predictable than in freshwater lakes where the removal of predators leads to cascading increases in prey abundance. Why might this be? It could be due to

difference in the size distribution and diversity of the biota. Within lake ecosystems the majority of biomass is aggregated into a few distinct size classes that loosely correspond to phylogenetic groupings. Moreover, phylogenetic diversity is lower, and organisms within size groups tend to have a relatively limited range of life-history traits and morphologies. Since size has a key role in determining the predators or prey of organisms within food webs, the organization of freshwater communities places strong constraints on community structure (Neill, 1994). Conversely, in species-rich marine systems the biomass spectrum is extended and there is more variance in size within phylogenetic groupings. This is particularly apparent on reefs. Moreover, the phylogenetic groupings tend to contain more species, with a wider range of life-history traits, behavioural differences and feeding strategies. Thus, 8–53% of species in reef fish communities eat other fishes, and many of these are generalists (Parrish *et al.*, 1986; Hixon, 1991). As a result, the overall effect of all piscivores on their prey can be substantial, although the impact of any individual species, or small group of species, is minor. This situation has been termed 'diffuse predation' by Hixon (1991) and differs markedly from the situation in lake ecosystems, and some marine systems such as the Barents Sea, where a few keystone species dominate the biomass within a trophic group (Jennings & Kaiser, 1998).

The best evidence for predator control of prey species is confined to the relationships between humans and their target fishes! There are numerous studies indicating that the abundance of target fishes decreases with increasing fishing effort and expands following the cessation of fishing (section 12.4.2). The strength of these relationships results from the conservative fishing strategies of humans who tend to target a relatively small proportion of the total fish fauna. Most predatory fish, conversely, are generalist feeders, often switching diet in response to prey abundance and cannibalizing their own species.

Invertebrates

The story is rather different for invertebrates, where keystone species are more common and density-dependent processes and competition are strong regulators of population dynamics. For example, the dynamics of competing red abalone *Haliotis rufescens* and red sea urchin *Strongylocentrotus franciscanus* populations on

the north-west coast of the United States are partly controlled by fishing effects. The fishery for red abalone is managed for recreational use, and diving with compressed air is banned. This means that populations in deeper areas are virtually unexploited. When a fishery for red sea urchins developed, the depletion of urchins led to the expected increase in algal recruitment. In deeper areas, this was followed by increases in the densities of red abalone that may have been released from competition (Tegner *et al.*, 1992).

Urchins and kelp forests

Kelp forests are found in cool, shallow and nutrient-rich coastal waters (section 2.4.1). They support invertebrate and fish fisheries, and the kelp may be harvested to produce alginates. Sea urchins feed on kelp and increases in urchin abundance lead to decreases in kelp cover. At high urchin densities, the kelp forest ecosystem may shift to an alternative state, known as '**urchin barren ground**'. The shift to barren ground reduces the habitat and primary production available to other species that usually use the kelp forest. If urchins are experimentally cleared from barren ground, then kelp will re-establish. Without such human intervention, barren grounds persist until disease, storms, sea otters or surge reduce urchin populations.

If fishing has direct or indirect effects on urchin abundance then it may lead to shifts between urchin and kelp dominated communities. In the *Macrocystis* forests off southern California (Fig. 2.9), the extent of kelp forests has shrunk since the mid-twentieth century. Although partly a response to changes in oceanographic conditions and increased sewage discharge, the fishing of urchin predators has also played a role. Both the spiny lobster *Panulirus interruptus* (Palinuridae) and the sheepshead *Semicossyphus pulcher* (Labridae) may once have eaten enough urchins to limit urchin population growth, but their abundance has now fallen following intensive fishing. Despite the loss of urchin predators, the overall impact of urchin grazing on the kelp forests may now be limited by a fishery for the red sea urchin *Strongylocentrotus franciscanus*. This fishery was the largest and most valuable in southern California through most of the 1990s and has led to clear reductions in the abundance and distribution of the urchins (Dayton *et al.*, 1998; Tegner & Dayton, 2000).

In the *Laminaria*-dominated kelp beds of the eastern

Canada and north-eastern United States, the ecosystem also shifts between kelp and urchin-dominated states. The shift had been associated with decreases in the abundance of the heavily fished Amercian lobster *Homarus americanus* that was known to prey on urchins (Mann & Breen, 1972; Mann, 1982). However, detailed analysis of feeding rates, stomach contents and abundance of lobsters indicated that they could not have controlled the population structure of urchins and it seemed that urchin populations expanded following increased larval recruitment (Miller, 1985; Hart & Scheibling, 1988). More recently, declines in the abundance of species such as wolf-fish *Anarhichas lupus* (Anarchichadidae) that feed on urchins were suggested to cause the shift away from a kelp-dominated community, but the relative impacts of these declines and oceanographic processes still need to be resolved (Vadas & Steneck, 1995). The indirect effects of fishing on kelp forests are more consistent in California than in eastern Canada and the United States. However, the importance of oceanographic factors should not be forgotten. Large area of kelp forests off Chile, Peru, southern and Baja California also died when they were bathed by warm and nutrient poor waters during El Niño events.

The removal of urchin predators by fishers on coral reefs may also lead to dramatic increases in the abundance of urchins and bioerosion of the reef matrix. We describe these effects in section 14.7. Once again, prey release appears to occur because very few species perform keystone roles.

Specific and aggregate responses
While changes in the abundance of keystone species have critical effects on community structure and ecosystem function, such species are probably rather rare. We say this because the diversity of species in the marine environment is often very high, because the numbers of known keystone species can be measured in tens rather than thousands, and because marine ecologists have tried very hard to find them. In general, it has proved very difficult to link the dynamics of small groups of species. This does not mean that they do not interact, just that the interactions may be minor or transient in space and time, and have a smaller relative impact on dynamics than other aspects of the environment.

However, at some level the removal of many different species with similar trophic functions must begin to influence the aggregate dynamics of other groups. Thus, in many fished ecosystems, fishers selectively remove many of the predatory fishes. If the marine ecosystem is not entirely inefficient then we would expect other groups to use the production that is no longer eaten by fish predators. Due to the scale on which such changes would occur, they are easier to examine with ecosystem models (section 8.7) and landings data than by experimental manipulation. One change in global landings that may be a response to the reduction in abundance of large predatory fish is the rapid rise in landings of predatory cephalopods. This may, of course, be the result of fishers targeting new cephalopod stocks, but a range of evidence now suggests these species are more abundant, and therefore more likely to support fisheries, in ecosystems where fish predators are overfished (Caddy & Rodhouse, 1998). Cephalopods are likely to sustain relatively high fishing mortality due to their fast life histories.

Summary

- Behaviour and life histories determine vulnerability to fishing.
- Fishers use their knowledge of behaviour to increase catches, and managers can use it to improve resource conservation.
- Species with slow life histories (slow growth, late maturity, large body size, low potential rates of population increase) are more vulnerable to fishing than species with faster life histories.
- Size-selective fishing can change the sex ratios of sequential hermaphrodites and curtail reproductive life span. This may reduce reproductive success.
- The life-history traits of fished species are often heritable, so populations will evolve in response to harvesting.

Continued

Summary *(Continued)*

- Fishing has extirpated a few species, but extinctions are rare. The most vulnerable species have limited geographical distribution, depend on specific habitats, and are accessible to fishers.
- Broad-scale patterns in diversity are usually explained by biogeographic factors and are modified by the direct and indirect effects of fishing. Fishing reduces species richness and can reduce ecosystem stability.

- Heavily fished communities are increasingly dominated by smaller species with faster life histories from lower trophic levels.
- Keystone species are relatively rare and most interactions between species in a community are weak. However, when keystone species exist, changes in their abundance due to fishing have critical effects on community structure.

Further reading

Several general texts explore links between diversity and ecosystem processes. Gaston (1996) is a good example. Magurran (1988) provides a useful summary of diversity indices and their uses. The ICES symposium volume on the ecosystem effects of fishing (Gislason & Sinclair, 2000) contains many papers that explore issues of vulnerability, genetic effects, community effects and changes in trophic interactions due to fishing.

13 Bycatches and discards

13.1 Introduction

The food demands of an ever-increasing world population and the industrialization of some fisheries have led to the overexploitation of many marine species. The bycatches associated with many fisheries have been highlighted in the media by conservation organizations and scientists. As a result, bycatches of species such as marine mammals, seabirds and turtles have become dominant factors in the management strategies of some fisheries. In some cases bycatches are so great that they are unsustainable, yet in other populations the effects may be negligible. Occasional bycatches may be a serious threat when they affect populations of rare and endangered species. Regardless of the ability of different species to withstand bycatch mortality, public perception of the species involved has a strong influence on the outcome of management measures to limit the effects of the fishery concerned.

In this chapter we discuss why many animals are removed from the sea and subsequently discarded. We explore the economic and other incentives that encourage discarding. We discuss the trophic and biomass composition of catches and discards in world fisheries and look at the differences between food fisheries and industrial fisheries that convert marine organisms into animal feeds and fuels.

13.2 Catches, bycatches and discards

13.2.1 Definitions

It is important to understand the terms used when discussing catches, bycatches and discards. Although this may seem rather pedantic, the use and misuse of the terminology has the potential to cause much confusion. As a result, Alverson *et al.* (1994) spent several pages clarifying the definitions of these terms. Probably the best way to understand them is to envisage the point at which they become relevant in the capture process.

The aim of most fishers is to capture species that have financial or energetic value. Thus, for an artisanal fisher the aim may be to catch anything that is edible, whereas those involved in high-technology fisheries may be interested in only one or a suite of species (e.g. shrimp, flounder, cod, hake). These species are defined as the **target species**. Target species are usually associated with other organisms that may not be the intended catch of that fishery. Once their gear is deployed, fishers have little or no control over the fate of the **non-target species** that can become part of the catch known as the **incidental catch**. Incidental catches can include benthic invertebrates, fishes, marine reptiles, birds and marine mammals. As soon as the catch is hauled in it is sorted, target species and valuable bycatch species are removed and the residual organisms (known as the **discarded catch** or **discards**) are returned to the sea. The **bycatch** is composed of the sum of the discarded catch and the incidental catch (Box 13.1).

13.2.2 Reasons for discarding

At this point, one might ask why should discarding occur, what is the point of returning edible protein back to the sea if it is going to die in any case? This is unlikely to happen in artisanal fisheries in which the emphasis is gathering enough protein for daily survival. However, the world's commercial fleets are motivated by financial reward. Larger commercial fishing vessels (> 30 m length) rarely go to sea for less than 1 week at a time. Often they travel long distances to reach productive fishing grounds. Each vessel has a limited fish-holding capacity. If these vessels were to retain all their bycatches, their fish-holds would fill up with low-value or (unfortunately named) **trash species** making the fishing trip uneconomically viable. The decision to reject non-target

Box 13.1

The composition and fate of catches, bycatches and discards.

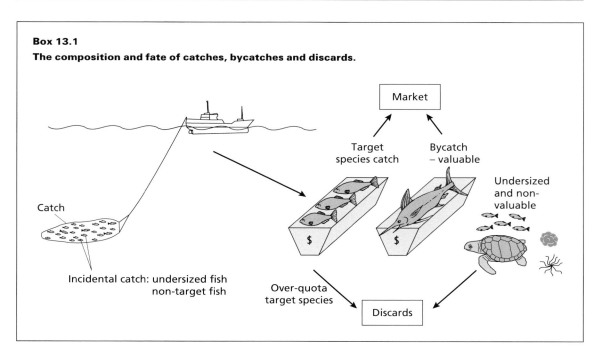

species is imposed by the fishers and the commercial market that pays low prices for bycatch species or by the imposition of regulations that prohibit their landing. However, it is economically feasible for industrial fisheries to target low-value species because they are removed in bulk. The identity of the fish harvested in industrial fisheries is largely irrelevant as they all yield protein or lipids; products that are the main objective of that fishery.

The influence of economic forces is taken to extremes when fishers discard target species that are above the minimum landing size in order to make room for more valuable larger fish in the fish-hold. This is a process known as **high-grading**. Gillis *et al.* (1995) used diet choice theory to investigate the mechanisms that underlie high-grading behaviour. They treated fishers as foragers that had to decide how much of each net's haul to 'ingest' before searching for more prey. They devised a model of discarding within each fishing trip that considered the availability of fish with different market values (financial rewards), constant factors such as trip quotas set by enforcement agencies, and the risk of premature trip termination due to gear loss or injury. Fish-hold size was considered to be analogous to

gut capacity, hence as catch increases, so the 'gut' fills up (assuming no digestion). Their model indicated that high-grading should be more common towards the end of a fishing trip and that it should increase with overall fish availability. High-grading was also predicted at the beginning of a fishing trip when the probability of the trip's quota being filled by the most valuable fish was highest. Gillis *et al.* (1995) were also able to determine that the imposition of more severe trip quotas would probably increase the incidence of high-grading. Thus, the conservation benefits gained by setting more severe quotas could be counteracted by the behavioural tendency of the fishers to discard perfectly marketable fish. Gillis *et al.* (1995) emphasized the need for fisheries managers to incorporate a consideration of the way fishers might alter their behaviour in response to the imposition of different restrictive management regimes.

The quota system (section 17.2.2) is particularly inefficient in mixed fisheries where several quota species coexist. For example, consider a fisher that has been allocated an annual quota for 20 tonnes of plaice *Pleuronectes platessa* and 10 tonnes of sole *Solea solea*. Both of these fish tend to be caught in similar grounds. Once the 10 tonne quota of sole has been caught, the fisher can continue fishing for plaice, but must discard any sole bycatches (note the sole has now shifted from the status of target to bycatch species), regardless of whether the fish are dead or not. It is almost impossible to avoid catching sole when fishing for plaice and vice versa. This seemingly senseless waste of perfectly good fish leads fishers to land **black fish** in their frustration. Black fish are those landed and sold illegally and do not occur in declared records of landings. In fact, many of the sole returned to the sea would survive as they are particularly resilient to the effects of retention within trawls (Kaiser & Spencer, 1995). This problem is more acute in fisheries for those fish with swimbladders that invariably die with the decrease in water pressure as they are hauled to the surface. Thus, current research to reduce bycatches is aimed at inducing bycatch species to escape the net before it is hauled (Chapter 5).

As long as management of mixed fisheries is based on quotas, the problems outlined above will persist, unless more radical action is taken. Norway is the first country to take initial steps towards reducing the practice of discarding. High-grading has been outlawed and fishers must land all fish above the minimum landing size. This could have several effects. Fishers will earn less money because they can no longer select only larger-sized fishes (although it should be noted that the Norwegian government subsidized the affected fisheries). Fishers might not travel as far in search of fish because they have less control over the size of the fishes retained. Quotas will be achieved more quickly, hence less fuel will be used thereby benefiting the environment.

13.3 Alternatives to discarding

The discarding of undersized fish is seemingly wasteful and is a fairly obvious underutilization of marine resources. This aside, the redirection of large amounts of energy into marine ecosystems can itself lead to population changes in scavenger species that are equally undesirable (Chapter 14). Alternatively, bycatches of undersized fish could be processed into fishmeal and fish oils. Fishmeal is used to produce animal and fish feeds while fish oils are used by the health industry and to fuel power stations. While there is some economic benefit to be gained from marketing these products, the fish would have a much higher value in its unaltered state if a consumer market existed for it, as far less money would be required to process the product. Hence,

this is an economically wasteful process that requires additional labour, infrastructure, transport and the consumption of energy to render fish down into other products. Processing fish into other products is also ecologically wasteful as each stage of the process requires not only financial investment, but also the loss of energy as protein and fats are converted from one form into another. Hall (1996) describes this as the 'fishmeal food chain' that incurs losses in nutritional value at each stage of the process.

13.4 Fisheries and bycatches

In a comprehensive review of world bycatches, Alverson *et al.* (1994) extracted data from over 800 published records. Although their text is the most authoritative on this subject, it is biased with respect to the quality of the data available, a point acknowledged by the authors. Bycatch and discard data are well documented for fisheries of the north-east Pacific and north-east and north-west Atlantic. The nations that monitor fisheries in these regions have sufficient resources to operate observer programmes on fishing vessels that are beyond the means of poorer nations. Trawl fisheries accounted for most of the records of bycatches (966) followed by drift nets and gill nets (232), line fisheries (150), pot (83) and purse seine (82) fisheries. An investigation of which target species or target groups were associated with most records of bycatches revealed that shrimp and groundfish species headed the list (Table 13.1; Fig. 13.1). The most commonly recorded animals taken as bycatch were salmon, marine mammals and halibut which had almost twice the number of records for all other bycatch species (Table 13.1). Since these species are either financially valuable or associated with high-value fisheries, the number of records is no doubt biased as a result of extra sampling effort (Alverson *et al.*, 1994).

Alverson *et al.* (1994) estimated that 27 million tonnes of material are discarded each year in comparison with the annual landed catch of around 100 million tonnes (Table 13.2). In their analysis they grouped the discards on the basis of the Food and Agriculture Organization of the United Nations (FAO) International Standard Statistical Classification of Aquatic Animals and Plants (ISSCAAP) species groups. Shrimp fisheries accounted for approximately 35% of the global commercial fisheries discards. In eight of the 18

Table 13.1 The number of records associated with the world's top 10 target and bycatch taxa or groups of animals. Number of records is the number of records of target or bycatch taxa in the database of bycatch studies compiled by Alverson *et al.* (1994). Groundfish are bottom dwelling or demersal species.

Taxa	Number of records
Target species	
Shrimp	255
Groundfish	247
Squid	130
Tuna	102
Pacific cod	94
Salmon	71
Atlantic cod	70
Whiting	68
Pollock	67
Nephrops	54
Bycatch species	
Salmon	281
Marine mammals	117
Halibut	113
Atlantic cod	63
King crab	48
Haddock	44
Tanner crab	41
Shark	35
Whiting	31
Birds	31

(a)

(b)

Fig. 13.1 Bycatch rates are high in groundfish fisheries that target species such as cod (a), but low in many pelagic fisheries that target species such as herring (b). Photographs copyright S. Jennings.

FAO areas evaluated, shrimp fisheries accounted for more than one-third of the total discards, and in four areas they accounted for over two-thirds of the discards. The ratio of discards to catch for shrimp fisheries was on average 5.2 : 1 (Table 13.2). Individual shrimp trawlers in the Australian northern prawn fishery discard approximately 1.5 tonnes or 70 000 individual organisms per night of fishing. This fishery generates roughly 30 000 tonnes of discards every year which is composed of 240 species including 75 families of fish and 11 families of shark. Shrimp fisheries off Brazil discard billions of fish each year while in the Gulf of Mexico shrimp fishery discards included 19 million red snapper and 3 million Spanish mackerel in 1989. It is for these reasons that shrimp fisheries are regarded as being among the least environmentally acceptable. The second highest discard ratio (2.49 : 1) was for crab fisheries, although their lower commercial landings meant that they produce the fourth highest amount

of discarded material. Crab fisheries tend to use very specific gear, hence it is not surprising that in the North Pacific crab fisheries the majority of discards are immature crabs or those below the minimum legal landing size. Many of these discarded crabs survive once returned to the sea as they are relatively undamaged by the fishing gear. A more detailed analysis of discards generated by trawl fisheries revealed discard to landed catch ratios of between 14.7 : 1 and 2 : 1 for 20 fisheries, of these, 14 were either shrimp or prawn fisheries (Table 13.3). The fisheries with the lowest discard to

Table 13.2 Global marine discards (includes bycatch landed but unreported by species in industrial fisheries) on the basis of the FAO International Standard Statistical Classification of Aquatic Animals and Plants species groups. From Alverson *et al.* (1994).

Species group	Mean discard weight (tonnes)	Landed weight (tonnes)	Ratio of discarded weight to landed weight	Ratio of discarded weight to total weight
Shrimps, prawns	9 511 973	1 827 568	5.20	0.84
Redfishes, basses, congers	3 631 057	5 739 743	0.63	0.39
Herrings, sardines, anchovies	2 789 201	23 792 608	0.12	0.10
Crabs	2 777 848	1 117 061	2.49	0.71
Jacks, mullets, sauries	2 607 748	9 349 055	0.28	0.22
Cods, hakes, haddocks	2 539 068	12 808 658	0.20	0.17
Miscellaneous marine fishes	992 356	9 923 560	0.10	0.09
Flounders, halibuts, soles	946 436	1 257 858	0.75	0.43
Tunas, bonitos, billfishes	739 580	4 177 653	0.18	0.15
Squids, cuttlefishes, octopuses	191 801	2 073 523	0.09	0.08
Lobsters, spiny-rock lobsters	113 216	205 851	0.55	0.35
Mackerels, snooks, cutlassfishes	102 377	3 722 818	0.03	0.03
Salmons, trouts, smelt	38 323	766 462	0.05	0.05
Shads	22 755	227 549	0.10	0.09
Eels	3 359	9 975	0.84	0.46
Total	27 012 099	76 999 942	0.35	0.26

landed catch ratios tended to be mid-water trawls, drift nets or certain types of traps (Table 13.3).

13.5 Incidental captures

13.5.1 Seabirds

There are numerous reported incidents of seabirds captured and drowned in fishing gears (Dayton *et al.*, 1995). In general, set net, drift net and long-line fisheries take the most birds. These fisheries often target large species such as adult tuna and salmon that do not occur in the diet of seabirds. So why do seabirds become caught in these fishing gears? Fish such as tuna pursue smaller prey fishes that are driven to the surface of the sea as they try to evade capture. Seabirds are able to recognize areas in which tuna are feeding by the disturbance created at the water's surface by the fleeing prey fishes. Seabirds capture these fish by either diving from the air or from the water's surface. The nylon meshes of the net are designed to be invisible underwater, hence birds have little chance of avoiding any nets set in the area where they are feeding. Surface feeding birds such as kittiwakes *Rissa tridactyla* and gulls *Larus* spp. are less likely to become entangled in subsurface

nets. Seabird mortality can be particularly high if nets are deployed in close proximity to breeding colonies.

The number of seabirds killed annually is quite staggering for some fisheries. It is not possible for us to list all the recorded instances here, but it is sufficient to say that the examples given are probably repeated elsewhere around the world. For example, an estimated 210 000–760 000 seabirds were killed every year by Japanese drift nets fishing in the North Pacific Ocean (Northridge, 1991). Subsequently, the use of these nets was banned in the early 1990s. Although such high incidental catches are undesirable, they are not necessarily unsustainable if they constitute a relatively small (< 1%) proportion of the population. However, in some circumstances bird deaths associated with nets can constitute the greatest source of mortality for some local populations. For example, a small population of 1600 razorbills *Alca torda* breeds in Newfoundland, Canada, and from 1981 to 1984 approximately 200 birds were killed in nets each year. This was slightly more than the 10% of the population that was estimated to die each year through natural causes. As a result, this population was considered under threat of extinction as nearly a quarter of the adult population was lost each year to a combination of natural and

Table 13.3 The 20 fisheries with the highest recorded discard ratios by weight (discarded weight per landed target catch weight) and the 10 fisheries with the lowest observed weight-based discard ratios. After Alverson *et al.* (1994).

Fishery description	Kg discarded per kg landed
Highest discard ratios	
Trinidadian shrimp trawl	14.71
Indonesian shrimp trawl	12.01
Australian northern prawn trawl	11.10
Sri Lankan shrimp trawl	10.96
US Gulf of Mexico shrimp trawl	10.30
Sea of Cortes shrimp trawl	9.70
Brazilian shrimp trawl	9.30
West Indian shrimp trawl	8.52
US Southeast shrimp trawl	8.00
Northwest Atlantic fish trawl	5.28
Persian Gulf shrimp trawl	4.17
Southwest Atlantic shrimp trawl	4.10
East Indian shrimp trawl	3.79
Bering Sea sablefish pot	3.51
Malaysian shrimp trawl	3.03
Senegalese shrimp trawl	2.72
Bering Sea rock sole trawl	2.61
British Columbia cod trawl	2.21
Gulf of Alaska flatfish trawl	2.08
Northeast Atlantic dab trawl	2.01
Lowest discard ratios	
Northwest Atlantic hake trawl	0.011
West Central Atlantic menhaden seine	0.029
Bering Sea cod pot	0.041
Northeast Pacific whiting trawl	0.043
Northwest Atlantic cod trawl	0.058
Bering Sea Pelagic pollock trawl*	0.062
Northwest Atlantic redfish trawl	0.063
Northeast Atlantic groundfish trawl	0.083
Gulf of Alaska midwater pollock trawl	0.086
Northwest Atlantic plaice trawl	0.118

* Includes some on-bottom fishing.

fishing mortality (Piatt & Nettleship, 1987). Albatrosses, and in particular the wandering albatross *Diomedea exulans*, have been severely affected by the activities of the southern bluefin tuna *Thunnus maccoyii* and Patagonian toothfish *Dissostichus eleginoides* long-line fisheries (Fig. 13.2). These fisheries deploy lines with thousands of hooks per set. Birds are attracted to the baited lines as they are paid out from the vessel and enter the water or when they can see the bait at shallow water depths. The Southern Ocean bluefin tuna fleet deploys 107 million hooks annually and is estimated to have killed 44 000 albatrosses up to 1989 when conservation measures were introduced. Nevertheless, modelling studies of wandering albatross populations predict that unless the number of hooks deployed per annum is reduced to 41 million in the Southern Ocean, this species will eventually decline to extinction (Weimerskirch & Jouventin, 1987; Brothers, 1991).

There are both conservation and financial incentives to reduce bird bycatches in long-line fisheries. For example, the southern bluefin tuna fishery lost about $US3.5 million in potential fish catches as a result of bait loss to birds in the 1980s. Seabirds in the north-east Atlantic were reported to remove up to 70% of the bait from long-lines which is costly in terms of lost bait, catch and time resetting the lines (Løkkeborg, 1998). Such high financial losses have led to the development of bird-scaring devices and setting funnels that set the

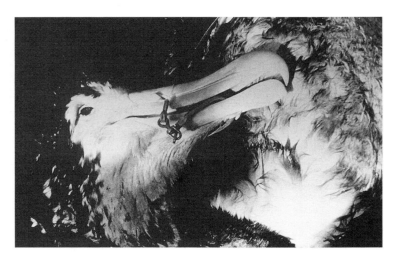

Fig. 13.2 A wandering albatross hooked on a tuna long-line. Photograph copyright N. Brothers.

line beneath the surface of the water out of the diving range of surface-feeding birds. Japanese tuna long-liners were able to reduce seabird catches by 88% by using bird deterrent devices (Fig. 5.16). Nevertheless, these measures may come too late for some endangered species of seabird. Currently, there are only 13 breeding pairs of the Amsterdam albatross *Diomedia amster-damensis*, and only 158 pairs of the short-tailed alba-tross *D. albatrus*. Both of these species are known to encounter long-line fisheries, and the latter have already been caught as incidental catch in Hawaiian waters. With so few remaining breeding adults in these popula-tions, even the loss of one bird will have a large impact on their chances of maintaining population numbers (Weimerskirch *et al.*, 1997).

13.5.2 Sea turtles

Captures of sea turtles have been recorded in long-line, set-net and towed gear fisheries. The survival of some turtle species is also threatened by loss of suitable hab-itat in which to lay their eggs through coastal tourist developments, egg collecting and illegal fishing. Marine litter is also a hazard for some species, particularly leatherback turtles *Dermochelys coriacea* that are indiscriminate feeders. Leatherback turtles have even died as a result of eating plastic bags that they pre-sumably mistake for jellyfishes that are their normal prey (Carr, 1987). Turtles are particularly vulnerable to entanglement and drowning in gill nets and associated

ghost-fishing gear, as the rough skin on their head and flippers catches easily on the meshes of these nets (Carr, 1987). Baited long-lines set for swordfish in the Mediterranean take approximately 20 000 endangered turtles every year. Similarly, shrimp trawl fisheries are a significant source of mortality for turtles (Poiner *et al.*, 1990; Fig. 13.3). In the shrimp fishery off North Carolina, five species of turtle had a mortality rate of up to 10%. Fewer turtles die when caught in towed gears, especially if the tow duration is less than 1 h. However, once tow duration rises to 200 min, mortality rate increases to 50% (Robins-Troeger, 1995).

Bycatches of turtles have been of concern to conser-vation organizations and there has been increased pub-lic awareness of this issue through media campaigns. Public opinion, in addition to scientific advice, has added urgency to the need to reduce bycatches of turtles in towed gear fisheries. This led to the development of trawl efficiency devices (TEDs) (Tucker *et al.*, 1997; Brewer *et al.*, 1998). Most TEDs are metal grids fixed within the net that deflect large organisms out of the net through an escape panel (Chapter 5). Threats of trade embargoes imposed by consumer nations, such as the US, have forced shrimp fisheries in the Far East to adopt TEDs. Fishers remain reluctant to use TEDs as they claim that they are difficult to use, but this is a particu-larly good example of how government restrictions and consumer choice can lead to the adoption of better fishing practices. Such initiatives tend to be driven by western cultures that are major consumers of shrimp.

Fig. 13.3 Shrimp fisheries have some of the highest discard to landed catch ratios of all fisheries. Long-lived large-bodied species are particularly vulnerable such as (a) sea turtles, and (b) large fishes such as rays and sharks. Photographs copyright T. Wassenberg.

(a)

(b)

Clearly the priority given to turtle conservation is likely to be viewed differently through the eyes of a shrimp fisher (section 13.8).

13.5.3 Sea snakes

Most sea snakes are highly venomous. Not surprisingly, there is relatively little work on their biology or population dynamics. As a result, it is difficult to gauge whether the estimated 120 000 sea snakes, comprising 10 different species, caught in the prawn trawl fisheries in the Gulf of Carpentaria, Australia, represents a significant threat to the survival of these species. Post-capture mortality varies from 10 to 40% depending on species, and, as with sea turtles, increases with tow duration. Handling on deck is no doubt a significant factor (Ward, 1996; Brewer *et al.*, 1998). Faced with a highly venomous animal, it is unlikely that the well-being of the sea snake will be foremost in the minds of most fishers.

13.5.4 Marine mammals

The high number of reported instances of incidental captures of marine mammals and the amount of research directed at quantifying this phenomenon and devising ways to reduce such bycatches is almost certainly linked to a 'sentimentality' factor. If one undertook a poll of the general public to determine which of

marine mammals, seabirds, sea turtles and sea snakes warranted greatest protection from human interference, marine mammals would be ranked first and sea snakes last. This is an unfortunate situation, as while we are well informed regarding interactions between fishing gears and marine mammals, we know very little with regard to sea snakes (see above).

It is an inescapable fact that wherever fishers pursue prey fish species they are likely to encounter larger predatory organisms such as sharks, swordfish and marine mammals. In many small-scale fisheries, the entanglement of a large animal in the net can lead to destruction of the gear and loss of catch which incurs financial losses for the fishers involved. This factor is less relevant in large-scale fisheries that use more robust fishing gear and power haulers with which to retrieve the gear. In this case, incidental captures are regarded as a nuisance rather than a financial impediment. Such was the case in the eastern Pacific tuna purse seine fishery that began in the late 1950s. This fishery replaced the highly selective pole-and-line fishery that had operated previously. It is thought that schools of tuna follow herds of dolphin as they pursue smaller fish species, but the exact reason for the association is not entirely understood (Hall, 1998). In the early days of this tuna fishery, dolphin were inevitably captured since their presence within the purse seine net ensured the presence of the tuna. Crude estimates of the number of dolphin killed during the 1960s indicated a mortality

Table 13.4 Bycatches (numbers of individuals per 10 000 sets) and tuna discards (tonnes per 10 000 sets) in tuna purse seine fisheries that set on logs, schools and dolphins (*n* = sample size). Combined data for 1993–1996 from Hall (1998).

Bycatch taxa	Log sets (*n* = 10 607)	School sets (*n* = 13 112)	Dolphin sets (*n* = 19 570)
Dolphins	6	11	4521
Turtles	232	100	64
Billfishes	4121	1708	894
Sharks and rays	105 632	30 258	7760
Large pelagic fishes	2 611 312	202 159	2608
Triggerfishes	1 735 960	11 714	1474
Other fishes	2 651 856	169 842	73 414

rate of hundreds of thousands per year, and this caused a population decline until the late 1970s. Knowledge of such high dolphin mortality sparked the **'tuna–dolphin' debate** that culminated in the *Marine Mammal Protection Act* of 1972 passed by the US Congress. This act necessitated the adoption of good fishing practices that involved **'backing-down'** (also known as the **back-down manoevre** or **back-down procedure**), which is a process to release dolphins from the net, and the provision of independent observers aboard tuna vessels to record dolphin mortality (Hall, 1996).

There are three methods of purse-seining for tuna: setting on free-swimming tuna that are detected by their feeding activities close to the water's surface; setting on tuna associated with floating objects such as dead trees and logs; and setting on dolphins. The latter is the cause of the majority of dolphin mortalities. When fishers use this technique, they use helicopters to spot herds of dolphin species that are associated with tuna. Speedboats are then launched that pursue and tire the dolphin herd. At this stage both the tuna and dolphin are surrounded with the net and the purse rope is then drawn closed. Only when the tuna are secured in the net are the dolphins released. At this point in the operation, a combination of current regime, equipment malfunctions, lack of fishers' expertise or motivation can lead to the death of dolphins (Hall, 1998). The fisher's skill in coordinating the retrieval of the purse seine net, several boats working simultaneously, releasing the dolphins and landing the catch is probably the most important factor that determines the level of dolphin mortality. Although this technique is the greatest source of dolphin mortality, it has the lowest bycatch of other species of fish and marine reptiles (Table 13.4).

Gill nets are also used to catch tuna in inshore areas.

United Nations legislation has banned the use of high-seas gill nets. Although highly selective with respect to the size class of animals captured, gill nets are associated with high numbers of incidental captures of cetaceans. In the Sri Lankan gill-net fishery, one dolphin is caught for every 1.7–4.0 tonnes of tuna landed. This compares very poorly with one dolphin for every 70 tonnes landed in the eastern Pacific purse-seine fishery. Hence, while the use of gill nets may have conservation benefits for target fish stocks, the negative aspects of bycatches of larger organisms may be counterproductive (Hall, 1998).

Seals are notorious for attacking fish reared in cages and damaging catch on long-lines and set nets (Pemberton & Shaughnessy, 1993). However, this tendency to steal the fisher's catch also leads to seals becoming entangled in netting or the anchor and buoy ropes (Croxall *et al.*, 1990). Seals are relatively intelligent and soon learn to associate a free meal with fishing activities. Evidence for this comes from the fact that it was more common for older Cape fur seals (*Arctocephalus pusillus pusillus*) to become entangled in nets than young individuals that were still learning how to find food (Meyer *et al.*, 1992).

While none of us is happy about the death and discarding of marine mammals taken in bycatches, it may not necessarily have an adverse effect on the persistence of a particular population. Bycatches of marine mammals are often bloody and gruesome affairs but they are no worse than the suffering of animals killed by traffic. The latter seems to evoke little concern and few have suggested banning the use of cars to reduce roadkill. Biologically acceptable levels of bycatch mortality can be calculated for individual populations of marine mammals and appropriate management taken when these levels are exceeded (Wade, 1998).

13.6 Methods to reduce bycatches

One of the most obvious methods of reducing bycatches is to reduce the total amount of fishing effort. This can be achieved directly through national or international legislation, or indirectly through trade embargoes, consumer orientated campaigns, product boycotts and the imposition of exorbitant tariffs. The United States imposed embargoes on shrimp harvested by fisheries that did not use bycatch reduction devices (BRDs, Chapter 5) and tuna from fisheries that did not make provisions to release dolphins from purse-seine nets (Joseph, 1994). The latter has resulted in tins of so-called **dolphin-friendly tuna** on supermarket shelves. However, while it may appear that public opinion is a powerful weapon to achieve reductions in bycatch, public opinion is also vulnerable to manipulation by large corporations that may have selfish motives. For example, a large international company recently publicized its policy to use marine resources that are harvested in a sustainable manner. At face value, this is an admirable stance, but the sceptic might suspect ulterior motives behind this policy. Companies that adopt such policies may gain an advantage over their competitors, by marketing seafood products under environmentally friendly slogans. Nevertheless, if the policy leads to sustainable management of the wild resource it may not matter what motivated it.

Technological improvements in gear design have been helpful in reducing bycatches (Chapter 5). Bycatch reduction devices include the use of grids or panels of rigid mesh that deflect organisms out of nets or encourage them to escape in some other manner (Chapter 5). Changes in fishing practice have also helped to reduce mortality of bycatch species, e.g. the back-down manoeuvre in purse-seine sets on tunas associated with dolphins (Francis *et al.*, 1992b). Acoustic 'pingers' can be attached to set nets to alert cetaceans to their presence (Lien *et al.*, 1995). Bird-scaring devices have been used to great effect to deter seabirds from attacking baited long-lines as they are shot away from fishing vessels (Fig. 5.16). These devices are basically strips of brightly coloured material that flutter in the wind as used by farmers to deter birds from eating their crops (Løkkeborg, 1998). Inexpensive devices such as bird scarers are likely to be adopted readily by the fishing industry especially when there are clear financial and gear deployment benefits. However, acoustic devices may be costly and require additional maintenance, and are unlikely to be popular with fishers.

A number of alternative regulatory measures might achieve a reduction in bycatch per unit effort (Hall, 1998). Perhaps the most familiar is the imposition of restrictions on gear use and operation. For example, increasing net mesh size will result in lower bycatches of smaller species of fishes and invertebrates, although there is usually a concomitant reduction in the catch of target species. Reducing tow duration may decrease the likelihood of encountering rarer bycatch species and will increase their chances of surviving the experience as shown for sea turtles and sea snakes (section 13.4). However, this decreases fishing efficiency, as it increases the number of times the gear has to be hauled and reset. For example, to reduce the number of turtle deaths in shrimp fisheries, tow duration would have to be restricted to 1 h, which is half that in most fisheries. This would be inconvenient for fishers, since halving tow duration would double the time spent hauling and resetting their gear.

Bycatch quotas can be used to cap the total number of animals caught incidentally. Eastern Pacific tuna purse-seiners have an annual limit on the number of dolphin they can catch while fishing for tuna. This limit is enforced by enforcement agency observers placed on fishing vessels. Once that limit has been reached, the fishers must turn their attention to other species that are not associated with dolphins. Bycatch quotas have the advantage that they put the onus on fishers to improve their fishing techniques such that they can continue to prosecute that fishery for as long as possible. This could lead to the weeding out of those fishers that are unable to reduce their bycatches. Bycatches may vary seasonally. For example, the incidence of whale captures in inshore waters may coincide with a certain point along a migration route, e.g. humpback whales *Megaptera novaeangliae* occur in the inshore waters of Peru during spring when they might be vulnerable to interactions with fishing gear. Partial closures of the affected areas of the fishery can be used effectively in such circumstances without reducing overall fishing effort.

13.7 Ghost fishing

Static fishing gear, such as pots and set nets, are usually deployed and then left unattended for periods of between 12 h and several days. During this period the gear is vulnerable to bad weather or can be snagged accidentally by passing vessels. Although fishers mark static gear with dhan buoys, they are difficult to see among waves unless fitted with radar reflectors. In bad weather, the marker buoy may come adrift making the gear more difficult to retrieve. If snagged by passing fishing vessels, the gear can be dragged many miles from its original position before it is finally cut free. No matter how the gear is lost, it could continue fishing indefinitely. This phenomenon is termed '**ghost fishing**' (Fig. 13.4). A questionnaire survey of set-net fishers in northern Spain revealed that fishers exploiting offshore grounds for species such as anglerfish *Lophius piscatorius* mostly lost gear as a result of trawling activity, while inshore fishers that targeted sea bream (Sparidae) most frequently lost gear as a result of bad weather (Puente, 1997).

Just how long a gear is likely to continue ghost fishing depends upon the circumstances of its loss. If the marker buoys are lost from a set gear it may remain anchored securely to the seabed in its original configuration. Gear that has been dragged off by trawlers is likely to be tangled and will fish less effectively when it is eventually cut free and allowed to sink back to the seabed. In very rough weather, set gears may be rolled across the seabed such that they become effectively useless. In other circumstances, the gear may be held in an open fishing position when it becomes snagged on rocks or similar seabed features. In sheltered or deepwater situations, lost gear is likely to continue fishing

for a long time, as it will not be subjected to damage by wave action and abrasion (Carr *et al.*, 1992). Kaiser *et al.* (1996a) found that a net that was snagged on rocky substrata in the Irish Sea continued to catch crustacea, fish and birds for over a year (Fig. 13.5). In contrast, Puente (1997) reports that a similar net lost off the north-western coast of Spain was completely destroyed by the action of the Atlantic swell after 1 month. In the shallow clear waters off the Algarve, Portugal, Erzini *et al.* (1997) reported that lost nets were rapidly overgrown by epiphytic algae, increasing their visibility and reducing their ability to catch fish. Eventually, these nets became so overgrown with biota that they became a microhabitat used by small fishes.

It is not easy to make general statements about the longevity of ghost-fishing nets or their ability to continue fishing based on experimental studies. This is largely because the circumstances of each gear loss will be unique. Nevertheless recent case studies have revealed some general trends. For most lost nets, a typical pattern of capture is observed. Over the first few days, catches decline almost exponentially as the increasing weight of the catch causes the net to collapse. Then, for the next few weeks, the cadavers within the net attract large number of scavenging crustaceans, many of which are valuable commercial species and also become entangled in the net. After this initial period there follows a continuous cycle of capture, decay and attraction for as long as the net has some entanglement properties (Carr *et al.*, 1992; Kaiser *et al.*, 1996a).

Fig. 13.4 A 'ghost-fishing' trammel net off the coast of Wales, UK. A tangled gadoid probably lured this shag, *Phalocracorax aristotelis*, into the net. Photograph copyright B. Bullimore, Countryside Council for Wales.

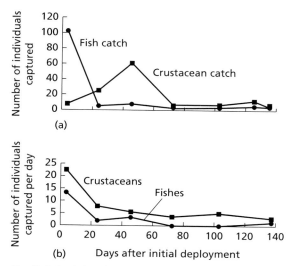

Fig. 13.5 Catch data for a 'ghost-fishing' gill net set on the seabed. (a) The change in the number of animals caught in the net between successive observation periods. Note how the catch of crustacea increases considerably after the initial deployment. These scavenging crustacea (crabs and lobsters) are presumably attracted by the corpses of fish tangled in the netting. (b) The decline in the catch rate of fishes and crabs in the same net which shows that these nets continue to fish for crustacea much longer than for fishes. After Kaiser *et al.* (1996a).

Pots or traps tend to be constructed of robust man-made materials and incorporate a rigid structure. This means that pots are likely to maintain their shape and hence capture efficiency for much longer than lost nets. Not surprisingly, ghost-fishing mortality rates of up to 55% of the mortality rates recorded in attended pots have been reported (High, 1976; Miller, 1977). As in lost nets, a re-baiting cycle occurs, which suggests that an intact pot could fish almost indefinitely. The 'ghost-fishing' potential of pots also varies according to the design for different fisheries. For example, Parrish and Kazama (1992) found that the majority of Hawaiian spiny lobster *Palinurus marginatus* and slipper lobster *Scyllarisdes squammosus* were adept at escaping traps, whereas parlour-type traps lead to mortalities of 12–25% of trapped individuals for American lobster, *Homarus americanus* (Smolowitz, 1978).

Little is known about the frequency of net or pot loss. This results from the reluctance of fishers to report such incidents, and the difficulty in undertaking quantitative surveys to estimate the prevalence of lost gear. However, the few available estimates of gear loss

indicate that it can be substantial in some fisheries. Approximately 7000 km of drift nets were lost per year in a north Pacific fishery, which is equal to 25% of the total length of netting set on a daily basis (Eisenbud, 1985). That fishers have complained about the problem of ghost fishing gives an indication of the scale of the problem. These complaints prompted a grapnel survey of the seabed on Georges Bank which yielded 341 actively fishing ghost nets from 286 tows (Brothers, 1992). The phenomenon of ghost fishing was clearly perceived to have negative effects on commercial stocks of Greenland halibut *Reinhardtius hippoglossoides* (Pleuronectidae) by commercial fishers involved in this fishery. As a result, they instigated their own voluntary clean-up programme (Bech, 1995). Considerable numbers of pots are also lost each year in North America. It was estimated that the 31 600 pots lost in the Bristol Bay king crab *Paralithodes camtschatica* fishery removed ≈ 80 000 kg of crabs from the stock (Kruse & Kimker, 1993). In another study, Breen (1987) reports an annual loss of ≈ 11% of the traps used in the Dungeness crab *Cancer magister* fishery in British Columbia. These fisheries tend to be highly localized, leading to a concentration of lost gear within relatively small areas. Potentially, the proportion of local stocks removed from the fishery could be significant. Furthermore, many of the targeted species have a high individual value and hence represent a large economic loss to the local fishing industry.

Many of the pots used in North America are made entirely of metal and will last for a considerable period of time in the marine environment. As a result, escape panels are now fitted to many pots used in North America, and biodegradable materials are used to ameliorate losses from 'ghost fishing' (Chapter 5). However, while it is easy to argue that metal traps or pots should include such devices, fishers that use gear covered with netting made of twine are more reluctant to incorporate escape panels, as they wrongly perceive that crustacea will be able to cut themselves free with their chelae.

Fragments of lost gear are spread far and wide by ocean currents and can cause high mortalities of marine mammals. Seals are particularly susceptible to entanglement in lost netting as they can be attracted to the catches within these nets. Seals are sometimes seen with pieces of netting around their neck and upper body. While the seals may not drown, cuts caused by the net material are easily infected, leading to the animal's eventual death. For example, 13% of the dead fur seals recorded off the coast of Japan were found entangled in net debris of one sort or another. Four thousand net fragments were found within 400 km of Japan's northern coast and these were estimated to have killed 533 fur seals. The high rate of net entanglement in northern fur seals *Callorhinus ursinus* accounted for their population decline between the 1970s and 1980s. Perhaps of most concern are the reported entanglements of the rare Hawaiian monk seal *Monachus schauinslandi* in nets off the coast of Hawaii.

13.8 Sociocultural differences

Hopefully, most readers will, by now, have formed the impression that discarding, bycatches and incidental catches are undesirable aspects of fishing activities. Our views are typical of those of people in developed countries that have the resources to implement bycatch reduction measures, the costs of which are absorbed by the consumer at the point of sale. At this point, it is useful to reflect that resource use in developing countries is markedly different to that in developed nations. Less than 3% of protein consumed in the United States is derived from fish, whereas this figure rises to over 40% for more than half the remaining world population. One billion people in Asia derive their entire animal protein intake from fish. Any reductions in the availability of fish for consumption in these countries would cause the average diet to become protein deficient. How might this chapter have been written from an Asian perspective?

Bycatches of marine mammals and turtles have been the subject of high-profile campaigns by organizations such as the Worldwide Fund for Nature and Greenpeace that have sought to influence governments to take measures to mitigate against these captures. However, this assumes that the protection of marine mammals is viewed equally by all nations. Western cultures have given whales and other marine mammals special status, segregating biological life into three catagories: humans, whales and 'the rest' (Lynge, 1992). It is easy to forget that these animals are regarded as an important source of energy, protein and other products (furs and oils) in countries such as Peru, Chile, Sri Lanka, Greenland, the United States (native harvest) and many other nations. The consumption of the meat of certain

(a)

(b)

Fig. 13.6 Cetacean deaths in bycatch are increasingly seen as repugnant in many of the developed western societies that once promoted commercial whaling (a). However, in other countries, cetaceans remain a highly prized source of marine produce such as this whale meat on sale in a Japanese market (b). Photographs copyright British Crown (a) and B. Glencross (b).

marine mammals has been part of the culture of some nations for centuries and the meat is highly prized for its gastronomic quality (Fig. 13.6).

The attitude of different societies towards bycatch issues is a fertile ground on which to generate conflict. Conservation groups and western societies demand embargoes on shrimp caught in Indonesia because the turtle bycatch is so high. Yet to Indonesian fishers, such measures seem totalitarian when the shrimp provide their major source of income.

In this chapter we have tried to stay on the scientific side of the bycatch debate, and simply review bycatches, their impacts on populations and the ways of reducing them. The ultimate fate of bycatch species and their fishers will depend on how such scientific information is used by people with different ethical viewpoints.

Summary

- Bycatches of invertebrates, fish, birds and mammals are taken by fisheries world-wide. Even the most selective of commercial fishing gears take some bycatch, and the weight of bycatch exceeds catch in many demersal trawl and shrimp fisheries.
- Some species are so abundant and productive that the proportion of the population lost to bycatches has insignificant effects on their dynamics. In sharp contrast, a number of albatross and dolphin species have small populations with slow rates of increase, and even a few deaths can cause long-term declines in abundance and place them at risk of extinction.
- In many cases it is possible to reduce bycatch by modifying gear design or changing the mechanism of gear deployment and fishers' behaviour. Many bycatch problems can be avoided by the application of common sense and codes of good practice, e.g. 'don't fish with gill nets near to nesting seabird colonies'.

Further reading

For a comprehensive account of bycatches we recommend Alverson *et al.* (1994). Hall (1998) provides an excellent review of the tuna–dolphin debate and Northridge (1991) reviews global bycatches in drift nets. Bycatches of sea snakes and turtles are documented in more specialist papers (sections 13.5.2 and 13.5.3).

14 Impacts on benthic communities, habitats and coral reefs

14.1 Introduction

Fishers use a great array of techniques and equipment to capture their prey. No matter which fishing technique is used, fishers and/or their gear interact with the habitat to some degree, for example, trampling intertidal fauna while collecting snails, lost fishing lines entangling non-target species, pots and traps landing on top of benthic fauna, and nets dragging across the seabed. The interaction of fishing gear with the environment either disturbs the habitat directly (**physical disturbance**) or indirectly, by removing competitors and predators from the system (**biological disturbance**). The environmental effects of fishing gears that are deployed at the surface of the sea or in mid-water are caused by gear loss and the direct and indirect consequences of removing bycatch and prey species. In contrast, most methods of fishing for bottom-dwelling species affect the seabed habitat and its inhabitants to some degree.

The aim of this chapter is to describe the ecological consequences of fishing on the seabed, both intertidally and subtidally. We consider the role of fishing activities as an agent of physical disturbance acting against a background of natural disturbance, the direct impacts of fishing methods and their effects on habitat structure, benthic communities and non-target species. We investigate long-term changes that might be linked to the effects of physical disturbance or the redirection of energy within the ecosystem and discuss the indirect effects of fisheries on habitat structure and production processes. Finally, we consider the direct and indirect effects of fishing on coral reefs and associated ecosystems.

14.2 Fishing disturbance

14.2.1 Fishing vs. natural disturbance

It is important at the outset to acknowledge that fishing is a form of physical disturbance as defined by Pickett and White (1985) (Box 14.1). Of course fishing is not the only form of disturbance in the marine environment, it is one of many such anthropogenic activities that include oil and aggregate extraction, dumping and the discharge of wastes and effluents. These activities occur against a background of naturally occurring physical disturbances that vary in scale and frequency from seasonal hurricanes or storm events (> 100 km^2) to daily wave or tidal action (> 1 km^2) and the feeding and burrowing activities of individual animals (< 10 m^2) (Van Blaricom, 1982; Hall, 1994) (Fig. 14.1). Although the latter occur at a relatively small scale, the frequency of bioturbation and predators' feeding disturbances may have a considerable additive effect on benthic communities. Thus, what may appear to be a relatively uniform seabed is probably a patchwork of similar substrata in various states of recolonization and succession (Grassle & Saunders, 1973; Connell, 1978). However, while it is possible to detect the short-term effects of predator disturbance, large-scale additive

---

Box 14.1

Pickett and White's (1985) definition of disturbance.

'Any discrete event in time that disrupts ecosystem, community, or population structure and changes resources, substrate availability, or the physical environment'.

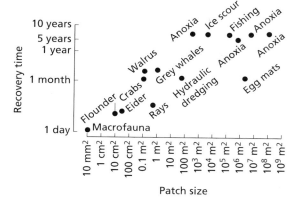

Fig. 14.1 Benthic disturbances occur over a wide range of scales. For example, at small scales, predators disturb sediments as they uproot or dig for prey (e.g. crabs, eider ducks, walrus). At larger scales, intense disturbances may occur as the result of scouring of the substratum by glaciers, through mass kills of benthos as a result of anoxic events (e.g. beneath suspended fish farm cages or as a result of stratification of the water column in enclosed areas such as the Baltic Sea) or hurricanes that affect many thousands of square kilometres of seabed. Adapted from Hall (1994).

effects have yet to be demonstrated (Hall *et al.*, 1993). This implies that small-scale disturbance events are masked by a background of large-scale disturbances. Alternatively, the scale of a small disturbance may permit such rapid recolonization that these effects are undetectable (Hall *et al.*, 1993). Presumably, as the scale and frequency of disturbance events increase, lasting ecological effects become apparent, despite the background of naturally occurring disturbances. The additive effects of an entire fishing fleet trawl-

ing the seabed would seem to be a good candidate to reach this threshold. As we will see in section 14.2.2, the global occurrence of bottom-fishing effort means that it can cause widespread changes to the composition of seabed communities.

As can be seen from Fig. 14.2 there are a number of interacting factors that need to be taken into account when trying to predict whether fishing disturbance is likely to have a significant long-term effect in a particular environment. These factors include habitat stability (e.g. rocky vs. loose sandy substrata), frequency of natural disturbance (related to depth, exposure and current regimes), the type of fishing gear used, and the scale, intensity and frequency of fishing activities. We might use this simple model (Fig. 14.2) to make predictions about the effects of fishing disturbance in different environments. For example, the effects of physical disturbance on communities of short-lived free-living worms that inhabit mobile sediments in shallow shelf seas (highly disturbed environment) will be harder to detect than the effects on communities that are associated with coral reefs (stable environment). The same model can be used to predict the impact of different gears. For example, there will be a large difference between the effects on the seabed of a light seine net hauled ashore by hand and a 10-tonne beam trawl towed at speeds greater than 15 km h^{-1}. Later in this chapter, we describe a study that tests some of these predictions (section 14.3.4).

14.2.2 Distribution of fishing disturbance

Fishing effort is not homogeneously distributed across the seabed. Fishers tend to concentrate their effort in

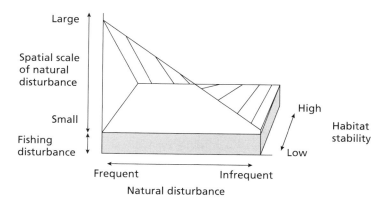

Fig. 14.2 A conceptual model demonstrating the relationship between the relative effects of fishing disturbance, natural disturbance and habitat stability. The effects of fishing become proportionately greater as natural disturbance occurs on smaller spatial scales and less frequently, and as habitat stability increases.

grounds that yield the best catches of commercial species, and avoid areas with obstructions and rough ground that would damage their fishing gear. This knowledge is either gained through personal experience, exchanged between fishers or traditionally passed from one generation to the next. The advent of satellite navigational aids and automated position logging has meant that fishers are able to relocate productive sites with high precision. In addition, fishing is severely restricted in some areas, such as shipping lanes and around oilrigs. So while some areas of the seabed may be intensely disturbed others may remain untouched. Knowledge of the distribution of fishing effort is important to enable us to predict the likely consequences of trawling for benthic habitats and communities. However, this information does not presently exist at a scale appropriate for such predictions (< 10 000 m²).

Continental shelf seas tend to be the areas of highest productivity, hence it is here that fishing intensity is highest. Most fishable grounds will be impacted several times per year. Early estimates of the area swept by bottom gears were unintentionally misleading as they implied that physical disturbance was spread homogeneously. Latterly, it became clear that effort was very patchy with some areas heavily impacted and others untouched. In the Middle Atlantic Bight, 18% of a 259 km² area was trawled in a 6-day period of intense fishing activity (Churchill, 1989) and up to 20 trawl tracks per 100 m² were observed in the New York Bight, at a depth of 100 m, where current action was weak (Twichell *et al.*, 1981). Similarly, Krost *et al.* (1990) found that trawl tracks occupied 19% of their muddy and relatively deep study area. However, these studies are not comparable because the persistence time of trawl marks will differ depending on the nature of the substratum and the local hydrographic regime. For example, beam trawl marks are detectable for less than 4 days in loose sandy sediment, whereas marks made by otter trawl doors were detectable for over 12 months in more stable muddy sediments (Krost *et al.*, 1990; de Groot & Lindeboom, 1994). In order to use direct observations of trawl marks on the seabed as an index of trawling intensity, more experimental studies need to be undertaken to determine their longevity in different sediments and under different environmental conditions.

More recently, 'black box' recorders have been fitted to a proportion of the Dutch beam trawl fleet that have

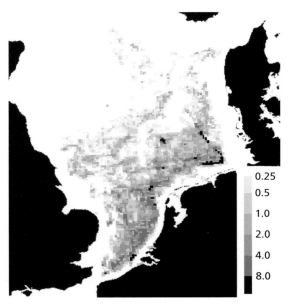

Fig. 14.3 Mean beam-trawling intensity by Dutch beam trawlers with engines > 300 horse-power from 1 April 1993 to 31 March 1996 in 3 × 3 nautical mile areas of the North Sea. Beam-trawl intensity within each area is expressed as the mean number of times that each area metre in that area is trawled each year. Data were collected from loggers which recorded the location of vessels at 6-min intervals. From Rijnsdorp *et al.* (1998).

allowed accurate tracking during fishing operations. The Dutch fleet accounts for 50–70% of the total beam-trawling effort in the North Sea and hence the data give a good idea of the distribution of beam trawl effort (Rijnsdorp *et al.*, 1998). As anticipated, these records indicated that beam-trawling effort is very patchily distributed in the North Sea; while it is estimated that some areas are visited > 400 times per year, others are never fished (Fig. 14.3). This study has given us a useful insight into the microdistribution of fishing effort, and currently provides the most accurate means of determining how often small areas of the seabed have been trawled.

There seems little hope of improving the resolution of fishing effort data until the tracking of fishing vessels becomes more widely accepted. However, it is possible to use indirect methods to measure fishing intensity. The distribution of bottom-trawling disturbance can be ascertained by the occurrence of anatomical injuries caused by physical damage in populations of animals that are able to withstand such injuries. Some bivalve species, such as the clams *Arctica islandica* and *Glycymeris glycymeris*, live in excess of 30 years and remain *in situ* throughout most of their life. Both of these species live close to the surface of the seabed and are vulnerable to damage as fishing gears pass across the seabed. Those bivalves that are not fatally damaged by the trawl or eaten by scavengers are able to repair their shell, leaving a mark within the shell matrix that records this disturbance event (Gaspar *et al.*, 1994; Witbaard & Klein, 1994). It should be possible to work out the minimum number of times a small area of seabed has been disturbed by counting the number of such marks in the shells of suitable bivalve species collected on the seabed.

Similarly, many starfishes are able to withstand the loss of several arms and regenerate them with time. Thus, the occurrence of starfish with missing arms can also be used as an index of fishing intensity. Starfish are relatively mobile, hence the resolution of fishing intensity derived from starfish damage frequency would apply to larger areas than data derived from bivalves. In addition, the record of disturbance will only last as long as it takes the starfish to regenerate the lost arms (6–24 months depending on the species). Kaiser (1996) investigated the validity of this technique by comparing the occurrence of damaged starfish in the Irish Sea. In heavily beam-trawled areas, up to 55% of the starfish *Astropecten irregularis* had lost arms compared with only 7% in a less intensively fished area (Fig. 14.4).

One of the advantages of using bivalves compared with echinoderms is that long-lived species will provide a disturbance history of a particular piece of ground that may cover a time span of more than 30 years. This is particularly important, as both the magnitude and distribution of fishing effort has changed through time. For example, fishers off the coast of north-east Scotland moved further away from their homeports as stocks declined in the early part of this century (Greenstreet *et al.*, 1999). These changes also coincided with a change in fishing practices as herring stocks declined and fishers targeted the grounds of the more lucrative Norway lobster (also known as Dublin bay prawn) *Nephrops norvegicus* (infraorder Astacidea). Improving our knowledge of the distribution and patterns of fishing effort is important to enable managers of the environment to target conservation measures such as area closures or gear restrictions.

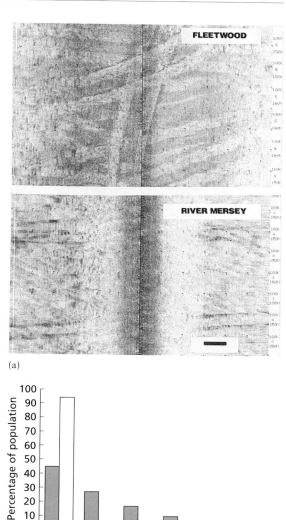

Fig. 14.4 (a) A portion of a side-scan sonar record showing pairs of 12-m-wide beam-trawl marks on muddy ground off Fleetwood, England, and pairs of 4-m-wide beam-trawl marks off the River Mersey, England. The trawl marks run in a north/south direction. Scale bar = 25 m. (b) The mean proportion of the starfish (*Astropecten irregularis*) populations with different numbers of damaged or regenerating arms at each site. Shaded bars, Fleetwood (heavily fished); open bars, Mersey (lightly fished). After Kaiser (1996).

14.3 Direct effects of fishing gear on the seabed

Fishing techniques that affect benthic fauna and habitats can be grouped into two categories: active and passive. Active bottom-fishing methods usually involve towing trawls or dredges across the seabed. However, artisanal fishers operating on tropical coasts also use a range of active techniques such as drive netting, spearing and fishing with chemicals or explosives. Passive fishing techniques include the use of pots or traps, baited hooks on long-lines, gill nets and drift nets (Chapter 5). Fishing gears come in all shapes and sizes, each with its own particular method of deployment. Not surprisingly, the impacts of each of these gears on non-target biota and habitats will also vary between different habitats and according to the manner in which they are used.

Most of the research centred on the environmental impacts of bottom fishing has focused on answering fundamental questions such as 'what happens when a fishing gear interacts with the seabed?'. The answer to this question appeared to be straightforward based on the observations of British fishermen who, in 1376, petitioned for the banning of beam trawls (Box 14.2). While it might seem intuitively obvious that a heavy trawl towed at high speed across the seabed will wreak havoc on animal and plant communities in its path, intuition does not provide a sound basis on which to make management decisions that may affect the livelihood of fishing communities.

Box 14.2
A long history of trawl disturbance.

Public concern about the long-term effects of bottom fishing can be traced back to the fourteenth century when the Commons petitioned the King of England 'that the great and long iron of the wondyrchoun [a beam trawl] runs so heavily and hardly over the ground when fishing that it destroys the flowers of the land below water there' (Graham, 1955).

14.3.1 Towed fishing gear

Although both towed and static fishing gears interact with the environment, towed fishing gears tend to be large, heavy, and sweep large areas of the seabed as they are towed at speeds up to 7 knots (approximately 13 km h^{-1}). Furthermore, towed bottom-fishing gears are used on every shelf sea throughout the world. As a result, towed bottom-fishing gears have the greatest potential to cause widespread ecological change to benthic community structure. Before discussing the way in which different towed bottom-fishing gears interact with the benthic environment, it is worth spending a few moments considering some of the factors that affect how fishers deploy their gear (Chapter 5). Some of this information cannot be found in reference books, and it can only be gleaned by talking directly with fishers as they work.

The primary objective of most commercial fishers is to maximize their catch for the least amount of effort. Large bycatches are viewed as a nuisance, as sorting the catch can be time consuming. Hence, fishers try to reduce bycatch if possible. Many commercial bottom trawls are towed for periods of up to 2 h. In areas where the bycatch consists of invertebrates or there is a risk of catching a lot of inert matter such as shell and rock, meshes are cut out of the belly (underside) of the net to eliminate as much of this material as possible before it enters the codend. A large bycatch of starfish can greatly reduce the financial value of a catch as starfish abrade scales and mucous from the commercial fish species and reduce their aesthetic appeal for consumers. Debris and animals such as starfish also block the meshes of the net which lowers their sorting efficiency and increases drag through the water. In extreme situations this can lead to 'boil back' where water is unable to pass through the net as quickly as the trawl is towed through the water. This effectively means that water ahead of the net is forced sideways and greatly reduces the catch efficiency of the trawl. Fishers can further reduce bycatches of invertebrates by reducing the length of warp between the vessel and the gear. This will increase the upward lift on the gear and will cause it to dig less deeply into the seabed. Although this can reduce catches of commercial species it may enable fishers to operate in areas that would otherwise be unworkable.

14.3.2 Direct effects on the substratum

Physical disturbance of the seabed results from direct contact with fishing gear. In soft sediments this will lead to the turbulent resuspension of surface sediments which may remobilize contaminants and radionucleides and expose the anoxic lower sediment layers. On hard substrata, boulders may be physically moved or rock reef and biogenic structures may be destroyed (Auster *et al.*, 1996). The magnitude of the impact on the seabed is determined by the towing speed, the physical dimensions and weight of the gear and type of substratum. The resultant changes to the habitat may persist for only a few hours in shallow waters with strong tides and continuous wave action, several years in muddy sediments found in sheltered areas such as fjords, or for decades in the relatively undisturbed deep sea (section 14.2.1). Alternatively, habitat alteration may be permanent, e.g. when bedrock formations are fragmented. Interactions with the habitat usually occur as the unavoidable consequence of using towed fishing gears in a particular environment. However, fishers are known to intentionally remove seabed structures that interfere with their fishing activities, a practice known as 'preparing the ground'. Faroese fishermen deliberately set out to destroy reefs of giant sponges off the Faroe's Bank by dragging chains attached between two boats across the seabed. Thankfully, this practice ceased when scientists revealed the importance of this habitat for juvenile cod *Gadus morhua* (ICES, 1994a).

Trawling and dredging (but not hydraulic dredges) typically reduce the surface roughness of the seabed in sandy habitats. It is important to appreciate the scale of roughness in this context, i.e. the microtopographic relief of the sediment. The presence of sessile fauna, biogenic features such as feeding pits, shell fragments and small rocks that protrude from the seabed all contribute to the microtopography of the habitat. As trawls pass across the seabed, surface sediments are resuspended. Heavier particles sink quickly to the bottom, while finer particles are winnowed away by tidal currents. The surface of freshly trawled sediments typically have a large proportion of shell debris (Kaiser & Spencer, 1996b; Schwinghamer *et al.*, 1996). Acoustic data, collected on trawled experimental sites on the eastern Grand Banks, Canada, showed that the effects of trawling were detectable to a depth of at least 4.5 cm within the sediment (hard packed sand). Hence bottom trawling reduced the complexity of both the surface and internal structure of soft-sediment habitats (Schwinghamer *et al.*, 1996).

14.3.3 Effects on infauna

Fishers that use bottom-fishing gears target species that live in or on the seabed or those that feed in that environment. As we have already seen, these fishing gears have been designed to remain in close contact with the seabed and inevitably catch a large proportion of non-target bycatch species. This bycatch gives a clue to the wider impacts that these gears have on the benthic community. Studies of the effects of fishing activities on benthic fauna have tended to deal with either **infauna** or **epifauna** (Box 14.3). The majority of studies to date have examined infauna (section 14.3.4). This preference is probably driven by the fact that infauna are more easily sampled quantitatively with standardized sampling devices such as grabs and corers. By default, studies of infauna are restricted to soft sediments whereas the nature of rocky habitats dictates the study of epifauna.

What can we learn from experiments?
Until recently, the wider ecological effects of fishing on benthic communities were relatively unknown. Increasing media attention from the 1980s onwards sparked over 40 experimental studies worldwide of the effects of bottom trawls on the seabed. Here, we highlight the

---

Box 14.3
Infauna and epifauna.

Infauna are defined as those animals living entirely within the sediment, whereas **epifauna** are defined as those animals living on, protruding from, anchored in, or attached to, the substratum. Both groups are commonly known as **benthic fauna**, **bottom-living fauna** or **benthos** (as opposed to **pelagic fauna** that live in the water column).

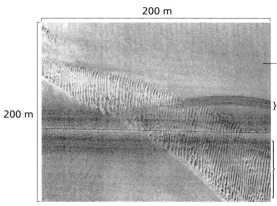

200 m

200 m

— Gravely seabed

} Beam trawl marks

Sandy megaripple
area

Fig. 14.5 A side-scan sonar image of an area of seabed in the Irish Sea approximately 200 m wide and 200 m long that demonstrates the patchy nature of seabed environments. A band of megarippled sand runs across the image; this habitat is highly mobile and has an impoverished faunal community. The sediment either side of the megaripple area is more stable and composed of coarse sand, gravel and shell fragments. This area has relatively high diversity and supports a high biomass of filter-feeding and emergent fauna.

most important findings of a selection of these studies as each one gives a different perspective depending on the gear and habitat studied.

Kaiser and Spencer (1996b) examined the effects of beam-trawl disturbance on benthic fauna in the Irish Sea. Their experimental site covered an area of 5.25 km² shelving from relatively shallow water (≈ 25 m) to deeper water further offshore (≈ 40 m). Two distinct habitats were found at this site. The first was composed of a relatively stable sediment of coarse sand, gravel and shell debris, characterized by a rich epifaunal filter-feeding community of soft corals and hydroids, and an amphipod dominated infauna. The second was characterized by mobile sediments with ribbons of megaripples and sand waves dominated by small polychaete species with little epifauna (Fig. 14.5). Trawling had no detectable effects in the mobile sand habitat, which is perhaps not surprising given the levels of natural disturbance experienced in such an environment (section 14.2.1; Shepherd, 1983).

Mobile sand environments are notoriously variable, and the fauna is often sparse. As a result, de Wolf and Mulder (1985) found that they were unable to make an accurate estimate of the abundance of benthic species in a megaripple habitat. Faunal abundance and composition varies according to the position sampled on the sand wave; for example, fauna tend to be most dense in the trough between sand waves where organic matter accumulates, while mobile species such as amphipods are prevalent on the slope of the sand wave (Shepherd, 1983). Animals living within the troughs of megaripples are less likely to be disturbed by fishing gear that will tend to ride over the crest of each sand wave.

Brylinsky *et al.* (1994) examined infaunal changes that occurred as a result of otter trawling on intertidal mudflats at high tide. They were able to collect samples directly from the otter trawl marks at low water. Despite the precise nature of their sample collection, they were also unable to detect any adverse effects of otter trawling. These intertidal mud flats are regularly exposed to large-scale natural disturbance such as ice-scour that probably maintain the benthic community in the initial stages of recovery. Thus several studies demonstrate that the effects of bottom fishing are relatively insignificant in habitats exposed to high levels of natural disturbance. Such habitats tend to be dominated by small-bodied opportunistic fauna that rapidly recolonize areas that have been physically disturbed.

In contrast to studies in highly disturbed habitats, the effects of fishing are much more noticeable in less disturbed areas. Kaiser and Spencer (1996b) tested the effects of beam-trawl disturbance in a stable sediment that supported a much more diverse benthic fauna than the mobile sand habitat previously described. They found that beam trawling reduced the number of species and individuals by two- and threefold, respectively. Detailed analyses of changes in particular types of fauna revealed that less common species were most severely depleted by beam trawling. Tube-building worms, amphipods and bivalves were among the most sensitive animals affected by trawl disturbance. However, 6 months later the benthic community in the experimentally trawled areas was indistinguishable from adjacent control sites (Kaiser *et al.*, 1998b). In a similar study undertaken in New Zealand, Thrush *et al.* (1995) studied the effects of scallop dredging on communities inhabiting coarse sand at a depth of 27 m. Dredging led to changes in the composition of the benthic infaunal community that lasted for at least 3 months after initial disturbance. The results of these experimental studies (Thrush *et al.*, 1995; Kaiser *et al.*, 1998b) should be regarded as conservative as it is impossible to simulate the area and intensity of disturbance created by an entire fishing fleet. This implies that, at larger scales of disturbance, recolonization probably takes longer.

Although both Thrush *et al.* (1995) and Kaiser *et al.* (1998b) selected experimental sites in areas of relatively low fishing intensity, it is impossible to be certain that they were not working in an environment that had been previously altered by fishing activities. This lack of knowledge of the disturbance history of a particular piece of seabed has dogged all but a few of the experimental studies undertaken to date. Tuck *et al.* (1998) were able to overcome this problem by conducting a similar experiment in Loch Gareloch, Scotland, which had been closed to all fishing activities for 25 years due to the presence of a nuclear submarine base within the loch. In contrast to the other experimental studies, Loch Gareloch is a sheltered site with a muddy seabed. Tuck *et al.* (1998) trawled a patch of the seabed completely once a month for 16 months using an otter trawl rigged with rock-hopper gear. They surveyed this and a control area throughout this period and at intervals up to 18 months after fishing disturbance was discontinued. They found that bivalve abundance diminished during the period of fishing disturbance whereas opportunistic polychaetes, such as cirratulids, increased in abundance. More importantly, some differences between the benthic assemblage in the fished and control area were still apparent after 18 months of recovery. These results suggest that even a confined period of fishing disturbance once per year would be able to maintain a muddy sediment community in an altered state (Tuck *et al.*, 1998).

Probably the best approach to investigating whether long-term changes have occurred within fished benthic systems is to exclude fishing activity from selected sites and study their response. The response is sometimes termed 'recovery' but this term should be avoided as it implies that we know the original composition of the community. Opportunities to undertake such studies are rare, largely due to the opposition from fishers that are denied access as a result of any closure. We know of three instances where the effects of closure have been studied, two of which we discuss a little later in this chapter.

The Bradda scallop ground off the Isle of Man in the Irish Sea is probably the most intensively fished scallop ground in Europe and has been exploited for over 50 years (Brand *et al.*, 1991). A small inshore closed area extending approximately 500–1500 m from the shore was established in 1989. Surveys undertaken several years after closure indicated that the variation between infaunal samples within the closed area was greater than in the adjacent dredged areas. In other words, intensive dredging activities appeared to have reduced the patchiness of the benthic community outside the closed area (Bradshaw *et al.*, 2000). Scallops are relatively sedentary, hence fishers tend to systematically fish adjacent tows in a manner analogous to a tractor ploughing a meadow. Such dredging would be expected to create a more homogeneous environment. Bradshaw *et al.* (2000) tested this expectation by experimentally scallop dredging some of the areas within the closed area to see if dredging would drive the areas towards an assemblage structure similar to that in the heavily fished areas close by. As predicted, the benthic assemblage in the experimentally scallop dredged areas became similar to that in the adjacent unprotected areas.

The results of these studies appear to concur with the conceptual model introduced earlier in this chapter (Fig. 14.2). As sedimentary habitats become more stable, so the effects of fishing disturbance are more dramatic in the short term and last for longer. This principle applies both to the structure and composition of the benthic assemblage and the topography and physical structure of the sediment. Infauna that live in the top few centimetres of coastal sediments at depths < 30 m tend to be small opportunistic species (e.g. spionid and capitellid polychaetes and amphipods) that are quick to recolonize areas after physical disturbance events (Dauer, 1984; Levin, 1984). As we have seen, the effects of trawling on this component of the infaunal community are unlikely to last more than 6 months. Furthermore, a recent study suggests that small body size may reduce the vulnerability of animals to damage from passing fishing gear. Gilkinson *et al.* (1998) constructed a life-size model of an otter door that they dragged across sediment in a flume tank. They took great care to ensure that the granulometric structure of the sediment was representative of a real seabed. Different species and sizes of bivalves were then added

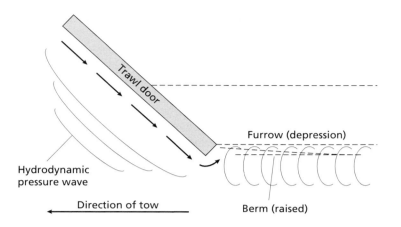

Fig. 14.6 Sediment and small bivalve displacement during otter trawling (plan view). The trawl door is being towed from right to left and the arrows in front of the door indicate the movement of sediment and small bivalves. After Gilkinson *et al.* (1998).

to the sediment and their response to the passage of the otter door noted. Gilkinson *et al.* (1998) were able to deduce that the otter door (and no doubt other fishing gears) created a pressure wave that fluidized sediments ahead of it. Small bivalves were fluidized with the sediment and were pushed aside unharmed in the wake of the otter door (Fig. 14.6). However, larger bivalves that were more firmly anchored within the sediment were damaged severely as they were too heavy to be moved aside by the pressure wave. Many of the opportunistic polychaetes that inhabit mobile sand areas are highly motile and would be able to re-bury themselves once disturbed from the sediment. Thus, in mobile sedimentary enviornments larger species, especially bivalves and sea urchins, may be the most vulnerable to trawling disturbance.

A recent study by Posey *et al.* (1996) suggested that deeper burrowing fauna are not adversely affected by severe natural disturbances such as episodic storms that only affect the upper layers of the sediment. They collected infaunal samples to a depth of 15 cm from inshore shallow water (13 m) sediment. Fauna living in the top few centimetres of the sediment were greatly depleted in the short term, whilst deeper burrowing species remained undisturbed. 'Deeper burrowing' was not defined, but it implies animals living at a depth of 7–15 cm, which is well within the depths disturbed by trawls and dredges (Bergman & Hup, 1992). This study implies that deeper burrowing species are protected from natural physical perturbations (except predatory events) and consequently may be less tolerant of fishing disturbances.

Bycatches of non-target infaunal species such as the heart urchin *Echinocardium cordatum* and the large infaunal bivalve *Arctica islandica* indicated that beam-trawl gear dug to a depth of at least 6 cm in a compact sandy seabed (Bergman & Hup, 1992). Heart urchin vulnerability varied with age as a consequence of the position occupied by the animal within the sediment. The position of small urchins within the sediment column, and not their size, made them vulnerable to bottom trawls. Smaller and younger sea urchins were found closest to the sediment surface. Consequently they were more vulnerable to physical damage than older and larger sea urchins that were buried at greater depths.

Factors affecting recovery rate

Judging by the experimental studies discussed so far, we would predict that the effects of physical disturbance would be short-lived for organisms adapted to frequent natural disturbance. This is in contrast to the effects within a habitat exposed to fewer disturbances (see Fig. 14.2). An example of fishing effects in a highly perturbed environment is Hall and Harding's (1997) study of the effects of mechanical and suction dredging and the scale of disturbance on an intertidal benthic community in the Solway Firth, Scotland. Their study is a particularly good example because they fished plots that were representative of commercial activities. Nevertheless, the infaunal community sampled within disturbed areas was comparable to that in control undisturbed areas after only 8 weeks. This rapid recolonization was achieved through the combination of seasonal larval settlement and large-scale bed-load transport leading to the passive immigration of adult fauna into the disturbed plots (Hall & Harding, 1997).

Suction dredging for manganese nodules on the abyssal plain of the Pacific Ocean provides a good example of the effects of physical impact in a relatively undisturbed environment (Theil & Schriever, 1990). Trenches created by the suction dredge head at depths of approximately 4000 m persisted for at least 2 years and probably longer. Thus, there appears to be strong evidence that the persistence of disturbance effects may be approximately correlated to the level of background natural disturbance experienced in a particular habitat. However, we do not wish the reader to assume that all sandy habitats are highly unstable and therefore immune to the effects of fishing activities. This point is well illustrated by a recent study that examined how sediment defaunation at different scales affected subsequent recolonization. Thrush *et al.* (1996) undertook their study on an intertidal mudflat in New Zealand characterized by dense mats of spionid worms. These worms live in tubes constructed of sand grains held together by mucus. They occur at densities of hundreds of thousands per square metre and help to increase the stability of the sediment. Removal of these animals destabilized the sediment and exacerbated the effects of disturbance by increasing the recovery time of the fauna.

While small, short-lived opportunistic species are able to recolonize disturbed areas relatively quickly, populations of larger, longer-lived organisms recover more slowly. Beukema (1995) reported on a series of studies that had examined community changes in the Dutch Wadden Sea in response to intensive commercial lugworm dredging activities. Areas of tidal flats were licensed to harvesters on an annual rotational basis. After 12 months of lying fallow, the complement of polychaete, crustacean and small bivalve species in previously dredged areas was restored to its previous state. However, gaper clams *Mya arenaria* were a major component of the infaunal biomass prior to harvesting. These clams are relatively long-lived and spawn and recruit infrequently, hence it took more than 2 years before their population began to recover from the effects of dredging. Later in this section, we will discuss the importance of the structural role of long-lived benthic organisms within benthic communities.

So far, we have considered mainly the effects of fishing on benthic fauna that inhabit coarse substrata.

Most animals are found within the upper 15 cm of these sediment habitats. However, fauna that inhabit soft mud and muddy sand environments live in burrows up to 2 m deep (Atkinson & Nash, 1990). As no fishing gear penetrates more than 20 cm into sediment, how might fishing gears affect these animals? As fishing gear is dragged across muddy seabeds, the upper portions of burrows constructed by bioturbating fauna will be collapsed. Nevertheless, the burrows are rapidly reconstructed. These organisms are continually reworking their burrows, building new chambers and opening new entrances, hence we assume that they are little affected by fishing activities. However, the energetic costs of repeated burrow reconstruction may have long-term implications for the reproductive output and survivorship of individuals. Periodically, the vulnerability of these animals to damage or capture by fishing gear may alter with tidal or light conditions. For example, the burrowing shrimp *Jaxea nocturna* remains deep within its burrow during the day and then moves to the entrance to feed at night when it is more likely to be caught by fishing gear (Nickell & Atkinson, 1995). Bioturbators such as these maintain the structure and increase the oxygenation of muddy sediment habitats. Consequently, if these organisms are adversely affected by fishing activities, changes in habitat complexity and community structure might occur. To date there is no evidence to suggest whether or not this is the case.

14.3.4 Effects on epifauna

We have already seen how bottom-fishing gears affect benthic fauna that live within the seabed. Consequently, we would expect that animals that live attached to, anchored in, or protruding from the surface layers of the seabed would be even more vulnerable to fishing disturbance. Not surprisingly, changes in the occurrence or abundance of epifaunal species were among the first indications that fishing may adversely affect benthic communities. The tube-building polychaete worm *Sabellaria spinulosa* forms colonies that eventually become reefs if left undisturbed. These were prevalent in the Wadden Sea off the coast of the Netherlands and Germany in the 1920s. However, a comparative survey conducted 55 years later found no evidence of these reefs. The benthic community had become dominated by small polychaete species, which is a typical response to stress in benthic communities. Riesen and Reise (1982) attributed these community changes to dredging and trawling activities.

Data collected from research surveys on the Australian north-west shelf between 1962 and 1983 indicated that the abundance of high-value commercial species declined with time. Catch rates of epibenthic organisms such as sponges also greatly decreased between 1963 and 1979 (Sainsbury, 1987). Sainsbury (1987) hypothesized that the degradation of the sponge community and associated soft corals and sea fans was partly responsible for the decline in the catches of emperors *Lethrinus* spp. and snappers *Lutjanus* spp. which sheltered and fed among these emergent fauna. As habitat structure was reduced the snappers and emperors were replaced by fish species of lower commercial value (*Nemipterus* spp. and *Saurida* spp.) (Sainsbury, 1991). Further experiments have shown that closing areas to fishing activities permits habitat regeneration and an improvement in the catch of more valuable species such as emperors and snappers (Sainsbury *et al.*, 1998) (Fig. 14.7). However, habitat regeneration was much slower than expected, and it was predicted that the largest epibenthic organisms (> 25 cm diameter) would only attain their full size after 15 years' growth.

Collie *et al.* (1997) identified comparable cobble substrata that were scallop dredged at different intensities at Georges Bank, north-west Atlantic. They located these areas by referring to fishing effort data, and then on a more refined scale using direct observations of the occurrence of dredging marks on the seabed with side-scan sonar. Infrequently fished areas were characterized by abundant growths of bryozoans, hydroids and tube worms that increased the three-dimensional complexity of the habitat (Fig. 14.8). These structural organisms dominated the biomass of the community, which is a typical feature of relatively undisturbed benthic communities. Dredged areas were very different, they had lower species diversity, reduced habitat complexity, and were dominated by species resistant to fishing activities: hard-shelled bivalves (e.g. *Astarte* spp.), echinoderms and scavenging crabs. There was a high diversity of animals living among the emergent epifauna including polychaetes, shrimp, brittle stars, mussels and small fishes (Collie *et al.*, 1997). Many of these species are important prey for commercially exploited fishes such as cod (Bowman & Michaels, 1984).

Auster *et al.* (1996) reported a reduction in habitat

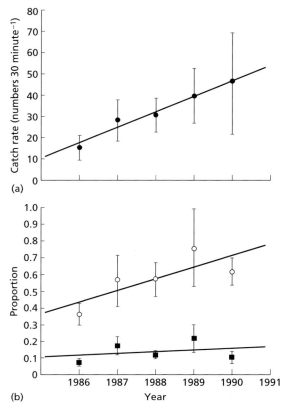

Fig. 14.7 (a) Total catch rates of *Lethrinus* plus *Lutjanus* (mean ± SE) during research trawl surveys in a zone otherwise closed to trawling on the Australian north-west shelf. The zone was closed to trawling in 1985. (b) Changes in the proportion of seabed with large (squares) and small (circles) benthic fauna (mean ± SE) in the zone closed to trawling (based on records of fauna in research trawl catches). After Sainsbury *et al.* (1998).

complexity as a result of trawling and scallop dredging activity at three sites in the Gulf of Maine, USA. The Swans Island conservation area has been closed to fishing with mobile gears since 1983. A video survey of the seabed using a remotely operated vehicle (ROV) revealed areas that had been cleared of emergent fauna by fishing activity close to the border of the conservation area. In addition, a localized reduction in some of the target species, such as scallops and crabs, reduced the number of habitat features that they created, e.g. feeding pits created by crabs. Fishing effects were even more obvious at Jeffrey's Bank. This site was surveyed by submersible in 1987 when boulders, 2 m in diameter, were a prominent feature of the habitat. It was

(a)   (b)

Fig. 14.8 Photographs of the seabed in (a) a heavily fished scallop ground, and (b) a similar habitat in a lightly fished area on the Georges bank. The fauna in the lightly fished area is typically dominated by foliose bryozoan and hydrozoan species that add to the structural complexity of the habitat. Photographs copyright D. Blackwood and P. Valentine.

thought that these large seabed obstructions would effectively exclude fishing mobile fishing gears from operating in this area. However, when the site was resurveyed in 1993, the overlying silt layer had been scoured away leaving an exposed gravel bed, sponges were far less common, and the boulders showed evidence of having been moved across the seabed. An ROV survey undertaken on the Stellwagen Bank revealed that dense aggregations of the hydrozoan *Corymorpha pendula* occurred on the seabed. Shrimp *Dichelopandalus leptoceros* shelter around these large hydrozoans, possibly using them for cover from predators, or perhaps feeding on rejected particles of food. Evidence of recent fishing activity was deduced from the wide linear swathes through benthic microalgal cover on the seabed. Within these areas there was an absence of hydrozoans and consequently shrimps were also absent (Auster *et al.*, 1996).

Recovery from disturbance may be rapid (< 6 months) for some epibenthic assemblages (Kaiser *et al.*, 1998b). In cases where communities are dominated by sessile emergent fauna, restoration of the habitat will take much longer. Collie *et al.* (1997) were able to study the recovery trajectory for several years after areas on the Georges Bank were closed to scallop dredging. They found that the emergent fauna had recolonized these areas 2 years later. However, the individual colonies were small, and it would take several more years before the bryozoans, hydroids and tube worms re-established their dominance in that community.

In special cases, benthic organisms become the substratum in which infaunal communities live. Maerl beds are formed of calcareous algae of the genus *Lithothamnion*, amongst the oldest living marine plants in Europe. These beds can take hundreds of years to accumulate, with live material at the surface of the bed overlying dead thalli (Potin *et al.*, 1990). The branched structure of the thalli provides a unique and complex habitat that supports a diverse community of animals. Commercially important species such as scallops inhabit maerl beds and have attracted fishers. Not surprisingly, scallop dredging severely disrupts this habitat by breaking down the interstices between the thalli and thereby causes long-term changes to the composition of the associated benthic fauna (Hall-Spencer & Moore, 2000).

Fisheries that occur in shallow, clear waters are likely to impact marine plant communities. In particular, seagrass meadows are vulnerable to physical disturbance by bottom-fishing gears that tear up individual plants, reduce biomass by shearing off fronds, and expose rhizomes (Fig. 14.9). Water turbidity may be increased locally through sediment resuspension by trawls and dredges, leading to regression in the seagrass meadow. The dense nature of seagrass meadows encourages sedimentation and accumulates organic matter. The plant's rhizomes help to stabilize sediment in the same way that the roots of trees bind soil. As a result, seagrass meadows are highly productive, support complex trophic food webs, and act as breeding and nursery areas for species of commercial importance.

Fig. 14.9 Sorting the catch from large quantities of seagrass (*Posidonia*) taken by a trawl net off the Italian coast. Photograph copyright G.D. Ardizzone.

In south-eastern Spain up to half the area of *Posidonia oceanica* meadows has been damaged by bottom fishing. Destruction of parts of the seagrass meadow has meant that sediment is no longer bound effectively, which has reduced the stability of this habitat (Fonseca *et al.*, 1984; Guillén *et al.*, 1994; Short & Wyllie-Echeverria, 1996).

14.3.5 Meta-analysis

As we have seen, the magnitude of the impact of bottom-fishing activities on benthic fauna and their habitat varies according to factors such as habitat stability, depth, disturbance frequency and presence of biogenic structures. Thus, the results from any one experiment or comparative study are limited in their scope for extrapolation to other situations. A far better approach is to take all the data from these studies and undertake what is known as a meta-analysis of the data. Such an analysis considers each separate study as

a replicate and therefore has great power for the analysis of general trends and patterns that might provide the basis of predictive models (Collie *et al.*, 2000). In addition, the effects of many different factors can be examined, something which would be unfeasible in a manipulative study. Collie *et al.* (2000) found that the effects of gear type were broadly consistent with expectations, i.e. intertidal dredging had more marked **initial effects** (immediate effects of impact) than scallop dredging, which in turn had greater effects than trawling (Fig. 14.10). The most consistently interpretable result was with respect to faunal vulnerability, with a ranking of initial impacts that matched expectations based on morphology and behaviour. For example, the sea anemones and their relatives (Anthozoa) were much more vulnerable than starfish (Asteroidea).

Collie *et al.* (2000) also employed a tree-based regression modelling approach, in which the data set was progressively split into increasingly homogenous subsets until it was unfeasible to continue (Fig. 14.11). This approach is becoming increasingly popular as a means for devising prediction rules and for summarizing large multivariate data sets in ways that are sometimes more informative than linear models. This regression tree analysis provided the first quantitative basis for predicting the relative impacts of fishing under different situations. Following the tree from its root to the branches we can make predictions, for example, about how a particular taxon would be affected initially by disturbance from a particular fishing gear in a particular habitat in different regions of the world. Collie *et al.* (2000) were also able to model recovery rate and speculated about the level of physical disturbance that was sustainable in a particular habitat. Their results suggested that sandy sediment communities were able to recover within 100 days, which implied that these habitats could perhaps withstand two to three incidents of physical disturbance per year without changing markedly in character. This, for example, is the average predicted rate of disturbance for parts of the southern North Sea. If the recovery rate estimates for sandy habitats are realistic then we would predict that the communities in heavily fished areas are held in a permanently altered state (Fig. 14.12).

14.4 Effects of static fishing gears

Static bottom gears are anchored to the seabed and left to fish passively (Chapter 5). The most commonly used

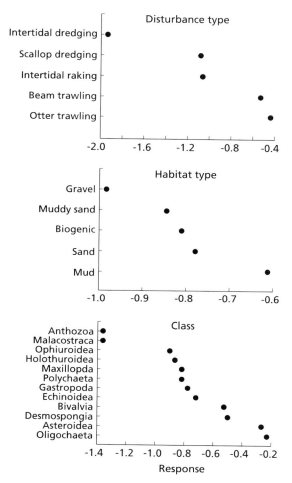

Fig. 14.10 The mean initial responses of all organisms, included in the Collie *et al.* (2000) analysis, to trawling by different gears (disturbance type) in different habitats (Habitat type) and then further broken down into the responses of different taxonomic groups (class). An initial response of 0 indicates no change in abundance, while a positive response indicates an increase and a negative response a decrease. In all cases, it is clear that the initial response to fishing disturbance was a decrease in abundance, although the extent of this decrease varies considerably for different disturance types, habitats and taxonomic groupings. After Collie *et al.* (2000).

cess. The introduction of mechanized hauling devices has increased the power with which gear is hauled. Thus, rocks and organisms are more easily sheared off if ropes, netting or traps become snagged on them (Munro *et al.*, 1987). Nevertheless, these effects are relatively minor in comparison with those associated with other active fishing techniques. By their nature, static gears do not affect great areas of seabed as in towed gear fisheries. A 12-m wide trawl towed for 1 h at 7 knots will affect approximately 160 000 m^2 of the seabed, compared with 100 m^2 covered by a single gill net panel or fleet of pots deployed for 1 day. However, in some fisheries, effort may be significant if concentrated in relatively small areas. In particular, many species targeted by static gear live in relatively complex habitats with a predominance of emergent fauna.

Eno *et al.* (1996) used divers to observe the possible interactions between static gear and fragile epifauna in close proximity to areas of conservation interest. Pots fished on rocky grounds came into contact with the fragile bryozoan *Pentapora foliacea* either as they landed on the seabed or when they were hauled across it. Although the bryozoan colonies were highly susceptible to damage by this gear, the chances of contact were relatively low because of the sparse nature of their distribution. In a similar habitat, sea fans *Eunicella verrucosa* flexed as pots were dragged over them and were not ripped off the rock substratum as anticipated (Eno *et al.*, 1996).

Divers also observed the interaction between Norway lobster *Nephrops norvegicus* creels and sea pens in a Scottish sea loch. As pots descended towards the seabed they created a pressure wave that caused the sea pens *Pennatula phosphorea*, *Virgularia mirabilis* and *Funiculina quadrangularis* to bend and lay flat on the seabed. When the pots were lifted off the seabed, the sea pens were able to re-establish themselves in the sediment even if they had been uprooted. Although these animals are considered to be highly vulnerable to towed fishing gear (MacDonald *et al.*, 1996), the direct observations made by divers suggest that they are far less sensitive to static fishing gear techniques. However, these observations are of occasional contacts with individual organisms and it is difficult to predict whether repeated disturbance would increase their mortality rate.

14.5 Long-term effects

While it has been relatively simple to demonstrate

static nets are gill, trammel or tangle nets. These gears have relatively little effect on seabed life although there are problems associated with bycatches in 'ghost-fishing' nets (Chapter 13). Traps and pots are deployed in fleets, each trap is baited to attract crustaceans, gastropods or fishes. Reefs can be damaged as nets or pots are dragged over the seabed during the retrieval pro-

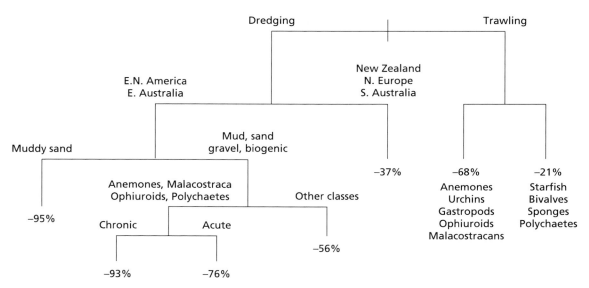

Fig. 14.11 A regression tree that summarizes the initial responses of organisms to fishing disturbance. The vertical height of a branch indicates the importance of a split in contributing to the overall variability in the initial response. Thus the difference in the initial response of the organisms to dredging and trawling was one of the most important factors accounting for variability in the initial response. The number beneath each node is the mean initial response for that combination of variables. For example, dredging in eastern North America or eastern Australia on mud, sand, gravel or biogenic habitats leads to a 93% (–93%) reduction in the abundance of anemones, malacostraca, ophiuroids and polychaetes under a chronic disturbance regime. After Collie *et al.* (2000).

short-term localized responses to fishing disturbance, it has been more difficult to infer long-term changes in all but the most extreme examples for which data have been collected over a long time scale (Riesen & Reise, 1982; Sainsbury, 1987; Sainsbury *et al.*, 1998). The few long-term datasets that exist of bycatches of benthic species have yielded strong evidence of fishing effects. In the 1930s, the Netherlands Institute for Sea Research (NIOZ) began to pay fishers to return a predetermined list of benthic and bycatch species from their bycatches (Philippart, 1999). Between 1930 and 1970, most of the fishers used otter trawls. Beam trawling started to become popular in the 1960s and finally superseded otter trawling as the main Dutch fishery from about 1970 onwards. As otter trawls are less effective at catching benthos than beam trawls, landings returns of benthic species to the NIOZ laboratory might have been expected to increase after 1970. In contrast, returns of many epifaunal species decreased steadily throughout the study period. In particular, elasmobranchs and whelks were among the species that demonstrated the greatest decrease in occurrence. The decline in the whelk population was previously attri-

buted the effects of ship antifouling paint that contained tri-butyl tin (TBT), but more recent studies strongly suggest that bottom trawling is the factor responsible (Fig. 14.13). There is a high incidence of damage to whelk shells collected from the North Sea which makes them vulnerable to predation by starfish and hermit crabs (Ramsay & Kaiser, 1998; Mensink *et al.*, 2000). A decline in whelk populations may have repercussions for other species. Hermit crabs *Pagurus bernhardus* are most commonly found in the shell of the common whelk in northern European waters. A reduced whelk population will restrict the availability of shells for use by hermit crabs. Whelk shells are most commonly inhabited by larger adult hermit crabs. As a result, hermit crab populations in areas where whelk populations become extinct may also decline or become dominated by smaller individuals.

Plankton communities tend to be dominated by the larvae of invertebrates. Consequently, alteration of benthic communities through fishing activities may cause changes in plankton community composition. Continuous plankton recorder samples collected from the central North Sea were numerically dominated by

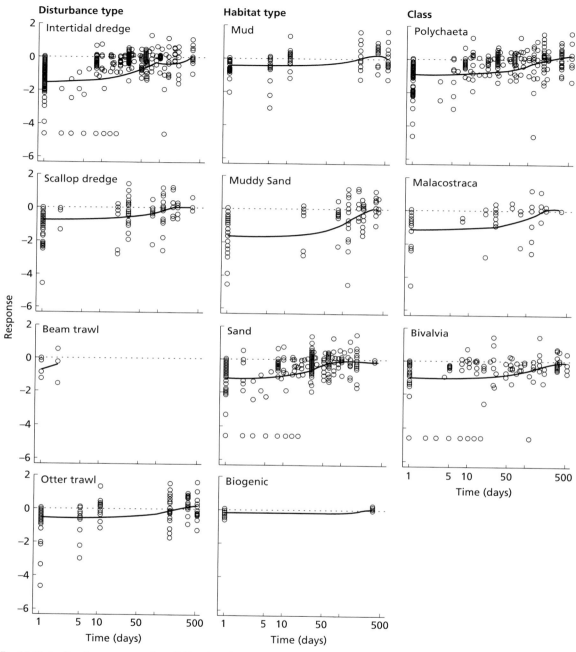

Fig. 14.12 Predicted recovery rates from fishing disturbance for different disturbance types, habitat types and taxonomic groups (class). Each circle shows a single response, with negative responses indicating decreased abundance. Fitted curves suggest mean recovery trajectories with time. It is assumed that recovery has occurred when the fitted curve reaches a response of zero. After Collie *et al.* (2000).

(a)   (b)

Fig. 14.13 These whelks *Buccinum undatum* from the Dutch Wadden Sea have been damaged by physical contact with bottom-fishing gear, probably shrimp trawls. The whelk (a) has been repeatedly damaged in the past, but has managed to repair its shell on at least three occasions. Whelk (b) has sustained very severe damage to the shell such that it would be unable to withdraw fully into its shell and would be very vulnerable to predation. Photographs copyright B. Mensink.

calanoid copepods from 1958 to the late 1970s. A shift in the plankton community then occurred. Samples that were collected after the late 1970s were dominated by the larvae of echinoderms such as sea urchins and brittlestars. This trend reflected the increase in the abundance of brittlestars observed in the North Sea, which was attributed to the eutrophication of coastal waters and possibly the effects of fishing disturbance (Lindley *et al.*, 1995).

14.6 Fishing as a source of energy subsidies

Although undesirable, most fisheries produce a proportion of bycatch. Bycatches are discarded dead or dying either for legal reasons or because there is insufficient money to be gained from sorting or landing them (Chapter 13). It is estimated that the equivalent of approximately a quarter (27 million tonnes) of current global landings are discarded (Alverson *et al.*, 1994). This estimate does not account for those animals that are never landed on-board vessels but remain dead on the seabed. Thus, fishing activities provide two main sources of food for benthic scavengers: discards rejected by fishers and animals killed on the seabed, damaged or exposed by passing trawls and dredges.

Food-falls have a profound effect on local diversity and production processes in the deep sea. Carrion arriving on the floor of the abyssal plain will attract scavengers over great distances, which indicates the importance of this opportunistic resource for deep-sea

organisms (Dayton & Hessler, 1972; Stockton & DeLaca, 1982). The responses of animals to carrion in continental shelf areas may be equally rapid. Trawlers in the North Sea are frequently observed fishing one behind the other. As the first trawls pass over the seabed, worms, sea urchins and molluscs are dug up and damaged, attracting fish into the area of disturbance (Kaiser & Spencer, 1996a; Kaiser & Ramsay, 1997). The fish respond to the chemical odours given off by the damaged animals and the feeding activities of other scavengers, or may be attracted by sediment plumes in clear water. Examination of gut contents reveals that several species of fish eat animals damaged by fishing activity (e.g. Fig. 14.14). Often, the prey will be organisms that are not available to the fish under normal conditions. For example, Kaiser and Spencer (1994) found that whiting *Merlangius merlangus* ate the gonads of scallops in recently trawled areas. Similar responses to fishing disturbance were also recorded for dab *Limanda limanda*. Under normal circumstances, dab feeding on a muddy seabed consume mainly the arms of the brittlestar *Amphiura* spp. which protrude from the surface of the sediment. In contrast, dabs sampled from heavily beam-trawled areas fed predominantly on the oral discs of brittlestars. This suggests that trawling had exposed entire brittlestars and made them more vulnerable to predation (Kaiser & Ramsay, 1997).

A large number of invertebrate epifauna are facultative scavengers (Britton & Morton, 1994) that have physiological features or behavioural adaptations that

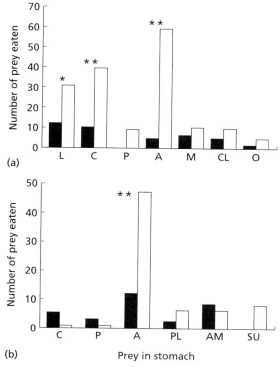

(a)

(b) Prey in stomach

Fig. 14.14 The change in the mean number of prey eaten by
(a) gurnards *Eutrigla gurnardus* and (b) whiting *Merlangius
merlangus* which were collected both before (shaded bars), and
3 h after (open bars) fishing an area with a 4-m commercial
beam trawl. Both the amount of food in the gut and the
composition of diets were different for fish feeding in areas of
trawl disturbance. Significant differences for particular prey
types are indicated by the asterisks above pairs of bars. Key to
prey: L, *Liocarcinus depurator* (swimming crab); C, *Crangon*
spp. (shrimps); P, *Pandalus* spp. (prawns); A, *Ampelisca
spinipes* (amphipod); M, *Macropodia* spp. (crabs); CL,
Callionymus spp. (dragonets); PL, polychaetes; AM,
Ammodytes spp. (sandeels); SU, sea urchin gonads; O, others.
After Kaiser and Spencer (1994).

enable them to survive the capture and discarding processes. These animals include starfishes, crabs and molluscs that are either able to regenerate damaged body
parts or live in protective shells. Invertebrate scavengers locate animals damaged by fishing by following
the odours released from their injured body tissues
(Sainte-Marie, 1986; Nickell & Moore, 1992; Ramsay
et al., 1996). Some animals arrive more quickly than
others, depending on their speed and mobility. By using
baited time-lapse and video cameras, Ramsay *et al.*

(1997) observed that starfish *Asterias rubens* and hermit crabs *Pagurus bernardus* were the two most abundant scavengers in offshore gravel and inshore sand
habitats. Under normal conditions, the ratio of starfish
to hermit crabs was 3 : 2 and 5 : 2 at the offshore and
inshore habitats, respectively. Hermit crabs always
arrived first at a food-fall and their numbers built up
rapidly, occasionally reaching densities of > 300 m^{-2} at
the offshore site. At such high densities, hermit crabs
competitively excluded most species, with a ratio of one
starfish to 10 hermit crabs feeding on the carrion. In
contrast, a greater diversity of scavengers gathered at
carrion at the shallower inshore study site. Hermit crabs
and starfish occurred in equal numbers on the carrion.
So what could explain the difference in patterns of scavenger aggregation at the different localities? Tidal currents were stronger at the offshore site and these would
have spread the odours of the carrion more quickly,
alerting greater numbers of hermit crabs to its presence.
As hermit crabs move more quickly than starfish, they
would rapidly attain high densities at the site of the
food-fall. In addition, there may have been fewer
food resources normally available to hermit crabs at
the deeper-water offshore site (Ramsay *et al.*, 1997).
These observations are directly analogous to the
behavioural interactions between different species of
seabirds observed at trawlers, where the feeding success
of fulmars is directly related to their numerical dominance in relation to other larger species (Furness *et al.*,
1988; Camphuysen *et al.*, 1993).

As we can see, food-falls often lead to multispecies
aggregations of different scavenging species. In some
cases group foraging can enhance food acquisition
(Brown & Alexander, 1994). Whelks *Buccinum undatum* are attracted to the feeding activities of starfish,
despite the risk of becoming prey themselves, as they
are able to feed on uneaten food. Although hermit crabs
gather quickly at carrion, they struggle to deal with it
as they have relatively weak chelae that are poorly
adapted for cutting flesh. As other scavengers arrive,
their feeding activities, enzymes secreted by feeding
starfish and bacterial decay will help to break down
flesh, making it easier to consume. Certain scavenging
species appear to have quite strong food preferences.
Whelks are most strongly attracted to traps baited with
dead crab, whereas hermit crabs strongly preferred
dead fish. As fishing gears pass across the seabed they
produce a trail of patchily distributed food resources of

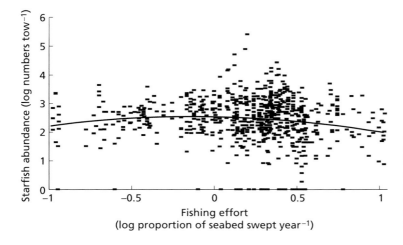

Fig. 14.15 The relationship between starfish numbers and fishing effort in the North Sea. Although the quadratic model accounted for only 5% of the variance, environmental variables only accounted for 1% and thus fishing effects would appear to be ecologically significant. After Ramsay *et al.* (2000).

different types. The food preferences of different scavengers may be a behavioural mechanism to avoid competitors and/or predators (Ramsay *et al.*, 1997).

14.6.1 Have population changes occurred?

The vast expansion in the size of populations of breeding seabirds in the North Sea has been linked partly to the increase in food resources generated by the discarding practices of many fisheries (Furness, 1996; section 15.2.3). However, the responses of fish and invertebrate scavenger populations to these energy subsidies are not clear. Populations of scavenging fish species, such as the North Sea dab *Limanda limanda* and long rough dab *Hippoglossoides platessoides*, have increased from 1970 to 1995 while populations of cod and industrially fished species have decreased (Heessen & Daan, 1996). In addition, the growth rates of plaice and sole have increased (Millner & Whiting, 1996; Rijnsdorp & Leeuwen, 1996). However, there are many alternative explanations for the proliferation of dabs and long rough dabs, and the increased growth of plaice and sole. For example, dabs and long rough dabs are small, grow rapidly and mature early and may have replaced slower growing species. Moreover eutrophication has increased populations of polychaetes and brittlestars in coastal waters, thereby increasing the food supply for juvenile flatfishes.

So far, evidence that populations of benthic invertebrate scavengers have increased in response to fisheries-generated carrion is weak. Starfishes are the scavenging taxon most likely to proliferate as a result of trawl

disturbance as they are able to consume fisheries-generated carrion and they are highly resilient to physical damage. Ramsay *et al.* (2000) found a quadratic relationship between starfish density and trawling intensity in the North Sea. This relationship suggested that at low to medium levels of fishing disturbance, starfish population size increased with increasing fishing intensity. However, at extremely high levels of fishing disturbance, starfish population size began to decline, presumably as a result of starfish fishing mortality (Fig. 14.15). While seabirds are able to actively search for fishing vessels over large areas (Camphuysen *et al.*, 1993), benthic invertebrates rely on the chance occurrence of fisheries carrion within areas restricted to a few 100 m. Thus, this source of carrion is probably too unpredictable to give significant benefit to most invertebrate populations.

14.7 Indirect effects on habitats

14.7.1 Loose seabeds

The direct physical contact of fishing gear with the substratum can lead to the resuspension of sediments and the fragmentation of rock and biogenic substrata. The resuspension, transport and subsequent deposition of sediment may affect the settlement and feeding of the biota in other areas. Sediment resuspended as a result of bottom fishing will have a variety of effects including: release of nutrients held in the sediment, exposure of anoxic layers, release of contaminants, increasing biological oxygen demand, and smothering of feeding and respiratory organs. The quantity of sediment resuspended by trawling depends on sediment grain size and the degree of compaction, which is higher on mud and fine sand than on coarse sand. Transmissiometers that measure background light levels in water, were reported to frequently record the highest levels of turbidity during periods of trawling activity (Churchill, 1989). In deeper water where storm-related bottom stresses have less influence, otter trawling activity contributed significantly to the resuspension of fine material. Churchill (1989) calculated sediment budgets for areas of the mid-Atlantic Bight and concluded that trawling was the main factor initiating the offshore transport of sediment at depths of 100–140 m. However, the transport of sediment resulting from fishing activities would not produce significant large-scale erosion over a period of a few years. The effects of sediment resuspension are clearer in deep-water environments that are relatively unperturbed. Thiel and Schriever (1990) experimentally harrowed an area of seabed at a depth of 4000 m. These observations revealed that 80% of their study site was covered by fine material that had settled out from the resultant sediment plume. This may have a similar effect to the seasonal settlement of organic material that occurs in deep-sea regions (Angel & Rice, 1996).

The sediment/water interface of marine sediments is an important site of benthic primary production. Brylinsky *et al.* (1994) found that benthic diatoms bloomed within otter door tracks 1 month after they had been created. They reasoned that the bloom was triggered by the release of nutrients from the sediment following trawling. The intensive trawling of *Posidonia oceanica* meadows in the Mediterranean Sea may lead to reductions in littoral primary productivity since large areas of *P. oceanica* are reported to have been killed by the mechanical action of fishing gears and the deposition of resuspended sediment (Guillén *et al.*, 1994). These meadows are known to be important sources of primary production although consequences of losses in production are not known. It is unlikely that large-scale changes in primary production could be reliably correlated with changes in fishing intensity using existing data.

14.7.2 Coral reefs

Coral reefs are formed by an actively growing veneer of living tissue, and are responsible for shaping and creating around one-third of tropical coastlines and whole archipelagos of atolls such as the Maldive Islands in the Indian Ocean. The veneer of living tissue, consisting of colonies of polyps and symbiotic algae known as zooxanthellae, is very susceptible to environmental change and the indirect effects of fishing. Actively growing corals can deposit 3–5 kg $CaCO_3$ m^{-2} y^{-1} and reefs can grow around 5 m in height every 1000 years. This allows growing coral to remain in shallow and well-illuminated water needed for the zooxanthellae to photosynthesize, and for corals to persist during slow increases in sea level. The limestone matrix of the reef is an important carbon sink. The reef habitat harbours a diverse fauna and flora and provides coastal protection.

As corals grow, other organisms will be eroding or weakening their calcareous skeletons. These are known as **bioeroders**. There are external bioeroders such as fish and sea urchins, and internal bioeroders living within the skeleton. Urchins are often the most important bioeroders and, when abundant, can erode reefs faster than they grow. Some examples of bioerosion rates for urchin species are given in Table 14.1. Rates of bioerosion can equal or exceed rates of reef accretion. The indirect effects of fishing can determine the abundance of bioeroding species and shift reef communities into a state that favours bioerosion and leads to loss of the reef matrix. This has important implications given the key role that reefs play in protecting coasts from storm damage, as a source of tourist revenue, as hot spots for diversity and as a source of fish production.

We can now look in some detail at the factors that control the abundance of bioeroders. Changes in the environment, especially temperature increases that occur during events such as the El Niño (section 2.3.2),

Table 14.1 Some examples of rates of coral bioerosion by sea urchins in different localities. After Glynn (1997).

Species	Location	Erosion rate (g $CaCO_3$ m^{-2} y^{-1})
Diadema antillarum	US Virgin Islands	4600
Diadema antillarum	Barbados	5300
Diadema mexicanum	Panama	139–10 400
Diadema savignyi	Moorea	3400
Echinometra lucunter	US Virgin Islands	3900
Echinometra mashei	Enewetak Atoll	70–260
Echinothrix diadema	Moorea	803
Eucidaris thouarsii	Galápagos	3320–22 300

can cause corals to lose their zooxanthellae and turn white, a process known as **coral bleaching**. If the temperature rises and bleaching lasts more than a few weeks, then the corals will die and be colonized by algae. Dead corals with algal growth are very susceptible to bioerosion since the bioeroding species wear them away as they graze. For example, since the 1982–83 El Niño in Galápagos the ecosystem has entered a phase of net bioerosion. Bioerosion rates are $20-40$ kg m^{-2} y^{-1} while coral growth is only 10 kg m^{-2} y^{-1}. Over the last few years, this bioerosion has led to the loss of quite substantial reef formations and their reduction to flat plains of coral rubble (Glynn, 1997). In other localities, coral death may be due to coral predators such as the snails *Drupella* spp. or crown-of-thorns starfish *Acanthaster planci*, and the dead coral again provides an ideal substratum for algal growth and colonization by bioeroders.

Fishing reduces the abundance of target fishes on tropical reefs (section 12.4.2). These fishes have different roles in the reef ecosystem. Some, such as the parrotfishes (Scaridae) are herbivores that graze algae and erode the reef matrix. Others feed on invertebrates, such as urchins and juvenile crown-of-thorns starfish. Studies in Kenya and the Caribbean have shown how changes in the fish community due to fishing can determine rates of bioerosion.

Gut contents analysis, video and diver observations have shown that fished species such as triggerfishes (Balistidae) and emperors (Lethrinidae) feed on urchins. Indeed, some emperor species are distinctive because they have clusters of black spots on their snouts. These spots are the remains of sea-urchin spines that penetrated the flesh and broke off as the emperors fed. On many Kenyan reefs, fishing has reduced populations of urchin predators by an order of magnitude or more and the abundance and behaviour of urchins has changed in response.

The urchin *Echinometra mathaei* is widely distributed on Kenyan reefs. When present in low numbers, this species usually lives in reef crevices that are gradually enlarged by grazing and the abrasive action of their spines. The urchins are thought to shelter in crevices as a defence from fish predation, and they feed in and around the crevices. On heavily fished reefs, it seems that the reduced predation on *Echinometra* has led to increases in their abundance. At high densities and with low predation pressure, *Echinometra* are found on exposed areas of reef rather than in crevices. Here, they outcompete other herbivores and erode the reef matrix. This ecological release of urchins has also led to the competitive exclusion of herbivorous fish because the urchins can persist at low levels of algal biomass and productivity, outcompeting herbivorous fishes and reaching maximum biomass levels an order of magnitude higher. Moreover, the ecological release of *Echinometra* has led to increased substrate bioerosion, reduced algal biomass, loss of topographic complexity and an associated decrease in reef fish biomass and production. Corals can no longer recruit to these reef areas due to continued bioerosion by urchins (McClanahan & Shafir, 1990; McClanahan, 1992). The processes that affect the relationship between sea urchins, fishes, algae and coral growth on Kenyan reefs are summarized in Fig. 14.16.

In the Caribbean, the principal urchin species is *Diadema antillarum*. This species is eaten by triggerfishes (Balistidae), wrasses (Labridae) and other fishes. At low densities, and when herbivorous fishes are not numerous, the grazing activities of *Diadema* do not prevent corals being overgrown by algae. Conversely, at high densities, grazing leads to bioerosion and the mortality of young corals (Sammarco, 1980). This means that changes in the density, and grazing pressure exerted by *Diadema* can regulate the relative abundance of corals and algae. When herbivorous fish such as parrotfishes are abundant, these graze algae and prevent them from growing over corals. Other herbivorous fishes such as the damselfishes may maintain the algal community in a state that favours coral settlement (Hixon, 1997). However, if herbivorous fishes are fished to low abundance and *Diadema* are also scarce then overall grazing pressure will be low and algal biomass will increase dramatically (Carpenter, 1997). This effect was dramatically demonstrated by a natural 'experiment' when there was mass mortality of *Diadema* throughout much of the Caribbean, including many areas that were intensively fished. This was thought to be due to a water-borne pathogen (Lessios, 1988). Algal biomass increased rapidly and an algal community that was once dominated by a low biomass and highly productive community of filamentous and turf species was increasingly dominated by macroalgae such as *Sargassum*. In the longer term this has prevented coral recruitment and growth and the reef ecosystem appears to have entered an alternate stable state.

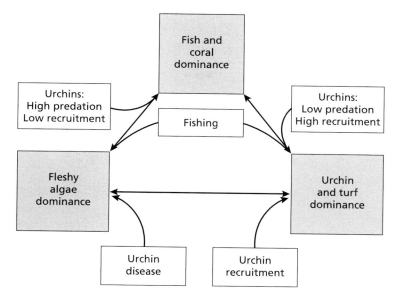

Fig. 14.16 A conceptual model of the coral reef ecosystem. The model indicates those processes that cause shifts between three equilibrium states (shaded) and which have been demonstrated by simulation and empirical studies. After McClanahan (1995).

Summary

• Fishing activities lead to short-term alteration in infaunal and epifaunal community structure, resuspension of sedimentary material, long-term changes in populations of vulnerable species, reduction of habitat complexity and the reallocation of energy resources.

• The relative impacts of fishing depend on the relative magnitude of fishing and natural disturbance.

• Degradation of habitat complexity is probably the most important effect of fishing as this can reduce the supply of important prey species, increase predation risk for juvenile commercial species and affect subsequent recruitment to the adult stocks.

• Intensive fishing of urchin predators and herbivorous fishes may cause reef communities to shift to alternate stable states that favour bioerosion rather than reef accretion.

Further reading

The effects of fishing on complex habitats are discussed by Watling and Norse (1998) and Auster and Langton (1999). Chapters in Hall (1999) and Kaiser and De Groot (2000) discuss impacts of trawling, dredging, discarding and ghost fishing in the marine environment. Hall (1994) reviews the factors that affect organisms living in soft sediments and places the influence of fishing disturbance in an ecological context. Several chapters in Birkeland (1997) review the indirect effects of fishing on coral reefs.

15 Fishery interactions with birds and mammals

15.1 Introduction

Birds and mammals are killed during fishing activities and may compete with fishers for prey. Low levels of mortality may cause decreases in abundance because bird and mammal populations have low intrinsic rates of natural increase. Moreover, any mortality may be significant for those concerned with the welfare of individuals, and this is reflected in the use of emotive terms such as 'slaughter' to describe their killing. Some fishers see birds and mammals as competitors and have been known to cull them in an attempt to boost fisheries yield.

Birds and marine mammals are hunted directly but, following the cessation of most commercial whaling, rates of mortality are low and rarely have a significant effect on population dynamics. Birds and marine mammals are also killed as bycatch during fishing activities (Chapter 13). However, the aim of this chapter is to describe indirect interactions. We show how fishers can compete with birds and mammals for prey, and how bird and mammal populations respond to the additional food that fishery discards provide. We also consider whether the removal of mammal predators has released prey that can then be caught by fishers, or provided additional food for other birds and mammals.

Interactions between fisheries, birds and mammals are amongst the most politically sensitive issues that fisheries managers face. In several countries fishery managers are expected to produce management plans that ensure the well being of birds and marine mammals, rather than maximizing fish production at any cost.

15.2 Birds

Birds that spend all or part of their life in the marine environment often feed on fished species. In particular, the pelagic seabirds that nest in colonies on isolated

Fig. 15.1 Blue-footed boobies feed on pelagic fishes in the Galápagos Islands. Photograph copyright S. Jennings.

coasts and islands (Fig. 15.1) often feed on oily plank-tivorous 'forage' fishes such as sandeels *Ammodytes* spp., sardines *Sardina* spp., horse mackerel *Trachurus* spp., capelin *Mallotus* spp. and anchovy *Engraulis* spp. The same fishes are intensively targeted by reduction (fishmeal) fisheries that frequently remove 20% or more of their production on an annual basis. Wading shorebirds and sea ducks also compete with fishers for shellfish such as clams, mussels and cockles, and the disturbance caused by fishers can interrupt the birds' feeding activity.

Birds that interact with fisheries (Table 15.1) can be divided into broad functional groups based on foraging behaviour. Foraging behaviour determines how and when birds interact with fisheries and how susceptible they are to changes in the distribution and abundance of prey. Some gulls, for example, may live and breed

Table 15.1 The main groups of birds that interact with fisheries. Based on Cramp (1977) and Perrins (1987).

Taxonomic group	Common name	No. of species
Order Sphenisciformes		
Family Spheniscidae	Penguins	18
Order Procellariiformes		
Family Procellariidae	Shearwaters	53
Family Hydrobatidae	Storm petrels	20
Family Pelecanoididae	Diving petrels	4
Family Diomedeidae	Albatrosses	14
Order Pelecaniformes		
Family Sulidae	Gannets and boobies	7
Family Phalacrocoracidae	Cormorants and shags	30
Family Pelecanidae	Pelicans	7
Family Fregatidae	Frigate birds	5
Family Phaethontidae	Tropic birds	3
Order Anseriformes		
Family Anatidae	Swans, geese, ducks	151
Order Charadriiformes		
Family Haematopodidae	Oystercatchers	6
Family Recurvirostridae	Stilts, avocets	7
Family Stercorariidae	Skuas	5
Family Scolopacidae	Sandpipers, snipes	81
Family Laridae	Gulls	45
Family Sternidae	Terns	41
Family Alcidae	Auks, guillemots, murres	21

in colonies on cliffs, isolated islands or rocks and forage locally (typically within 100 km of the colony) throughout the year. The limited feeding areas used by these species make them susceptible to local fluctuations in prey density. Some gannets and albatrosses, conversely, live an almost exclusively pelagic life and only return to colonies for breeding. Even when breeding, they may make trips of several hundred kilometres over a 24-h period to bring back food for the chicks. Colonies of these species are usually found on rocks and islands free from all land predators, since predators would eat chicks while the adults were foraging. Breeding constrains the range over which seabirds can forage, so they may miss breeding years when food supply is poor. This allows them to range more widely in search of food. Wading birds such as oystercatchers and stilts may migrate between breeding and wintering areas. Wintering areas provide warmer conditions than breeding areas at a given time of year. While wintering, waders feed on bivalves and polychaete worms living on tidal mudflats, salt marshes or mangals. Because their feeding areas are restricted, birds may not feed

successfully if fishers compete for their food or cause disturbance.

Vast colonies of screeching seabirds on cliffs and islands, the deep deposits of white and grey faeces they produce, and the frequency with which adults return from foraging trips suggest that fish consumption by seabirds is considerable. This is often true. The seabirds of the North Sea, for example, are thought to consume around 600 000 tonnes of fish each year. This includes 200 000 tonnes of sandeel, 30 000 tonnes of sprat and small herring, 22 000 tonnes of small gadoids and 13 000 tonnes of large herring and mackerel. The seabirds also consume considerable quantities of discards (Chapter 13; Furness & Tasker, 1997). The consumption rate of 600 000 tonnes can be compared with total predation on fishes by fishes of 6.6 million tonnes, total fish production of around 16.2 million tonnes and total catches of approximately 3 million tonnes (Sparholt, 1990). These estimates will vary considerably from year to year, but suggest that birds are important consumers of fish production. The figures exclude the 10 million sea ducks and 1 million waders that forage on

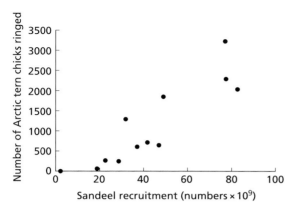

Fig. 15.2 The relationship between the number of Arctic terns marked with rings (successfully fledged juveniles) in the Shetland Islands and the recruitment of young-of-the-year sandeels on 1 July each year from 1976 to 1987. After Monaghan *et al.* (1989).

shellfish in north-west Europe during winter. Common eider ducks *Somateria mollisima* are the most abundant species, numbering around 3 million. Ducks and waders have major impacts on shellfish populations, probably consuming around 20% of mussel and cockle production in the Wadden Sea in each year. By comparison, fishers may remove 25% of production in heavily exploited areas.

15.2.1 Competition between birds and fisheries

Most seabirds eat small schooling pelagic fishes that live within a few metres of the sea surface. In upwelling and productive coastal systems, seabirds can consume 5–30% of fish production. Seabirds appear to prefer small pelagic fishes because they are relatively easy to catch and have oily tissues that are rich in energy and help to meet the high metabolic requirements of the birds. The consumption of so much fish oil also accounts for the remarkably smelly breath of seabirds.

Numerous studies show convincing and consistent linkages between prey availability and the reproductive success of seabirds. Thus, reproductive success is positively correlated with the availability of favoured prey fishes and prolonged periods of low prey biomass may lead to significant decreases in seabird population size (Furness, 1982; Powers & Brown, 1987). An example of the relationship between the breeding success of Arctic terns in the Shetland Islands (northern North Sea) and food supply is presented in Fig. 15.2.

In the upwelling ecosystems off south-west Africa and Peru, where the abundance of pelagic fish is highly variable (section 2.3.2), there are strong correlations between bird population size (often measured as the rate of faeces (guano) production) and sardine or anchovy abundance (Crawford & Shelton, 1978; Duffy, 1983). Birds that produce deposits of guano are collectively termed **'guano birds'** and typically include gannets and boobies (Sulidae). Rates of guano production can be so high that the guano is mined as a phosphate and nitrate rich agricultural fertilizer. Figure 15.3 shows how the abundance of bird populations that feed in upwelling systems can vary. Some of the seabirds found in upwelling systems produce several broods of young

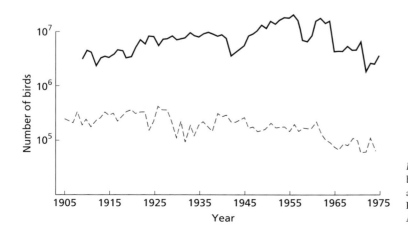

Fig. 15.3 Fluctuations in the numbers of breeding seabirds using the Peruvian (——) and South African (– – –) sectors of the Humboldt and Benguela upwellings. After Duffy and Siegfried (1987).

each year, allowing their population size to grow rapidly when feeding conditions are good. In less variable systems, trends in bird abundance are correspondingly less variable. Because fish production and the reproductive success of seabirds are so closely linked, fishing that reduces prey availability will cause population declines in seabirds. Thus, much of the argument for interactions between fisheries and seabirds depends on the extent to which fishing can be proven to influence population trends in prey fishes. For example, seabird colonies at the margin of a prey species range may not have accessible prey sources if the range of the prey fish population shrinks when its abundance is reduced by fishing (section 4.3).

Two good examples of studies that link fisheries, fish population structure and trends in seabird reproductive success and abundance come from the Shetland Islands in the northern North Sea and from the Barents Sea.

Seabirds and sandeels in the Shetland Islands

The Shetland Islands are an internationally important seabird breeding area, with colonies of 13 species containing 25–100% of their respective populations in the North Sea. Many of these seabirds feed on sandeels, predominantly the lesser sandeel *Ammodytes marinus*. In the 1970s, an industrial fishery for sandeels started in the North Sea, and by the 1980s annual landings exceeded 1 million tonnes, the largest single species fishery in the North Sea at that time. The sandeels were used to produce fishmeal and fish oil for fertilizers, food manufacture or fuelling power stations. Sandeels were a key component of the diet for seabirds in this region, since other suitable prey fishes were scarce (Wright, 1996). Seabirds were estimated to eat around 200 000 tonnes of sandeels each year, or 4% of the stock. This suggested that the potential for the fishery to affect birds was much greater than the potential of birds to affect the fishery (Furness & Tasker, 1997).

The development of the industrial fishery for sandeels during the 1980s coincided with rapid declines in the reproductive success of many Shetland seabirds. Birds such as skuas *Stercorarius* spp., terns *Sterna* spp. and kittiwakes *Rissa tridactyla* that fed on sandeels close to the sea surface were most affected and there was widespread concern that the sandeel fishery was limiting the sandeel supply for breeding seabirds.

Annual estimates of sandeel abundance were made with virtual population analysis (Chapter 7) since 1974. These estimates suggested that the decline in sandeel abundance around the Shetland Islands resulted from poor recruitment. The seabirds that suffered population declines generally fed on small, young-of-the-year sandeel. Fishing mortality first affects young-of-the-year sandeel in July, at the end of the chick rearing period. As such, the seabirds could not have competed directly with fisheries for young-of-the-year sandeels during the chick-rearing period. However, if fishing reduced the spawning stock biomass of sandeel, and low stock sizes were associated with low recruitment (section 4.2.1), then intensive fishing of sandeels would have indirect effects on seabird reproductive success. There was no evidence for a relationship between recruitment and parent stock in sandeels and, during a year of good recruitment in 1991, the reproductive success of seabirds improved when sandeels recolonized their feeding grounds. The initial decline in sandeel recruitment preceded any change in size of the spawning stock, and when the fishery was closed in 1991 because sandeel biomass was excessively low, recruitment was the strongest since records began. As a result, seabird reproductive success increased. Despite strong evidence for the indirect effects of fishing on seabird populations in some other ecosystems, changes in the sandeel stocks around Shetland appeared to reflect variability in the pre-recruit immigration and survival of sandeels rather than changes in stock size caused by fishing (Wright, 1996).

Barents Sea

The collapse of fish stocks in the Barents and Norwegian Sea ecosystems had profound effects on bird populations. Around 3 million pairs of seabirds breed in northern Norway, mostly Atlantic puffins *Fratercula arctica*, kittiwakes and common murres (guillemots, *Uria* spp.). Puffin and murre populations declined with the collapse of the Norwegian herring *Clupea harengus* and Barents Sea capelin *Mallotus villosus* stocks. The largest puffin colony was on the Lofoten Islands in northern Norway. Here, puffins fed on young-of-the-year herring passing the islands. Once the herring stock collapsed there were virtually no alternate prey sources and from 1969 to 1987, when herring stock biomass was low, less than 50% of chicks fledged in all but 3 years and in most years there was complete breeding failure. Chicks often lost mass and died in the nests only 10–20 days after hatching. The herring stock started to recover in 1988 after the fishery had been closed. Puffins showed an immediate response with four successful breeding seasons from 1989 to 1992.

Intensive fishing caused the collapse of the Norwegian herring stock, although recruitment fell because sea temperatures were abnormally low at that time. At first, the massive reduction in herring biomass reduced herring predation on larval and juvenile capelin and, coupled with lower sea temperatures, led to stronger capelin recruitment (Blindheim & Skjoldal, 1993). However, in the 1960s, abnormal amounts of polar water joined currents that circulate in the north Atlantic and produced a slug of cold and low-salinity water. The 'Great Slug' was carried south and east along the coast of Greenland and arrived off Newfoundland in 1971 and 1972 before going east across the Atlantic and turning north towards the Barents and Norwegian Seas (Dickson et al., 1988). Temperature and salinity fell to a minimum in 1978 and 1979, but when temperature rose again in the early 1980s, cod recruitment increased. Warmer conditions did not favour capelin and their recruitment failed in 1984 and 1985. This poor recruitment, coupled with the effects of intensive fishing, led to the collapse of the capelin stock (section 12.4.4). Murre populations on the Norwegian coast fed on capelin rather than herring and fell by 80% in 1985–87 (Anker-Nilssen et al., 1997). Moreover, since herring biomass was already very low, those species of birds that fed on pelagic fishes had few alternate prey and failed to reproduce or starved.

So, in the Barents Sea, the effects of fishing, coupled with environmental change, led to population declines of seabirds. These seabirds were particularly susceptible to the collapse of herring and capelin stocks because there were no abundant alternative food sources. As is often the case, ecosystems showing the strongest system-wide responses to fishing are those where a few species fulfil key functional roles.

Role of life histories

Responses of seabirds to changes in prey abundance are closely linked to characteristics of the birds' life history and feeding behaviour. Table 15.2 summarizes life-history data for a number of families that interact with fisheries. For all groups, adult survival and clutch sizes are low. Changes in population size tend to be gradual and follow changes in reproductive success rather than adult mortality. Mass adult mortality is generally rare, with the starvation of murres in the Barents Sea and 'guano' birds in highly variable upwelling ecosystems being notable exceptions.

The abundance of seabird prey varies in space and time (Chapter 4), and the slow life histories of many seabirds buffer their populations against occasional breeding failures. Seabirds are most susceptible to variations in food supply during the breeding season because the need to return to their offspring, at intervals of hours or days, reduces their foraging range. Moreover, chicks may need a smaller range of species and sizes of fish than adult birds. Since breeding birds forage in the vicinity of the colony, local fluctuations in prey recruitment and distribution can be major determinants of breeding success. Some birds omit breeding seasons altogether if adequate food is not available. This removes the constraint of having to forage near the colony and adults can feed wherever they find sufficient prey.

Changes in capelin abundance affect the behaviour of feeding seabirds. In Newfoundland, comparisons of capelin and seabird abundance showed capelin had to reach a threshold density for birds to feed effectively. Above the threshold, fluctuations in capelin density were compensated for by behavioural responses, but below the threshold density, no amount of foraging would compensate for the lack of prey. One analysis showed that competition for prey between birds, whales and cod was weak because their different feeding

Family	Adult weight (kg)	Age at first breeding (years)	Adult survival (% y⁻¹)	Clutch size (number)
Penguins	4–40	4–8	75–95	1–2
Albatrosses	2–12	7–13	92–97	1
Gannets/boobies	1–3.5	3–5	90–95	1–3
Cormorants	1–5	4–5	85–90	2–3
Auks	0.1–1.0	2–5	80–93	1–2

Table 15.2 Life histories of some birds that interact with fisheries. From Croxall (1987).

strategies led to spatial separation. The key factor determining seabird reproductive success was whether the total size of suitable prey stocks, as determined by fishing and environmental effects, exceeded threshold levels (Burger & Piatt, 1990; Piatt, 1997).

Seabirds switch prey rapidly if preferred prey are depleted and alternatives are available. A good example of prey switching is provided by the Cape gannet and African penguin which feed in the Benguela upwelling off south-west Africa. Anchovy stocks in the upwelling collapsed during the late 1980s and the Cape gannet *Morus capensis* and African penguin *Spheniscus demersus*, which had fed primarily on anchovies *Engraulis japonicus*, began to eat the increasingly abundant pilchard *Sardinops ocellatus* (Fig. 15.4).

Surface and nearshore foraging birds often show greater interannual variability in reproductive success than diving and offshore foraging birds that can seek alternative prey in deeper or offshore areas. Seabirds such as the *Larus* gulls, terns and kittiwake *Rissa tridactyla* cannot dive more than a few metres, and have to feed on fishes shoaling at the surface. Surface feeders often disperse widely in search of alternative food following the breeding season and are also major scavengers of fishery discards (Chapter 13). Conversely, seabirds such as the auks (guillemots, razorbills and puffins) swim and dive effectively. Piatt and Nettleship (1985) recorded diving depths of auks (Alcidae) off Newfoundland based on incidental capture in bottom set gill nets. These birds get trapped in nets because they dive to feed on capelin that form large dense schools in inshore areas. The maximum recorded depths to which murres, razorbills, puffins and guillemots would dive were 180 m, 120 m, 60 m and 50 m, respectively. Murres can swim at up to 2 m sec^{-1} while underwater and remain there for up to 4 min. As such, they can feed on a variety of prey and are less likely to be affected by the population collapse of one prey species.

The role of fishing

In our example from the Shetland Islands, there was little evidence for a fishing effect since prey availability was largely independent of fish population size. Conversely, in the Barents Sea, fishing contributed to the collapse of the herring and capelin stocks and was an indirect cause of mass seabird mortality.

Populations of short-lived pelagic forage fishes are highly variable in space and time, because total stock

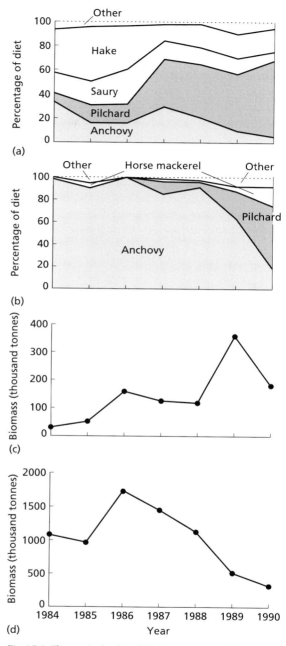

Fig. 15.4 Changes in the diet of (a) Cape gannets and (b) African penguins feeding in the Benguela upwelling ecosystem, and the abundance of (c) pilchard and (d) anchovy prey. After Adams *et al.* (1992).

biomass is not buffered against recruitment failure when only one or two age classes contribute to that biomass. Moreover, because the range of shoaling pelagic fishes is often linked to their abundance, their abundance at a given location changes dramatically over time (section 4.3). Much of the variance in abundance can be related to changes in the environment, for example during periods of low productivity associated with cessation of upwelling or as a result of mismatch between spawning time and larval food production (section 4.2.2). However, fishing increases the susceptibility of pelagic fish populations to environmental effects by reducing the age structure and increasing dependence on individual recruitment events. Thus, it can be an indirect cause of declines in seabird abundance.

Fig. 15.5 Birds feeding on fish discards from a trawler in the northern North Sea. Photograph copyright A.V. Hudson, by courtesy of R.W. Furness.

15.2.2 Benefits of discarding

Fishers routinely discard unwanted fish, invertebrates and offal at sea. Fishes may be discarded because they are more costly to sort than to sell, because fishers have exceeded a catch quota or because they are below a minimum landing size. In addition, fish offal is discarded as fishers gut and otherwise prepare catches for storage and sale (Chapter 13). Much discarded material floats on the sea surface, or sinks slowly, and is easily consumed by seabirds. Seabirds follow fishing boats by day and night in search of discards (Fig. 15.5).

Globally, around 27 million tonnes of bycatch are discarded every year; 25% of the current landings (Alverson *et al.*, 1994). The effects of discards on seabirds have been studied extensively in the North Sea where 475 000 tonnes of fish, offal and benthic inverte-

brates are discarded annually (Camphuysen *et al.*, 1993). Seabirds consume approximately 90% of offal, 80% of roundfish, 20% of flatfish and 10% of the invertebrates discarded. This level of food availability would maintain around 2.2 million seabirds; more than the total estimated population of scavenging seabirds in the North Sea. The effects of discarding are reflected in seabird population trends, with a 10-fold increase in the number of breeding seabirds from 1900 to 1990 (Furness, 1996). The main scavengers, such as fulmars and kittiwakes, have become particularly abundant (Fig. 15.6).

There is intense competition for offal and discards between scavenging birds (Hudson & Furness, 1988; Camphuysen *et al.*, 1993; Garthe *et al.*, 1996). Fulmars *Fulmarus glacialis* and gulls *Larus* spp. are the main consumers of offal in the northern and southern

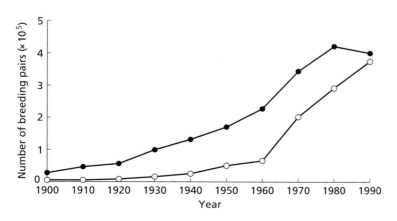

Fig. 15.6 Changes in the number of pairs of kittiwakes (open circles) and fulmars (closed circles) breeding on the North Sea coasts of the United Kingdom. After Furness (1992).

North Sea, respectively, and their feeding success is positively correlated with numerical dominance at fishing boats. Feeding success is also related to bird size and handling ability; thus, kittiwakes consume small fish while gannets take larger ones. The inability of smaller gulls to swallow whole large fish makes them vulnerable to kleptoparasitism by large gulls, great skuas and gannets (Furness, 1996). **Kleptoparasitism** is the deliberate stealing of food by one animal when that food has been captured by another. No seabirds are obligate kleptoparasites, but some Arctic skua (appropriately called Parasitic Jaegers in North America) are known to subsist from kleptoparasitism. Rates of kleptoparasitism depend on food availability, since kleptoparasitism allows birds to access resources exploited by other species (Furness, 1987).

Many seabird populations have grown as a result of feeding on discards, and a fishery practice regarded as wasteful and damaging to fish populations may be beneficial to seabirds. In some cases, plans to reduce discarding could lead to reductions in seabird populations (Furness *et al.*, 1992) and changes in mesh-size regulations which reduce the proportion of catch discarded should be introduced gradually to avoid adverse effects for competitively inferior seabirds (Camphuysen *et al.*, 1993). An interesting conflict between the costs and benefits of discarding has arisen in north-east Spain. The Audouin's Gull *Larus audouinii* is an endangered seabird, and two colonies hold 74% of the world population. The larger of these colonies, on the coast, holds 70%, and the smaller colony, on an island 600 km from the shore, holds 4%. When a trawling moratorium was imposed near the colonies, the larger colony shifted diet and maintained reproductive success by feeding at inland sites such as rice fields, coastal lagoons, dunes and beaches. Such feeding areas were not accessible to the smaller colony, and the birds failed to reproduce successfully. If development on the Spanish coast leads to the loss of inland feeding sites, the Audouin's gull could be threatened by discarding bans (Ore *et al.*, 1996).

15.2.3 Waders and shellfish

Sea ducks and wading birds feed on fished invertebrates such as mussels *Mytilus edulis* (Mytilidae) cockles *Cerastoderma edule* (Cardiidae) and trough shells *Spisula subtruncata* and *S. solida* (Mactridae). Thus, the overexploitation of shellfish may reduce the quantity of food available for birds such as common eider duck *Somateria mollissima* or Eurasian oystercatcher *Haematopus ostralegus*, and fishers may disturb feeding birds. Conversely, fishers have complained that birds eat large quantities of shellfish that could otherwise be harvested. Some of the most impressive and well-validated studies of interactions between fishers and other predators were conducted with wader populations, and these can provide quantitative predictions of fishing effects on seabird populations.

Many wading birds winter in the estuaries of northern Europe, and feed on mussel and cockle beds. These beds are also fished. The activities of fishers affect birds in two ways. First, fishers disturb birds and drive them away from areas where they would otherwise feed. This reduces feeding time and increases energy use because the birds have to fly. Disturbance also increases competition between birds as it forces them to forage in smaller areas. Second, fishers compete with birds for shellfish. The effects of different shellfishery management regimes on the mortality rates of oystercatchers were explored with a model that predicts population biology from individual behaviour (Goss-Custard, 1996; Stillman *et al.*, 1996).

The model was developed using data from the Exe estuary in southern England, where oystercatchers feed primarily on mussels. The model uses knowledge of foraging decisions, interactions between birds, and the energetic requirements of individual birds to predict the population level consequences of fishing on shellfish beds (Fig. 15.7). When fishers are on the mussel beds, oystercatchers must relocate to feed, and each bird has to decide between feeding in areas with many mussels, where many competitors are encountered, or feeding in areas with few mussels where few competitors are encountered. The decision made by each bird depends on the decisions made by other birds and competitively dominant birds tend to feed in areas where mussels are abundant, while subdominants feed elsewhere. The model accounts for individual decisions and for the metabolism (hence food requirements) of the birds, tidal cycles (which affect the availability of mussels), gut processing rates and the availability of other feeding grounds. Gut processing rates of overwintering oystercatchers on the Exe estuary are such that birds disturbed by fishing activities (and also by people walking themselves and their dogs) may not meet their energetic

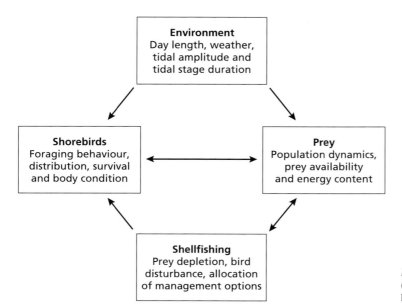

Fig. 15.7 The structure of the Stillman et al. (1996) model that describes interactions between wading birds and shell fisheries.

demands and switch to feeding on earthworms in fields surrounding the estuary (Stillman *et al.*, 1996).

The first run of the model used the known population of overwintering oystercatchers and the density of mussels on the beds. Every day, each bird would adjust its feeding strategy with respect to all other birds. The oystercatchers were required to meet their energetic demands on a daily basis, and if they could not do this by feeding in the estuary or surrounding fields they would die. A comparison of the results with empirical data (counts of the birds feeding on adjacent fields) indicated that the model provided a good description of population processes, and so it was run to investigate the effects of different mussel fishing strategies on the population. Three fishing strategies were compared, low tide thinning, low tide stripping and high tide stripping. Low tide thinning involved the removal of adult mussels at low tide and left bed size unchanged, low tide stripping involved the removal of all mussels for sorting after collection and could reduce bed area to zero, and high tide stripping is equivalent to low tide stripping, but bed area can only be reduced by 75% because all the beds are not accessible at high tide. With all fishing strategies, the fishers' collection rate depended upon the density of mussels, so this affects the time the fishers stay on the beds and disturb the birds. Of the three fishing methods considered, low tide strip-

ping led to the lowest adult mortality at low fishing effort while low tide thinning was better at the highest fishing effort (Stillman *et al.*, 1996).

The model can be used to select fishing strategies and alert managers to the impending starvation of overwintering oystercatchers. It showed that risk of starvation is correlated with the numbers of birds feeding upshore or leaving to feed on fields adjacent to the estuary (Fig. 15.8). When such changes in behaviour are observed, fisheries could be regulated to minimize their effects on bird populations. This form of adaptive management is already in use on wildfowl reserves where shooting is banned during cold weather because feeding birds must not be disturbed if they are to meet their energetic requirements.

The Stillman *et al.* (1996) model has also been parameterized for the Loughor Estuary (Burry Inlet) in south Wales where oystercatchers feed on cockle beds exploited by fishers. Within the Burry Inlet, oystercatchers did not feed in fields and, on average, they spent 40% of the day feeding as opposed to 85% in the Exe. These preliminary analyses suggest that fishing is likely to have more effects on overwintering oystercatcher populations on the Exe estuary (Fig. 15.9).

Stillman *et al.* (1996) conclude that shellfish stocks can be exploited without increasing wintering mortality of shorebirds, provided that shellfish are

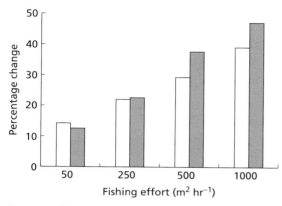

Fig. 15.8 Relationship between fishing effort and the percentage change in the number of Eurasian oystercatchers feeding in upshore areas (open bars) and field areas (shaded bars). Fishing effort is expressed as rate of low tide stripping. After Stillman *et al.* (1996).

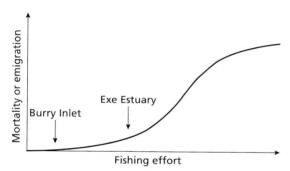

Fig. 15.9 Hypothetical relationship between fishing effort in shellfisheries and the mortality or emigration of shorebirds. Arrows indicate the status of the relationship between fishing effort and mortality or emigration in two British estuaries.

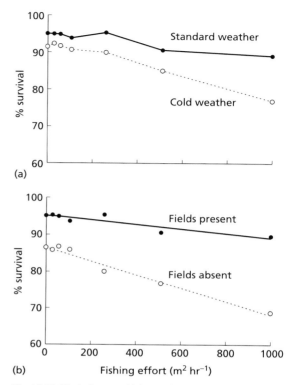

Fig. 15.10 The influence of fishing effort on the overwintering mortality of Eurasian oystercatchers (a) in the event of a 2-week spell of cold weather (0°C) at the end of January, and (b) in the event that no fields are available for supplementary feeding. After Stillman *et al.* (1996).

abundant and alternative food sources are present. The oystercatchers become more vulnerable during colder weather when energy demands are higher and supplementary food sources on land are inaccessible because they are frozen (Fig. 15.10). Cumulative effects of small increases in mortality over a period of years can have large effects on population size because shellfishing has disproportionately large effects on inexperienced and subdominant birds that would otherwise contribute to future generations (Fig. 15.11).

The wader and shellfish model provides an excellent example of how to predict the indirect effects of fishing and how to make management decisions that allow fish-

eries and shorebirds to coexist. Unfortunately, the data requirements are prodigious, and it is unlikely that many such models will be developed for other ecosystems.

15.3 Mammals

Marine mammals are widely distributed and may forage on species targeted by fishers or consume prey eaten by fished predators. Those groups that interact with fisheries are listed in Table 15.3. Marine mammals can feed on larger fish than birds, either by catching them in the open sea or by removing them from set nets. Particular controversy has arisen when mammals feed on adult salmon migrating through estuaries on their way to riverine spawning grounds. Human hunters always hate competition, whether the competitors are wolves, birds of prey or seals. Fishers are no exception

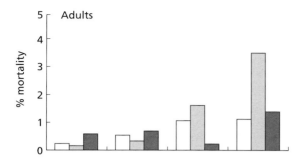

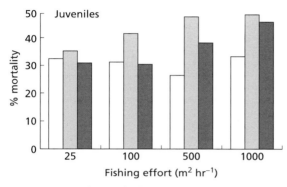

Fig. 15.11 The influence of different shell fishing strategies on the mortality of adult and juvenile Eurasian oystercatchers. Open bars, low tide thinning; light shade, low tide stripping; dark shade, high tide stripping. After Stillman *et al.* (1996).

and many people feel that marine mammals are consuming fish they would otherwise catch. There are many examples of fishers culling marine mammals in an attempt to boost fisheries yield (Earle, 1996).

When mammals are culled for food or fur, their reduced abundance may affect prey species. For example, reductions in whale populations may lead to changes in krill populations and changes in the abundance of sea otters *Enhydra lutris* have affected shellfisheries.

15.3.1 Competition between mammals and fisheries

Theory
In the simplest case, mammals and fishers compete for the same species, at the same stage of its life history, in the same location. We will describe this competition mathematically as it provides a basis for (i) estimating

changes in fishery yield when the abundance of mammals changes and (ii) estimating changes in the abundance of mammalian prey when fishing mortality changes (Beverton, 1985). We assume that instantaneous mortality rates are constant across the range of observed prey sizes, that recruitment can be expressed as a mean rate and that the parameters are not density dependent.

If instantaneous natural mortality due to predation by the marine mammal is M^* and that due to all other causes is M^1, then the total mortality Z is

$$Z = F + M^* + M^1 \qquad (15.1)$$

This breakdown of natural mortality is the same as presented in Chapter 8 for multispecies models. If the mean biomass of prey is $\bar{P}$, the amount caught by fishers is Y and the amount consumed annually by the marine mammal is C^* then

$$Y = \frac{F\bar{P}}{F + M^* + M^1} \qquad (15.2)$$

$$C^* = \frac{M^*\bar{P}}{F + M^* + M^1} \qquad (15.3)$$

and the ratio of the amount caught by the fishery to the amount consumed by mammals is

$$\frac{Y}{C^*} = \frac{F}{M^*} \qquad (15.4)$$

The quantity Y/C^* can be used to establish the relative impact, in terms of mortality, of fishing and predation on the prey species. We can also ask how the yield in a fishery that is currently producing yield Y_1 in the presence of a marine mammal that generates predation mortality M_1^*, will change if the size of the mammal population changes and generates predation mortality M_2^*. We denote the yield in numbers prior to the change as Y_1 and that after as Y_2. From equation 15.2, and given that $\bar{P} = \bar{R}\,\bar{W}$, where $\bar{R}$ = average recruitment and $\bar{W}$ = mean weight of fish:

$$Y_1 = \frac{F\bar{R}\,\bar{W}_1}{F + M^1 + M_2^*} \qquad (15.5)$$

$$Y_2 = \frac{F\bar{R}\,\bar{W}_2}{F + M^1 + M_2^*} \qquad (15.6)$$

Table 15.3 Mammals that interact with marine fisheries. Based on King (1964) and Jefferson *et al.* (1993).

Taxonomic group	Common name	Number of species
Order Mysticeti	**Whales**	
Family Balaenidae	Baleen/right whales	3
Family Neobalanidae	Pygmy right whale	1
Family Eschrichtidae	Grey whale	1
Family Balaenopteridae	Fin whales	6
Order Odontoceti	**Whales and dolphins**	
Family Ziphiidae	Beaked/toothed whales	20
Family Monodontidae	Beluga/narwhal	2
Family Physeteridae	Sperm whales	1
Family Kogiidae	Pygmy sperm whale	2
Family Phocoenidae	Porpoises	6
Family Delphinidae	Dolphins/pilot/killer	32
Order Pinnipedia	**Seals and sealions**	
Superfamily Phocoidea		
Family Phocidae	True seals (no ears)	17
Superfamily Otarioidea		
Family Otariidae		
subfamily Otariinae	Sealions	5
subfamily Arctocephalinae	Fur seals	8
Family Odobenidae	Walruses	1
Order Carnivora	**Otters/Martens**	
Family Mustelidae		
Subfamily Lutrinae		
*Enhydra lutris**	Sea otter	1

* Other species of otter may interact with fisheries, but this is the only species that consistently spends most of its life in the marine environment.

so

$$\frac{Y_2}{Y_1} = \frac{\overline{W}_2}{\overline{W}_1} \frac{F + M^1 + M^*}{F + M^1 + M_2^*} \qquad (15.7)$$

then the relative increase in yield in the absence of mammal predation ($M_2^* = 0$) is:

$$\frac{Y_2 - Y_1}{Y_1} = \frac{\overline{W}_2}{\overline{W}_1} \frac{F + M^1 + M^*}{F + M^1} - 1 \qquad (15.8)$$

To make a general comparison between yields in numbers we can assume $\overline{W}_1 = \overline{W}_2$, and equation 15.8 can be rearranged:

$$\frac{Y_2 - Y_1}{Y_1} = \frac{M^*}{F + M^1} \qquad (15.9)$$

Figure 15.12 shows how the ratio $(Y_2 - Y_1)/Y_1$ changes with $F + M^1$ for different values of M^*. The plot shows

that the expected enhancement of yield in the absence of mammal predation is relatively larger if predation were previously high. Figure 15.12 also gives some idea where given predator–prey relationships may lie and whether a fall in marine mammal populations would have any discernible fisheries benefit. Thus, a small mammal population consuming a fraction of the fishes caught by the fishery would fall at the base of the plot (where enhancement would be minimal) and a large marine mammal population consuming as much as the fishery would fall in the centre of the plot.

Although models of this type provide a basis for understanding fishery marine mammal interactions, the reality is rarely that simple. For example, reductions in the abundance of a marine mammal could decrease fisheries yield because the mammal usually feeds on a fish predator which feeds on the fished species. Examples of alternative relationships are given in Fig. 15.13. These include situations when the fish predator is

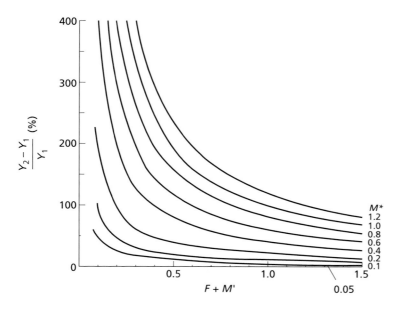

Fig. 15.12 Potential enhancemnent of yield ($(Y_2 - Y_1)/Y_1$) as a function of the total mortality from all causes other than marine mammal predation ($F + M^1$) for given levels of marine mammal predation (M^*). After Beverton (1985).

eaten by the marine mammals and fished (15.13b), the fishery exploits only prey species but the marine mammal feeds on both prey and fish predator (15.13c), the marine mammal feeds only on the fish predator and the fishery exploits only the prey (15.13d), and the the marine mammal feeds only on the prey and the fishery exploits only the predator (15.13e). In reality the situation is often a mix of these models, the balance shifting with the abundance, age structure and size of the marine mammals and prey species (Beverton, 1985; Yodzis, 1998). As a result, there are few empirical examples of marine mammals affecting the abundance of fished stocks and few cases where populations of fish-eating mammals were adversely affected by interaction with the fishery. An elegant example concerns proposals to cull Cape fur seals *Arctocephalus pustillus pustillus* in order to increase yields of fish which the seals eat (Yodzis, 1998). Yodzis analysed a complex food web using a population dynamics model that incorporated predation and nonpredatory interference. The analyses indicated that because of trophic interactions, a cull of fur seals was more likely to *reduce* the total yield from fish stocks than to increase it. This work provides a clear warning about making simplistic inferences from complex systems, and highlights the need for further research on multispecies management.

Empirical evidence for interactions

The collapse of the Norwegian capelin stock (section 15.2.1) not only had implications for seabirds, but also for grey seals *Phoca vitulina vitulina* (Hamre, 1991, 1994). When the capelin stock collapsed, herring biomass was already low and there were no abundant and alternative prey for seals (Dragesund & Gjosaeter, 1988). Some seals died of starvation, but others made an atypical migration out of the Barents Sea, presumably in search of food. During the course of this migration many seals died of starvation or were trapped in fishing nets (Haug *et al.*, 1991). Since fishing and environmental change contributed to the collapse of the herring and capelin stocks, fishing had indirect effects on the seal population.

The abundance of common porpoise *Phocoena phocoena* in the north-east Atlantic has declined in recent years. Since their distribution overlapped that of North Sea herring, it was thought that the demise of herring stocks was responsible. However, porpoises appear to feed on both demersal and pelagic fishes, and a number of gadoids increased in abundance as the herring declined. Moreover, in some areas, porpoise remained abundant following the collapse of herring stocks. Hutchinson (1996) concluded that the incidental catch of porpoises in nets was more likely to affect porpoise

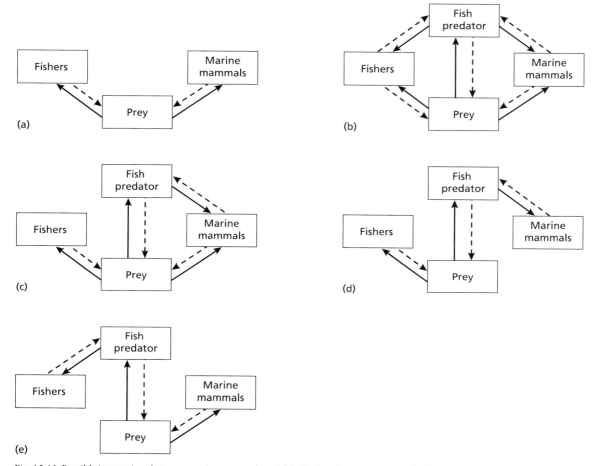

Fig. 15.13 Possible interactions between marine mammals and fish. Broken lines represent predation and continuous lines the flow of energy from prey to predator. Arrows indicate predation or capture. After Beverton (1985).

populations than any impact of fishing on their prey. The porpoises in the north-east Atlantic, in common with other marine mammal species that can switch diets and feed in ecosystems where the choice of prey is varied, are not dramatically affected by the overfishing of some prey species.

In general, only those marine mammal populations with a limited choice of prey (such as the seals feeding on capelin and herring in the Barents Sea) will be affected by fluctuations in prey abundance. If fishing causes the declines in prey populations then it is indirectly responsible for the reduced abundance of marine mammals.

15.3.2 Prey release

Reductions in the abundance of marine mammals will reduce their total food consumption. Will this lead to the release of prey that can be eaten by other species or fished by humans?

Krill surplus hypothesis
Baleen whales were the main consumers of krill in Antarctic waters. When their populations were dramatically reduced by commercial whaling, they were expected to release krill production for their competitors. This was known as the 'krill surplus hypothesis'.

Fig. 15.14 Crabeater seals *Lobodon carcinophagus* on Antarctic pack ice. Photograph copyright D.K.A. Barnes.

Some support for the hypothesis was provided by increases in the abundance of chinstrap penguins *Pygoscelis antarctica* and southern fur seals *Arctocephalus gazella*. However, at the same time, other krill feeders with similar geographical range had become scarce. With hindsight it seems that the population trends of chinstrap penguins and fur seals were driven by climate change rather than krill availability. Both these species favour ice-edge and open-water habitats, while the species that declined, the Adélie penguin *Pygoscelis adelie* and crab-eater seal *Lobodon carcinophagus* (Fig. 15.14) favour pack-ice habitats. The periods of low whale abundance have coincided with climatic changes that led to reductions in pack ice habitat (Fraser *et al.*, 1992).

Moreover, if whales and penguins formerly competed for krill they would have to have fed in the same areas. Large baleen whales are seasonally present in the Antarctic from November to May, but there is limited overlap between their feeding grounds and those of the penguins. Only the small minke whales *Balaenoptera acutorostrata* winter in the marginal ice zone where they could compete with penguins, and they have actually become more abundant (Fraser *et al.*, 1992).

Competition between krill feeders is likely if their distributions and feeding behaviour overlap in space and time. Krill production in the south-west Atlantic has been estimated as 28.6–97.6 million t y^{-1}, while seabird, seal, whale and fish consumption in the same area was estimated as 32.6 million t y^{-1} (Trathan *et al.*, 1995). However, many krill predators cannot compete

strongly because they feed in different areas at different times. This does not mean that there is no scope for density-dependent effects or competition, it is just that the effects are relatively minor in comparison with the oceanographic factors that determine krill abundance on feeding grounds. Models that are based on a homogeneous view of food chain assume that krill are equally vulnerable to predators when they are not (Murphy, 1995).

Temporal and spatial variation in krill abundance drives the dynamics of krill consumers. This variation is due to recruitment fluctuations and the differential transport of krill by ocean currents. On the Antarctic Peninsula, for example, krill recruitment is highest following seasons of extensive and prolonged sea-ice cover (Siegel & Loeb, 1995). On South Georgia, krill abundance fluctuates by a factor of 20 between years, and in years of low abundance krill predators suffer catastrophic breeding failures (Croxall *et al.*, 1988; Brierley *et al.*, 1997; Murphy *et al.*, 1998).

Sea otters

Sea otters *Enhydra lutris* prey on benthic invertebrates that inhabit shallow temperate and boreal waters of the North Pacific. The history of the relationships between these otters and fisheries provides an excellent example of the ways in which fishers can alter relationships between marine mammals and their prey. Otters were hunted heavily during the eighteenth and nineteenth centuries, and local populations were rapidly extirpated. On coasts once used by the otters, a number of shell fisheries developed. Otters were increasingly scarce by the early 1900s, and in 1911 further hunting was banned. Following the ban, otter populations increased in abundance and range, and started to compete with shellfisheries for food. Otters mostly feed at depths of 60 m or less and eat many commercially exploited shellfish such as crabs *Cancer* spp., spiny lobster *Panulirus interruptus*, abalone *Haliotus* spp., sea urchins *Strongylocentrotus* spp. and various clams. In most cases, there was little evidence that consumption rates had significant effects on shellfisheries, but there were some notable exceptions. These included the valuable abalone species whose abundance consistently declined when sea otters reappeared in a given locality. However, with hindsight, it was probably the hunting of otters that reduced shellfish predation and allowed shellfisheries to develop in the first place (Estes & Van Blaricom, 1985).

Summary

• Birds and mammals may compete with fishers for prey, or be killed as bycatch during fishing activities.
• Seabird reproductive success depends on prey availability. Since fishing has contributed to the collapse of some prey stocks, it can cause seabird population declines.
• Many seabirds feed on pre-recruits, while fishers target older fish. The effects of fishing on pre-recruit abundance are determined by the spawner–recruit relationship.
• Scavenging seabirds feed on discards and their populations have grown as discarding has increased.
• Several bird populations are now dependent on discards and rapid reductions in discarding could threaten competitively inferior species.

• Sea ducks and wading birds feed on fished invertebrates. Fishers can disturb these birds and compete with them for prey, leading to increased overwintering mortality.
• In some estuaries, interactions between wading birds and fishers are well understood, and provide a basis for management that allows fishers and shorebirds to coexist.
• Marine mammals may compete with fishers for prey, but only those species with restricted diets are usually affected.
• If marine mammal populations are reduced by hunting they eat less. However, the proliferation of their prey may not benefit other predators that are separated from them in space or time.

Further reading

Hunt and Furness (1996) review interactions between seabirds and fisheries in northern Europe, and Beddington *et al.* (1985) contains several reviews that describe interactions between fishers and marine mammals.

16 A role for aquaculture?

16.1 Introduction

Present global yields from capture fisheries are not taken in a sustainable fashion and many stocks have become progressively overexploited or fished to collapse. Hence, there is an increasing need to look to alternative methods of obtaining protein from the sea. We are increasingly fishing down food webs and harvesting smaller, faster-growing and more productive species from lower trophic levels that are converted into animal feeds or oils rather than being used for human consumption (Chapter 12). As a result, many have indicated that aquaculture is a panacea that will make up the shortfall in food production from the sea. Could aquaculture even remove the need to fish wild stocks altogether? It was hoped that as the world's fleets are slowly decommissioned, some of the unemployed fishers could retrain as farmers of the fish they used to catch.

Despite this optimistic outlook, aquaculture has its own social and environmental problems. In some ways aquaculture is rather like electricity generation. While we are very good at producing electricity using a variety of fuels (mainly coal, gas, oil and uranium), in the past no one really considered the associated problems of pollution and environmental degradation associated with these activities (global warming, acid rain, open-caste mining, oil spills). Similarly, humans have invested considerable time and finance in the development of techniques to rear aquatic organisms with little consideration of the environmental damage that might arise from the outputs from such systems and the localized habitat degradation that might occur. Reducing the environmental impact of aquaculture practices has become one of the main research problems that needs to be addressed before further expansion can occur in the industry. The aim of this chapter is to describe the history of aquaculture, the scale of aquaculture production, the environmental problems of aquaculture and the potential for expansion of this industry.

16.2 Aquaculture past and present

Humans have farmed aquatic organisms, including fish, molluscs, crustaceans and plants, for several thousand years, although the earliest examples of cultivated species are almost certainly from fresh waters. Aquaculture provided a ready and reliable source of protein irrespective of the fluctuations that occurred naturally in capture fisheries (Box 16.1). Despite its long history, the term 'aquaculture' was only invented about 40 years ago and implies controlled farming or intervention that will enhance production (e.g. stocking or feeding). In 1974, it was estimated that 6 million tonnes of fin fish and shellfish were produced worldwide. At the time, this equated to 12% of global landings for direct human consumption and 4% of the world's animal protein supply (excluding milk), although the importance of this protein source was much greater for land-locked Asian countries (Allsopp, 1993; 1997). The World Bank and the Food and Agriculture Organization of the United Nations (FAO) set up an aquaculture working group to assess the potential of aquaculture to meet the growing world demand for protein. This group estimated that there was potential to increase fin fish production 10 fold (from 2 to 20 million tonnes) whereas it was estimated that the potential increase for capture fisheries was less than twofold (up to 90 million tonnes). This prophecy has become reality as aquaculture now contributes over 20% of global production (Fig. 16.1). There are major differences in the role of aquaculture in marine and inland environments, however. In inland environments, aquaculture production consistently exceeds capture production, while in marine environments it is typically around 10% of capture production (Table 16.1).

Box 16.1
Varro on Roman Aquaculture

Marcus Terentius Varro (1126–27 BC) was renowned in Roman society for his intelligence and knowledge. He worked as a historian, agronomist and poet, and wrote several hundred volumes during his life. Julius Caesar asked him to supervise an intended national library. In this extract, from his work *Rerum Rusticarum* (On Agriculture), he describes marine and freshwater aquaculture.

'There are two kinds of fish-ponds, the fresh and the salt. The one is open to common folk, and not unprofitable, where the Nymphs furnish the water for our domestic fish; the ponds of the nobility, however, filled with sea-water, for which only Neptune can furnish the fish as well as the water, appeal to the eye more than the purse, and exhaust the pouch of the owner rather than fill it. For in the first place they are built at great cost, in the second place they are stocked at great cost, and in the third place they are kept at great cost. Hirrus used to take in 12000 sesterces from the buildings around his fish-ponds but he spent all that income for the food which he gave his fish. No wonder; for I remember that he leant Caesar on one occasion 2000 lampreys by weight; and that on account of the great number of fish his villa sold for 4000000 sesterces. Our inland pond, which is for the common folk, is properly called 'sweet', and the other 'bitter'; for who of us is not content with one such pond? Who, on the other hand, who starts with one of the sea-water ponds doesn't go on to a row of them.'

From a translation by Hooper & Ash (1934)

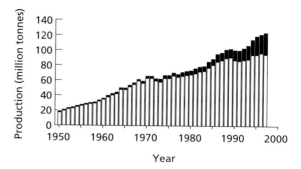

Fig. 16.1 Global production by capture fisheries (open bars) and aquaculture (shaded bars). After FAO (1999).

Table 16.1 Production (million tonnes) by capture fisheries and aquaculture in inland and marine environments. Data from FAO (1999).

Environment	1994	1995	1996	1997
Inland				
Aquaculture	12.11	13.86	15.61	17.13
Capture	6.91	7.38	7.55	7.70
% from aquaculture	63.7	65.3	67.4	69.0
Marine				
Aquaculture	8.67	10.42	10.78	11.14
Capture	85.77	85.62	87.07	86.03
% from aquaculture	9.2	10.8	11.0	11.5

As we saw in Chapter 2, the productivity of most fished species depends upon uncontrollable natural fluctuations in the ocean environment. In contrast, aquaculture production is more directly related to controlled inputs. Somewhat ironically, many aquaculture feeds are based on fishmeal, the production of which is unlikely to increase any further. For this reason, most aquaculture feed companies have been searching for viable plant-based feeds. This problem is currently most pressing for the cultivation of piscivorous species that have very particular protein and lipid requirements. While vegetable-based proteins and lipids can be added from plants such as soya beans, they are not composed of the essential amino acids and fatty acids required for fish growth.

In terrestrial systems, the development of intensive husbandry has concentrated mainly on four herbivorous mammals (cattle, pigs, sheep and goats) and four birds (chickens, turkeys, geese and ducks). This contrasts sharply with the development of husbandry techniques for hundreds of fish, mollusc, crustacean and latterly echinoderm species. Of these, the molluscs are most efficient in converting consumed food (phytoplankton) into body tissues. Furthermore, cultivated molluscs are filter feeders that require no additional input of feed. In contrast, fish and crustacea feed higher up the food chain and the overall energy conversion efficiency is relatively low (i.e. we often end up feeding fish to fish in aquaculture systems).

16.3 What is cultivated?

Approximately 300 species are cultivated worldwide, of which 20% are predatory species that yield 10% of all production. However, these species tend to command the highest unit value in the marketplace and account for 40% of the total market value. In contrast, herbivorous and omnivorous fishes accounted for approximately 90% by weight of world production in 1974, but continue to command a much lower unit price (FAO, 1993b). These species have been identified as the most suitable subjects to fulfil basic world food requirements because they are relatively cheap to produce and do not require sophisticated cultivation systems. To this day, fish form the largest component of world aquaculture production both in terms of tonnage and financial value (Fig. 16.2), largely due to their importance in fresh water. In marine and brackish water environments, molluscs and crustaceans are dominant. Species that dominated global aquaculture production by weight and value in 1997 are listed in Tables 16.2 and 16.3. The figures we discuss in this chapter exclude kelp *Laminaria japonica* production. These algae now dominate marine aquaculture production by volume. In 1997, kelp production was 4.17 million tonnes with a value of $US2.70 billion.

Fish production is strongly dominated by carp cultivation which reflects the influence of the Asian region. The other main species are tilapia and salmonids. By the early 1990s, salmonid production in the Atlantic

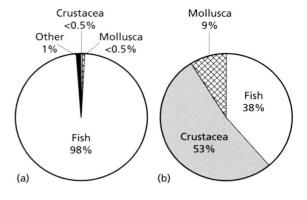

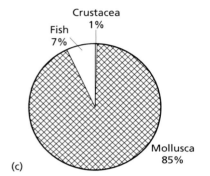

Fig. 16.2 Global aquaculture production by species groups in (a) freshwater (b) brackish and (c) marine environments. Data from FAO (1999).

and Pacific Oceans had exceeded the quantity of fish taken in salmonid capture fisheries (Fig. 16.3). Marine fish species account for only a small proportion of fish production. Farmed marine fishes are almost exclusively

Table 16.2 The species that contributed most to global aquaculture production by weight in 1997. Species are classified as marine if they are reared in marine or brackish environments for part of their life cycle. Data from FAO (1999).

Common name	Scientific name	Group	Environment	Production (million tonnes)
Pacific cupped oyster	*Crassostrea gigas*	Mollusc	Marine	2.92
Silver carp	*Hypophthalmichthys molitrix*	Fish	Freshwater	2.88
Grass carp	*Ctenopharyngodon idellus*	Fish	Freshwater	2.44
Common carp	*Cyprinus carpio*	Fish	Freshwater	1.99
Bighead carp	*Aristichthys nobilis*	Fish	Freshwater	1.41
Yesso scallop	*Pecten yessoensis*	Mollusc	Marine	1.27
Japanese carpet shell	*Ruditapes philippinarum*	Mollusc	Marine	1.12
Crucian carp	*Carassius carassius*	Fish	Freshwater	0.69

Table 16.3 The species that contributed most to global aquaculture production by value in 1997. Species are classified as marine if they are reared in marine or brackish environments for part of their life cycle. Data from FAO (1999).

Species	Common name	Group	Environment	Value ($US billion)
Giant tiger prawn	*Penaeus monodon*	Crustacea	Marine	3.93
Pacific cupped oyster	*Crassostrea gigas*	Mollusc	Marine	3.23
Silver carp	*Hypophthalmichthys molitrix*	Fish	Freshwater	2.79
Common carp	*Cyprinus carpio*	Fish	Freshwater	2.42
Grass carp	*Ctenopharyngodon idellus*	Fish	Freshwater	2.23
Atlantic salmon	*Salmo salar*	Fish	Marine	1.87
Yesso scallop	*Pecten yessoensis*	Mollusc	Marine	1.62
Japanese carpet shell	*Ruditapes philippinarum*	Mollusc	Marine	1.52

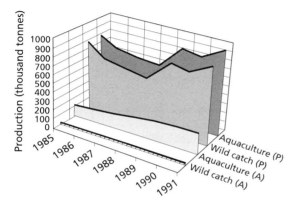

Fig. 16.3 World aquaculture production of Atlantic (A) and Pacific (P) salmon compared with landings by capture fisheries for the years 1985–91. After FAO (1993b).

carnivorous species that command a high market value. European Sea bass *Dicentrarchus labrax*, sea bream (Sparidae) and turbot *Scopthalmus maximus* are now commonly cultivated throughout southern Europe where warm waters stimulate high growth rates. In Southern Australia, southern bluefin tuna *Thunnus maccoyii* are captured at sea on rod and line and then towed to shore in cages behind the fishing vessel. The capture site can be hundreds of miles from the shore and the return voyage is a slow and tedious affair to ensure the fish are kept in pristine condition. Once they have returned to port, the fish are transferred underwater into holding pens for fattening up prior to slaughter. Sturgeon *Acipenser* spp. are now cultivated around the Caspian Sea for their caviar, which reduces the fishing pressure on dwindling wild stocks.

Nearly all cultivated molluscan species are bivalve molluscs such as mussels, oysters and scallops. The cultivation of crustacea is largely based on penaeid shrimp which are currently valued at around $US5 billion per year at first sale.

16.4 Production systems

The production of aquatic species can be split broadly into three techniques: extensive, semi-intensive and intensive systems. In **extensive systems** the organisms stocked into the water body rely almost entirely on naturally derived food with little or no human intervention. Brackish water lagoons that have high levels of primary productivity or the cultivation of molluscs by spreading collected juveniles on the seabed would be good examples of extensive cultivation systems. In **semi-intensive systems** productivity is enhanced with fertilizers or manures or organic waste. Supplemental feeding may occur and stocking density or species composition may be manipulated. **Intensive systems** are defined by the most extreme levels of human control. In intensive systems there will be high stocking density, extensive use of artificial feeds that can be supplemented with vitamins, essential elements, antibiotics, and close environmental control (Allsopp, 1997). Intensive production systems are energetically costly, and more energy is often required to produce protein than in marine capture fisheries (Box 16.2).

Two other important terms applied to aquaculture production systems are mono- and polyculture. **Monoculture** is typically undertaken in developed economies and is commonly used in the cultivation of salmonid, catfish and molluscan species. While highly productive, monoculture requires a large financial and time investment to ensure that the environmental and feeding conditions are closely controlled. Outbreaks of disease are the greatest threat to monoculture systems as the organisms are often much more vulnerable to infection due to their high stocking density and elevated stress levels. Monoculture systems are probably most common in intensive marine aquaculture systems. **Polyculture** traditionally has been practised with carp, tilapia or crustacea. In this case the aquatic organisms are cultivated in combination with plant or animal husbandry. Typically the byproducts of animals wastes increase the nutrient supply to enclosed systems, increasing primary productivity. Consequently the chosen aquatic organisms tend to be either herbivorous or consume prey low down the food chain (i.e. zooplankton or arthropod grazers).

16.5 Feeding constraints

Molluscs, cultivated in nutrient-rich coastal waters, are the most efficient organisms for aquaculture production as they feed on phytoplankton. The most cost-effective finfish for mass production are those that graze phytoplankton and macrophytes (e.g. tilapias and carps). Accordingly, these species tend to dominate nations with developing economies such as those on the Asian and African continents. Most marine species tend to be carnivorous (fish, crustacea and gastropods) and require a high protein content in their feed. These species are also the most valuable, but are also the most costly to produce. The conversion ratio of pelleted feeds can be as high as 2 : 1 feed weight/flesh weight. Nevertheless, even the best conversion ratio of carnivorous species are usually well below the production efficiency of herbivores.

Cultivation of high-valued carnivorous species depends upon the availability of high protein feeds. These feeds still depend on fishmeal or low value bycatch species for their protein component. Hence it is not surprising that 50% of the production costs for salmonids and Asian shrimp production are associated with feed requirements (New et al., 1993). Fish based products account for up to 64% of the cost of pelleted feeds in the salmon aquaculture industry (Prendergast, 1994). As a result research has been concentrated on finding plant-based substitutes for traditional sources of protein. Both soybean and canola (a type of rape-

seed) appear to be acceptable fish-based protein substitutes that yield comparable growth rates but are half as costly to produce. Cost savings are not the only consideration. The supply of plant-based products is far more reliable than fluctuating supplies of marine products, nutrient composition is more uniform and shelf life of these plant-based feeds is more stable. Another consideration is the introduction of pathogens through imports of feed. For example, in South Australia pilchards (Clupeidae) are fed to southern bluefin tuna kept in pens (section 16.4). Local supplies of pilchards could not meet the demand and growers imported frozen fish from South America. These fish carried a virus that remained unaffected by cold preservation, which then infected South Australian stocks of pilchard causing high mortalities (Thorpe et al., 1997).

The idea of catching fish to feed to fish (or terrestrial animals) seems ethically perverse and is energetically inefficient, especially when this food source would be acceptable to meet the protein shortages that exist in developing countries. However, such practices are likely to continue as long as they remain financially more rewarding than the use of alternative sources of protein.

16.6 Prospects for expansion

In general, the marine environment offers the greatest potential for the expansion of finfish and molluscan aquaculture. This is largely because of the wide range of potentially suitable species that are yet to be cultivated and the large extent of available cultivation sites. In contrast, the number of freshwater lakes, rivers, ponds and reservoirs suitable for aquaculture are severely limited as is the number of species that are appropriate for cultivation.

The productivity of both freshwater and marine aquaculture systems is limited by the rate of biotechnological advances. This is particularly relevant for those species that tend to have the highest unit value as these tend to be the most demanding in terms of husbandry or dietary requirements. Halibut Hippoglossus spp. require very cold water at elevated pressure for their eggs to hatch, while the formulation of an appropriate pelleted feed is currently a major problem for the cultivation of southern bluefin tuna in South Australian fish farms.

Aquaculture production has been improved by advances in a number of areas of research. Gonadotropic

Box 16.2

The energy costs of aquaculture.

The energy consumption of various fisheries (Table B16.2.1) is compared with the energy costs of selected fish-farming operations (Table B16.2.2). The energy costs are expressed as GJ t^{-1} (1GJ = 10^9 Joules) in either harvested (aquaculture) or landed (fisheries). These figures are based on data published in the early 1970s and

there is no doubt that they will have changed. Technological advances in the design of engines for fishing vessels mean that fuel consumption is now approximately 12.5% lower than that required to attain the same constant speed in the 1970s and 1980s (figures quoted by Wärtsilä NSD UK Ltd). However, vessels now have to travel a lot further to find sufficient fish, hence the gains in fuel savings are almost certainly outweighed by the extra energetic costs incurred while travelling.

Table B16.2.1 Energy consumption of various fisheries. Adapted from Pitcher and Hart (1982).

Species/taxon	Locality	Energy cost (GJ tonne^{-1})
Shrimp	Gulf of Mexico	359
All fish species	Adriatic	170
Demersal fish	North Sea trawl fleet	50
All fish species	UK average	35
Pilchards	Purse-seiner fleet Namibia	10
Herring, mackerel	Scottish purse seines	4

Table B16.2.2 Energy costs of selected fish-farming operations. Adapted from Pitcher and Hart (1982).

Species	Locality	Type of farm	Energy cost (GJ tonne^{-1})
Carps	Japan	Intensive recirculation	300
Carps	Czech Republic	Intensive ponds	54
Carps	Philippines	Extensive ponds	2
Tilapia	Israel	Intensive, aeration chambers	90
Tilapia	Thailand	Ponds, intensive	11
Tilapia	Congo	Ponds, extensive	0.2
Trout	UK	Intensive, recirculation	280
Trout	UK	Intensive ponds	55
Polyculture	Sri Lanka	Milkfish/prawns intensive	170
Polyculture	Israel	Carps/mullet/tilapia intensive	65
Polyculture	Thailand	Carps/tilapia extensive	0.13

extracts were initially used to induce breeding, but these have been replaced by refined synthetic hormones. Genetic selection, gene-transfer, gamete storage by cryopreservation, fertilization techniques, sex control, incubation and larval rearing procedures are all subjects of past improvement and current research. The cryopreservation of viable gametes has eliminated the constraints of species that have seasonal reproductive cycles and ensures a reliable supply of larvae year round or when conditions are most favourable. This is par-

ticularly important for markets in developed countries where retailers now expect a reliable and constant source of products throughout the year. Controlled sex differentiation has enabled the production of sexually sterile fish. This can be achieved by a number of techniques including the induction of triploidy in females that yield monosex stocks of females. Sexually sterile fish direct a greater proportion of their metabolized energy into somatic growth that gives a higher proportion of consumable tissues at market size. Genetic selection has been used to reduce head and fin size which further reduces waste in the final product. Artificial diets now include growth stimulants that are peptide or protein based and improve feed conversion rations. These diets are often dosed with antibiotics to reduce the risk of infection in intensive finfish and crustacean aquaculture systems. Finally, in common with terrestrial farmed systems, marine organisms are now the subject of trials using recombinant DNA methodologies to produce fish with altered body shape and size characteristics.

While the technological improvements listed above have improved yields and provided better control of the quality and consistency of organisms produced, many are also a cause for concern. Escapes of fish into the wild from cultivation sites have the potential to reduce the genetic diversity of wild stocks and large releases of hatchery individuals can competitively exclude wild fish. Dilution of the genetic diversity of a stock of fish has the potential to weaken that stock's resistance to environmental change and pathogens. This issue has been particularly prominent in the salmon industry of North America. The release of hatchery-reared salmon has been used to mitigate the effects of habitat loss as a result of dam building programmes. However, these fish compete with the existing wild fish for habitat and food. Hilborn (1992) argues that large-scale hatchery programmes for salmonids in the Pacific North-west pose the greatest single threat to the long-term maintenance of salmonid stocks. He further argues that any attempt to add hatchery-reared fish to existing healthy stocks of wild fish is ill advised.

While artificial feeds loaded with antibiotics reduce infection rates in intensive cultivation systems, these substances can affect microbial communities in the immediate vicinity leading to a reduction in microbial diversity and the possible development of drug resistance in potentially harmful species. The International Council for the Exploration of the Sea Working Group on the Environmental Interaction of Mariculture listed some of the concerns associated with the use of antimicrobials in fish feeds, these include:
- development of drug resistance in fish pathogens;
- spread of drug resistant plasmids to human pathogens;
- transfer of resistant pathogens from fish farming to humans;
- presence of antimicrobials in wild fish;
- impact of antimicrobials in the sediment beneath cages on the rates of microbial processes, bacterial population composition and the relative size of the resistant subpopulation of bacteria.

Among the many environmental concerns associated with aquaculture, the introduction of alien species is one of the most difficult to control. Introduced species may compete with native species for the same resources, and they may carry pests, predators and diseases to which native species are more vulnerable. Specific examples of species introductions linked with bivalve mariculture are dealt with in section 16.7.3, but Box 16.3 gives an idea of the range of organisms introduced to the North Sea through aquaculture practices.

16.6.1 Cage cultivation

Freshwater fishes tend to be cultivated in pond systems. The most commonly cultivated species, carp and tilapia, are highly tolerant of low oxygen levels, poor water quality and high stocking densities. In contrast, marine fishes require high quality water, high oxygen saturation and are often sensitive to high stocking densities. The cultivation of marine fishes tends to occur either in lagoons or on the open coast in cages. Lagoons are a relatively scarce resource and require careful management to ensure that they are not overstocked or polluted by the waste products of cultivated fish species and uneaten feed. Cultivation along the coastline requires the use of suspended cages. However, these are vulnerable to damage from wave and storm action and tend to be sited in sheltered bays, lochs and fjords. Such

---

Box 16.3

Introduced species

Table B16.3.1 shows species introduced into the North Sea region presumably through the transfer of organisms for the aquaculture industry. Pest organisms such as the American oyster drill created severe problems for the aquaculture industry in the UK (Reise *et al.*, 1999).

Continued.

Box 16.3 *(Continued)*

Table B16.3.1 Species introduced into the North Sea region.

Taxon	Origin	First record	Transport
Bacillariophyceae (diatoms)			
Thalassiosira punctigera	NP	1978	A/S
Thalassiosira tealata	NP	1950	A/S
Coscinodiscus wailesii	NP	1977	A/S
Phaeophyceae (brown algae)			
Sagassum muticum	NP	1960s	A
Undaria pinnatifida	NP	1986	A/S
Colpomenia peregrina	IP	1905	A
Corynophlaea umbellata	NP	1990	A
Rhodophyceae (red algae)			
Bonnemaisonia hamifera	NP	1890	A
Asparagopsis armata	SP	1950	A
Grateloupia doryphora	NP	1969	A
Grateloupia luxurians	IP	1947	A
? Agardiella subulata	?	1973	A
Dasya baillouviana	WA?	1950	A
? Dasysiphonia sp.	NP	1994	A
Anotrichium furcellatium	NP	1976	A
Polysiphonia senticulosa	NP	1993	A
Polysiphonia harveyi	NP	1908	A
Chlorophyceae (green algae)			
Codium fragile ssp. *atlanticum*	NP	1839	A
Codium fragile ssp. *tomentosoides*	NP	1900	A/S
Codium fragile ssp. *scandinavicum*	NP	1919	A/S
Ascetospora (parasitic protozoans)			
Bonamia ostreae	WA	1982	A
Marteilia refringens	EA	1970s	A
Haplosporidium armoricanum	EA	1970s	A
Hydrozoa (hydroids)			
Gonionemus vertens	NP	1913	A/S
Anthozoa (sea anemones)			
Diadumene cincta	?NP	1925	A/S
Bivalvia (lamellibranchs)			
Crassostrea gigas (incl. *C. angulata*)	NP	1964	A
Mercenaria mercenaria	WA	1864	A
Petricola pholadiformis	WA	1890	A
Gastropoda (snails)			
Crepidula fornicata	WA	1887	A
Urosalpinx cinerea	WA	1900	A
Polychaeta			
Clymenella torquata	WA	1936	A
Hydroides dianthus	WA	1970	A/S
Hydroides ezoensis	NP	1976	A/S
Crustacea			
Eusarstella (Sarsiella) zostericola	WA	1940	A
Mytilicola orientalis	NP	1992	A
Mytilicola ostreae	NP	1992	A
Nematoda (here: swimbladder nematode)			
Anguillicola crassus	NP	1982	A

Origin: EA Eastern Atlantic outside North Sea; WA Western Atlantic; NP northern Pacific; SP southern Pacific; IP Indo-Pacific. Transport: known or assumed to have ocurred with aquaculture (A) or ships (S).

sites are a finite resource and each bay, loch or fjord can only sustain a limited number of fish cages (Gowen & Bradbury, 1987; Gowen *et al.*, 1989). The rain of faeces and uneaten fish food that descends to the seabed beneath fish cages leads to locally anoxic conditions and degradation of the benthic community in the immediate area. Eutrophication may occur in confined waters with excessive inputs from fish farms and such cases have been linked to outbreaks of toxic algal blooms. Many countries now insist that fish cages are re-sited after several years to permit the affected seabed to recover through natural recolonization processes (ICES, 1996). As the suitable coastal sites for fish-cage cultivation have been used up, fish-cage technologists have developed automated cage systems that are moored further offshore and feed the fish automatically. These cage systems are automatically raised to the surface of the sea for servicing and harvesting. The advantage of these systems is that they are less prone to wave disturbance, do not present a hazard for vessels on passage and can be re-sited whenever required to reduce environmental degradation (Dahle, 1995).

16.6.2 Stock enhancement and ranching

The idea of increasing the stocks of wild fish by releasing large numbers of recently hatched fish into the wild is not new. Scientists in the United States, Norway and the United Kingdom were debating the efficacy of stock enhancement in the latter part of the nineteenth century. Scientists in the United States and Norway succeeded in hatching cod eggs and those of several other species. The methodology of hatching cod eggs was displayed at the 1883 Fisheries Exhibition in London followed quickly by the release of millions of yolk-sac larvae into Naragansett Bay, Rhode Island. Miraculously, young cod were caught the following year, and while the scientists made no claims as to the origin of these fish, fishers were quick to place their faith in the restocking as the solution to declining stocks of wild fish (Smith, 1994). Of course, the chances of any of the yolk-sac larvae surviving to be caught the following year were remote given the massed ranks of predators waiting to eat them. Although scientists were able to hatch fish larvae, they were unable to rear them beyond the non-feeding stage during which mortality is > 95% and this precluded further attempts at stock enhancement (Shelbourne, 1964). In

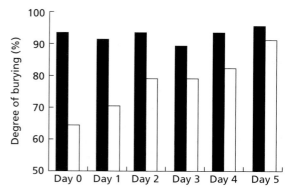

Fig. 16.4 The incidence of burying among experienced (shaded) and naïve sole following transfer to tanks containing a sandy substratum. After Howell (1994).

recent decades, advances in rearing techniques have considerably increased the range of fish that can be cultivated and has provided a new impetus for stock enhancement programmes worldwide. Examples of such species include red sea bream *Pagrus major* (Sparidae) and Japanese flounder *Paralichthys olivaceus* (Paralictithyidae, order Pleuronectiformes) in Japan, and Atlantic cod in Norway (Svånstad & Kristiansen, 1990; Watanabe & Nomura, 1990). The behaviour and physiology of reared juvenile fish is significantly different or inferior to equivalent aged fish in the wild and their survival rate is so low that stock enhancement is still not economically viable (Howell, 1994). Nevertheless, selective breeding and pre-release conditioning and exposure to mock predators are techniques that certainly improve post-release survival (Olla & Davis, 1989; Howell, 1994). Hatchery-reared fish that have never encountered a natural habitat, prey or predators will be naïve and hence may not behave appropriately in the natural environment. Survivorship can be improved however, by a period of learning prior to release (Fig. 16.4).

In order to assess the success of hatchery-release programmes it is necessary to have an effective and workable means of tracking, observing and identifying the released organisms at a later date. Failure to address this need meant that it was impossible to assess whether the 42 million fingerling red drum *Sciaenops ocellatus* (Sciaenidae) released into Texas bays and estuaries had a significant positive effect on the local populations. However, in Kaneohe Bay, Hawaii, hatchery-reared striped mullet *Mugil cephalus* (Mugilidae) accounted for 75% of the sampled fish population in nursery areas. The success of this programme was attributed to thorough pilot studies that identified the best release size, sites and season. The hatchery release of molluscs is complicated by considerations of the minimum density required to permit effective fertilization at spawning. Not only must the density at which the animals are released be considered, but also local hydrography that affects larval dispersal. Stocks of pink and green abalone *Haliotis corrugata* and *H. fulgens* (Haliotidae) increased dramatically after nearly 4500 mature adult animals were transplanted to the coastal waters of California, demonstrating that such releases can have a beneficial effect if properly planned (Bohnsack, 1996).

Reseeding and stock enhancement programmes are more likely to be financially viable if the species in question have a high financial or resource value. This is certainly the case with species such as pearl oysters and lobsters. Pearl oysters, as their name suggests, are harvested and cultured for the pearls that they produce. Two species are commercially significant; the black-lipped pearl oyster *Pinctada margaritifera* which produces black pearls, and the gold- or silver-lipped pearl oyster *P. maxima*. Although natural pearls are rare in these oysters, both produce large cultured pearls. Pearl oysters have suffered high levels of overfishing and a suite of management techniques have been applied to maintain a sustainable population. Among these, reseeding depleted areas with spat collected from the wild and from hatcheries has proved successful as long as the animals are permitted to reach maturity before they are havested (Sims, 1993). If the oysters are allowed to mature and spawn before they are harvested then culture may be unnecessary in the future.

Lobsters command a high unit value at market and their populations are easily overfished and hence lend themselves to stock enhancement programmes. Habitat suitability is a critical factor that determines the number of lobsters that can be supported in a given area of seabed. In particular, lobsters require shelter in the form of either burrows or crevices throughout their life history. One method of increasing the available habitat is to introduce artificial reefs into the environment (Jensen *et al.*, 1994). When released into a suitable habitat, juvenile lobsters *Homarus gammarus* have been shown to contribute to the commercial fishery (Bannister *et al.*, 1994). However, juvenile lobsters are costly to rear as they must be maintained in individual compartments to restrict their cannibalistic tendencies. As for hatchery-reared fish that are released into the wild, lobsters also need to be trained to avoid natural predators before they are released.

16.7 Case studies

So far we have talked in general terms about the contribution of aquaculture to the world production of marine products and discussed the benefits and less positive aspects. At this point in the chapter we look more specifically at two case studies: shrimp farming and bivalve mariculture. Both activities have brought both financial benefits and environmental problems to the areas in which they are undertaken.

16.7.1 Shrimp farming

World landings of shrimp (including prawns) from capture fisheries reached a plateau of approximately 2 million tonnes in the 1980s. Worldwide production of farmed shrimp rose from 84 000 to 891 000 tonnes between 1982 and 1994 (FAO, 1996). Despite the fact that shrimp farming has become a multibillion dollar business worldwide, the sustainibility of this sector of the aquaculture industry is currently affected badly by widespread production failures and associated negative environmental impacts that result from the farming practices. As a result, production has been unpredictable and varied between 789 000 and 891 000 tonnes from 1992 to 1994. Disease appears to be the major cause of recent production declines. Despite mitigating measures, the first sign of major problems occurred in 1988 when Taiwan's production dropped dramatically, shortly followed by a similar occurrence in China. Much of Asia's shrimp production is now seriously affected by rapidly spreading viral epidemics. To date, there are at least 20 identified shrimp viruses of which the white-spot virus in Asia and Taura syndrome virus in the Americas are the most threatening.

Disease outbreaks

Taiwan was among the earliest leaders in shrimp production that was based mainly on the output of small, family-owned intensive farms that produced 115 000 tonnes of *Penaeus monodon* (superfamily Penaeoidea) in 1987. However, a combination of industrial pollution, bacterial and viral diseases and the recirculation of

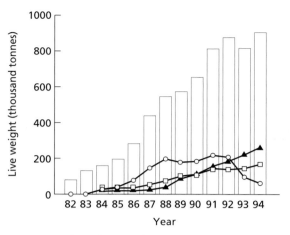

Fig. 16.5 World production of farmed penaeid shrimp (open bars) for the years 1982–94, showing the rise and decline in the production of shrimp from China. Key: China (circles); Indonesia (squares); Thailand (triangles). After Chamberlain (1997).

pond effluents between farms led to a 66% decrease in production in 1988. Despite attempts to revive the industry by cultivating alternative *Penaeus* species, shrimp production collapsed to 25 000 tonnes in 1994 and most Taiwanese shrimp farmers have now moved into marine fish cultivation (Rosenberry, 1994). This collapse in the Taiwanese shrimp industry left a gap in the world market that was quickly filled by production from China which rose at a rate of 80% per year and reached 199 000 tonnes in 1988. However, as in Taiwan, production became erratic and dropped to 88 000 tonnes in 1993 (Fig. 16.5). This decline was attributed to deterioration in water quality linked to industrial, agricultural and domestic pollution, and organic pollution from adjacent shrimp farms. The situation was further aggravated by outbreaks of disease and the occurrence of red tides. While the latter and pollution were at first blamed for increased mortality rates, it was later discovered that white-spot virus was the main cause of mortality, causing 100% death 2–3 days after infection (Huang *et al.*, 1995). This virus spread rapidly from China to other Asian countries causing mass mortality in a wide range of shrimp species (Nakano *et al.*, 1994).

Although Thailand has a long history of shrimp cultivation, it was not until 1985 that intensive cultivation

techniques were introduced from Taiwan. Most of these shrimp farms were small family owned and operated outfits that occupied less than 2 ha. The shrimp are cultivated in ponds which exchange water with the adjacent estuary or ocean to maintain water quality. However, the rapid spread of shrimp farms lead to a situation whereby wastewater from adjacent farms was recirculated among neighbouring shrimp operations. The resulting decrease in water quality leads to elevated levels of stress for the shrimp and causes a decline in growth rate, food conversion ratio and an increase in disease transmission. These problems were exemplified by the development of over 5000 ha of intensive ponds in the northern Gulf of Thailand. This is an excellent example of poor site location. The Gulf receives the runoff from four river systems that create extensive mudflat systems that extend for up to 15 km at low water. As the tide inundates these mudflats, sediment is resuspended and was transported into the shrimp ponds. The extreme sediment load lowered water quality to such an extent that shrimp farming was abandoned in this area (Chamberlain, 1997).

As in China, shrimp farming in Thailand was also affected by outbreaks of disease. Yellow-head disease caused total mortality within 5 days in some farms. The pathogen responsible for causing this disease was identified as a cytoplasmic virus that is carried by brackish water shrimp *Palaemon styliferus* and *Acetes* sp. (Wongteerasupaya *et al.*, 1995). However, this virus is only viable for 72 h outside the host in sea water. The Thai Department of Fisheries were able to instigate a management programme as a result of these studies. Shrimp farmers were instructed to erect filters on their seawater intakes to prevent the entrance of the host shrimp into their ponds. The use of trash fish feed that often contains the host shrimps was prohibited. Incoming water had to be held in reservoirs for 72 h prior to use and in the event of infection occurring, neighbouring farms had to be warned to prevent them pumping contaminated water for a period of 72 h. These measures contributed to a decline in this particular disease problem although this may have coincided also with changes in host susceptibility or the viral pathogen. White-spot disease has also occurred in Thailand and was particularly prevalent from 1994 to 1996. Measures to control this disease have included viral screening at water intakes, disinfection of pond water with chlorine or formalin and pumping of waste-

Fig. 16.6 Fishers catching shrimp larvae in push nets on the coast of Ecuador. The wild larvae are sold to coastal shrimp farms where they are grown-on to a marketable size. Photograph copyright FAO.

water offshore and pumping of clean oceanic water ashore. Nevertheless, despite these expensive efforts to limit the effect of the disease, production of shrimp in Thailand dropped by 25–40% in 1996 although the long-term impact of white-spot virus is unclear (Chamberlain, 1994, 1997).

Similar problems have occurred in the shrimp cultivation industry in the western hemisphere. Ecuador is the largest producer of shrimp in this region and peaked at 113 000 tonnes in 1992. In Ecuador, wild shrimp larvae are used to stock shrimp farms (Fig. 16.6). These farms obtain their water supply from coasts and estuaries. However, extended drought caused increases in salinity and nutrient concentrations in the Guayas River estuary, which supported a large number of farms, and these environmental conditions favoured the proliferation of *Vibrio* spp. that then infected the shrimp causing them to swim erratically at the surface of the ponds. This shrimp behaviour attracted large numbers of gulls that fed on the disorientated shrimp—a phenomenon that gave its name to the syndrome, *Gaviota*, which is the Spanish word for seagull. These outbreaks of disease were treated with antibiotics or by encouraging the growth of harmless bacteria that outcompeted the harmful *Vibrio* spp. but had mixed results. Fortunately, El Niño rains lowered the salinity in the affected estuaries and diluted nutrient concentrations and eliminated the disease in 1993 (Chamberlain, 1997). Other areas in Ecuador were affected by Taura syndrome virus which caused 80–90% mortalities in some pond systems (Chamberlain,

1994). Nevertheless, despite the continuing presence of the Taura syndrome virus, high market prices have continued to drive Ecuador's shrimp production to a new peak of 115 000 tonnes (Rosenberry, 1996).

Environmental issues

Water exchange between shrimp farms and their surrounding environment is a standard practice to avoid excessive build-up of waste products, avoid excessive eutrophication within ponds, and to maintain healthy planktonic blooms. Extensive cultivation systems require a daily water exchange rate of up to 5% whereas intensive systems require an exchange rate of up to 30% (Clifford, 1985). Calculations of the nutrient budgets for such systems indicated that more than 76% of the nitrogen and more than 87% of the phosphorous input is retained within the pond water and sediments (Robertson & Philips, 1995). These nutrients are discharged into the surrounding ecosystem whenever the pond water is exchanged, when harvesting occurs or when the ponds are dredged out. However, the receiving body of water will have a limited capacity to assimilate concentrated and persistent pulses of highly nutrient-enriched waters and sediments. At some point, the critical load of nutrients will be attained beyond which the water quality will begin to deteriorate with negative consequences for the local flora and fauna. Mangroves are an obvious sink for such nutrient outputs and Robertson and Philips (1995) calculated that approximately 3 ha of mangrove was required to assimilate the nutrient load generated by a 1-ha semi-intensive shrimp farm. This rose to 22 ha of mangrove for 1 ha of intensive shrimp pond. While the outputs from isolated shrimp farms do not tend to critical levels, a successful farming operation often attracts the development of other farms in close proximity. As the number of farms in a restricted area increases so the water quality begins to deteriorate as the concentration of nutrient discharges increases. This inevitably leads to elevated stress levels in the shrimp and increased susceptibility to infection by pathogens. Proactive management can help to avoid these situations. For example in Canada the capacity of proposed salmon farm sites to assimilate nutrients is examined prior to consent for planning, while within enclosed bodies of water, salmon farms are required to be spaced at least 3 km apart to minimize environmental impacts (Black & Truscott, 1994).

Mangrove destruction

Mangroves are an important habitat, acting as nursery areas for estuarine fishes and invertebrates, and providing feeding and breeding grounds for birds and mammals. They provide protection against storm activity and prevent soil erosion and provide an important source of income for poor coastal communities (Bailey, 1988). In the initial stages of the shrimp industry, mangroves were considered to be areas of low commercial value that were ideal for development for shrimp farming. Furthermore, mangroves are the natural habitat for many of the cultivated shrimp species, hence they seemed to be the ideal location for shrimp farms (Fegan, 1996). However, mangroves are actually very poor sites for shrimp farms as their acid sulphate soils have acidity of pH 3–4 when dried out.

On a global scale, shrimp farming has been responsible for less than 5% of mangrove destruction to date. However, on a localized scale the impact of shrimp farming may be far more severe (Phillips *et al.*, 1993). Aquaculture pond construction projects have destroyed 20% of the mangrove forests in some areas of Ecuador (Snedaker *et al.*, 1986) while in Indonesia the majority of the 300 000 ha of mangrove forest cleared to date is currently used for shrimp cultivation.

Most shrimp-producing countries now recognize the ecological importance of mangroves and have legislation to protect them from development. However, as with the enforcement of fisheries regulations, many of these countries do not have the resources to monitor and police those regulations (Bailey, 1988). While there may be pressure on governments to turn a blind eye to small-scale shrimp farms within areas of protected mangrove, there is increasing pressure from environmental lobby groups in developed countries that advocate banning imports of shrimp farmed at the expense of mangrove forests (Woodhouse, 1996). Nevertheless, the high revenues earned from shrimp cultivation are likely to maintain the pressure for development of some areas of mangroves. Recent attempts to model the social and biological benefits of mangrove conservation indicated that only 12% of the mangrove outside Thailand's mangrove conservation areas should be given over to shrimp farming and that 61% of these mangroves should be maintained in their natural state (Pongthanapanich, 1996).

Shrimp farming has a number of other impacts on the environment such as the intrusion of saltwater into neighbouring agricultural areas and pollution of groundwater supplies. Many of the chemical treatments used to control outbreaks of disease are occasionally used in excess and pollute surrounding waters. Alternatively, they may build up within sediments to high concentrations (e.g. copper compounds). The collection of post-larvae and reproductive adults from the wild for the aquaculture industry supports a large artisanal fishery but also threatens wild populations of shrimp (Fig. 16.6). As virtually all shrimp cultivation is supported by the collection of wild juveniles, excessive collection of shrimp may also have effects on other biota as shrimp are major predators of larval fishes and benthos in coastal waters.

16.7.2 Bivalve mariculture

Bivalve mariculture is mainly restricted to coastal areas from the shoreline to depths of no more than approximately 20 m. The choice of locations for cultivation sites will include considerations of water quality and contaminant load, the occurrence of disease organisms and conflicts with other coastal users (Laing & Spencer, 1997). The cultivation of bivalves can be a method of alleviating adverse environmental impacts arising from other activities in the coastal zone. As we have already seen, intensive fish farming practices can have undesirable environmental impacts, particularly as the effluents are highly nutrient enriched, promoting the development of algal blooms, some of which are toxic.

It has been proposed that integrated fish and bivalve mariculture systems can ameliorate this effect, as the bivalves reduce algal densities and nutrients which are effectively removed when the bivalve product is harvested (Folke & Kautsky, 1989; Shpigel *et al.*, 1993). Mussels are extremely effective at removing excess particulate-bound nutrients from eutrophic systems (Haamer, 1996), and have been used for the restoration of enclosed water masses that are heavily polluted (Russell *et al.*, 1983). While these mussels are unfit for human consumption because of their assimilation of heavy metals and other persistent pollutants, they represent a cost-effective and self-perpetuating means of maintaining water quality.

One of the biggest attractions of cultivating bivalve species is that they require no inputs of food by the farmer as they feed on phytoplankton that occur naturally. Despite these obvious advantages, there has been an increase in the awareness of the environmental effects that may result from the various stages of bivalve cultivation processes. Most notably, adverse effects have been associated with mussel and oyster farms in Spain and France located at sites where hydrographical conditions are unsuitable for high-density cultivation (Tenore *et al.*, 1985; Castel *et al.*, 1989).

The commercial cultivation of bivalves involves three distinct processes: seed collection, seed nursery and on-growing, and harvesting. Each stage of the cultivation process has the potential to cause environmental changes, e.g. through overexploitation of seed, exceeding local carrying capacity, or the damaging effects of harvesting. We now go on to examine the various environmental problems associated with each stage of cultivation.

Seed production

There are three main methods of producing or collecting young bivalves for on-growing: spat collection, mechanical harvesting and hatchery production.

Spat (young bivalve molluscs at the time when they first settle from the plankton) can be obtained from the wild with the use of collectors that are made from a wide variety of natural and man-made materials. These are generally laid out on the shore or suspended in the water from ropes. There are few environmental effects of these collectors apart from the removal of the spat of both target and non-target species. The use of broken shell material, otherwise known as **cultch**, to encourage

spatfalls often entails modifying the habitat by laying this material on muddy or sandy shores.

While the adults of many bivalve species are harvested subtidally with dredges (e.g. clams *Arctica islandica*, *Mercenaria mercenaria*, *Ensis* spp. and scallops *Pecten maximus*, *Aequipecten opercularis*), we know of only one species, the mussel *Mytilus edulis*, which is dredged as **seed** (small juveniles) for relaying and on-growing. As we have already seen in Chapter 14 an increasing number of experimental studies have examined the ecological effects of disturbance created by bivalve dredging activities. Whereas scallops and clams are found in relatively low densities ($< 1 \text{ m}^{-2}$) seed mussels form dense aggregations in discrete areas of the seabed (Dankers & Zuidema, 1995). As a result, the extent of physical disturbance caused while dredging for mussel seed is confined to relatively small areas of the seabed, and the benefits of harvesting a resource that is lost regularly as a result of natural perturbations probably outweigh the limited negative environmental effects that occur. However, in extensive mussel fisheries the depletion of the seed mussel stocks has the greatest effect on trophic interactions within this system. Dankers and de Vlas (1992) estimated that from 1984 to 1990 between 30 and 130×10^6 kg of mussel seed were harvested from an estimated subtidal stock of 165×10^6 kg in the Dutch Wadden Sea. One consequence of the heavy exploitation of these seedbeds is that they are not permitted to develop into mature adult mussel beds. In the early 1990s the entire intertidal mussel stock was removed from the Wadden Sea by commercial dredgers. This resulted in increased mortalities of common eider duck and reduced breeding success for Eurasian oystercatchers that depend on mussels as their main food source. Dankers and Zuidema (1995) suggested that these ecological effects would persist for many years until new mussel beds develop and mature, indeed there is no guarantee that mussels will resettle in the future.

Bivalve seed are often produced in hatcheries from broodstocks and then sold on to farmers. Hatcheries offer the advantage of environmental control such that spawning can be induced by increasing food supply, and altering temperature or day length. As in finfish hatcheries, antibiotics are used in some bivalve hatcheries. To date there is little information on the environmental effects of using antimicrobial products in bivalve hatcheries. Drug resistance in fish pathogens is now well established as are the effects of antibiotics on microfaunal communities beneath fish cages (ICES, 1996). However, only a few bivalve hatcheries supply a much larger on-growing industry (e.g. two hatcheries supply the entire United Kingdom seed requirements), hence any effects are likely to be highly localized. As in fish culture, efforts have been made to produce sterile spat that will have better somatic growth, but these attempts are still in their infancy.

Introductions of alien species

Many bivalve species are exotic to the locations in which they are grown, e.g. Asian Manila clams *Tapes philippinarum* and Pacific oysters *Crassostrea gigas* are grown in Europe. Imports of adult and bivalve seed have been the cause of introductions of alien marine organisms to countries where formally they did not occur (Reise *et al.*, 1999; Box 16.3). These include competitors of bivalves, such as the slipper limpet *Crepidula fornicata*. Slipper limpets can reach high biomasses and compete with bivalves and other biota for food and space. Other introductions include bivalve predators, such as the American whelk tingle *Urosalpinx cinerea* and diseases, such as *Bonamia*, which infects the blood cells of flat oysters, causing high mortalities under conditions of intensive cultivation. Introductions of alien organisms can be limited by the use of quarantine arrangements, but their use varies from country to country.

Many countries also have additional national legislation to control the introduction of exotic bivalve species for cultivation. In the United Kingdom, for example, release of exotic species into the wild is only permissible by licence under the Wildlife and Countryside Act (1981). The International Council for the Exploration of the Sea (ICES) has produced a Code of Practice entitled 'The Introductions and Transfers of Marine Organisms 1994'. This is designed to prevent inadvertent co-introductions of harmful organisms associated with the target species, the ecological and environmental impacts of introduced and transferred species, and the genetic impact of these species on indigenous stocks. While the genetic impacts of escapee salmon are well recognized, the problems associated with transferring bivalve stocks from one area to another are yet to be addressed.

Introductions of algae, including toxic dinoflagellates, blooms of which can have a significant impact on commercial bivalve mollusc culture, have generally been attributed to the transportation of resting cysts in ships' ballast water (Hallegraeff & Bolch, 1991). However, normal trading, involving transport of shellfish stocks from one area to another followed by relaying or storage in open basins, can provide another mechanism of transfer. The faeces and digestive tracts of bivalves can be packed with viable dinoflagellate cells or can contain resting cysts. In the Netherlands, recirculating storage systems are used to quarantine mussels and oysters as a precaution against such introductions (Dijkema, 1995). Invasive alien seaweeds, including *Sargassum muticum*, *Undaria pinnatifida* and *Laminaria japonica* are also thought to have been introduced into European waters through transport of the sporophyte stage in juvenile oysters, or as small plants attached to bivalve shells (Rueness, 1989).

On-growing

Unless bivalves are cultivated directly on the substratum or ranched within confined bodies of water, it is often necessary to grow them on structures that are introduced into the marine environment. The introduction of such structures has an immediate effect on local hydrography and provides a new substratum upon which epibiota can settle and grow. In addition, the introduction of high densities of cultivated organisms increases local oxygen demand and elevates the input of organic matter into the immediate environment. Where there is a high density of bivalve stock the larval settlement of other benthic species may be reduced. It has been shown that the larvae of other benthic species are filtered and digested by adult bivalves. Most of these larvae either pass through the digestive system alive or are rejected by the adults, but they then become bound in the faeces or pseudofaeces (Baldwin *et al.*, 1995).

Despite the ecological effects associated with bivalve cultivation, it should be remembered that natural beds of these animals were once extensive and played a vital part in the functioning of the local ecosystem where they occurred, processing excess phytoplankton and cycling organic material. Oysters in Chesapeake Bay, on the north-eastern coast of the USA, were historically so abundant that they filtered most of the Bay's water once per week, whereas in recent decades their numbers have fallen to such low levels that the remaining oysters would take nearly a year to filter the same amount of water (Newell, 1988). Overharvesting, disease and to

Fig. 16.7 Oyster trestles close to the low water mark. Note the heavy fouling by green algae on the net bags that contain the oysters. Algal growths such as these further reduce water flow around the trestles and increase the input of organic matter into the surrounding sediments as the algae die and become detached from the netting. Photograph copyright M.J. Kaiser.

some extent pollution have often almost destroyed these native populations, and the introduction of commercially cultivated bivalves may help to restore some ecosystems (Dame *et al.*, 1989; Gottlieb & Schweighofer, 1996).

Intertidal cultivation

In Arcachon Bay, France, 10 km^2 of the lower intertidal zone is occupied by beds of cultivated areas, which constitutes approximately 7% of the total intertidal area (Castel *et al.*, 1989). Castel *et al.* (1989) found that the presence of densely stocked oyster beds elevated organic carbon levels in the adjacent sediments. This elevated oxygen demand and produced anoxic conditions and a decrease in macrofaunal abundance. Nugues *et al.* (1996) examined environmental changes associated with mariculture at a relatively small oyster farm in the River Exe, England. They found that water currents were significantly reduced in close proximity to oyster trestles which consequently doubled the sedimentation rate and increased the organic content of the underlying sediments, and led to a reduction in the depth of the oxygenated layer of sediment. Nevertheless, the changes observed in the benthic fauna were restricted to the area immediately beneath the trestles. Hence, at low stocking densities, the effects of oyster cultivation are relatively benign and highly localized (Fig. 16.7).

In some areas of Europe and North Amercia, oysters are cultivated directly on the ground on a variety of substrata including mud, sand and gravel (Simenstad & Fresh, 1995). In North America, preparation of ground cultivation plots can involve severe levels of disturbance. In some areas, such as Willapa Bay, Washington, the insecticide carbaryl is sprayed on intertidal areas to kill populations of burrowing shrimp (*Neotrypaea californiensis* and *Upogebia pugettensis*) which destabilize the sediment and smother oysters by their burrowing activities. Spraying is carried out every 6 years and is strictly regulated due to its controversial nature. In addition to the application of chemicals, oyster grounds are harrowed to level the ground prior to cultivation, and raked and dredged to distribute and thin oysters during on-growing (Simenstad & Fresh, 1995). As we have seen in Chapter 14, the removal of burrowing fauna and physical disturbance are likely to induce habitat and community changes similar to those attributed to dredge harvesting techniques.

In North America, many native clam species are cultivated including Manila clam, *Tapes philippinarum* and hard-shelled clam *Mercenaria mercenaria*. Cultivation usually involves some form of habitat modification in the form of adding gravel or gravel and crushed shell to the substratum, placing protective plastic netting over seed clams, or laying the clams directly onto the sediment in net bags called poches. Not surprisingly, such habitat modifications lead to alterations in the local environment and, consequently, faunal composition. Simenstad and Fresh (1995) reported that the application of gravel to intertidal sediments resulted in a shift from a polychaete to a bivalve and nemertean dominated community, but emphasized that changes are likely to be site-specific. Such shifts in community composition could have repercussions at other trophic levels, e.g. changes in the abundance of certain harpacticoid copepod populations which are important prey for juvenile salmon and flatfish species.

Suspended raft cultivation

Superficially, suspended raft culture of bivalves has little visual impact on the landscape. However, the large biomass of cultivated and fouling organisms suspended beneath rafts and buoys has a major effect on phytoplanktonic, benthic and hydrographic processes in close proximity to the cultivation site. Mussels provide a complex surface area on which dense epifaunal communities consisting of over 100 species can develop (Tenore & Gonzalez, 1976). Small portunid crabs *Pisidia longicornis* were found to be abundant among fallen mussels beneath rafts in the Spanish rias. These, in turn, were fed upon opportunisitically by several fish species that normally consume polychaete worms (Lopez-Jamar *et al.*, 1984). *P. longicornis* are so abundant in areas of mussel cultivation that their larvae dominate the zooplankton community that is normally characterized by copepods (Alvarez-Ossorio, 1977). Mussels excrete high levels of ammonia which promotes high levels of productivity in algae attached to mussel lines which is equivalent to algal production in local intertidal systems (Lapointe *et al.*, 1981). So great is the productivity associated with mussel lines in the Spanish rias, that Tenore *et al.* (1982) speculated that inshore fisheries were potentially enhanced by the bedload transport of organic rich sediment into coastal areas.

The relatively beneficial effects that occur in the Spanish rias contrast sharply with the effects observed by Dahlbäck and Gunnarsson (1981) in Sweden. They demonstrated that organic sedimentation rates beneath mussel ropes were twice as much as those found in adjacent uncultivated areas. This excessive organic enrichment was associated with anoxic sediment and the growth of bacterial mats, *Beggiatoa* spp., beneath the mussel ropes. In this situation, the benthic infauna had low diversity and biomass which is a well documented response at polluted sites (Pearson & Rosenberg, 1978). Similarly, the productivity of densely stocked Japanese oyster grounds was detrimentally affected by the generation of large quantities of pseudofaeces and high filtration rates. Pseudofaeces production was so great beneath oyster cultivation rafts that it was at least equivalent to natural sources of sedimentation (Mariojouls & Kusuki, 1987).

Summary

- The contribution of aquaculture to protein production from the sea is increasingly important.
- Technological advances are improving product quality and the efficiency with which marine organisms are bred and grown. However, aquaculture can have environmental costs.
- The ethics of harvesting fish to make animal feed is clearly driven by economic considerations and not by the necessity to feed the human population.
- Few herbivorous fishes are suitable for the mariculture industry which contrasts sharply with the dominance by carp and tilapia in freshwater.

- The financial success of shrimp farming has led to a gold rush mentality, and farms have been established with little environmental planning.
- The disastrous consequences of poor planning and environmental management have severely limited the growth of the shrimp farming industry.
- Environmental concerns need to be addressed in parallel with the technological obstacles that hinder the expansion of the aquaculture industry.
- While aquaculture provides an alternative to fishing wild stocks of marine species, aquaculture is associated with its own suite of ecological problems.

Further reading

Basic statistics of world aquaculture production and trends can be found in the FAO publications series (e.g. FAO, 1999). Bohnsack (1996) provides a good overview of attempts to reintroduce and improve the stocks of certain species in the tropics, while Howell (1994) considers some of the major behavioural and physiological problems associated with hatchery-released fishes. A comprehensive treatment of ecological problems associated with aquaculture can be found in Volume 18 (1A) of the journal *Estuaries*.

17 Management and conservation options

17.1 Introduction

Throughout this book we have described the biological, social and economic diversity of fisheries, the selection and application of assessment methods and how fishing affects species, habitats and ecosystems. In this final chapter, we explore how the fisheries scientist can inform the management process and what can realistically be achieved through management.

Governments and other authorities manage fisheries because the biological, social and economic consequences of an unregulated fishery are undesirable. Their management objectives may be intended to ensure the economic and social well being of future generations or to protect habitats and species of conservation concern. Effective fisheries management requires clear objectives supported by the best scientific advice and appropriate management actions.

We begin by showing how management objectives are translated into an operational management strategy and how management actions, such as catch or effort controls, can be used to implement the strategy. We discuss how different management objectives are reconciled, the benefits of involving fishers in the management process, and the problems caused by uncertainty. We focus on single-species management that is often supported by explicit and quantitative scientific advice and how such management may be improved.

Fisheries managers increasingly need scientific advice on multispecies and ecosystem issues. We ask whether the fisheries scientist can usefully advise them and what can be achieved. Given that concerns about the effects of fishing on non-target species of birds, mammals, reptiles, fishes and invertebrates have started to dominate the political and scientific agenda, protection of these species is likely to become a key objective in many management plans. We consider how endangered species and habitats can be protected from fishing and the

potential uses of protected areas and no-take zones. In the final section of this chapter, we predict the future for fisheries science and fisheries management in the developed and the developing world.

17.2 Management objectives, strategies and actions

17.2.1 From objective to action

The preceding chapters have emphasized that the biological, social and economic consequences of an unregulated fishery are undesirable. For this reason governments and other authorities, including fisher's unions and cooperatives, intervene to manage fisheries. If they do not intervene then there is little reason why an individual fisher should act with regard to other fishers, society and the environment. Self interest is paramount in all but a few societies.

We listed the possible objectives of management in the Introduction to this book (section 1.5). These could loosely be defined as biological, social, economic and political. Although there are many possible objectives, some clearly dominate existing management decisions. In Chapters 7 and 8 we saw that the main biological objective is to protect the resource from overexploitation that jeopardizes future production. Given that many stocks are now overexploited, this means stock rebuilding. Other biological objectives of increasing significance involve protecting species and habitats of conservation concern (e.g. Chapters 14 and 15). Economic objectives include the maximization of economic benefits to harvesters, processors, distributors, marketers and consumers (Chapter 11). Social objectives include high employment, stability of coastal communities and safety at sea. Political objectives may include the avoidance of conflict (section 1.5).

The first stage of the management process is to

Catch controls	Effort controls	Technical measures
Total allowable catches (TAC)	Limited licences	Size and sex selectivity
Individual quotas (IQ)	Effort quotas	Time and area closures
Catch limits	Gear or vessel restrictions	

Table 17.1 Types of management action. From OECD (1997).

express the objective as a management strategy such as a target escapement. Quantitative strategies will be difficult or impossible to set in relation to ecosystem-based objectives until we gain a better understanding of ecosystem function. If the fishery scientist uses an appropriate assessment technique when formulating strategies, and accounts for many of the potential uncertainties and errors we have emphasized throughout this book, the strategy will work if effective management action is taken to implement it. Management actions can be divided into catch controls, effort controls and technical measures. Catch controls limit the catches of individual fishers or the fleet as a whole, effort controls limit the numbers of fishers in the fishery and what they can do, while technical measures control the catch that can be made for a given effort (Table 17.1) (OECD, 1997). Box 17.1 shows an example of how objectives are translated into actions in the western Australia abalone fishery.

Outcomes of a given management action in a given fishery depend on the biological, social and economic environment. Management action has to be taken with an understanding of how the action will affect the behaviour of fishers and whether it can be implemented. This understanding comes from interaction between fishers, managers and scientists. Some management actions have undesirable and unexpected consequences. Thus, in Chapter 13 we saw how catch quotas could lead to high-grading, where fishers discarded smaller individuals of quota species dead or dying because a quota would be more valuable if larger individuals were landed. Contrary to the manager's expectation, the effect of quota management was to maintain rather than decrease fishing mortality.

In the ideal world, there would be clear management strategies, it would be easy to measure whether or not they had worked, and the management actions would be supported by all fishers and other interest groups. Unfortunately, the world is not ideal. A Food and Agriculture Organization (FAO) review of the status of global fishery resources suggested that 57 of the 80 major stocks investigated were already fully or over-exploited (FAO, 1994). These stocks are all 'managed'. There are lots of theories about management, but there is much to be said for looking at existing management systems and seeing how well they operate rather than guessing what might happen. In 1997, the Organization for Economic Co-operation and Development (OECD) adopted this approach and looked at the value of catch controls, effort controls and technical measures. Their study applied to the larger and predominantly single-species fisheries in developed countries, but gives a very useful assessment of the successes and failures of different management actions (OECD, 1997).

17.2.2 Catch control

Catch controls, also known as **output controls**, are intended to control fishing mortality by limiting the weight of catch that fishers can take. These include **total allowable catches** (TAC) or **quotas** (Q), which are limits on the total catch to be taken from a specified stock, as well as **individual quotas** (IQ) and vessel catch limits where the TAC is divided between fishing units. In many cases, catch controls are really landings controls, since fishers may kill and discard unseen large numbers of fish of a size or quality that do not attract the highest prices in order to make the most money from their IQ (Chapter 13). Clearly, if such high-grading takes place, the IQ will not directly control fishing mortality.

Catch controls are amongst the most widely used management regulations. TAC management tends to encourage a race to fish amongst fishers because the resource is still common property and they race to maximize their share of the TAC (the fishery is closed when the aggregate catch equals the TAC). This is manifest as shorter fishing seasons, reduced fish quality, higher bycatch and more dangerous working conditions because fishers will take greater risks to keep fishing. TACs also encourage overcapacity because larger and

Box 17.1
From objective to action in the Western Australia abalone fishery.

The abalone fishery of Western Australia is small but profitable, benefiting from the considerable demand for abalone on international markets. Fishers target three species, the greenlip *Haliotis laevigata*, brownlip *H. conicopora* and Roe's abalone *H. roei*. The fishers dive to depths of around 50 m using surface supply equipment and prise the abalone from their rocky habitats using a flattened steel bar.

Western Australian Fisheries (WA Fisheries) specify the management objectives for the abalone fishery. These are:

1 to maintain sustainability of abalone stocks through the maintenance of the breeding stock and habitat
2 to maximize the economic return from the abalone resource to the community while maintaining sustainability of the stock and the habitat

3 to ensure cost-effective management of the fishery
4 to encourage maximum commercial flexibility and administrative simplicity for industry participants.

To meet these management objectives, the management strategy of WA Fisheries is to maintain a target abundance and size composition for each abalone species. The target is based on scientific advice (stock assessment) from the Abalone Management Advisory Committee.

To meet the target abundance and size composition for each species, management action is taken using catch controls and technical measures. First, total allowable catches (TACs) are set for each species, and these are divided among the fishers with licences to give them individual quotas (IQs) for the season. TACs and IQs are allocated by fishing zones. Second, minimum size limits are set for each species.

In the 1998/99 season, the TACs and IQs were set as shown in Table B17.1.1. The minimum sizes were 140 mm shell length for greenlip and brownlip abalone and 70 mm shell length for Roe's abalone.

Table B17.1.1 Abalone catch controls by species and fishing zone in the 1998/99 fishing season.

Fishing Zone	Number of Licensees	Greenlip/Brownlip (tonnes meat weight)		Roe's abalone (tonnes whole weight)	
		TAC	IQ	TAC	IQ
1	6	40.2	6.7	9.96	1.66
2	8	40.0	5.0	8.0	1.0
3	12	0	0	108.0	9.0

Source: Western Australian Fisheries (www.wa.gov.au/westfish).

more powerful vessels will compete more effectively for a share of the catch. TACs often result in gluts of supply and markets and processors have to invest in the capacity to deal with them. This increases marketing and processing costs. TACs may be good biologically but are poor economically.

IQs restrict the catches of individual fishers or boats. The sum of all IQs will equal the TAC. If IQs can be bought and sold by fishers then they are known as **individual transferable quotas** (ITQ). IQs allow fishers to catch their quota at a rate that suits them since they

are guaranteed a share of the TAC rather than having to compete for it. This more relaxed approach to fishing tends to improve catch quality, increase stability of supply to markets and processors and increase safety for fishers who no longer have to risk working in marginal conditions (OECD, 1997).

The main problem with IQs is that they increase the risk of fishers high-grading their catch. Moreover, the initial allocation of IQs to fishers can present many problems because fishers will want the greatest possible proportion of the TAC. IQs are usually allocated based

Characteristic	Total Allowable Catch (TAC/Q)	Individual Quota (IQ)	Individual Transferable Quota (ITQ)
Race to fish	↑	↓	↓
Profitability	—	↑	↑
Employment	↑	↓	↓
Capacity	↑	—	↓
Fishers safety	↓	—	↑
Stability of catch	↓	↑	↑
Quality of catch	↓	↑	↑
Discarding	↑	↑	↑
High-grading	↑	↑	↑
Resource conservation	↑	↑	↑
Misreporting of catch	↑	↑	↑
Assessment costs	↑	↑	↑
Administrative costs	↑	↑	↑
Enforcement costs	↑	↑	↑

Table 17.2 Impacts of three methods of catch control on social, economic and biological characteristics of a fishery. Arrows summarize how each method of catch control most frequently affected fishery characteristics; compared with the period before it was introduced. Note that the fisheries studied by OECD were not all affected in the same way because they were influenced by different social, biological and economic factors and because the initial management regimes were different. Based on OECD (1997).

on some measure of activity within the fishery during preceding years. Enforcement of IQs has to be effective or there will be a race to fish. This means that the system is better suited to managing the activities of a few large vessels that land catches in major ports than hundreds of fishers who sell their catches to local markets.

An ITQ system gives fishers property rights in the fishery and allows them to trade those rights with other fishers. This means that fishers who decide to leave the fishery by selling their ITQs will benefit financially and are compensated for their action. In general, the most efficient fishers buy ITQs from those who are less efficient. In fisheries where ITQ systems have been introduced, they generally have desirable impacts. Thus, operating costs start to fall, fleet capacity falls and profitability increases. Employment is likely to fall in a more efficient industry, but the remaining fishers will make more money. Since fishers benefit from effective enforcement they may help to police their own fishery or contribute to enforcement costs. ITQs may not be appropriate in countries where the concept of trading such rights is culturally unacceptable. The likely impacts of catch controls are summarized in Table 17.2.

TACs are set to meet the target levels of fishing mortality determined by stock assessment (Chapter 7). They may be fixed, or they may change from year to year because fish stocks fluctuate and the future is unpredictable. In order to adjust catches from year to year, the government or regulatory authority could buy

and sell ITQs. Thus, when the TAC had to be set at a low level, the government would buy any excess ITQs such that the sum of all ITQs remaining in the fishery was equal to the new TAC. This would give the industry a relatively stable income from year to year. An alternative is to allow the ITQ to change from year to year by fixing it at a set proportion of the TAC. In this case the industry bears the risk of uncertainty and will have good and bad years (Clark, 1985; OECD, 1997). In other cases ITQs are fixed from year to year, and the TAC set at a level that is considered to be sustainable in the long term. This approach has been adopted in some crustacean and mollusc fisheries, and has worked successfully. However, because there is an inherent degree of uncertainty in the system, there is always the risk of overexploitation.

Another way of controlling catch is to tax it. Taxes are already imposed on the extraction of other natural resources such as gas, minerals and oils, and would be expected to control catches. We saw in Chapter 11 that overcapacity develops in fisheries because fishers do not pay to fish even when fish are scarce. A tax on landings would increase the costs of fishing and mean that revenues would have to be higher if fishing was to be profitable. Taxes on catches do not have the undesirable effects of taxes on income or profit because they are not an incentive to discourage work or investment (Hannesson, 1993). Taxes would be hard to impose on real fisheries because, unlike oil and gas, the abundance of the resource varies from year to year and taxes would have to be adjusted accordingly (Clark, 1985). This would cause stress and hardship for fishers.

17.2.3 Effort control

Effort controls, also known as **input controls**, limit the number of boats or fishers who work in a fishery, the amount, size and type of gear they use, and the time the gear can be left in the water. Effort controls may also limit the size or power of vessels and the periods when they fish. The aim of effort control is to reduce the catching power of fishers and thus reduce fishing mortality. The response of fishing mortality to effort control is difficult to predict because it depends on the way that fishers change their behaviour in response to the regulations. All we can safely say is that the removal of all fishing effort will reduce fishing mortality to zero!

Effort controls can be divided into licences, **individual effort quotas** (IEQ) and vessel or gear restrictions. Limited licences restrict the number of boats or fishers in the fishery. Licences can be transferable. Effort quotas limit the amount of time spent working by a given unit of gear, a vessel or a fisher. For example, a fisher may only be allowed to set a given number of pots or go to sea for a given number of days in the year. An **individual transferable effort quota** is a tradable IEQ. Vessel or gear restrictions try to limit the catching capacity of vessels or fishers. These may control the size and design of pots or nets, the dimensions of a fishing vessel or ban specific gears that are seen as too effective (OECD, 1997).

Licences, also known as permits or concessions, are given or sold to each fisher or vessel participating in the fishery. Long-term licences allow access to the fishery and short-term licences allow fishing at specified times in specified places. The initial allocation of licences may be a problem if they are used to effect an immediate reduction in effort. This can be overcome by issuing licences to everyone already in the fishery and not renewing them as fishers leave or, if a faster reduction in effort is needed, by buying back licences after issue. If licensing does lead to a real reduction in effort, then stock size and the profitability of the fishery should increase in the longer term. In reality, however, limited licences do not control fishing mortality because the race to fish between licence holders persists, and this leads to overcapitalization and increased harvesting costs. Licence regulations are relatively simple to

enforce, but enforcement costs may be high (OECD, 1997).

If IEQs are to control fishing effort then the number of fishers or vessels (fishing units) also has to be controlled. There is little benefit in limiting every fisher to 2 days' fishing with one pot if any number of fishers can join the fishery. Moreover, fishers are very good at adapting their gear, vessels and behaviour to make up for the loss of catch when effort restrictions are first imposed, a process known as **technological creep**. Thus, the form of effort control should not give fishers the scope to modify gear and fishing behaviour to increase catch per unit effort. If this occurs, then effort controls will not control fishing mortality. The initial allocation of IEQs, as with ITQs, will be problematic, and enforcement of complex regulations applying to many fishers may be difficult and expensive (OECD, 1997).

Gear and vessel restrictions generally stop fishers from using the gears that would be most effective and therefore profitable. This will increase fishing costs. The continued evolution of more powerful vessels and gears may limit the impact of restrictions. Thus, limitations to vessel size have simply resulted in the design of smaller and more powerful vessels. If regulators respond to this by cutting vessel size further then fishers will lose all their investment unless they pay to modify vessels to comply with regulations. Vessel restrictions may also compromise safety by forcing fishers to work further out to sea in smaller vessels. A good example of the consequences of such measures is the changes that have occurred in the European beam-trawl fleet. Various gear and area restrictions were applied to vessels over a certain length. Almost immediately, ship yards were designing shorter and fatter (beamier) vessels with the same engine power.

While it may be easier to control effort instead of catch, effort control rarely has the desired consequences. The control of effort is unlikely to be effective, biologically or economically, if catch controls and technical measures are not imposed as well. Table 17.3 summarizes some of the expected consequences of effort control.

17.2.4 Technical measures

Technical measures restrict the size and sex of fished species that are caught or landed, the gears used and the times when, or areas where, fishing is allowed. The size

Characteristic	Licences	Individual Effort Quotas (IEQ)	Vessel/gear restrictions
Race to fish	↑	↓	↑
Profitability	↓	↓	↓
Employment	↑	↓	↓
Capacity	↑	↑	↓
Fishers safety	↓	↑	↓
Stability of catch	↓	↑	↓
Quality of catch	↓	↑	↓
Discarding	↓	↓	↓
High-grading	↓	↓	↓
Resource conservation	↓	—	—
Misreporting of catch	↓	↓	↓
Assessment costs	↑	↑	↑
Administrative costs	↑	↑	↑
Enforcement costs	↑	↑	↑

Table 17.3 Impacts of three methods of effort control on social, economic and biological characteristics of a fishery. Arrows summarize how each method of effort control most frequently affected fishery characteristics; compared with the period before it was introduced. Note that all the OECD fisheries studied were not affected in the same way because they were influenced by different social, biological and economic factors, and because initial management regimes were different. Based on OECD (1997).

of individuals that are landed may be controlled by minimum landing sizes (MLS). These will be most useful as a conservation measure if individuals below the minimum landing size can be measured *in situ* (as in crustacean and mollusc fisheries where divers carry a measure) or returned to the sea alive. In cases where individuals are dead or dying at the time of capture, and where the MLS has not deterred fishers from targeting small fish, minimum landing sizes will not affect fishing mortality. Gear restrictions, such as the size of mesh in traps and nets, control the minimum sizes at which fished species are caught. If used in conjunction with an MLS, this reduces the number of unwanted individuals that fishers have to sort and discard and ensure that most smaller individuals are never landed. These restrictions should reduce fishing mortality for a given fishing effort, provided that fish do not receive fatal injuries as they escape. Mesh size restrictions can have unintended effects in multispecies fisheries (OECD, 1997; Chapter 8). Sex restrictions usually restrict the capture of mature or egg-bearing females. They are used in crustacean fisheries. Empirical evidence from crustacean fisheries that have persisted despite very high fishing effort suggests they can be effective. Size- and sex-selective measures do not reduce the race to fish and would be expected to encourage overcapitalization and increased employment. They are likely to increase enforcement costs and levels of discarding (OECD, 1997).

Time and area closures can protect fished species at specific phases of their life history. Examples are the protection of juvenile nursery areas or adult spawning grounds. Time closures can protect annual stocks until their production and quality is high, but also lead to market gluts at the start of the fishing season. These can cause prices to fall, and force processors to invest in capacity that is idle for much of the year. Area closures may stimulate effort redistribution and increase fishing costs without reducing fishing mortality. Time and area closures have been most effective when used in conjunction with other measures such as catch and effort controls. Their impacts are summarized in Table 17.4. We return to the use of protected areas for fisheries and general conservation in section 17.5.3.

17.2.5 Management in action

Optimization

Optimization is used to find the combination of management actions that provide the best means of achieving a strategy (section 1.5.2). Optimization could, for example, be used to determine the combination of mesh size and fishing effort that would maximize the profits from a gill-net fishery. In this case, the profits that result from different combinations of fishing effort and mesh size (based on desired age and size at first capture coupled with knowledge of selectivity, section 9.3.1) would be calculated using a bioeconomic model. The results could be presented graphically on a plot of mesh size against effort, with contours being used to show

Table 17.4 Impacts of technical measures on social, economic and biological characteristics of a fishery. Arrows summarize how each technical measure most frequently affected fishery characteristics; compared with the period before it was introduced. Note that all the OECD fisheries studied were not affected in the same way because they were influenced by different social, biological and economic factors and because the initial management regimes were different. Based on OECD (1997).

Characteristic	Sex and size limits	Time and area closures
Race to fish	↑	↑
Profitability	↓	↓
Employment	↑	↑
Capacity	↑	↑
Fishers safety	↓	—
Stability of catch	—	—
Quality of catch	↑	—
Discarding	↑	—
High-grading	—	—
Resource conservation	—	—
Misreporting	↑	↑
Assessment costs	↑	—
Administrative costs	↑	—
Enforcement costs	↑	—

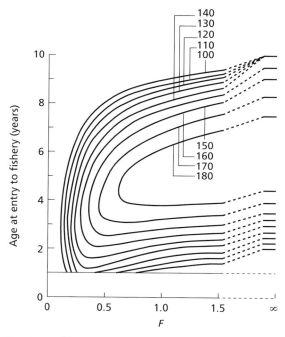

Fig. 17.1 Yield per recruit (Y_w/R) diagram showing mean Y_w/R for the North Sea haddock at different fishing mortalities (F) and ages of entry to the fishery. Y_w/R is given in grams with isopleths at intervals of 10 g. After Beverton and Holt (1957).

the profits that result from different management options. The choice of mesh size and effort would be optimal when profit is highest.

When the effects of two or three management actions have to be optimized, 2-D and 3-D contour plots are a useful and accessible way of presenting the results. For example, Fig. 17.1 shows the combinations of fishing mortality and ages of entry to the fishery that result in given yields per recruit for North Sea haddock. However, if more than three actions are considered, then other approaches such as non-linear search procedures and stochastic dynamic programming are used (Mangel, 1985; Mangel & Clark, 1988; Hilborn & Walters, 1992; Quinn & Deriso, 1999).

What has worked?

When used independently, catch controls, effort controls or technical measures are unlikely to provide a basis for meeting management objectives. The OECD assessment of the effects of different management actions in developed countries showed that ITQs, used in conjunction with technical measures, had worked most effectively to optimize resource conservation and

economic performance of the fisheries they considered. ITQs were one of the most effective ways of controlling exploitation, reducing the race to fish, reducing overcapacity and increasing profits. Technical measures helped to reduce discarding, control bycatches and protect target species at vulnerable stages of the life history. However, the biological characteristics of the fishery can greatly complicate management. Management using ITQs and technical measures was generally most effective for single-species and selective fisheries, such as those for many crustaceans and molluscs (OECD, 1997). Conversely, management was particularly ineffective for multispecies demersal fisheries where gears were highly unselective and bycatch abundant. Because there were many species on the fishing grounds, no single management action, or combination, would result in optimal mortality for each species (OECD, 1997). The result is that target and non-target species with slower life histories are overexploited (section 12.4.2).

If it is possible to identify effective management actions, then why are these actions not widely adopted or enforced? Many strategies and actions are never based on assessment advice because the assessment scientists did not consider all the objectives in their analysis. Thus, a scientist may be asked how to stabilize catches in order to increase the profitability of a processing plant while the government also has an implicit, but unstated, objective of maximizing catches this year to avoid hardship and dissent in coastal towns. A second problem is that the capacity of fishers to adapt their behaviour and fishing methods to management action is often overlooked. If 10 units of effort catch 10 tonnes of fish today, a regulation that limits fishers to five units of effort will not reduce catches to 5 tonnes for long. Enforcement is also a problem. Complex regulations are costly and difficult to enforce and many of the most detailed action plans cannot be followed in the real world.

Thus far we have mostly dealt with larger fisheries that are fished for profit. Many small subsistence fisheries scattered along the world's coastlines are very different. There are few resources for assessment or management, and complex regulations cannot be enforced unless the fishers help. Most management problems result from common access or people's desperate need for food and income when nothing else is available (Chapters 6 and 11). Tenure systems giving communities rights to specific fishing grounds offer many benefits (section 6.4.5), particularly if supported by the government. We discuss the potential benefits of tenure systems and co-management in section 17.3.2.

Management feedback

Fishery management is not a static process, and feedback is needed to adjust management actions and ensure that management objectives are met. For example, the management strategy for Icelandic capelin is to leave 400 000 tonnes to spawn each year (Jakobsson & Stefánson, 1998; Chapter 1). Management action, in this case a TAC, is used to control fishing mortality, and periodic assessments are used to confirm the size of the spawning stock. Should the spawning stock exceed the target size, then there is feedback from the assessment scientists to the managers, and they modify their management actions accordingly. In most fisheries, there is a continuous cycle of setting management strategies, implementing management actions, assessing the impacts of those actions, and modifying the actions in accordance with strategies.

Reducing capacity

Many fisheries are heavily overexploited, and a little tinkering with existing management practices will be insufficient to realize long-term biological, social or economic benefits. As such, major structural readjustment is needed. This is likely to include extensive decommissioning of boats to remove excess capacity, bans on new entrants to the fishery, and allocation of property rights. Change would have undesirable consequences in the short term, including job losses and reductions in income. However, in the long term capacity reduction would produce a more efficient industry and happier and wealthier fishers.

Displaced fishers will need to be supported by the government until they find new jobs. Decommissioning of vessels and removal of capacity is best done quickly as a 'one-off' exercise. Repeated and long-term decommissioning schemes encourage excess capacity because they reduce investment risks for fishers. Capacity reduction will be expensive, but is preferable to providing an inefficient industry with long-term subsidies. Voluntary and cash-starved decommissioning schemes tend to remove the least effective capacity (Munro, 1998).

The fishery for the black-lipped abalone *Haliotis rubra* (Haliotidae) off western Victoria in Australia shows how capacity may be reduced (Sanders & Beinssen, 1996). The fishery began in 1967 and abalone were collected by divers from inshore reefs. Capacity grew very quickly as fishers realized the profits that could be made, and by 1968 the government had banned new fishers, increased licence fees and raised the minimum landing size. Biological assessments showed that capacity still had to be reduced, and in 1984, fishers were allowed to sell their licence entitlements to people who wanted to become fishers. However, these new entrants had to buy two licences to enter the fishery and then 'retire' one of them. This reduced number of divers. In 1988, managers decided to control catch rather than effort. They set the annual TAC at 280 tonnes with each of 14 divers now working in the fishery getting an ITQ of 20 tonnes. This was 20% reduction in catch from the preceding years of effort regulation. The lower capacity fishery is now flourishing and the increased value of ITQs has partly compensated divers for the initial reduction in catches.

Capacity reduction would greatly improve the economic efficiency of many fisheries. We might expect that this would be in a nation's interest, and yet there are few schemes that achieve real capacity reduction. Hannesson (1993) has an interesting perspective on the problem. He suggests that the cost of subsidizing fisheries is spread across many groups in society and so it does not upset anyone too much because the financial burden is not too onerous for the individual. However, the loss of jobs in the fishing industry is seen as a further erosion of a traditional way of life. As an organized and coherent group, fishers can generate trouble and bad publicity for the government of the day. As such, governments tend to serve fishers at the expense of society as a whole.

Developed and developing countries

The removal of capacity in developed countries can have harsh social and economic consequences in the short term but, in the longer term, these are often ameliorated by support from the government and job opportunities in other sectors of the economy. In the poorer countries of the world, fishing is often the forced occupation of families who have no other opportunities and it provides their main source of food and income. These countries often have huge international debt and are not in a position to buy excess capacity from fisheries, even if, in the longer term, this would increase protein supply. The scale of poverty and reliance on fisheries in the developing world has been widely described elsewhere (Kent, 1998). The extent of the problem is highlighted by the fact that 40% of the 4.4 billion people that live in developing countries lack basic sanitation and 30% have no access to clean water. Worse still, estimates suggest that 1 billion people in 40 developing countries may lose access to their primary source of protein as a result of overfishing (UNDP, 1999).

While an objective such as 'maximizing protein supply' may seem desirable, the action needed to meet that objective, notably a dramatic reduction in capacity, will leave some people to starve. We should not expect attempts to control fishing in many parts of the developing world to be any more successful than attempts to prevent the felling of rain forest unless the developed world is willing to pay. This does not mean through fisheries subsidies that have encouraged overcapacity and overfishing, but by assisting with debt relief, development and social restructuring that will give fishers alternative livelihoods. Once this type of support is

available it may be possible to give specific and limited groups of fishers access to fishing grounds, to limit their fishing capacity and to provide government support for enforcement.

This gloomy situation does not apply to all artisanal fisheries, especially where population density is low and income is available from farming or tourism. Empirical evidence from fisheries that were shown to be sustainable suggests that 1 km^2 of actively growing reef, fished for a variety of algal, fish and invertebrate species, can support approximately 320 people if no other protein sources are available. Shallow reef and lagoon areas in general would support about one-fifth of this number (Jennings & Polunin, 1996c). At these human population densities, the fishery is unlikely to collapse if the fishers are catching solely for subsistence. Capacity can be constrained by balancing the number of fishers who have fishing rights with the size of the area of tenure. It can further be constrained by limiting fishery development programs to new (usually pelagic) resources and educating fishers about the limits to resources in the area of tenure. Traditional management has generally broken down in those parts of tropics where population growth and urbanization are occurring but employment prospects for the majority are limited (Chapter 6). Moreover, no fishery development is advisable without a recognition of the limits of sustainability within the fishery and an assessment of existing catch rates. While these comments may seem obvious, they are frequently ignored by those promoting fishery development that ties fishers into schemes that force them to overfish to meet loan repayments and fuel costs.

17.3 Improving management

17.3.1 Enforcement and compliance

Management actions only help to meet an objective if they are accepted or enforced. In reality, many fisheries with management plans have been overexploited and their social, biological or economic potential has not been realized. This is because fishers do not comply with management actions or managers do not enforce them.

Enforcement is easier and cheaper if fishers can see that management actions benefit them. Indeed, in a few well-managed fisheries, fisheries scientists and managers are now treated as though they are working for the fishers and not against them. If fishers actively oppose management actions, then enforcement is less likely to be effective. More powerful enforcement will help, but creates resentment, is costly, and can develop into an arms race. It is worth remembering that even in military dictatorships, some of the population continue to behave in accordance with their own beliefs rather than cower to the threat of violence. Some level of industry support is essential for management to work, as draconian measures are almost impossible to enforce effectively, and governments rarely allocate significant resources to fisheries enforcement.

In general, managers could do a great deal more to help fishers understand why regulations are imposed and to tailor those regulations to help and encourage fishers to comply with them. In the most simplistic form this could mean that the manager uses scientific advice to ensure that mesh sizes are compatible with minimum landing sizes. Fishers with exclusive rights to a fishery are likely to see the benefits of allowing small fish to escape alive but do not see benefits of discarding many dead, dying and valuable fish back into the sea to feed seabirds and crabs. Fishers will generally value the information that is based on their experience. It is no good telling a fisher who is catching lots of cod that there are none left in the sea without further explanation of density-dependent habitat use and other factors that may mask the collapse of a stock. Similarly, the choice of management action should be sympathetic to the fisher's culture (Chapter 6). ITQs which provide effective ownership of part of a stock may suit the market economies of the developed world but may not be appropriate for fishers who work together to provide for their whole community. These fishers are more likely to support a tenure agreement that gives them shared rights to a fishing ground.

Fishery regulations are enforced in many ways. Large vessels such as tuna purse-seiners may have to carry an observer to record landings and bycatch. Many fisheries are patrolled by surveillance aircraft. These can direct fishery or military patrol vessels to fishing boats that may not be complying with the regulations. Other enforcement is conducted by fisheries officers at ports and fish markets. In larger commercial fisheries there is increasing use of modern technology. Larger otter and beam trawlers fishing in European waters are now satellite tracked to provide their positions in real

time. In some Australian fisheries, vessels have to log their catches electronically at sea and are subject to spot checks while fishing and when landing. If there is a discrepancy between data logged electronically and the catches on board the fishers are subject to heavy fines. Penalties for breaking fishery regulations are very variable within and between nations. In Chapter 6 we saw that death and beating were not unusual punishments for poaching in early Pacific Island cultures. Nowadays, monetary fines and the confiscation of gear are common penalties. There has been concern that the fines imposed for illegal fishing are often too small to act as deterrents, but some countries are quite proactive in defending their fishery rights. In recent years for example, the Canadian navy impounded a Spanish vessel that was fishing with illegal nets, and the Australian navy has captured Indonesian fishing boats (Chapter 6).

When enforcement is needed in smaller fisheries it is often carried out at 'bottlenecks' in the capture and sale process. Thus, size or species restrictions may be enforced by watching the activities of fish traders, or ice that is needed to preserve catches may only be supplied to fishers that possess a fishing licence (Adams, 1996). Enforcement can be difficult because enforcers are often poorly paid and may be tempted by bribes. In Papua New Guinea, one secretary at the Department of Fisheries and Marine Resources was offered, and rejected, $US23 000 in bribes during the first 3 months in the job (Adams, 1996). Compliance is preferred to enforcement and this is encouraged when government managers and fishers work together to manage the fishery (co-management). This is especially important in smaller fisheries since there will never be sufficient resources to police the activities of every fisher.

Enforcement is usually more difficult when the management system promotes a race to fish (section 11.3.1). Management systems that give property rights to stocks or provide access rights to fishing grounds are quite often supported by fishers and they will take some responsibility for enforcement. Under ITQ systems, fishers who flout the regulations or fish without an ITQ will reduce the value of the quotas and will not be tolerated by legal quota owners. Under tenure systems, fishers who fish illegally (poach) will not be tolerated because they are depriving the owners of potential food and income. Governments can help to empower fishers to police their own fishery, something that is

Type of management	Role of fishers	Role of State
Self management	Manage	Not involved
Co-management	Consult	Advise
Co-management	Co-operate	Co-operate
Co-management	Advise	Consult
State management	Informed	Instruct

Fig. 17.2 The role of fishers and the state in different types of fisheries management. Based on Pinkerton (1994).

Table 17.5 Conditions that promote successful co-management. Based on Pinkerton (1994).

- Fishers already deal with management problems collectively in cooperatives, villages or ports
- Boundaries of the fishery are clearly defined and stocks are not shared with other fishers
- Fishery has not collapsed and a race to fish or Malthusian over-fishing are not in progress
- Community has alternative sources of income and support
- State supports local institutions and supports devolved power
- Locals can police the fishery and State can provide legal backup

encouraged in Japan and several Pacific Island nations (Chapter 6).

Clearly, all management will not be supported by fishers. Most will not approve of being displaced from fishing grounds to protect a species of conservation concern or having to use bycatch reduction measures that reduce their catches and increase running costs. Similarly, they may not want to reduce catch rates in the short term, even though this may increase profitability in the long term, because their perception of discount rates differs from that of managers (section 11.3.1). In these cases, enforcement is needed. Enforcement should be equitable and comprehensive, so fishers do not see other fishers consistently flouting regulations and not being punished. If they do, then the fishers' contempt for the enforcers will increase, and a race to fish may ensue. Inequitable enforcement has created distrust and conflict in internationally managed fisheries for shared stocks where some countries try to ensure their fleets follow management regulations while others do not.

17.3.2 Co-management

Management and enforcement are usually more effective if managers and fishers work together. **Co-management** means that fishers and the state share responsibility for fisheries management. It lies between extremes where the state either dictates or leaves the fishers to their own devices. At other levels, the state and fishers can consult, advise or cooperate fully (Fig. 17.2). Co-management helps to minimize conflicts

that impede management and allows fishers to indicate when regulations, such as quotas that encourage high-grading or landing of illegal fish, are inappropriate, even if they work in theory.

Co-management has worked best when the fishery is under the control of fishers, either through ITQs or tenure arrangements. There are other characteristics of fisheries that also promote successful co-management (Table 17.5).

As pressures on a fishery increase, co-management can help communities to preserve their traditional conservation ethic. Co-management has also helped to improve understanding between fishers, managers and scientists that were formerly at odds with each other (Pinkerton, 1994).

Co-management was introduced to deal with the collapsed Chilean loco (muricid gastropod) fishery shortly after it was closed in 1989 (section 1.3.1). Co-management worked well, the fishery was reopened, and stocks are now rebuilding (Castilla & Fernandez, 1998). Under Chilean Fishing and Aquaculture law, exclusive fishing rights to benthic resources are only granted to organized fisher unions. The unions, *sindicatos*, played a key role in the establishment of Management and Exploitation Areas (MEAs). In El Quisco in central Chile, for example, the Union jointly developed the loco management plan with scientists following collaborative research, and the plan included closure of part of their fishing ground. The government provided support through legislation that protected the Unions rights. The benefits of co-management were seen by fishers and scientists when loco stocks recovered faster than in adjacent areas (Castilla & Fernandez, 1998).

Co-management is a nice concept and has worked well in a few areas. While it is an excellent idea to promote better communication and understanding between fishers, fishery scientists and regulators, it is unrealistic to expect that co-management will solve all management problems. Ultimately, effective management may require that catches and fishing capacity are reduced, and this makes it impossible to please everyone.

17.3.3 Ownership of resources and harvesting rights

Public ownership of marine resources and freedom of access to the sea has contributed to the race to fish and ultimately to overfishing (section 11.3.1). Management systems that restrict the ownership of resources or the right to harvest can help prevent overfishing. For example, an ITQ system usually restricts the right to harvest while the ownership of resources remains with the public or government. Tenure systems, conversely, may allow fisheries to own the resources. ITQ systems and tenure arrangements have helped to mitigate the race to fish and encourage resource conservation for stocks that do not cross management or national boundaries (**straddling stocks**). For highly migratory species, management is much more difficult as the demands of different nations have to be reconciled. For example, purse-seine vessels may catch small yellowfin tuna in one part of the Pacific while long-line vessels from a different nation catch adults in another. Improved international collaboration is needed. All countries must agree on regulations and enforce them effectively and equitably (section 17.3.1).

Ownership of resources or harvesting rights can provide many biological, economic and social benefits for fishers and society as a whole. As such, both ITQ and tenure systems are increasingly promoted. Moreover, ownership of resources or harvesting rights creates some intriguing possibilities for conservation groups that could even start to buy, but not use, these rights to reduce fishing mortality and protect rare species.

17.3.4 Uncertainty and the precautionary approach

If there is one thing you are likely to remember after reading this book it will be the uncertainty that faces the fisheries scientist (and we do not mean job prospects). It was well known that the future of fisheries was uncertain in the late 1850s when the Norwegian government hired scientists to explain why catches of Lofoten cod fluctuated from year to year (section 1.2.2). In recent years the problems of uncertainty have been considered explicitly (Walters, 1984, 1986; section 7.9). An unpredictable climate means that future production is uncertain, and there is uncertainty about the integrity of stocks, model specification, model parameters and whether today's parameters will apply tomorrow. We hope this book will encourage you to look critically at assessment methods and to offer appropriate caution if giving, receiving or interpreting advice.

Fisheries scientists, managers or fishers may respond differently when faced with uncertainty. For example, fishers want to catch more now because the future is uncertain while managers want lower catches for the same reason (Clark, 1985). Managers generally propose conservative limits in the face of uncertainty because they feel that risks should not be taken with the stock until they know more about it. In some ways, this is strange, since it may be hard to judge the productivity of the stock until it has been exercised by exploitation (Walters, 1986).

One approach is to adopt **adaptive management**, to learn more about the effects of fishing and the potential of the fishery. This would consist of a series of experimental manipulations of the fishery. Different harvesting strategies could, for example, be applied to different stocks of the same species. Adaptive management is an interesting idea, but is not practical in most fisheries. While most would accept that existing fisheries represent the worst forms of experimental design (one way trips to collapse), fishers may not have alternative stocks to harvest and can be disadvantaged by experimentation unless someone compensates them. Occasionally, we can learn more about the effects of fishing and the potential of fisheries by using 'experimental' designs that have been established inadvertently by fishers. Thus, in subsistence reef fisheries, islands or reefs of different sizes that are fished by different numbers of people will have different harvest rates and may, in effect, provide a spatially replicated experiment. Deliberate experimentation could also be worthwhile if benefits are being dissipated under the status quo.

From the manager's perspective, the response to uncertainty has been to adopt a precautionary approach to management. This avoids a situation where, to quote

John Gulland, 'overfishing continues until all doubt is removed and so are the fish' (Sissenwine & Rosenberg, 1993). We have seen examples of the many precautionary biological reference points that can be used for single species management (Chapter 7; FAO, 1995b; Caddy, 1998), but precaution applies at all levels from fishery development to technology transfer. The precautionary approach as used today is based on the application of prudent foresight, but from the perspective of the manager rather than fisher. In general, the precautionary approach is based on the idea that management actions in the present should consider the needs of future generations, avoid changes that are not reversible, identify undesirable outcomes of management action in advance, and identify measures to correct them. The manager should also conserve the productive capacity of the resource where fishing effects are uncertain, link fishing and processing capacity to the expected magnitude and variability of production and ensure that all fishing is authorized and monitored. The precautionary approach also applies to the management of impacts on habitats and non-target species (FAO, 1995b).

In a new or developing fishery, the manager acting in accordance with the precautionary approach would:
1 control access to the fishery at the outset;
2 place a conservative limit on fishing mortality or impacts on the environment and non-target species;
3 control investment;
4 establish precautionary biological reference points; and
5 initiate regular monitoring and research.

In smaller multispecies and artisanal fisheries area closure may provide a more appropriate means of precautionary control (section 17.5.3) and poorly targeted development aid should be discouraged. The precautionary approach should not stop fishers gaining any access to new sources of protein and income. The real aim is to avoid the usual problems of overcapacity, inefficiency and undesirable biological effects.

In existing fisheries, the manager acting in accordance with the precautionary approach would:
1 ensure that mechanisms for the control of further fishing effort and investment are operative;
2 monitor the fishery in relation to reference points; and
3 act on warnings of over-exploitation to protect the reproductive capacity of the resource.

In overexploited fisheries, the manager would:
1 impose immediate limits on capacity;
2 develop and institute a recovery plan to rebuild spawning stock biomass; and
3 monitor recovery.
Clearly this requires a lot of cash (FAO, 1995b).

Some fisheries are now managed in accordance with the precautionary approach. Biological reference points (Chapter 7) are used to provide quantitative targets for managers of single-species fisheries and it is increasingly possible to set minimum biologically acceptable levels for those species of conservation concern that can be censured effectively. However, it is more difficult to adopt a precautionary approach to ecosystem-based concerns because quantitative targets are hard to set.

17.3.5 Role of science

Quantitative understanding of the responses to fishing is desirable because it allows managers to set strategies that are clearly understood. Of course, if the models on which strategies are based are wrongly specified, and the parameters invalid, then they are of little use without some explicit understanding of where the uncertainty arises and what effect it has on the output. The role of science is to inform the management process and to monitor the effectiveness of management. At present, fishery science has provided a useful analytical framework for single-species management but not for multispecies and ecosystem-based management where we still have much to learn. With the exception of studies on fished species, most of the quantitative science that provides a basis for setting management strategies has been applied to interactions between birds and fisheries. For example, existing models can be used to assess the impacts of shellfish harvesting on oystercatcher populations (section 15.2.2).

The dangers of fire fighting
With increasing concern for the ecosystem effects of fishing, threatened species and the viability of fishing communities, fishery scientists could take a stronger lead in setting an appropriate research strategy. Indeed, parts of research agenda are now run by conservation or pressure groups because some scientists have been slow or unwilling to treat the wider aspects of fisheries as important. This has engendered a fire-fighting

approach where governments and scientists rush from subject to subject in an attempt to placate pressure groups. This may provide short-term fixes for problems as they are encountered, but does not necessarily solve many long-term problems or resolve conflicts between groups with different objectives. Moreover, actions to resolve single issues may have unexpected and undesirable consequences.

The conservation and ecosystem-based issues that we described in Chapters 12–15 will increasingly dominate fisheries management in coming years. Fisheries are likely to be managed to conserve non-target species and habitats. However, short-term mitigation can cause long-term problems. A nice example of the dangers of the fire-fighting approach comes from shipping. In the 1970s, there was growing public concern that the oils and contaminants that entered the environment when ships discharged their ballast water were having negative impacts on the environment. The effects, although significant, were actually rather localized. Legislation to control ballast water discharge was hurriedly put in place, and the shipping industry responded by improving the cleanliness of ballast water. In fact, they did such a good job that ships' ballast tanks became giant aquaria, and the living organisms that entered tanks with the ballast water could be transported across the world's oceans. When the ballast water was discharged, thousands of miles from the location of collection, new species were introduced to the local environment. Some of these, such as toxic phytoplankton, have had far more extensive and harmful impacts than the oil and contaminants that used to be discharged (Richardson, 2000).

If you are not convinced that fire fighting could be a real problem in fisheries, then the story of the Audouins gull may be a warning (Chapter 15). Here, the simple and seemingly obvious policy of reducing waste in fisheries by trying to limit discarding could endanger a rare species of seabird. Seventy-four per cent of the global population of the endangered Audouins gull (*Larus audouinii*) lives on the Ebro Delta and a small island off the Spanish coast. With loss of coastal feeding grounds due to development, these gulls have supplemented their diets by feeding on discards. However, in 1993, there was a seasonal closure of the local trawl fishery, and the breeding success of a gull colony fell dramatically because they could not find alternative food sources in the absence of discards (Oro *et al.*,

1996). In this case, sustained fishery closure or the use of gears that reduced discarding could threaten an endangered seabird.

There are other cases in which management measures based on inadequate information may not have the desired effects. The design of more selective fishing gear may be desirable, but may also encourage the selective removal of keystone groups, and increase the probability of ecosystem shifts. As we saw in Chapter 16, aquaculture may, in some cases, provide an alternative source of protein that reduces our reliance on fishing but the environmental impacts of aquaculture may be worse than those of fishing. For these reasons it is important that management decisions are based on the best available science and that management problems are considered from different perspectives. In the absence of good science, insurance through the precautionary use of no-take zones may be preferable to reactive use of 'band-aids'.

Data poor management

Notwithstanding our comments on fire fighting, there is much information on fisheries and their responses to exploitation that can be gleaned from existing studies and may help to set initial and precautionary harvesting strategies. This knowledge is rarely used when the development of new fisheries is planned. While many view the approach of successively adding new capacity at long intervals to be unacceptably slow, general limits to capacity can quickly be determined with an estimate of abundance and some knowledge of the life history of the species. Past experience provides a good guide to fisheries yield but is often ignored. Witness the development and investment programmes on tropical reefs which quickly increase capacity well beyond that which the most simplistic analysis would show to be sustainable. For example, numerous empirical studies on reefs where multispecies fisheries exist and were shown to be sustainable, suggest that yields of 2–3 t km^{-2} y^{-1} can readily be taken from reef areas, and yields of 10 t km^{-2} y^{-1} can be sustained from small areas of actively growing reef (Dalzell, 1996). This knowledge could easily be used as a guide for initial development, rather than the traditional *ad hoc* investment in capacity (Munro & Fakahua, 1993).

Empirical studies can also provide a basis for planning harvest strategies that minimize the probability of ecosystem shifts on tropical reefs. First, fishery managers should seek to identify and protect the few species of fishes, such as urchin predators, which have keystone roles and maintain the ecosystem in a state that favours fish production. Second, fishers should catch the remaining species from many trophic levels. This increases the probability that a given total yield can be sustained and reduces the risk of deleterious shifts in ecosystem function. Moreover, if fishers are used to catching, eating and marketing many types of fish, they are more likely to accept changes in catch composition which follow changes in recruitment success and less likely to target specific species to economic extinction. At present the increased targeting of a wider range of species tends to be a reactive response to overfishing favoured species rather than a proactive approach to sustainable management (Jennings & Polunin, 1996c).

17.4 Multispecies and ecosystem-based management

17.4.1 What are the objectives?

The formulation of strategy and action for multispecies and ecosystem-based management is particularly difficult because interactions are complex or poorly understood. Thus, an objective such as 'maintaining a healthy ecosystem' cannot necessarily be translated into a harvest strategy and management action. As such, contemporary multispecies and ecosystem management means dealing with specific problems as they emerge. This approach is not holistic. For example, if the impacts of trawling are leading to reductions in benthic diversity then managers try to limit trawling impacts in order to reverse this trend. Ecosystem-based management does not, at the present time, mean adjusting the pattern of fisheries exploitation to regulate productivity or other fundamental processes which many would view as the real meaning of the term. Rather, it just means taking a wider view than the single species approach.

17.4.2 What can be achieved?

There have been various examples of attempts to manage multispecies communities, but few were based on the levels of scientific understanding equivalent to those used in single-species fisheries. For example, anchovy were intensively fished off Namibia in an attempt to shift the ecosystem to a phase dominated by the more valuable pilchard (Shelton, 1992) and predatory fishes have been cleared from a portion of the Spanish continental slope with the hope of establishing a shrimp fishery (Sugihara et al., 1984). There was also a proposal to fish for spiny dogfish and winter skate on the Georges Bank to encourage the recovery of more valuable demersal species (Rothschild, 1991). Other proposals have focused on marine mammal–fishery interactions. Thus there have been calls to limit krill fishing in case it delayed the recovery of whale populations (Lyster, 1985) and calls to safeguard catches in south-west African fisheries by killing seabirds and fur seals. As we saw in section 15.3.1, a detailed analysis of the south-west African fisheries suggests a cull of fur seals was more likely to reduce yield because of the complex interactions between species (Yodzis, 1998; Crawford et al., 1992).

In most cases the scientific understanding of relationships between species is too weak to make realistic assessment of indirect fishing effects, and management would be unlikely to have the desired effects. Intuition is not a good substitute for scientific understanding, and the idea that valuable predatory fish at the highest trophic levels should be caught to boost fisheries for their prey may result in many fishers trying to do their best for the ecosystem (Jennings & Kaiser, 1998).

At present, multispecies models are not widely used to provide management advice, but they do help to inform decisions on mesh size in the North Sea, to provide improved mortality estimates for single-species assessments, and to set TACs for capelin *Mallotus villosus* in the north-west Atlantic. The capelin are managed on a conservative basis so as to preserve them as forage species for the more valuable cod *Gadus morhua*. Multispecies analyses will never be applied to many fisheries, because their data requirements are so great. Moreover, while scientists still argue over the factors that control the dynamics of 'simple' predator–prey systems on rocky shores, the short-term prospects of understanding the relative roles of environment, competition and predation in most multispecies fisheries are not promising.

In a few exceptional circumstances, models of ecosystem interactions have been empirically validated and can provide the basis for managing ecosystem function. On East African coral reefs, relationships between fishers, fish, urchins, corals and algae are well known (section 14.7), and deliberate removal of sea urchins can be used to promote the recovery of reefs that shifted to alternate stable states following fishing (McClanahan *et al.*, 1996). After removing sea urchins from unfished experimental plots, the predicted increases in fish abundance, fish species richness and algal cover followed within 1 year. Here, the system was relatively simple and well understood, so quantitative sea-urchin reduction targets could be implemented. The manipulation of ecosystem state is likely to be a lot harder in most systems and managers are unlikely to favour ecosystem-based approaches without good evidence for system level responses to fishing. Thus, in the North Sea, where strong links between predator and prey dynamics are rarely observed, there has been greater emphasis on managing species individually and in small groups.

The largest step forward in ecosystem management will occur when we can start to set strategies based on quantitative indicators. On reefs, for example, we might aim to maintain their function as carbon sinks by setting acceptable accretion rates. For this to happen, research on the ecosystem effects of fishing will have to focus on the quantitative understanding of process rather than the description of effects. Some of the ecosystem models described in section 8.7 may provide a basis for doing this, although their validity will depend on the degree of simplification that can be achieved without losing realism.

17.5 Managing fisheries for conservation

For many years, the overriding objective of fishery management was to maximize yield. However, species and habitat conservation have become increasingly important objectives; a reflection of public and scientific concern about the wider effects of fishing. Throughout this book we have seen examples of non-target species such as dolphins, skates and seabirds that are impacted by fishing. In coming years we feel that species and habitat conservation will become the dominant management objectives in several fisheries. In much of the developed world, fishing is now seen as a threat to the environment rather than a source of protein and income. Many international treaties have begun to identify the environmental impacts of fishing as undesirable.

17.5.1 Endangered species

Species of fish, marine mammals, invertebrates and seabirds have been identified as endangered by fishing. As approaches for listing endangered species improve and publicity for the lists increases, the protection of endangered species will become a primary management objective (Norse, 1993; Hawkins *et al.*, 2000; Musick, 2000). The white abalone *Haliotis sorenseni*, once common on the coast of Baja California, is now close to extinction following exploitation. The barndoor skate *Dipturus laevis* of the north-west Atlantic, once abundant on demersal fishing grounds from Cape Hatteras to George's Bank, is now relatively rare, and the Banggai cardinalfish *Pterapogon kauderni*, a small species that lives in shallow water is threatened with extinction following the activities of aquarium fish collectors (Casey & Myers, 1998; Roberts & Hawkins, 1999). Many other marine species have been locally extirpated by fishing (section 12.4.1). Although species with restricted ranges, low rates of intrinsic population increase and subject to Allee effects may be particularly vulnerable, fishing is unlikely to cause extinctions on a scale comparable with, for example, those due to the hunting of terrestrial birds on the Pacific or Indian Ocean Islands. Habitat development, reclamation, pollution and the indirect effects of fishing on habitats, such as coral reefs, remain far greater threats. However, because other threats exist, the impacts of fishing may have greater relative significance. Thus, catch rates that might be acceptable in a pristine ocean environment are not acceptable for species impacted by other activities.

Many fisheries scientists have been sceptical about the potential of fisheries to drive exploited marine species to extinction. Indeed, there was an uproar from some scientists and administrators when Atlantic cod *Gadus morhua* and haddock *Melanogrammus aeglefinus* were listed as vulnerable to extinction the 1996 Red List of Threatened Animals (IUCN 1996). It is hard not to take it personally when a respected organization suggests that the species you are managing with the goal of maximizing long-term yields actually has a good chance of becoming extinct! The IUCN assesses species according to a rigid set of criteria, including rates of population decline. Thus, cod and haddock easily met the thresholds of declining by 20% in the previous 10 years or 3 generations. Indeed, southern bluefin tuna *Thunnus maccoyii* met the 80% decline threshold, triggering a listing of critically endangered. Some biologists argue that the criteria of population decline are inappropriate for many fish stocks, because their structure is highly variable in space and time (Chapter 4) and because management principles dictate that stocks should be reduced deliberately to some extent, in order increase production (Chapter 7). Furthermore, it has been argued that high potential rates of increase of most fishes give them a much greater ability to recover than the terrestrial birds and mammals for which declines deserve to be taken more seriously. Other biologists counter that declines of the order of 95–99% (as in the case of southern bluefin tuna) indicate that management is out of control. Furthermore, they worry about the continuing failure of some stocks with high reproductive potential to recover from collapses, despite moratoria on fishing. At the time of writing, cod stocks around eastern Canada are an example. The jury is still out on this debate, which is leading to re-evaluations of the likelihood of driving marine species to extinction, as well as how to assess such risks (Beverton, 1990; ICES, 1997b; Matsuda *et al.*, 1997; Musick, 1999; Reynolds & Mace, 1999; Roberts & Hawkins, 1999; Pope *et al.*, 2000).

In multispecies fisheries, fishing leads to high mortality of non-target species that share grounds with target species. Many of the larger skates and rays can only tolerate low rates of fishing mortality because their potential rates of population increase are also low (section 13.2.2). The large size of these fish at hatching, in many cases so large that they will immediately be retained in fishing nets used to catch other species, means that fishing is a real threat. In some cases, bans on unselective gears such as trawls or closure of areas to fishing may be the only way to reduce fishing mortality rates and prevent extirpation.

Some bird species are taken as bycatch at rates exceeding those that are predicted to drive them to extinction (Chapter 13). Prior to the introduction of mitigation measures in 1989, the southern bluefin *Thunnus maccoyii* tuna fleet set 107 million hooks and was estimated to kill 44 000 albatrosses as bycatch. In 1995, long-lining was listed as a threat to wandering albatross under the Australian Endangered Species Protection Act. Models suggest that the number of hooks set has to be reduced to 41 million per year to prevent decline to extinction (Chapter 13). Given that

many international organizations with a strong political following are powerful lobbyists for bird conservation, it is likely that severe restrictions will be imposed on fisheries that ultimately threaten bird populations.

17.5.2 Habitats

Habitat conservation is now a management objective in several fisheries. In the United States, 1996 amendments to Magnuson–Stevens Fishery Conservation and Management Act require fisheries managers to describe and identify essential fish habitat, to minimize to the extent practicable adverse effects on such habitat caused by fishing, and to identify other actions to encourage the conservation and enhancement of such habitat (Kurland, 1998). **Essential fish habitat** (EFH) has been defined as 'those waters and substrate necessary for spawning, breeding, feeding or growth to maturity'. 'Necessary', at least to the National Marine Fisheries Service, means habitat required to support a sustainable fishery. EFH has now been designated for all species managed under federal fishery management plans in the United States, based on the best available information. As the habitat requirements of a large number of species are described it is likely that the impacts of fishing gears will be controlled on a much larger scale than ever before.

Habitat protection can be achieved using gear restrictions and protected areas. Dynamite fishing has already been banned, although not necessarily effectively, on many reefs. Habitats such as coral reefs may be valuable for other purposes such as dive tourism, and economic values are increasingly assigned to them (Spurgeon, 1992; Costanza *et al.*, 1997). This opens the possibility that fishers will be charged for habitat damage in the same way that ship owners are forced to pay compensation if they damage reefs in tourist areas such as the Caribbean and Red Sea.

Gear restrictions have the advantage that they do not preclude all fishing. A good example of this situation is the voluntary agreement that exists between the South Devon and Channel Fishermen Ltd (potters), the Trawler Owners Association (towed gear fishers) that operate off Salcombe in the English Channel (Chapter 6; Hart, 1998). Towed bottom-fishing gear is excluded from certain areas of the seabed that are reserved for use by pot fishers. Pots and traps have a much less severe impact on benthic communities and habitats than towed bottom-fishing gear. A comparative study of benthic fauna in the areas reserved for potting and those open to fishing, either all year or periodically, demonstrated that species sensitive to bottom-fishing disturbance were most abundant in the areas fished by pots (Kaiser *et al.*, 2000). This study demonstrates how it is possible to continue exploiting a fishery (the brown crab *Cancer pagurus* fishery in this case) using gear that causes little damage to the environment. The segregation of these two sectors of the local fishing industry also reduces the likelihood of trawlers snagging fixed gear and towing it away so that it becomes lost and may continue to fish for many years (section 13.7).

Gear restrictions can limit the type of ground that can be fished and create refuges from fishing. For example, bans on the use of large bobbins on trawl ground gear could prevent fishers from trawling on rocky substrates. However, as with most effort restrictions, fishers would doubtless modify gear over time to counteract them.

17.5.3 Protected areas and no-take zones

A **no-take zone** (NTZ) is an area where all fishing is prohibited. Many small NTZs have been established throughout the world, ranging in size from a fraction of 1 km^2 to 10 km^2 or more. In addition, there are other types of **marine protected areas**, loosely termed **marine reserves**, where fishing may be limited to local people, recreational anglers or banned on a seasonal basis. NTZs and protected areas used solely for fisheries management are a technical measure (section 12.2.4), but we discuss them here because they are increasingly used to protect habitats and non-target species from fishing.

The impetus to establish NTZs came from countries such as Australia, New Zealand and the Seychelles, where NTZs have provided many general benefits as a tool for conservation and marine environmental management. Within NTZs, the abundance of many invertebrate and fish species increased in comparison with adjacent fished areas and NTZs have provided income for local people by attracting divers and 'eco-tourists' (Roberts & Polunin, 1991, 1993; Rowley, 1994; Bohnsack, 1998; Mosquera *et al.*, 2000).

When used as an action for managing target species NTZs are a technical measure (section 17.2.4). Their value depends on the life history and dynamics of the target stock. Many of the stocks that support the

largest marine fisheries are highly migratory and fished by many fleets over thousands of square kilometres. The movements and dynamics of these stocks are well known. When the objective of management is to improve yields of target species, the NTZs have to reduce overall fishing mortality. For some stocks in some circumstances they will do this but their success depends on the life history of the stock and the extent to which fishing is controlled in those parts of the stocks' range outside the NTZ. This suggests that the utility of NTZs for fish stock management should be considered on a case-by-case basis and the benefits that NTZs may provide should be assessed against the benefits of other management options.

The impacts of NTZs on potential yields from fish stocks are usually predicted using dynamic pool models (section 7.7) with overall fishing mortality (F) dependent on the rate of movement between no-take and fished zones and F in the fished area (Guénette *et al.*, 1998). In the case of the North Sea cod, for example, closure of 25% of the North Sea would have had minimal impact on the rate of fishing mortality and spawning stock biomass because this wide-ranging fish would be accessible to displaced fishing effort outside the NTZ. The only way to control mortality would be through catch or effort controls. Conversely, the North Sea plaice spends the first few years of its life cycle in relatively restricted habitats, and large reductions in juvenile mortality could be achieved with a NTZ. This is because most juvenile plaice would remain within the NTZ and would not be accessible to displaced fishing effort. However, as they mature, the plaice would begin to migrate outside the NTZ, boosting yields in the fishery. A protected area of around 38 000 km^2 has been introduced as part of the management strategy for North Sea plaice. This has reduced fishing effort and increased yield to the fishery. At present, there are derogations for smaller beam trawlers to enter the box and fish for shrimp, but they also take plaice as bycatch. If these boats were excluded, and the box became a true NTZ, then potential benefits to the fishery should be even greater (ICES, 1994c; Piet & Rijnsdorp, 1998). For management of migratory stocks that are targeted by flexible fleets, the value of NTZs needs to be assessed on a case-by-case basis (Horwood *et al.*, 1998, 2000). In many cases, NTZs will only be effective when used in conjunction with catch and effort controls.

For smaller multispecies fisheries NTZs may provide

important safeguards against ineffective management, in accordance with the precautionary approach (section 17.3.4), and help with management in areas where there are no resources to implement complex management strategies. Moreover, in many inshore reef fisheries NTZs are a logical extension of existing tenure arrangements and likely to be accepted by fishers. Multispecies assessment has not provided the basis for management in most large and well-studied fisheries so there is little prospect of applying multispecies assessment in artisanal fisheries. As such, management will have to be implemented without perfect scientific knowledge. NTZs may provide a means of doing this. Empirical studies have shown that there are increases in the abundance and spawning stock biomass of a proportion of target species in NTZs. Although benefits cannot be expected for all species, because they have different life histories, many fished species are relatively site-attached after recruitment to the reef and a NTZ will help to reduce fishing mortality on part of the population. There is a growing consensus that networks of NTZs will provide sources and sinks for larval production, guard against fishery collapse and may maintain recruitment to heavily fished areas (Bohnsack, 1998).

If fish spill over from the reserve to adjacent fished areas, they may enhance yields. However, the empirical evidence for such effects is quite limited. At a Kenyan marine reserve, catches just outside the reserve boundary were higher but they did not compensate for those that were lost as a result of closure of part of the fishing area (McClanahan & Kaundaarara, 1996). Conversely, a study at Sumilon Island in the Philippines suggests that aggregate catch rates were higher when a marine reserve was in operation than when it was not (Russ & Alcala, 1996b). The differences between studies are not surprising as the extent of spill-over will depend on the species in the multispecies community and the size and siting of the reserve.

Temporary NTZs could be used to implement rotational harvesting strategies. Such strategies may have a useful role in fisheries management but are rarely used. This is unfortunate because they are consistent with traditional management methods that evolved without the input of modern fishery managers and because fishers who are not competing for access to a resource often choose to harvest on a rotational basis. Temporary NTZs are particularly suited to shellfish and crustacean fisheries, because closure allows stocks to

rebuild to levels that make harvesting cost effective and opening reduces the stocks to levels that remain in the most productive growth phase. In addition, time and area closures are relatively easy to enforce and likely to be understood and appreciated by fishers.

It is relatively easy to enforce fishing bans in NTZs if the will and resources exist: quite simply, any fisher working in the reserve is breaking the rules. In some cases, however, the biomass of fish in managed marine reserves has increased dramatically and they have become victims of their own success by becoming magnets for poachers. A particular concern is that the high biomass of large and relatively fecund spawning fish in the reserve is rapidly reduced by very low levels of poaching activity. Thus, some minor poaching and fishing activity in a Seychelles marine reserve dramatically reduced the biomass of target groupers and snappers (Jennings et al., 1996). The more effectively a reserve is working the more carefully restrictions must be enforced. This process is helped if the community sees the wider benefits of the reserve and supports its operation.

NTZs are widely used to address management objectives other than those based on yield and profit. They are likely to play an increasingly important role in protecting non-target species and habitats of conservation concern. NTZs protect habitats from destructive fishing methods, maintain diversity and prevent adverse ecosystem shifts. NTZs also provide important sites for study and experimentation, including control sites for studies of fishing effects. They have been widely used for education, tourism and to increase public awareness and understanding of environmental issues. Net economic benefits from some of these activities can far exceed those from fisheries. Thus, in the Seychelles, over one-third of all tourists arriving in the islands will pay to visit a marine protected area. NTZs are likely to play an increasing role in safeguarding non-target species. This could include closure of fisheries adjacent to seabird nesting colonies and closure of areas where vulnerable species are taken as bycatch. NTZs are particularly applicable to relatively site-attached species that are almost impossible to exclude from catches, such as skates, rays and groupers.

There have been many problems sustaining the effective management of NTZs, particularly when they were not imposed with full support and involvement of local fishers. Russ and Alcala (1994) described 20 years of hopes and frustrations on Sumilon Island in the central Philippines. An NTZ was established there in 1974, but there have been several major breakdowns of management. Remarkably, this was at a site where the aggregate fisheries yield from the island was higher when the NTZ was operating (Russ & Alcala, 1996a). One of the key problems with management on the island seemed to be that fishers were not convinced of the real benefits of the NTZ and were not fully involved in its management. There are many other examples of the repeated breakdown of NTZs in tropical reef fisheries. As we discussed in the section on co-management, management strategies that are imposed from outside the community are unlikely to work in societies where fisheries are prosecuted by many small-scale fishers and there are few, if any, resources for enforcement.

17.6 Future trends

17.6.1 Fisheries science

We have warned repeatedly about the problems of trying to predict the future, but we cannot resist trying to do it here. Fisheries science is changing rapidly. Bioeconomic analyses are increasingly important in the assessment and management of fisheries and there is more concern than ever before about the impacts of fishing on the marine environment. Many laboratories that formerly worked on stock assessment are now concerned with the impacts of fishing on seabirds, mammals, rare fishes, habitats and ecosystem processes. The rapid diversification of the field has resulted in a lot of science but it is not always clear how this can be used to meet specific management objectives. In the next decade there is likely to be much more emphasis on producing quantitative indicators of the impacts of fishing and setting quantitative targets for managers.

In the future, we predict that there will be more research on the ways in which fishing affects ecosystem processes such as productivity and stability, and how it alters the role of ecosystems as nutrient stores or carbon sinks. This may provide the basis for setting management strategies such as limits to the frequency and intensity of trawling. The impacts of fishing are likely to be measured in economic terms and costed against the potential benefits. We expect to see much more research on biological reference points and precautionary approaches to dealing with uncertainty. For species

of conservation concern, scientists are likely to identify better indicators of vulnerability to mortality and conservation biologists will direct more and more research effort to fisheries-related issues. Behavioural ecologists may start to have more input to gear design, gear operation and strategies for bycatch reduction. Even very basic knowledge, such as the recognition that turtles need to surface every hour or so to breathe can be usefully employed. Thus, turtle mortality in prawn trawls off Queensland is almost nil when the nets are towed for 60 minutes but rises to 50% with 200-minute tows (Chapters 5 and 13).

The burden of proof for showing that fishing does not have adverse environmental effects is likely to shift towards the fisher, though we should be aware that excessive restrictions on the fishing industry could encourge society to meet its protein requirements from less desirable sources. Since any human activity will impact the environment, the effects of fishing must be costed against the effects of obtaining similar protein yields elsewhere. Such cost analyses have hardly been attempted, and yet the results may show that fishing is a better way of obtaining protein than intensive farming and aquaculture.

17.6.2 Fisheries management

In the short term, we expect that fisheries will be increasingly over-exploited and the ecosystem effects of fishing will cause more and more concern to conservation and pressure groups with an interest in the marine environment. However, this concern is likely to increase pressure for major structural changes in fisheries, and significant capacity reduction will probably follow, whether driven by the removal of subsidies, the collapse of stocks, or the purchase of excess capacity. Ownership of fishing rights and fish stocks is likely to become the norm and fishers will become more actively involved in management decisions. We expect that many management decisions will be based on non-target species and habitats and that fisheries will be closed to protect species of conservation concern. Conservation groups may even start to buy fishing rights, but not use them, in order to protect species and habitats. Marine reserves are likely to be used more to protect species and habitats of conservation concern and to provide insurance against management failure. New technologies, such as satellite monitoring, will probably improve enforcement and compliance. Within a few decades we would expect that the fisheries of the developed world will be characterized by lower employment, higher profitability and less conflict.

In the developed world, we also expect that consumers will influence the actions of fishers. There will be increased demand for fish products of high quality that are 'ethically' sourced, or even 'humanely' killed, and these will fetch higher prices. Codes such as the FAO code of conduct for sustainable fisheries (FAO, 1995c), may be used as a basis for certifying fish from 'sustainable' sources. Supermarkets and buyers, under pressure from the public, may refuse to deal with fishers and nations that support activities such as whaling. Aquaculture is likely to become increasingly important

for enhancing the production of valuable species but we expect widespread concern about the impacts of genetically modified fish and shellfish.

With more groups interested in management of the marine environment, more objectives are likely to be considered and the trade-offs between them more widely explored. This will tend to be done at the outset and not as a result of *ad hoc* adjustments by lobbyists. We expect to see a growing dichotomy between the sustainability, profitability and capacity for species and habitat conservation in fisheries of the developing and developed world, unless the international community intervenes to provide people in the poorest countries with alternative (not fished) sources of food and income.

Summary

• Management actions are taken to meet a management objective. Actions can be categorized as catch controls, effort controls and technical measures. Management actions only help to meet an objective if they are accepted or enforced.

• Catch controls that give fishers rights to the resource, used in conjunction with technical measures, are often the most effective management actions.

• Stock assessments and the future are uncertain, so managers are increasingly adopting a precautionary approach to management.

• Quantitative understanding of the responses to fishing is desirable because it allows managers to set strategies that are clearly understood.

• Fisheries science has provided a useful analytical framework for single-species assessment and management, but multispecies and ecosystem management is more difficult because the supporting science is often not so well developed.

• In the future fisheries science is likely to become an increasingly broad discipline and quantitative approaches to ecosystem analysis will be developed. Fishery managers are likely to be more concerned with economic and social issues and the conservation of non-target species and habitats.

• In the future fishers will probably have to pay for access to the marine environment or to land catches. There are likely to be fewer fishers but their activities will be more profitable. New technologies are expected to improve enforcement and compliance.

• If the international community does not help people from poor countries to access alternative sources of food and income, we expect to see a growing gap between the sustainability, profitability and capacity for species and habitat conservation in fisheries of the developing and developed world.

Further reading

The variety and utility of management actions used within OECD member countries are described by OECD (1997). Chapters in Polunin and Roberts (1996) consider management and enforcement in reef fisheries and developing countries. Clark (1985), Gulland (1988), Hilborn and Walters (1992) and Quinn and Deriso (1999) all contain comprehensive discussions of management issues. Rosenberg (1998) predicts the future for fisheries science and management.

References

Adams, N.J., Seddon, P.J. & van Heezik, Y.M. (1992) Monitoring of seabirds in the Benguela upwelling system: can seabirds be used as indicators and predictors of change in the marine environment? *South African Journal of Marine Science* **12**, 959–974.

Adams, P.B. (1980) Life history patterns in marine fishes and their consequences for management. *Fishery Bulletin* **78**, 1–12.

Adams, T.J.H. (1996) Modern institutional framework for reef fisheries management. In: *Reef Fisheries* (eds N.V.C. Polunin & C.M. Roberts), pp. 337–360. Chapman & Hall, London.

Addison, J.T. (1995) Influence of behavioural interactions on lobster distribution and abundance as inferred from pot-caught samples. *ICES Marine Science Symposia* **199**, 294–300.

Admiraal, W. (1984) The ecology of estuarine sediment inhabiting diatoms. In: *Progress in Phycological Research* (eds F.E. Round & D.J. Chapman), pp. 269–322. Biopress, Bristol.

Alheit, J. (1987) Egg cannibalism versus egg production: their significance in anchovies. *South African Journal of Marine Science* **5**, 467–470.

Allee, W.C. (1931) *Animal Aggregations: a Study in General Sociology*. University of Chicago Press, Chicago.

Allen, K.R. (1963) Analysis of stock-recruitment relations in Antarctic fin whales. *Rapports et Procès-Verbaux des Réunions, Conseil International pour l'Exploration de la Mer* **164**, 132–137.

Allen, K.R. (1980) *Conservation and Management of Whales*. University of Washington Press, Seattle.

Allen, R.L. (1985) Dolphins and the purse-seine fishery for yellowfin tuna. In: *Marine Mammals and Fisheries* (eds J.R. Beddington, R.J.H. Beverton & D.M. Lavigne), pp. 236–252. George Allen and Unwin, London.

Allsopp, W.H.L. (1993) *Aquaculture in Southeast Asia*. International Development Research Centre, Ottawa.

Allsopp, W.H.L. (1997) World aquaculture review: performance and perspectives. In: *Global Trends: Fisheries Management* (eds E.L. Pikitch, D.D. Huppert & M.P. Sissenwine), pp. 153–165. American Fisheries Society Symposium, Bethesda, Maryland.

Alvarez-Ossorio, M. (1977) Un estudia de la Ria de Muros en Noviembre de 1975. *Boletín Instituto Español Oceanografía* **2**, 1–223.

Alverson, D.L., Freeberg, M.H., Pope, J.G. & Murawski, S.A. (1994) A global assessment of fisheries bycatch and discards. *FAO Fisheries Technical Paper* **339**, 233.

Andersen, K.P. & Ursin, E. (1977) A multispecies extension to the Beverton and Holt theory of fishing, with accounts of phosphorus circulation and primary production. *Meddelelser Fra Danmarks Fiskeri-Og Havundersogelser* **7**, 310–435.

Andre, C. & Lindegarth, M. (1995) Fertilisation efficiency and gamete viability of a sessile free-spawning bivalve, *Cerastoderma edule*. *Ophelia* **43**, 215–227.

Angel, M. & Rice, A. (1996) The ecology of the deep ocean and its relevance to global waste management. *Journal of Applied Ecology* **33**, 915–926.

Anker-Nilssen, T., Barrett, R.T. & Krasnov, J.V. (1997) Long- and short-term responses of seabirds in the Norwegian and Barents Seas to changes in stocks of prey fish. In: *Forage Fishes in Marine Ecosystems* (ed. Anon), pp. 683–698. University of Alaska Sea Grant College Program, University of Alaska, Fairbanks.

Anon. (1885) *Report of the commissioners appointed to inquire and report upon the complaints that have been made by line and drift net fishermen of injuries sustained by them in their calling owing to the use of the trawl net and beam trawl in the territorial waters of the United Kingdom*. Eyre and Spottiswoode, London.

Anon. (eds) (1992) Age determination and growth in fish and other aquatic animals. *Australian Journal of Marine and Freshwater Research* **43**, 879–1330.

Anon. (1997) *Estimation of Fish Biomass in the Irish Sea by Means of the Annual Egg Production Method*. MAFF Directorate of Fisheries Research, Lowestoft.

Appeldoorn, R.S. (1995) Covariation in life-history parameters of soft-shell clams (*Mya arenaria*) along a latitudinal gradient. *ICES Marine Science Symposia* **199**, 19–25.

Arnold, G.P. & Metcalfe, J.D. (1989) Fish migration: orientation and navigation or environmental transport. *Journal of Navigation* **42**, 367–374.

Arnold, G.P., Walker, M.G., Emerson, L.S. & Holford, B.H. (1994) Movements of cod (*Gadus morhua* L.) in relation to the tidal streams in the southern North Sea. *ICES Journal of Marine Science* **51**, 207–232.

Arntz, W.E. & Weber, W. (1970) *Cyprina islandica* (Mollusca: Bivalvia) als Nahrung von Dorsch und Kliesche in der Kieler Bucht. *Berichte der Deutschen Wissenschaftlichen Kommission für Meeresforschung* **21**, 193–209.

Atkinson, R.J.A. & Nash, R.D.M. (1990) Some preliminary observations on the burrows of *Callianassa subterranea* (Montagu) (Decapoda: Thalassinidea) from the west coast of Scotland. *Journal of Natural History* 24, 403–413.

Augustin, N.H., Borchers, D.L., Clarke, E.D., Buckland, S.T. & Walsh, M. (1998) Spatiotemporal modelling for the annual egg production methods of stock assessment using generalized additive models. *Canadian Journal of Fisheries and Aquatic Sciences* 55, 2608–2621.

Auster, P.J. & Langton, R.W. (1999) Indirect effects of fishing. In: *Fish Habitats: Essential Fish Habitat and Rehabilitation* (ed. L.R. Benaka), pp. 150–187. American Fisheries Society Symposium 22, American Fisheries Society, Bethesda, Maryland.

Auster, P.J., Malatesta, R.J., Langton, R.W., Watling, L., Valentine, P.C., Donaldson, C.L., Langton, E.W., Shepard, A.N. & Babb, I.G. (1996) The impacts of mobile fishing gear on seafloor habitats in the Gulf of Maine (Northwest Atlantic): implications for conservation of fish populations. *Reviews in Fisheries Science* 4, 185–202.

Axelrod, R. (1984) *The Evolution of Co-operation*. Basic Books, New York.

Azam, F., Fenchel, T., Field, J.G., Gray, J.S., Meyer-Reil, L.A. & Thingstad, F. (1983) The ecological role of water-column microbes in the sea. *Marine Ecology Progress Series* 10, 257–263.

Bagenal, T.B. (eds) (1974) *Ageing of Fish*. Unwin Brothers Ltd, Woking.

Bagenal, T.B. & Braum, E. (1978) Eggs and early life history. In: *Methods for Assessment of Fish Production in Freshwaters* (ed. T.B. Bagenal), pp. 165–201. Blackwell Scientific Publications, Oxford.

Bagenal, T.B. & Tesch, F.W. (1978) Age and growth. In: *Methods for the Assessment of Fish Production in Freshwaters* (ed. T.B. Bagenal), pp. 101–136. Blackwell Science, Oxford.

Bailey, C. (1988) The social consequences of tropical shrimp mariculture development. *Ocean and Shoreline Management* 11, 31–44.

Bailey, K.M. & Houde, E.D. (1989) Predation on eggs and larvae of marine fishes and the recruitment problem. *Advances in Marine Biology* 25, 1–83.

Bailey, R.F.G. & Dufour, R. (1987) Field use of an injected ferromagnetic tag on the snow crab (*Chionoecetes opilio* O. Fab). *Journal du Conseil, Conseil International pour l'Exploration de la Mer* 43, 237–244.

Baldwin, B., Borichewski, J. & Lutz, R.A. (1995) Predation on larvae: filtration by adult bivalves directly and indirectly kills bivalve larvae. *Proceedings of the Twenty Third Benthic Ecology Meeting*, Rutgers State University, New Brunswick.

Bannerman, N. & Jones, C. (1999) Fish-trap types: a component of the maritime cultural landscape. *International Journal of Nautical Archaeology* 28, 70–84.

Bannister, R.C.A., Addison, J.T. & Lovewell, S.R. (1994) Growth, movement, recapture rate and survival of hatchery-reared lobsters (*Homarus gammarus* Linnaeus, 1758) released into the wild on the English east coast. *Crustaceana* 67, 156–172.

Barbieri, L.R., Chittenden, M.E. Jr & Jones, C.M. (1997) Yield-per-recruit analysis and management strategies for Atlantic croaker, *Micropogonias undulatus*, in the Middle Atlantic Bight. *Fishery Bulletin* 95, 637–645.

Barnes, R.S.K. & Hughes, R.N. (1999) *An Introduction to Marine Ecology*. Blackwell Science, Oxford.

Basson, M., Rosenberg, A.A. & Beddington, J.R. (1988) The accuracy and reliability of two new methods for estimating growth parameters from length frequency data. *Journal du Conseil, Conseil International pour l'Exploration de la Mer* 44, 277–285.

Basson, M., Beddington, J.R., Crombie, J.A., Holden, S.J., Purchase, L.V. & Tingley, G.A. (1996) Assessment and management techniques for migratory annual squid stocks: the *Ilex argentinus* fishery in the Southwest Atlantic as an example. *Fisheries Research* 28, 3–27.

Baumann, M. (1995) A comment on transfer efficiencies. *Fisheries Oceanography* 4, 264–266.

Baumgartner, T.R., Soutar, A. & Ferreira-Bartrina, V. (1992) Reconstruction of the Pacific sardine and northern anchovy population over the past two millenia from sediments of the Santa Barbara basin, California. *California Co-operative Fishery Investigations Report* 33, 24–40.

Bax, N.J. (1991) A comparison of the fish biomass flow to fish, fisheries, and mammals in six marine ecosystems. *ICES Marine Science Symposia* 193, 217–224.

Bayliss-Smith, T. (1990) Atoll production systems: fish and farming on Ontong Java Atoll, Solomon Islands. *Occasional Papers in Prehistory* 18, 57–69.

Beamish, R.J. (ed.) (1995) *Climate change and northern fish populations*. Canadian Special Publications of Fisheries and Aquatic Sciences 171.

Bech, G. (1995) Retrieval of lost gillnets at Ilulissat Kangia. *NAFO Scientific Council Research Document* 95/6.

Beckman, D.W. & Wilson, C.A. (1995) Seasonal timing of opaque zone formation in fish otoliths. In: *Recent Developments in Fish Otolith Research* (eds D.H. Secor, J.M. Dean & S.E. Campana), pp. 27–44. University of South Carolina Press, Columbia.

Beddington, J.R. & Cooke, J.G. (1983) The potential yield of fish stocks. *FAO Fisheries Technical Paper*, 242.

Beddington, J.R., Beverton, R.J.H. & Lavigne, D.M. (eds) (1985) *Marine Mammals and Fisheries*. George Allen and Unwin, London.

Beddington, J.R., Rosenberg, A.A., Crombie, J.A. & Kirkwood, G.P. (1990) Stock assessment and the provision of management advice for the short fin squid fishery in Falkland Islands waters. *Fisheries Research* 8, 351–365.

Begon, M., Harper, J.L. & Townsend, C.R. (1996a) *Ecology: Individuals, Populations and Communities*. Blackwell Science, Oxford.

Begon, M., Mortimer, M. & Thompson, D.J. (1996b) *Population Ecology: a Unified Study of Animals and Plants* (3rd edn). Blackwell Science, Oxford.

Bell, G. (1982) *The Masterpiece of Nature: the Evolution and Genesis of Sexuality*. Croom-Helm. London.

Berger, W.H. & Herguera, J.C. (1992) Reading the sedimentary record of the ocean's productivity. In: *Primary Productivity and Biogeochemical Cycles in the Sea* (eds P.G. Falkowski & A.D. Woodhead), pp. 455–486. Plenum Press, New York.

Berger, W.H., Fischer, K., Lai, C. & Wu, G. (1987) *Ocean productivity and organic carbon flux. Part 1. Overview of maps of primary production and export production*. Scripps Institute of Oceanography, University of California, San Diego.

Bergman, M.J.N. & Hup, M. (1992) Direct effects of beamtrawling on macrofauna in a sandy sediment in the southern North Sea. *ICES Journal of Marine Science* **49**, 5–11.

Beukema, J.J. (1995) Long-term effects of mechanical harvesting of lugworms *Arenicola marina* on the zoobenthic community of a tidal flat in the Wadden Sea. *Netherlands Journal of Sea Research* **33**, 219–227.

Beverton, R.J.H. (1963) Maturation, growth and mortality of Clupeid and Engraulid stocks in relation to fishing. *Rapports et Procès-Verbaux des Réunions, Conseil International pour l'Exploration de la Mer* **154**, 44–67.

Beverton, R.J.H. (1985) Analysis of marine mammal–fisheries interactions. In: *Marine Mammals and Fisheries* (eds J.R. Beddington, R.J.H. Beverton & D.M. Lavigne), pp. 3–33. George Allen and Unwin, London.

Beverton, R.J.H. (1987) Longevity in fish: some ecological and evolutionary perspectives. In: *Ageing Processes in Animals* (eds A.D. Woodhead, M. Witten & K. Thompson), pp. 161–186. Plenum Press, New York.

Beverton, R.J.H. (1990) Small pelagic fish and the threat of fishing: are they endangered? *Journal of Fish Biology* **42**, 6–16.

Beverton, R.J.H. (1992) Patterns of reproductive strategy parameters in some marine teleost fishes. *Journal of Fish Biology* **41** (Suppl. B), 137–160.

Beverton, R.J.H. (1995) Spatial limitation of population size; the concentration hypothesis. *Netherlands Journal of Sea Research* **34**, 1–6.

Beverton, R.J.H. & Holt, S.J. (1956) A review of methods for estimating mortality rates in fish populations, with special reference to sources of bias in catch sampling. *Rapports et Procès-Verbaux des Réunions, Conseil International pour l'Exploration de la Mer* **140**, 67–83.

Beverton, R.J.H. & Holt, S.J. (1957) *On the Dynamics of Exploited Fish Populations*. Ministry of Agriculture, Fisheries and Food, London (republished by Chapman & Hall, 1993).

Beverton, R.J.H. & Holt, S.J. (1959) A review of the lifespan and mortality rates of fish in nature and their relationship to growth and other physiological characteristics. *Ciba Foundation Colloquim on Ageing* **5**, 142–180.

Beverton, R.J.H. & Iles, T.C. (1992) Mortality rates of O-group plaice (*Pleuronectes platessa* L.), dab (*Limanda limanda* L.) and turbot (*Scophthalmus maximus* L.) in European waters. III Density-dependence of mortality rates of O-group plaice and some demographic implications. *Netherlands Journal of Sea Research* **29**, 61–79.

Bhattacharya, C.G. (1967) A simple method of resolution of a distribution into Gaussian components. *Biometrics* **23**, 115–135.

Billet, D.S.M., Lampitt, R.S., Rice, A.L. & Mantoura, R.F.C. (1983) Seasonal sedimentation of phytoplankton to the deep sea benthos. *Nature* **302**, 520–522.

Birkeland, C. (ed.) (1997) *Life and Death of Coral Reefs*. Chapman & Hall, New York.

Bjorndal, K.A. (ed.) (1982) *Ecology and Conservation of Sea Turtles*. Smithsonian Institute, Washington DC.

Blaber, S.J.M. (1997) *Fish and Fisheries of Tropical Estuaries*. Chapman & Hall, London.

Black, E.A. & Truscott, J. (1994) Strategies for regulation of aquaculture site selection in coastal areas. *Journal of Applied Ichthyology* **10**, 294–306.

Blaxter, J.H.S. (ed.) (1974) *The Early Life History of Fish*. Springer-Verlag, New York.

Blaxter, J.H.S. (1986) Development of sense organs and behavior of teleost larvae with special reference to feeding and predator avoidance. *Transactions of the American Fisheries Society* **115**, 98–114.

Blaxter, J.H.S. (1987) Structure and development of the lateral line. *Biological Reviews of the Cambridge Philosophical Society* **62**, 471–514.

Blaxter, J.H.S. (1992) The effect of temperature on larval fishes. *Netherlands Journal of Zoology* **42**, 336–357.

Blaxter, J.H.S. & Ehrlich, K.F. (1974) Changes in behaviour during starvation of herring and plaice larvae. In: *The Early Life History of Fish* (ed. J.H.S. Blaxter), pp. 575–588. Springer-Verlag, Berlin.

Blaxter, J.H.S. & Fuiman, L.A. (1990) The role of sensory systems of herring larvae in evading predatory fishes. *Journal of the Marine Biological Association of the United Kingdom* **70**, 413–427.

Blaxter, J.H.S. & Hempel, G. (1963) The influence of egg size on herring larvae (*Clupea harengus* L.). *Rapports et Procès-Verbaux des Réunions, Conseil International pour l'Exploration de la Mer* **28**, 211–240.

Blaxter, J.H.S., Gamble, J.C. & Westernhagen, H.V. (eds) (1989) The early life history of fish. *Rapports et Procès-Verbaux des Réunions, Conseil International pour l'Exploration de la Mer* **191**.

Blindheim, J. & Skjoldal, H.R. (1993) Effects of climatic changes on the biomass yield of the Barents Sea, Norwegian Sea and West Greenland large marine ecosystems. In: *Large Marine Ecosystems: Stress, Mitigation and Sustainability* (eds K. Sherman, L.M. Alexander & B.D. Gold), pp. 185–189. American Association for the Advancement of Science, Washington.

Boehlert, G.W. (1996) Larval dispersal and survival in tropical reef fishes. In: *Reef Fisheries* (eds N.V.C. Polunin & C.M. Roberts), pp. 61–84. Chapman & Hall, London.

Boehlert, G.W. & Mundy, B.C. (1988) Roles of behaviour and physical factors in larval and juvenile recruitment to estuarine nursery areas. *American Fisheries Society Symposium* 3, 51–67.

Boehlert, G.W. & Mundy, B.C. (1993) Ichthyoplankton assemblages at sea-mounts and oceanic islands. *Bulletin of Marine Science* 53, 336–361.

Boerema, L.K. & Gulland, J.A. (1973) Stock assessment of the Peruvian anchovy (*Engraulis ringens*) and management of the fishery. *Journal of the Fisheries Research Board of Canada* 30, 2226–2235.

Bohnsack, J.A. (1982) Effects of piscivorous predator removal on coral reef fish community structure. In: *Gutshop '81: Fish Food Habits and Studies* (eds G.M. Caillet & C.A. Simenstad), pp. 258–267. University of Washington, Seattle.

Bohnsack, J.A. (1996) Maintenance and recovery of reef fishery productivity. In: *Reef Fisheries* (eds N.V.C. Polunin & C.M. Roberts), pp. 283–313. Chapman & Hall, London.

Bohnsack, J.A. (1998) Application of marine reserves to reef fisheries management. *Australian Journal of Ecology* 23, 298–304.

Bohnsack, J.A. & Bannerot, S.P. (1986) A stationary visual census technique for quantitatively assessing community structure of coral reef fishes. *NOAA Technical Report NMFS* 41.

Bone, Q., Marshall, N.B. & Blaxter, J.H.S. (1995) *Biology of Fishes*. Blackie, London.

Borchers, D.L., Buckland, S.T., Priede, I.G. & Ahmadi, S. (1997) Improving the precision of the daily egg production method using generalized additive methods. *Canadian Journal of Fisheries and Aquatic Sciences* 54, 2727–2742.

Borgmann, U. (1987) Models of the slope of, and biomass flow up, the biomass size spectrum. *Canadian Journal of Fisheries and Aquatic Sciences* 44 (Suppl. 2), 136–140.

Botsford, L.W., Moloney, C.L., Largier, J.L. & Hastings, A. (1998) Metapopulation dynamics of meroplanktonic invertebrates: the Dungeness crab (*Cancer magister*) as an example. *Canadian Special Publications of Fisheries and Aquatic Sciences* 125, 295–306.

Bowman, R.E. & Michaels, W.L. (eds) (1984) Food of seventeen species of northwest Atlantic fish. *NOAA Technical Memorandum. NMFS-NEFC*, 28.

Boyle, P.R. (1990) Cephalopod biology in the fisheries context. *Fisheries Research* 8, 303–321.

Boyle, P.R. & Bolettzky, S.V. (1996) Cephalopod populations: definition and dynamics. *Philosophical Transactions of the Royal Society* 351, 985–1002.

Bradshaw, C., Veale, L.O., Hill, A.S. & Brand, A.R. (2000) The effects of scallop dredging on gravely seabed communities. In: *The Effects of Fishing on Non-Target Species and Habitats: Biological, Conservation and Socio-Economic Issues* (eds M.J. Kaiser & S.J. de Groot), pp. 83–104. Blackwell Science, Oxford.

Brand, A.R., Allison, E.H. & Murphy, E.J. (1991) North Irish Sea scallop fisheries: a review of changes. In: *An International Compendium of Scallop Biology and Culture* (eds S.E. Shumway & P.A. Sandifer), pp. 204–218. World Aquaculture Society, Baton Rouge.

Brander, K. (1981) Disappearance of common skate, *Raia batis*, from the Irish Sea. *Nature* 290, 48–49.

Breder, C.M. & Rosen, D.E. (1966) *Modes of Reproduction in Fishes*. Natural History Press, New York.

Breen, P.A. (1987) Mortality of Dungeness crabs caught by lost traps in the Fraser River Estuary, British Columbia. *North American Journal of Fisheries Management* 7, 429–435.

Brett, J.R. (1979) Environmental factors and growth. In: *Fish Physiology 8: Bioenergetics and Growth* (eds W.S. Hoar, D.J. Randall & J.R. Brett), pp. 599–675. Academic Press, New York.

Brett, J.R. & Groves, T.D.D. (1979) Physiological energetics. In: *Fish Physiology 8: Bioenergetics and Growth* (eds W.S. Hoar, D.J. Randall & J.R. Brett), pp. 279–352. Academic Press, New York.

Brewer, D., Rawlinson, N., Eayrs, S. & Burridge, C. (1998) An assessment of bycatch reduction devices in a tropical Australian prawn trawl fishery. *Fisheries Research* 36, 195–215.

Brierley, A.S., Watkins, J.L. & Murray, A.W.A. (1997) Interannual variability in krill abundance at South Georgia. *Marine Ecology Progress Series* 150, 87–98.

Briggs, R.P. (1992) An assessment of nets with a square mesh panel as a whiting conservation tool in the Irish Sea Nephrops fishery. *Fisheries Research* 13, 133–152.

Britton, J.C. & Morton, B. (1994) Marine carrion and scavengers. *Oceanography and Marine Biology: an Annual Review* 32, 369–434.

Brothers, E.B., Williams, D.M. & Sale, P.F. (1983) Length of larval life in 12 families of fishes at One Tree lagoon, Great Barrier Reef, Australia. *Marine Biology* 76, 319–324.

Brothers, G. (1992) *Lost or Abandoned Fishing Gear in the Newfoundland Aquatic Environment*. Department of Fisheries and Oceans, St Johns, Newfoundland, Canada.

Brothers, N. (1991) Albatross mortality and associated bait loss in Japanese longline fishery in the Southern Ocean. *Biological Conservation* 55, 255–268.

Browder, J.A. (1993) A pilot model of the gulf of Mexico Continental Shelf. In: *Trophic Models of Aquatic Ecosystems* (eds V. Christensen & D. Pauly), pp. 279–284. ICLARM Conference Proceedings 26, ICLARM, Manilla.

Brown, K.M. & Alexander, J.E.J. (1994) Group foraging in a marine gastropod predator: benefits and costs to individuals. *Marine Ecology Progress Series* 112, 97–105.

Brylinsky, M., Gibson, J. & Gordon, D. (1994) Impacts of flounder trawls on the intertidal habitat and community of the Minas Basin, Bay of Fundy. *Canadian Journal of Fisheries and Aquatic Sciences* 51, 650–661.

Buckland, S.T., Anderson, D.R., Burnham, K.P. & Laake, J.L. (1993) *Distance Sampling: Estimating Abundance of Biological Populations*. Chapman and Hall, London.

Burger, A.E. & Piatt, J.F. (1990) Flexible time budgets in breeding common murres: buffers against variable prey availability. *Studies in Avian Biology* 14, 71–83.

Burke, W.T. (1983) Extended fisheries jurisdiction and the new law of the sea. In: *Global Fisheries: Perspectives for* the 1980s (ed. B.J. Rothschild), pp. 7–50. Springer-Verlag, Berlin.

Burke, W.T. (1994) *The New International Law of Fisheries: UNCLOS 1982 and Beyond*. Clarendon Press, Oxford.

Bustamante, R.H. & Castilla, J.C. (1987) The shellfishery in Chile: an analysis of 26 years of landings (1960–85). *Biologia Pesquera* **16**, 79–97.

Bustard, R. (1972) *Sea Turtles: their Natural History and Conservation*. Tapinger, New York.

Butterworth, D.S. & Andrew, P.A. (1984) Dynamic catch–effort models for the lake stocks in ICSEAF Divisions 1.3–2.2. *Collected Papers of the International Commission for the South-East Atlantic Fisheries* **11**, 29–58.

Buxton, C.D. (1993) Life-history changes in exploited reef fishes on the east coast of South Africa. *Environmental Biology of Fishes* **36**, 47–63.

Caddy, J.F. (ed.) (1983) Advances in assessment of world cephalopod resources. *FAO Fisheries Technical Paper* **231**.

Caddy, J.F. (ed.) (1989) *Marine Invertebrate Fisheries: Their Assessment and Management*. John Wiley & Sons, New York.

Caddy, J.F. (1998) A short review of precautionary reference points and some proposals for their use in data-poor situations. *FAO Fisheries Technical Paper* **379**.

Caddy, J.F. & Rodhouse, P.G. (1998) Cephalopod and groundfish landings: evidence for ecological change in global fisheries? *Reviews in Fish Biology and Fisheries* **8**, 431–444.

Cambridge, H.L. & Hocking, P.J. (1997) Annual primary production and nutrient dynamics of the seagrasses *Posidonia sinuosa* and *Posidonia australis* in south-western Australia. *Aquatic Botany* **59**, 277–295.

Camphuysen, C.J., Ensor, K., Furness, R.W., Garthe, S., Huppop, O., Leaper, G., Offringa, H. & Tasker, M.L. (eds) (1993) *Seabirds Feeding on Discards in Winter in the North Sea*. Netherlands Institute for Sea Research, Den Burg, Texel.

Carlson, C.A., Ducklow, H.W. & Michaels, A.F. (1994) Annual flux of dissolved inorganic carbon from the euphotic zone in the northwestern Sargasso Sea. *Nature* **371**, 405–408.

Carpenter, R.C. (1997) Invertebrate predators and grazers. In: *Life and Death on Coral Reefs* (ed. C. Birkeland), pp. 198–229. Academic Press, New York.

Carpenter, R.C., Hackney, J.M. & Adey, W.H. (1991) Measurements of primary productivity and nitogenase activity of coral reef algae in a chamber incorporating oscillatory flow. *Limnology and Oceanography* **36**, 40–49.

Carr, A. (1987) Impact of nondegradable marine debris on the ecology and survival outlook of sea turtles. *Marine Pollution Bulletin* **18**, 352–356.

Carr, H.A. & Milliken, H. (1998) Conservation engineering: options to minimize fishing's impacts to the sea floor. In: *Effects of Fishing Gear on the Sea Floor of New England* (eds E.M. Dorsey & J. Pederson), pp. 100–103. Puritan Press, Hollis, New Hampshire.

Carr, H.A., Blott, A.J. & Caruso, P.G. (1992) eds. A study of ghost gillnets in the inshore waters of southern New England. In: *MTS 92: Global Ocean Partnership*, pp. 361–367. Marine Technology Society, Washington DC.

Carvalho, G.R. & Hauser, L. (1994) Molecular genetics and the stock concept in fisheries. *Reviews in Fish Biology and Fisheries* **4**, 326–350.

Casey, J.M. & Myers, R.A. (1998) Near extinction of a large, widely distributed fish. *Science* **28**, 690–692.

Castel, J., Labourg, P.-J., Escaravage, V., Auby, I. & Garcia, M. (1989) Influence of seagrass beds and oyster parks on the abundance and biomass patterns of meio- and macrobenthos in tidal flats. *Estuarine, Coastal and Shelf Science* **28**, 71–85.

Castilla, J.C. & Fernandez, M. (1998) Small-scale benthic fisheries in Chile: on co-management and sustainable use of benthic invertebrates. *Ecological Applications* **8** (Suppl.), S124–S132.

Castonguay, M., Dutil, J.D. & Desjardins, C. (1989) Distinction between American eels (*Anguilla rostrata*) of different geographic origins on the basis of their organochlorine contaminant levels. *Canadian Journal of Fisheries and Aquatic Sciences* **46**, 836–843.

Chamberlain, G.W. (1994) Taura Syndrome and China collapse caused by new shrimp viruses. *World Aquaculture* **25**, 22–25.

Chamberlain, G.W. (1997) Sustainability of world shrimp farming. In: *Global Trends: Fisheries Management* (eds E.L. Pikitch, D.D. Huppert & M.P. Sissenwine), pp. 195–212. American Fisheries Society Symposium, Bethesda, Maryland.

Chambers, R.C. & Trippel, E.A. (eds) (1997) *Early Life History and Recruitment in Fish Populations*. Chapman & Hall, New York.

Chapman, V.J. (1977) *Wet Coastal Ecosystems*. Elsevier Science, Amsterdam.

Chapman, V.J. & Chapman, D.J. (1980) *Seaweeds and their Uses*. Chapman & Hall, London.

Charles, A. (1983) Optimal fisheries investment under uncertainty. *Canadian Journal of Fisheries and Aquatic Sciences* **40**, 2080–2091.

Charnov, E.L. (1976) Optimal foraging: the marginal value theorem. *Theoretical Population Biology* **9**, 129–136.

Charnov, E.L. (1993) *Life History Invariants: Some Explorations of Symmetry in Evolutionary Ecology*. Oxford University Press, Oxford.

Chittenden, M.E. Jr (1991) Operational procedures and sampling in the Chesapeake Bay pound-net fishery. *Fisheries* **16**, 22–27.

Choat, J.H. & Robertson, D.R. (1975) Protogynous hermaphroditism in fishes of the family Scaridae. In: *Intersexuality in the Animal Kingdom* (ed. R. Reinboth), pp. 263–283. Springer-Verlag, Berlin.

Christensen, V. (1996) Virtual population reality. *Reviews in Fish Biology and Fisheries* **6**, 243–247.

Christensen, V. & Pauly, D. (1992) ECOPATH II—a software for balancing steady-state models and calculating network characteristics. *Ecological Modelling* **61**, 169–185.

Christensen, V. & Pauly, D. (eds) (1993) *Trophic Models of Aquatic Ecosystems.* ICLARM Conference Proceedings 26, ICLARM, Manilla.

Churchill, J.H. (1989) The effect of commercial trawling on sediment resuspension and transport over the Middle Atlantic Bight continental shelf. *Continental Shelf Research* 9, 841–864.

Clark, C.W. (1976) A delayed-recruitment model of population dynamics, with an application to baleen whale populations. *Journal of Mathematical Biology* 3, 381–391.

Clark, C.W. (1985) *Bioeconomic Modelling and Fisheries Management.* John Wiley & Sons, New York.

Clark, C.W., Edwards, G. & Frielaender, M. (1973) Beverton-Holt model of a commercial fishery: optimal dynamics. *Journal of the Fisheries Research Board of Canada* 30, 1629–1640.

Clark, K.R. & Warwick, R.M. (1998) A taxonomic distinctness index and its statistical properties. *Journal of Applied Ecology* 35, 278–289.

Clarke, M.R. (1996) The role of cephalopods in the world's oceans: an introduction. *Philosophical Transactions of the Royal Society* 351, 979–983.

Clifford, H.C. (1985) Semi-intensive shrimp farming. In: *Texas Shrimp Farming Manual* (eds G.W. Chamberlain, M.G. Haby & R.J. Miget), pp. IV-15-IV-42. Publication of Texas Agricultural Extension Service, College Station, Texas.

Clough, B.F., Ong, J.E. & Gong, W.K. (1997) Estimating leaf area index and photosynthesis production in canopies of the mangrove *Rhizophora apiculata. Marine Ecology Progess Series* 159, 285–292.

Cobb, J. & Caddy, J.F. (1989) The population biology of decapods. In: *Marine Invertebrate Fisheries: Their Assessment and Management* (ed. J.F. Caddy), pp. 327–374. Wiley Interscience, New York.

Cohen, J.E. & Briand, F. (1984) Trophic links of community food webs. *Proceedings of the National Academy of Science* 81, 4105–4109.

Cole, J.J., Findlay, S. & Pace, M.L. (1988) Bacterial production in fresh and saltwater ecosystems: a cross-system overview. *Marine Ecology Progess Series* 43, 1–10.

Collie, J.S. & Sissenwine, M.P. (1983) Estimating population size from relative abundance data measured with error. *Canadian Journal of Fisheries and Aquatic Sciences* 40, 1871–1879.

Collie, J.S. & Walters, C.J. (1991) Adaptive management of spatially replicated groundfish populations. *Canadian Journal of Fisheries and Aquatic Sciences* 48, 1273–1284.

Collie, J.S., Escanero, G.A. & Valentine, P.C. (1997) Effects of bottom fishing on the benthic megafauna of Georges Bank. *Marine Ecology Progress Series* 155, 159–172.

Collie, J.S., Hall, S.J., Kaiser, M.J. & Poiner, I.R. (2000) A quantitative analysis of fishing impacts on shelf sea benthos. *Journal of Animal Ecology* 69, 785–798.

Collins, S.L. & Benning, T.L. (1996) Spatial and temporal patterns in functional diversity. In: *Biodiversity: a Biology of Numbers and Difference* (ed. K.J. Gaston), pp. 253–280. Blackwell Science, Oxford.

Compagno, L.J.V. (1984a) FAO Species Catalogue 4: Sharks of the world. An annotated and illustrated catalogue of shark species known to date. Part 1. Hexanchiformes to Lamniformes. *FAO Fisheries Synopses*, 125.

Compagno, L.J.V. (1984b) FAO Species Catalogue 4: Sharks of the world. An annotated and illustrated catalogue of shark species known to date. Part 2. Carcharhiniformes. *FAO Fisheries Synopses*, 125.

Compagno, L.J.V. (1990) Alternative life-history styles of cartilaginous fishes in time and space. *Environmental Biology of Fishes* 28, 33–75.

Conand, C. (1989) *Les holothuries Aspidochirote du lagun de Nouvelle-Caledonie: biologies, ecologie et exploitation.* PhD Thesis, University of Western Brittany, Brest.

Cone, R.S. (1989) The need to reconsider the use of condition indices in fishery science. *Transactions of the American Fisheries Society* 118, 510–514.

Connell, J.H. (1978) Diversity in tropical rain forests and coral reefs. *Science* 199, 1302–1310.

Conover, D.O. & Present, T.M.C. (1990) Counter-gradient variation in growth rate: compensation for length of the growing season among Atlantic silversides from different latitudes. *Oecologia* 83, 316–324.

Conover, D.O. & Schultz, E.T. (1995) Phenotypic similarity and the evolutionary significance of countergradient variation. *Trends in Ecology and Evolution* 10, 248–252.

Cook, R.M., Sinclair, A. & Stéfansson, G. (1997) Potential collapse of North Sea cod stocks. *Nature* 385, 521–522.

Costanza, R., d'Arge, R., de Groot, R., Farber, S., Grasso, M., Hannon, B., Limburg, K., Naeem, S., O'Neill, R.V., Paruelo, J., Raskin, R.G., Sutton, P. & van der Belt, M. (1997) The value of the world's ecosystem services and natural capital. *Nature* 387, 253–260.

Cotter, A.J.R. (1998) Method for estimating variability due to sampling of catches on a trawl survey. *Canadian Journal of Fisheries and Aquatic Sciences* 55, 1607–1617.

Cramp, S. (eds) (1977) *Handbook of the Birds of Europe, the Middle East and North Africa The Birds of the Western Palaearctic.* Oxford University Press, Oxford.

Crawford, R.J.M. & Shelton, P.A. (1978) Pelagic fish and seabird interrelationships off the coasts of southwest and South Africa. *Biological Conservation* 14, 85–109.

Crawford, R.J.M., Shannon, L.V. & Pollock, D.E. (1987) The Benguela ecosystem: 4. The major fish and invertebrate resources. *Oceanography and Marine Biology: an Annual Review* 25, 353–505.

Crawford, R.J.M., Underhill, L.G., Raubenheimer, C.M., Dyer, B.M. & Martin, J. (1992) Top predators in the Benguela ecosystem—implications of their trophic position. *South African Journal of Marine Science* 12, 675–687.

Creech, S. (1992) A multivariate morphometric investigation of *Atherina boyeri* Risso, 1810 and *A. presbyter* Cuvier, 1829 (Teleostei, Atherinidae): morphometric evidence in support of the two species. *Journal of Fish Biology* 41, 341–353.

Croxall, J.P. (1987) Introduction. In: *Seabirds: Feeding Ecology and Role in Marine Ecosystems* (ed. J.P. Croxall), pp. 1–5. Cambridge University Press, Cambridge.

Croxall, J.P., McCann, T.S., Prince, P.A. & Rothery, P. (1988) Reproductive performance of seabirds and seals at South Georgia and Signy Island, South Orkney Islands, 1976–1987: implications for Southern Ocean monitoring studies. In: *Antarctic Ocean and Resources Variability* (ed. D. Sahrhage), pp. 261–285. Springer-Verlag, Berlin.

Croxall, J.P., Rodwell, S. & Boyd, I.L. (1990) Entanglement in man-made debris of Antarctic fur seals at Bird Island, South Georgia. *Marine Mammal Science* 6, 221–233.

Cruetzberg, F., Duineveld, G.C.A. & van Noort, G.J. (1987) The effect of different numbers of tickler chains on beam trawl catches. *Journal du Conseil, Conseil International pour l'Exploration de la Mer* 43, 159–168.

Csirke, J. (1987) The Patagonian fishery resources and the offshore fisheries in the southwest Atlantic. *FAO Fisheries Technical Paper*, 286.

Cushing, D.H. (1971) The dependence of recruitment on parent stock in different groups of fishes. *Journal du Conseil, Conseil International pour l'Exploration de la Mer* 33, 340–362.

Cushing, D.H. (1973) Dependence of recruitment on parent stock. *Journal of the Fisheries Research Board of Canada* 30, 1965–1976.

Cushing, D.H. (1975) *Marine Ecology and Fisheries.* Cambridge University Press, Cambridge.

Cushing, D.H. (1982) *Climate and Fisheries.* Academic Press, London.

Cushing, D.H. (1988a) *The Provident Sea.* Cambridge University Press, Cambridge.

Cushing, D.H. (1988b) The study of stock and recruitment. In: *Fish Population Dynamics* (ed. J.A. Gulland), pp. 105–128. Wiley, Chichester.

Cushing, D.H. (1996) *Towards a Science of Recruitment in Fish Populations.* Ecology Institute, Oldendorf/Luhe.

Daan, N. (1979) Multispecies versus single-species assessment of North Sea fish stocks. *Canadian Journal of Fisheries and Aquatic Sciences* 44 (Suppl. 2), 360–370.

Daan, N. (1987) Multispecies versus single-species assessments of North Sea flatfish stocks. *Canadian Journal of Fisheries and Aquatic Sciences* 44 (Suppl. 2), 360–370.

Daan, N. & Sissenwine, M.P. (eds) (1991) *Multispecies models relevant to management of living resources.* ICES Marine Science Symposia 193.

Dadswell, M.J., Rulifson, R.A. & Daborn, G.R. (1986) Potential impact of large scale tidal power development in the Upper Bay of Fundy on fisheries resources of the northwest Atlantic. *Fisheries* 11, 26–35.

Dahle, L.A. (1995) Off-shore fish farming systems. *INFOFISH-International* 2, 24–30.

Dalhbäck, B. & Gunnarson, L. (1981) Sedimentation and sulfate reduction under a mussel culture. *Marine Biology* 63, 269–275.

Dalzell, P. (1996) Catch rates, selectivity and yields of reef fishing. In: *Reef Fisheries* (eds N.V.C. Polunin & C.M. Roberts), pp. 161–192. Chapman & Hall, London.

Dalzell, P., Adams, T.J.H. & Polunin, N.V.C. (1996) Coastal fisheries in the Pacific Islands. *Oceanography and Marine Biology: an Annual Review* 34, 395–531.

Dame, R., Spurrier, J. & Wolaver, T. (1989) Carbon, nitrogen and phosphorus processing by an oyster reef. *Marine Ecology Progress Series* 54, 249–256.

Dankers, N. & de Vlas, J. (eds) (1992) *Multifunctioneel beheer in de Waddenzee-Integratie van naturbeheer en schelpdiervisserij.* Netherlands Institute for Sea Research, Texel, The Netherlands.

Dankers, N. & Zuidema, D. (1995) The role of the mussel (*Mytilus edulis* L.) and mussel culture in the Dutch Wadden Sea. *Estuaries* 18, 71–80.

Dauer, D.M. (1984) High resilience to disturbance of an estuarine polychaete community. *Bulletin of the of Marine Science* 34, 170–174.

Davies, J.K. (1988) A review of information relating to fish passage through turbines: implications to tidal power schemes. *Journal of Fish Biology* 33, 111–126.

Dayton, P.K. & Hessler, R.R. (1972) Role of biological disturbance in maintaining diversity in the deep sea. *Deep-Sea Research* 19, 199–208.

Dayton, P.K., Thrush, S.F., Agardy, M.T. & Hofman, R.J. (1995) Environmental effects of marine fishing. *Aquatic Conservation* 5, 205–232.

Dayton, P.K., Tegner, M.J., Edwards, P.B. & Riser, K.L. (1998) Sliding baselines, ghosts, and reduced expectations in kelp forest communities. *Ecological Applications* 8, 309–322.

de Groot, S. & Lindeboom, H. (eds) (1994) *Environmental impact of bottom gears on benthic fauna in relation to natural resources management and protection of the North Sea.* Netherlands Institute for Sea Research, Texel.

de Jonge, V.N. & Colijn, F. (1994) Dynamics of microphytobenthos biomass in the Ems estuary. *Marine Ecology Progress Series* 104, 185–196.

de Jonge, V.N. & van Beusekom, J.E.E. (1992) Contribution of resuspended microphytobenthos to total phytoplankton in the Ems estuary and its possible role for grazers. *Netherlands Journal of Sea Research* 30, 91–105.

De Lury, D.B. (1947) On the estimation of biological populations. *Biometrics* 3, 145–167.

De Martini, E.E., Ellis, D.M. & Honda, V.A. (1992) Comparisons of spiny lobster *Panilirus marginatus* fecundity, egg size and spawning frequency before and after exploitation. *Fishery Bulletin* 91, 1–7.

de Wolf, P. & Mulder, M. (1985) Spatial variation of macrobenthos in the southern North Sea. *Estuaries* 8, (2B): A64.

Deriso, R.B. (1980) Harvesting strategies and parameter estimation for an age-structured model. *Canadian Journal of Fisheries and Aquatic Sciences* 37, 268–282.

Deriso, R.B. (1987) Optimal $F_{0.1}$ criteria and their relationship to maximum sustainable yield. *Canadian Journal of Fisheries and Aquatic Sciences* 44 (Suppl. 2), 339–348.

Deriso, R.B., Quinn, T.J.I.I. & Neal, P.R. (1985) Catch-age analysis with auxiliary information. *Canadian Journal of Fisheries and Aquatic Sciences* 42, 815–824.

Dery, L.M. (1988) American plaice *Hippoglossoides platessoides.* In: *Age Determination Methods for Northwest Atlantic Species* (eds J. Pentilla & L.M. Dery), pp. 111–118. NOAA Technical Report NMFS 72, US Department of Commerce, Springfield.

Dickey-Collas, M., Brown, J., Fernand, L., Hill, A.E. & Horsburgh, K.J. (1997) Does the western Irish Sea gyre influence the distribution of pelagic juvenile fish. *Journal of Fish Biology* **51** (Suppl.), 206–229.

Dickie, L.M., Kerr, S.R. & Boudreau, P.R. (1987) Size-dependent processes underlying regularities in ecosystem structure. *Ecological Monographs* **57**, 233–250.

Dickson, R.R., Meincke, J., Malmberg, S.A. & Lee, A.J. (1988) The great 'salinity anomoly' in the northern North Atlantic 1968–82. *Progress in Oceanography* **20**, 103–151.

Dijkema, R. (1995) Large-scales recirculations systems for storage of imported bivalves as a means to counteract introduction of cysts of toxic dinoflagellates in the coastal waters of the Netherlands. In: *Shellfish Depuration. Purification des Coquillages* (eds. R. Poggi & J.Y. LeGall), pp. 355–367. IFREMER Brest, Plouzane.

Dingle, H. (1996) *Migration: the Biology of Life on the Move.* Oxford University Press, Oxford.

Doubleday, W.G. (1976) A least squares approach to analysing catch at age data. *ICNAF Research Bulletin* **12**, 69–81.

Doubleday, W.G. & Rivard, D. (eds) (1983) Sampling commercial catches of marine fish and invertebrates. *Canadian Special Publications of Fisheries and Aquatic Sciences* **66**.

Dragesund, O. & Gjosaeter, J. (1988) The Barents Sea. In: *Continental Shelves (Ecosystems of the World 27)* (eds H. Postma & J.J. Zijlstra), pp. 339–361. Elsevier, Amsterdam.

Duarte, C.M. & Cebrián, J. (1996) The fate of marine autotrophic production. *Limnology and Oceanography* **41**, 1758–1766.

Duffy, D.C. (1983) Environmental uncertainty and commercial fishing-effects on Peruvian guano birds. *Biological Conservation* **26**, 227–238.

Duffy, D.C. & Siegfried, W.R. (1987) Historical variations in food consumption by breeding seabirds of the Humboldt and Benguela upwelling regions. In: *Seabirds: Feeding Ecology and Role in Marine Ecosystems* (ed. J.P. Croxall), pp. 327–346. Cambridge University Press, Cambridge.

Dugdale, R.C. & Goering, J.J. (1967) Uptake of new and regenerated forms of nitrogen in primary productivity. *Limnology and Oceanography* **12**, 196–206.

Dulvy, N.K. & Reynolds, J.D. (1997) Evolutionary transitions among egg-laying, live-bearing and maternal inputs in sharks and rays. *Proceedings of the Royal Society: Biological Sciences* **264**, 1309–1315.

Dulvy, N.K., Metcalfe, J.D., Glanville, J., Pawson, M.G. and Reynolds, J.D. (2000) Fishery stability, local extinctions, and shifts in community structure in skates. *Conservation Biology* **14**, 283–293.

Dunn, M.R., Potten, S., Radford, A. & Whitmarsh, D. (1994) *An economic appraisal of the fishery for bass* Dicentrarchus labrax L. *in England and Wales* CEMARE, Portsmouth.

Duplisea, D.E. & Bravington, M.V. (1999) Harvesting a size-structured ecosystem. *International Council for the Exploration of the Sea, Committee Meeting* 1999/Z: 01.

Duplisea, D.E. & Kerr, S.R. (1995) Application of a biomass size spectrum model to demersal fish data from the Scotian shelf. *Journal of Theoretical Biology* **177**, 263–269.

Dutil, J.D., Legare, B. & Desjardins, C. (1985) Discrimination d'un stock de poisson l'anguille (*Anguilla rostrata*), basee sur la presence d'un produit chimique de synthese et mirex. *Canadian Journal of Fisheries and Aquatic Sciences* **42**, 455–458.

Earle, M. (1996) Ecological interactions between cetaceans and fisheries. In: *The Conservation of Whales and Dolphins* (eds M.P. Simmonds & J.D. Hutchinson), pp. 167–204. John Wiley, Chichester.

Eaton, D.R. (1996) *The Identification and Separation of Wild-Caught and Cultivated Sea Bass* (Dicentrarchus labrax). Fisheries Research Technical Report 103, MAFF Directorate of Fisheries Research, Lowestoft.

Ebert, T.A. & Russell, M.P. (1992) Growth and mortality estimates for red sea-urchin *Strongylocentrotus franciscanus* from San Nicolas Island, California. *Marine Ecology Progress Series* **81**, 31–41.

Edley, M.T. & Law, R. (1988) Evolution of life histories and yields in experimental populations of *Daphnia magna*. *Biological Journal of the Linnean Society* **34**, 309–326.

Edwards, A.W.F. (1992) *Likelihood*. The Johns Hopkins University Press, Baltimore, Maryland.

Eggleston, D.B., Armstrong, D.A., Elis, W.E. & Patton, W.S. (1998) Estuarine fronts as conduits for larval transport: hydrodynamics and spatial distribution of Dungeness crab postlarvae. *Marine Ecology Progress Series* **164**, 73–82.

Eisenbud, R. (1985) The pelagic driftnet. *Salt Water Sportsman* **1985**, 65–72.

Elliott, J.M. (1994) *Quantitative Ecology and the Brown Trout*. Oxford University Press, Oxford.

Ellis Ripley, W. (1946) The soupfin shark and the fishery. *California Division of Fish and Game, Fish Bulletin* **64**, 7–37.

Emerson, L.S., Greer-Walker, M. & Witthames, P.R. (1990) A stereological method for estimating fish fecundity. *Journal of Fish Biology* **36**, 712–730.

Emlen, J.M. (1966) The role of time and energy in food preference. *American Naturalist* **100**, 611–617.

Engås, A. & Ona, E. (1990) Day and night distribution pattern in the net mouth area of the Norwegian bottom-sampling trawl. *Rapports et Procès-Verbaux des Réunions, Conseil International pour l'Exploration de la Mer* **189**, 123–127.

Eno, N.C., MacDonald, D. & Amos, S.C. (1996) *A study on the effects of fish (Crustacea/Mollusc) traps on benthic habitats and species*. Report to European Commission Directorate General XIV, Studies Contract 94/076.

Erzini, K., Monteiro, C., Ribeiro, J., Santos, M., Gaspar, M., Monteiro, P. & Borges, T. (1997) An experimental study of gill net and trammel net 'ghost-fishing' off the Algarve (southern Portugal). *Marine Ecology Progress Series* **158**, 257–265.

Estes, J.A. & VanBlaricom, G.R. (1985) Sea-otters and shellfisheries. In: *Marine Mammals and Fisheries* (eds J.R. Beddington, R.J.H. Beverton & D.M. Lavigne), pp. 187–235. George Allen and Unwin, London.

Estrella, B.T. & McKiernan, D.J. (1989) Catch-per-unit-effort and biological parameters from the Massachusetts coastal lobster (*Homarus americanus*) resource: description and trends. *NOAA Technical Bulletin NMFS* **81**.

FAO (1978) Some scientific problems of multispecies fisheries. *FAO Fisheries Technical Paper* **181**.

FAO (1993a) Marine fisheries and the law of the sea: a decade of change. *FAO Fisheries Circular* **853**.

FAO (1993b) Aquaculture production 1985–91. *FAO Fisheries Circular* **815**, Rev. 5.

FAO (1994) Review of the state of world marine fishery resources. *FAO Fisheries Technical Paper* **335**.

FAO (1995a) *World Fishery Production 1950–93*. FAO, Rome.

FAO (1995b) Precautionary approach to fisheries. Part 1. Guidelines on the precautionary approach to capture fisheries and species introductions. *FAO Fisheries Technical Paper* **350**, Part **1**.

FAO (1995c) *Code of Conduct for Sustainable Fisheries*. FAO, Rome.

FAO (1996) *Time Series on Aquaculture-Quantities and Values*. FAO, Rome.

FAO (1997a) Fishery statistics: catches and landings 1995. *FAO Fisheries Yearbook* **80**. FAO, Rome.

FAO (1997b) Fishery Statistics: Commodities. *FAO Fisheries Yearbook*, **80**. FAO, Rome.

FAO (1999) *The State of World Fisheries and Aquaculture 1998*. FAO, Rome.

Fegan, D.F. (1996) Sustainable shrimp farming in Asia: vision or pipe dream. *Asian Aquaculture* **2**, 22–24.

Felbeck, H. & Somero, G.N. (1982) Primary production in deep sea hydrothermal vent organisms. *Trends in Biochemical Sciences* **7**, 201–204.

Felsenstein, J. (1985) Phylogenies and the comparative method. *American Naturalist* **125**, 1–15.

Fernö, A. & Olsen, S. (1994) *Marine Fish Behaviour in Capture and Abundance Estimation*. Fishing News Books, Oxford.

Field, J.G., Crawford, K.L. & Villacastin-Herrero, C.A. (1991) *Network analysis of Benguela pelagic food webs*. Bengula Ecology Programme, Workshop on Seal–Fishery Biological Interactions. University of Cape Town. 16–20 September, 1991. BEP/SW91/M5a.

Field, K.D. & Clark, M.R. (1996) Assessment of the ORH 7A orange roughy fishery for the 1996–97 fishing year. *N.Z. Fisheries Assessment Research Document* **96/120**.

Foale, S. & Day, R. (1997) Stock assessment of trochus (*Trochus niloticus*) (Gastropoda Trochidae) fisheries at West Nggela, Solomon Islands. *Fisheries Research* **33**, 1–16.

Folke, C. & N. Kautsky (1989) The role of ecosystems for a sustainable development of aquaculture. *Ambio* **18**, 234–243.

Fonseca, M.S., Thayer, G.W. & Chester, A.J. (1984) Impact of scallop harvesting on eelgrass (*Zostera marina*) meadows: implications for management. *North American Journal of Fisheries Management* **4**, 286–293.

Forward, R.B., Burke, J.S., Rittschof, D. & Welch, J.M. (1996) Photoresponses of larval Atlantic menhaden (*Brevoortia tyrannus* Latrobe) in offshore and estuarine waters: implications for transport. *Journal of Experimental Marine Biology and Ecology* **199**, 123–135.

Fossum, P. Kalish, J. & Moksness, E. (eds) (2000) Second international symposium on fish otolith research and application, Bergen, Norway 20–25 June 1998. *Fisheries Research*, **46**.

Fournier, D. & Archibald, C.P. (1982) A general theory for analyzing catch at age data. *Canadian Journal of Fisheries and Aquatic Sciences* **39**, 1195–1207.

Fournier, D.A. & Doonan, I.J. (1987) A length-based stock assessment method utilizing a generalized delay-difference model. *Canadian Journal of Fisheries and Aquatic Sciences* **44**, 422–437.

Fournier, D.A., Sibert, J.R., Majkowski, J. & Hampton, J. (1990) MULTIFAN: a likelihood based method for estimating growth parameters and age composition from multiple length frequency data sets illustrated using data from southern bluefin tuna *Thunnus maccoyii*. *Canadian Journal of Fisheries and Aquatic Sciences* **47**, 301–317.

Fowler, A.J. (1995) Annulus formation in coral reef fish—a review. In: *Recent Developments in Fish Otolith Research* (eds D.H. Secor, J.M. Dean & S.E. Campana), pp. 45–63. University of South Carolina Press, Columbia.

Fox, W.W. Jr (1970) An exponential surplus-yield model for optimizing exploited fish populations. *Transactions of the American Fisheries Society* **99**, 80–88.

Francis, R.I.C.C. (1988) Maximum likelihood estimation of growth and growth variability from tagging data. *New Zealand Journal of Marine and Freshwater Research* **22**, 42–51.

Francis, R.I.C.C. (1990) Back-calculation of fish length: a critical review. *Journal of Fish Biology* **36**, 883–902.

Francis, R.I.C.C. (1992) Use of risk analysis to assess fishery management strategies: a case study using orange roughy (*Hoplostethus altanticus*) on the Chatham Rise, New Zealand. *Canadian Journal of Fisheries and Aquatic Sciences* **49**, 922–930.

Francis, R.I.C.C. (1993) Estimation of risk: a response to Hilborn *et al*. *Canadian Journal of Fisheries and Aquatic Sciences* **50**, 1125–1126.

Francis, R.I.C.C. & Shotton, R. (1997) 'Risk' in fisheries management: a review. *Canadian Journal of Fisheries and Aquatic Sciences* **54**, 1699–1715.

Francis, R.I.C.C., Paul, L.J. & Mulligan, K.P. (1992a) Ageing of adult snapper (*Pagrus auratus*) from otolith annual ring counts: validation by tagging and oxytetracycline injection. *Australian Journal of Marine and Freshwater Research* **43**, 1069–1089.

Francis, R.I.C.C., Awbrey, F.T., Goudey, C.L., Hall, M.A., King, D.M., Medina, H., Norris, K.S., Orbach, M.K., Payne, R. & Pikitch, E. (1992b) *Dolphins and the Tuna Industry*. National Research Council. National Academy Press, Washington DC.

Fraser, W.R., Trivelpiece, W.Z., Ainley, D.G. & Trivelpiece, S.G. (1992) Increases in Antarctic penguin populations: reduced competition with whales or a loss of sea ice due to environmental warming? *Polar Biology* **11**, 525–531.

Fréon, P. (1993) Consequences of fish behaviour for stock assessment. *ICES Marine Science Symposia* **196**, 190–195.

Fretwell, S. (1972) *Populations in a Seasonal Environment.* Princeton University Press, New Jersey.

Fretwell, S.D. & Lucas, H.L. (1970) On territorial behaviour and other factors influencing habitat distribution in birds. *Acta Biotheoretica* **19**, 16–36.

Frid, C.L.J., Clark, R.A. & Hall, J.A. (1999) Long-term changes in the benthos on a heavily fished ground off the NE coast of England. *Marine Ecology Progress Series* **188**, 13–20.

Friedlander, A., Beets, J. & Tobias, W. (1994) Effects of fish aggregating device design and location on fishing success in the US Virgin Islands. *Bulletin of the of Marine Science* **55**, 592–601.

Fry, B. & Quinones, R.B. (1994) Biomass spectra and stable-isotope indicators of trophic level in zooplankton of the northwest Atlantic. *Marine Ecology Progress Series* **112**, 201–204.

Fuiman, L.A. & Higgs, D.M. (1997) Ontogeny, growth and the recruitment process. In: *Early Life History and Recruitment in Fish Populations* (eds R.C. Chambers & E.A. Trippel), pp. 225–249. Chapman & Hall, New York.

Furness, R.W. (1982) Competition between fisheries and seabird communities. *Advances in Marine Biology* **20**, 225–327.

Furness, R.W. (1987) Kleptoparasitism in seabirds. In: *Seabirds: Feeding Ecology and Role in Marine Ecosytems* (ed. J.P. Croxall), pp. 77–100. Cambridge University Press, Cambridge.

Furness, R.W. (1992) Implications of changes in net mesh size, fishing effort and minimum landing size regulations in the North Sea for seabird populations. *Joint Nature Conservaton Committee Report* **133**, Peterborough.

Furness, R.W. (1996) A review of seabird responses to natural or fisheries-induced changes in food supply. In: *Aquatic Predators and their Prey* (eds S.P.R. Greenstreet & M.L. Tasker), pp. 168–173. Blackwell Scientific Publications, Oxford.

Furness, R.W. & Tasker, M.L. (1997) Seabird consumption in sand lance MSVPA models for the North Sea and the impact of industrial fishing on seabird population dynamics. In: *Forage Fishes in Marine Ecosystems* (ed. Anon), pp. 147–169. University of Alaska Sea Grant College Program, Fairbanks.

Furness, R.W., Hudson, A.V. & Ensor, K. (1988) Interactions between scavenging seabirds and commercial fisheries around the British Isles. In: *Seabirds and Other Marine Vertebrates: Competition, Predation and Other Interactions* (ed. J. Burger), pp. 240–268. Columbia University Press, New York.

Furness, R.W., Ensor, K. & Hudson, A.V. (1992) The use of fishery waste by gull populations around the British Isles. *Ardea* **80**, 105–113.

Garthe, S., Camphuysen, K.C.J. & Furness, R.W. (1996) Amounts of discards by commercial fisheries and their significance as food for seabirds in the North Sea. *Marine Ecology Progress Series* **136**, 1–11.

Gaspar, M.B., Richardson, C.A. & Monteiro, C.C. (1994) The effects of dredging on shell formation in the razor clam *Ensis siliqua* from Barrinha, southern Portugal. *Journal of the Marine Biological Association of the United Kingdom* **74**, 927–938.

Gaston, K.J. (1996) *Biodiversity: a Biology of Numbers and Difference.* Blackwell Science. Oxford.

Gauldie, R.W. (1991) Taking stock of genetic concepts in fisheries management. *Canadian Journal of Fisheries and Aquatic Sciences* **48**, 722–731.

Gayanilo, F.C., Sparre, P. & Pauly, D. (1994) *The FAO-ICLARM Stock Assessment Tools (Fisat) User's Guide.* FAO, Rome.

Getz, W.M. & Swartzman, G.L. (1981) A probability transition matrix model for yield estimation in fisheries with highly variable recruitment. *Canadian Journal of Fisheries and Aquatic Sciences* **38**, 847–855.

Ghiselin, M.T. (1969) The evolution of hermaphroditism among animals. *Quarterly Review of Biology* **44**, 189–208.

Gilkinson, K., Paulin, M., Hurley, S. & Schwinghamer, P. (1998) Impacts of trawl door scouring on infaunal bivalves: results of a physical trawl door model/dense sand interaction. *Journal of Experimental Marine Biology and Ecology* **224**, 291–312.

Gilliam, J.F. & Fraser, D.F. (1987) Habitat selection under a predation hazard: test of a model with foraging minnows. *Ecology* **68**, 1856–1862.

Gillis, D., Pikitch, E. & Peterman, R. (1995) Dynamic discarding decisions: foraging theory for highgrading in a trawl fishery. *Behavioural Ecology* **6**, 146–154.

Gislason, H. & Sinclair, M.M. (eds) (2000) Ecosystem effects of fishing. *ICES Journal of Marine Science* **57**, 465–791.

Glass, C.W. & Wardle, C.S. (1995) Studies on the visual stimuli to control fish escape from codends. II. The effect of a black tunnel on the reaction behaviour of fish in otter trawl codends. *Fisheries Research* **23**, 157–164.

Glass, C.W., Wardle, C.S., Gosden, S.J. & Racey, D. (1995) Studies on the visual stimuli to control fish escape from codends. I. Laboratory studies on the effect of a black tunnel on mesh penetration. *Fisheries Research* **23**, 157–164.

Glynn, P.W. (1997) Bioerosion and coral reef growth: a dynamic balance. In: *Life and Death on Coral Reefs* (ed. C. Birkeland), pp. 68–95. Academic Press, New York.

Godø, O.R. & Engås, A. (1989) Swept area variation with depth and its influence on abundance indices of groundfish from trawl surveys. *Journal of Northwest Atlantic Fishery Science* **9**, 133–139.

Godø, O.R., Pennington, N. & Vølstad, J.H. (1990) Effect of tow duration on length composition of trawl catches. *Fisheries Research* **9**, 165–179.

Gordon, H.S. (1954) The economic theory of a common property resource: the fishery. *Journal of Political Economy* **62**, 124–142.

Goss-Custard, J.D. (ed.) (1996) *The Oystercatcher: from Individuals to Populations.* Oxford University Press, Oxford.

Gottlieb, S.J. & Schweighofer, M.E. (1996) Oysters and the Chesapeake Bay ecosystem: a case for exotic species introduction to improve environmental quality? *Estuaries* **19**, 639–650.

Gowen, R.J. & N.B. Bradbury (1987) The ecological impact of salmonid farming in coastal waters: a review. *Oceangraphy and Marine Biology: an Annual Review* **25**, 563–575.

Gowen, R.J., Bradbury, N.B. & Brown, J.R. (1989) The use of simple models in assessing two of the interactions between fish farming and the marine environment. In: *Aquaculture—a Biotechnology in Progress* (eds N. De Pauw, E. Jaspers, H. Ackefors & N. Wilkins), pp. 1071–1080. European Aquaculture Society, Bredene, Belgium.

Graham, M. (1955) Effect of trawling on animals of the sea bed. *Deep Sea Research (Supplement)* **3**, 1–16.

Graham, N.E. & White, W.B. (1988) The El Niño cycle: a natural oscillator of the Pacific Oceanatmosphere system. *Science* **240**, 1293–1301.

Grassle, J.F. (1986) The ecology of deep-sea hydrothermal vent communities. *Advances in Marine Biology* **23**, 301–362.

Grassle, J.F. & Saunders, H.L. (1973) Life histories and the role of disturbance. *Deep Sea Research* **20**, 643–659.

Green, E.P., Mumby, P.J., Edwards, A.J. & Clark, C.D. (1996) A review of remote sensing for the assessment and management of tropical coastal resources. *Coastal Management* **24**, 1–40.

Greenstreet, S.P.R. & Hall, S.J. (1996) Fishing and groundfish assemblage structure in the northwestern North Sea: an analysis of long-term and spatial trends. *Journal of Animal Ecology* **65**, 577–598.

Greenstreet, S.P.R., Spence, F.E. & McMillan, J.A. (1999a) Fishing effects in northeast Atlantic shelf seas: patterns in fishing effort, diversity and community structure. V. Changes in the groundfish species assemblage of the northwestern North Sea between 1925 and 1996. *Fisheries Research* **40**, 153–184.

Greenstreet, S.P.R., Spence, F.B., Shanks, A.M. & McMillan, J.A. (1999b) Fishing effects in northeast Atlantic shelf seas: patterns in fishing effort, diversity and community structure. II. Trends in fishing effort in the North Sea by UK registered vessels landing in Scotland. *Fisheries Research* **40**, 107–124.

Grimes, C.B., Idelberger, C.F., Able, K.W. & Turner, S.C. (1988) The reproductive biology of tilefish *Lopholatilus chamaeleonticeps* Goode & Bean, from the United States mid-Atlantic Bight, and the effects of fishing on the breeding system. *Fishery Bulletin* **86**, 745–762.

Groombridge, B. & Luxmoore, R. (1989) *The Green Turtle and Hawksbill (Reptilia: Cheloniidae): World Status, Exploitation and Trade*. CITES, Lausanne.

Gruber, S.H. & Stour, R.G. (1983) Biological materials for the study of age and growth in a tropical marine elasmobranch, the lemon shark *Negaprion brevirostris* (Poey). *NOAA Technical Report NMFS* **8**, 193–205.

Guénette, S., Lauck, T. & Clark, C.W. (1998) Marine reserves: from Beverton and Holt to the present. *Reviews in Fish Biology and Fisheries* **8**, 1–21.

Guillén, J., Ramos, A., Martinéz, L. & Sánchez Lizaso, J. (1994) Antitrawling reefs and the protection of *Posidonia oceanica* (L.) delile meadows in the western Mediterranean

Sea: demands and aims. *Bulletin of Marine Science* **55**, 645–650.

Guillory, V. (1993) Ghost fishing by blue crab traps. *North American Journal of Fisheries Management* **13**, 459–466.

Gulland, J.A. (1971) *The Fish Resources of the Ocean*. Fishing News Books, West Byfleet.

Gulland, J.A. (1983) *Fish Stock Assessment: a Manual of Basic Methods*. Wiley-Interscience Publications.

Gulland, J.A. (ed.) (1988) *Fish Population Dynamics*. John Wiley, London.

Gulland, J.A. & Holt, S.J. (1959) Estimation of growth parameters for data at unequal time intervals. *Journal du Conseil, Conseil International pour l'Exploration de la Mer* **25**, 47–49.

Gulland, J.A. & Rosenberg, A.A. (1992) A review of length based approaches to assessing fish stocks. *FAO Fisheries Technical Paper* **323**.

Gunderson, D.R. (1993) *Surveys of Fisheries Resources*. John Wiley, New York.

Gunderson, D.R. & Dygert, P.H. (1988) Reproductive effort as a predictor of natural mortality rate. *Journal du Conseil, Conseil International pour l'Exploration de la Mer* **44**, 200–209.

Haamer, J. (1996) Improving water quality in a eutrophicated fjord system with mussel farming. *Ambio* **25**, 356–362.

Haedrich, R.L. & Barnes, S.M. (1997) Changes over time of the size structure in an exploited shelf fish community. *Fisheries Research* **31**, 229–239.

Haedrich, R.L. & Merrett, N.R. (1992) Production/ biomass ratios, size frequencies, and biomass spectra in deep-sea demersal fishes. In: *Deep-Sea Food Chains and the Global Carbon Cycle* (eds G.T. Rowe & V. Pareinte), pp. 157–182. Kluwer Academic Publishers, Dordrecht.

Hall, M.A. (1996) On bycatches. *Reviews in Fish Biology and Fisheries* **6**, 319–352.

Hall, M.A. (1998) An ecological view of the tuna–dolphin problem: impacts and trade-offs. *Reviews in Fish Biology and Fisheries* **8**, 1–34.

Hall, S.J. (1994) Physical disturbance and marine benthic communities: life in unconsolidated sediments. *Oceanography and Marine Biology: an Annual Review* **32**, 179–239.

Hall, S.J. (1999) *The Effects of Fishing on Marine Ecosystems and Communities*. Blackwell Science, Oxford.

Hall, S.J. & Harding, M.J.C. (1997) Physical disturbance and marine benthic communities: the effects of mechanical harvesting of cockles on non-target benthic infauna. *Journal of Applied Ecology* **34**, 497–517.

Hall, S.J., Basford, D.J. & Robertson, M.R. (1990) The impact of hydraulic dredging for razor clams *Ensis* sp. on an infaunal community. *Netherlands Journal of Sea Research* **27**, 119–125.

Hall, S.J., Robertson, M.R., Basford, D.J. & Fryer, R. (1993) Pit-digging by the crab *Cancer pagurus*: a test for long-term, large-scale effects on infaunal community structure. *Journal of Animal Ecology* **62**, 59–66.

Hallegraeff, G.M. & Bolch, C.J. (1991) Transport of toxic dinoflagellate cysts via ships' ballast. *Marine Pollution Bulletin* **22**, 27–30.

Hall-Spencer, J.M. & Moore, P.G. (2000) Impacts of scallop dredging on maerl grounds. In: *The Effects of Fishing on Non-Target Species and Habitats: Biological, Conservation Socio-Economic Issues* (eds M.J. Kaiser & S.J. de Groot), pp. 105–117. Blackwell Science, Oxford.

Hamner, W.M., Jones, M.S., Carleton, J.H., Hauri, I.R. & Williams, D.M. (1988) Zooplankton, planktivorous fish and water currents on a windward reef face, Great Barrier Reef, Australia. *Bulletin of Marine Science* **42**, 459–479.

Hampton, I. (1996) Acoustic and egg production estimates of South African anchovy abundance over a decade: comparisons, accuracy and utility. *ICES Journal of Marine Science* **53**, 493–500.

Hamre, J. (1988) Some aspects of the interrelation between the herring in the Norwegian Sea and stocks of capelin and cod in the Barents Sea. *International Council for the Exploration of the Sea, Committee Meeting*, 1988/H:42, 15.

Hamre, J. (1991) Interrelation between environmental changes and fluctuating fish populations in the Barents Sea. In: *Long Term Variability of Pelagic Fish Populations and their Environment* (eds T. Kawasaki, S. Tanaka, Y. Toba & A. Taniguchi), pp. 259–270. Pergamon Press, Oxford.

Hamre, J. (1994) Biodiversity and exploitation of the main fish stocks in the Norwegian-Barents Sea ecosystem. *Biodiversity and Conservation* **3**, 473–494.

Hannesson, R. (1993) *Bioeconomic Analysis of Fisheries*. FAO and Fishing News Books, Oxford.

Hannesson, R. (1995a) Fishing on the high seas: co-operation or competition. *Marine Policy* **19**, 371–377.

Hannesson, R. (1995b) Sequential fishing: cooperative and non-cooperative equilibria. *Natural Resource Modeling* **9**, 51–59.

Hannesson, R. (1996) *Fisheries Mismanagement: the Case of the North Atlantic Cod*. Fishing News Books, Oxford.

Hannesson, R. (1997) Fishing as a supergame. *Journal of Environmental Economics and Management* **32**, 309–322.

Hanski, I. (1991) Single-species metapopulation dynamics: concepts, models and observations. *Biological Journal of the Linnaean Society* **42**, 17–38.

Hanski, I. (1999) *Metapopulation Ecology*. Oxford University Press, Oxford.

Harden-Jones, F.R. (1968) *Fish Migration*. Edward Arnold, London.

Harden-Jones, F.R. (1981) Fish migration: strategy and tactics. In: *Animal Migration* (ed. D.J. Aidley), pp. 139–165. Cambridge University Press, Cambridge.

Hardin, G. (1968) The tragedy of the commons. *Science* **162**, 1243–1248.

Hare, J.A. & Cowen, R.K. (1996) Transport mechanisms of larval and pelagic juvenile bluefish (*Pomatomus saltatrix*) from South Atlantic Bight spawning grounds to Middle Atlantic Bight nursery habitats. *Limnology and Oceanography* **41**, 1264–1280.

Harmelin, J.G., Bachet, F. & Garcia, F. (1995) Mediterranean marine reserves—fish indexes as tests of protection efficiency. *Marine Ecology* **16**, 233–250.

Harmelin-Vivien, M.L., Harmelin, J.G., Chauvet, C., Duval, C., Galzin, R., Lejeune, P., Barnabe, G., Blanc, F., Chevalier, R., Duclerc, J. & Lasserre, G. (1985) Evaluation visuelle des peuplements et populations de poissons: Methods et problemes. *Revue d'Ecologie (Terre Vie)* **40**, 467–539.

Harris, M. (1998) *Lament for an Ocean*. McClelland and Stewart, Toronto.

Hart, A.M. & Russ, G.R. (1996) Response of herbivorous fishes to crown-of-thorns starfish outbreaks III. Age, growth, mortality and maturity indices of *Acanthurus nigrofuscus*. *Marine Ecology Progress Series* **136**, 25–35.

Hart, M. & Scheibling, R. (1988) Heat waves, baby booms, and the destruction of kelp beds by sea urchins. *Marine Biology* **99**, 167–176.

Hart, P.J.B. (1998) Enlarging the shadow of the future: avoiding conflict and conserving fish. In: *Reinventing Fisheries Management* (eds T.J. Pitcher, P.J.B. Hart & D. Pauly), pp. 227–238. Kluwer Academic Publishers, London.

Harvey, P.H. & Pagel, M.D. (1991) *The Comparative Method in Evolutionary Biology*. Oxford University Press, Oxford.

Hatcher, B.G. (1981) The interaction between grazing organisms and the epilithic algal community of a coral reef: a quantitative assessment. *Proceedings of the Fourth International Coral Reef Symposium* **2**, 215–224.

Hatcher, B.G. (1988) The primary productivity of coral reefs: a beggar's banquet. *Trends in Ecology and Evolution* **3**, 106–111.

Hatcher, B.G. (1997) Organic production and decomposition. In: *Life and Death on Coral Reefs* (ed. C. Birkeland), pp. 140–174. Chapman & Hall, New York.

Haug, T., Kroyer, A.B., Nilssen, K.T., Ugland, K.I. & Aspholm, P.E. (1991) Harp seal (*Phoca groenlandica*) invasions of Norwegian coastal waters: age composition and feeding habits. *ICES Journal of Marine Science* **48**, 363–371.

Hawkins, J.P., Roberts, C.M. & Clark, V. (2000) The threatened status of restricted-range coral reef fish species. *Animal Conservation* **3**, 81–88.

Hay, M.E. (1991) Fish–seaweed interactions on coral reefs: effects of herbivorous fishes and adaptations of their prey. In: *Ecology of Fishes on Coral Reefs* (ed. P.F. Sale), pp. 96–119. Academic Press, San Diego.

Hay, S. (1995) Egg production and secondary production of common North Sea copepods: field estimates with regional and seasonal comparisons. *ICES Journal of Marine Science* **52**, 315–327.

Heath, M.R. (1992) Field investigations of the early life stages of marine fish. *Advances in Marine Biology* **28**, 1–174.

Heath, M.R. (1993) An evaluation and review of the ICES herring larval surveys in the North Sea and adjacent waters. *Bulletin of Marine Science* **53**, 795–817.

Heessen, H.J.L. & Daan, N. (1996) Long-term trends in ten non-target North Sea fish species. *ICES Journal of Marine Science* **53**, 1063–1078.

Helgason, Th. & Gislason, H. (1979) VPA—analysis with species interaction due to predation. *International Council for the Exploration of the Sea Committee Meeting* 1979/G: 52.

High, W.L. (1976) Escape of Dungeness crabs from pots. *Marine Fisheries Reviews* **38**, 19–23.

Hilborn, R. (1992) Hatcheries and the future of salmon in the Northwest. *Fisheries* **17**, 5–8.

Hilborn, R. (1996) Risk analysis in fisheries and natural resource management. *Human Ecological Risk Assessment* **12**, 655–659.

Hilborn, R. & Walters, C.J. (1992) *Quantitative Fisheries Stock Assessment: Choice, Dynamics and Uncertainty.* Chapman & Hall, New York.

Hilborn, R., Pikitch, E.K. & Francis, R.C. (1993a) Current trends in including risk and uncertainty in stock assessment and harvest decisions. *Canadian Journal of Fisheries and Aquatic Sciences* **50**, 874–880.

Hilborn, R., Pikitch, E.K., McAllister, M.K. & Punt, A.E. (1993b) Comment on 'Use of risk analysis to assess fishery management strategies: a case study using orange roughy (*Hoplostethus altanticus*) on the Chatham Rise, New Zealand' by R.I.C.C. Francis. *Canadian Journal of Fisheries and Aquatic Sciences* **50**, 1122–1125.

Hilborn, R., Pikitch, E.K. & McAllister, M.K. (1994) A Bayesian estimation and decision analysis for an age-structured model using biomass survey data. *Fisheries Research* **19**, 17–30.

Hill, K. & Hawkes, K. (1983) Neotropical hunting among the Ache of eastern Paraguay. In: *Adaptive Responses of Native Amazonians* (eds R. Hames & W. Vickers) pp. 139–188. Academic Press, New York.

Hill, A.S., Brand, A.R., Wilson, U.A.W., Veale, L.O. & Hawkins, S.J. (1996) Estimation of by-catch composition and the numbers of by-catch animals killed annually on Manx scallop grounds. In: *Aquatic Predators and Their Prey* (eds S.P.R. Greenstreet & M.L. Tasker), pp. 111–115. Fishing News Books, Blackwell Science, Oxford.

Hill, B. & Wassenberg, T. (1990) Fate of discards from prawn trawlers in Torres Strait. *Australian Journal of Marine and Freshwater Research* **41**, 53–64.

Hill, M.O. (1973) Diversity and evenness: a unifying notation and its consequences. *Ecology* **54**, 427–432.

Hinckley, S., Bailey, K.M., Picquelle, S.J., Schumacher, J.D. & Stabeno, P.J. (1991) Transport, distribution and abundance of larval and juvenile walleye pollock (*Theragra chalcogramma*) in the western Gulf of Alaska. *Canadian Journal of Fisheries and Aquatic Sciences* **48**, 91–98.

Hirshfield, M.F. (1980) An experimental analysis of reproductive effort and cost in the Japanese medaka. *Ecology* **61**, 282–292.

Hirth, H.F. (1993) Marine turtles. In: *Neashore Marine Resources of the South Pacific* (eds A. Wright & L. Hill), pp. 329–370. Institute of Pacific Studies, Suva.

Hislop, J.R.G. (1996) Changes in North Sea gadoid stocks. *ICES Journal of Marine Science* **53**, 1146–1156.

Hixon, M.A. (1991) Predation as a process structuring coral reef fish communities. In: *The Ecology of Fishes on Coral Reefs* (ed. P.F. Sale), pp. 475–508. Academic Press, San Diego.

Hixon, M.A. (1997) Effects of reef fishes on corals and algae. In: *Life and Death on Coral Reefs* (ed. C. Birkeland), pp. 230–248. Academic Press, New York.

Hjort, J. (1914) Fluctuations in the great fisheries of northern Europe viewed in the light of biological research. *Rapports et Procès-Verbaux Des Réunions, Conseil International pour l'Exploration de la Mer* **20**, 1–228.

Hoar, W.S., Randall, D.J. & Brett, J.R. (eds) (1979) *Fish Physiology VIII: Bioenergetics and Growth.* Academic Press, London.

Hoenig, J.M. (1983) Empirical use of longevity data to estimate mortality rates. *Fishery Bulletin* **81**, 898–903.

Holden, M.J. & Meadows, P.S. (1962) The structure of the spine of the spur dogfish (*Squalus acanthias* L.) and its use for age determination. *Journal of the Marine Biological Association of the United Kingdom* **42**, 179–197.

Holland, K.N., Peterson, J.D., Lowe, C.G. & Wetherbee, B.M. (1993) Movements, distribution and growth rates of the white goatfish *Mulloides flavolineatus* in a fisheries conservation zone. *Bulletin of Marine Science* **52**, 982–992.

Holling, C.S. (1959) The components of predation as revealed by a study of small-mammal predation of the European sawfly. *Canadian Entomologist* **91**, 293–320.

Holt, S.J. (1963) A method for determining gear selectivity and its application. *ICNAF Special Publication* **5**, 106–115.

Hooper, W.D. & Ash, H.B. (1934) Translation of *On Agriculture* by Marcus Terentius Varro. Harvard University Press, Cambridge, Massachusetts.

Horwood, J.W. (1987a) A calculation of locally optimal fishing mortalities. *Journal du Conseil, Conseil International pour l'Exploration de la Mer* **43**, 199–208.

Horwood, J.W. (1987b) *The Sei Whale: Population Biology, Ecology and Management.* Croom-Helm, Beckenham.

Horwood, J.W. (1990) Near-optimal harvests from multiple species harvested with several fishing fleets. *IMA Journal of Mathematics Applied in Medicine and Biology* **7**, 55–68.

Horwood, J.W. (1993) Stochastic Locally-Optimal Harvesting. *Canadian Special Publications of Fisheries and Aquatic Sciences* **120**, 333–343.

Horwood, J.W. (2000) No-take zones: a management context. In: *The Effects of Fishing on Non-Target Species and Habitats: Biological, Conservation and Socio-Economic Issues* (eds M.J. Kaiser & S.J. de Groot), pp. 302–311. Blackwell Science, Oxford.

Horwood, J.W. & Whittle, P. (1986) Optimal control in the neighbourhood of an optimal equilibrium with examples from fisheries models. *IMA Journal of Mathematics Applied in Medicine and Biology* **3**, 129–142.

Horwood, J.W., Bannister, R.C.A. & Howlett, G.J. (1986) Comparative fecundity of North Sea plaice (*Pleuronectes platessa* L.). *Proceedings of the Royal Society London* **B228**, 401–431.

Horwood, J.W., Nichols, J.H. & Milligan, S. (1998) Evaluation of closed areas for fish stock conservation. *Journal of Applied Ecology* **35**, 893–903.

Houde, E.D. (1987) Early life dynamics and recruitment variability. *American Fisheries Society Symposium* **2**, 17–29.

Houde, E.D. (1997) Patterns and trends in larval-stage growth and mortality of teleost fish. *Journal of Fish Biology* **51**, 52–83.

Howell, B.R. (1994) Fitness of hatchery-reared fish for survival in the sea. *Aquaculture and Fisheries Management* **25** (Suppl. 1), 3–17.

Huang, J.S., Xiaoling, Y.J. & Conghai, Y. (1995) Baculoviral hypodermal and hematopoietic necrosis—study on the pathogen and pathology of the explosive epidemic disease of shrimp. *Marine Fisheries Research* **16**.

Hudson, A.V. & Furness, R.W. (1988) Utilization of discarded fish by scavenging seabirds behind whitefish trawlers in Shetland. *Journal of Zoology* **215**, 151–166.

Hughes, R.N. (ed.) (1993) *Diet selection: An Interdisciplinary Approach to Foraging Behaviour*. Blackwell Scientific Publications, Oxford.

Hughes, R.N. & Croy, M.I. (1993) An experimental analysis of frequency-dependent predation (switching) in the 15-spined stickleback, *Spinachia spinachia*. *Journal of Animal Ecology* **62**, 341–352.

Hunt, G.L. & Furness, R.W. (1996) Seabird/fish interactions, with particular reference to seabirds in the North Sea. *ICES Co-operative Research Report* **216**, 1–87.

Hunter, J.R. & Goldberg, S.R. (1980) Spawning frequency and batch fecundity in northern anchovy *Engraulis mordax*. *Fishery Bulletin* **77**, 641–652.

Hunter, J.R. & Kimbrell, C.A. (1980) Egg cannibalism in the northern anchovy *Engraulis mordax*. *Fishery Bulletin* **78**, 811–816.

Hunter, J.R. & Lo, N.C.H. (1993) Icthyoplankton methods for estimating fish biomass: introduction and terminology. *Bulletin of Marine Science* **53**, 723–727.

Hunter, J.R. & Macewicz, B.J. (1985) Measurement of spawning frequency in multiple spawning fishes. *NOAA Technical Report NMFS* **36**, 79–94.

Hunter, J.R., Lo, N.C.H. & Leong, R.J.H. (1985) Batch fecundity in multiple spawning fishes. *NOAA Technical Report NMFS* **36**, 67–77.

Hunter, J.R., Argue, A.W., Bayliff, W.H., Dizon, A.E., Fonteneau, A., Goodman, D. & Seckel, G.R. (1986) The dynamics of tuna movements: an evaluation of past and future research. *FAO Fisheries Technical Paper* **277**.

Hutchings, J.A. (1996) Spatial and temporal variation in the density of northern cod and a review of hypotheses for the stock's collapse. *Canadian Journal of Fisheries and Aquatic Sciences* **53**, 943–962.

Hutchings, J.A. & Myers, R.A. (1994) What can be learned from the collapse of a renewable resource—Atlantic Cod *Gadus morhua*, of Newfoundland and Canada. *Canadian Journal of Fisheries and Aquatic Sciences* **51**, 2126–2146.

Hutchings, L., Verheye, H.M., Mitchell-Innes, B.A., Peterson, W.T., Huggett, J.A. & Painting, S.J. (1995) Copepod production in the southern Benguela system. *ICES Journal of Marine Science* **52**, 439–455.

Hutchinson, J. (1996) Fisheries interactions: the harbour porpoise—a review. In: *The Conservation of Whales and Dolphins* (eds M.P. Simmonds & J.D. Hutchinson), pp. 128–165. John Wiley, Chichester.

ICES (1994a) Report of the Benthos Ecology Working Group. *International Council for the Exploration of the Sea, Committee Meeting*, 1994/L: 4.

ICES (1994b) Report of the mackerel/horse mackerel egg production workshop. *International Council for the Exploration of the Sea, Committee Meeting*, 1994/H: 4.

ICES (1994c) Report of the study group on the North Sea plaice box. *International Council for the Exploration of the Sea, Committee Meeting*, Assess: 14.

ICES (1995) Report of the study group on the ecosystem effects of fishing activities. *ICES Co-operative Research Report* **200**.

ICES (1996) Report of the Working Group on Environmental Interactions of Mariculture. *International Council for the Exploration of the Sea, Committee Meeting*, 1996/F: 5.

ICES (1997a) Report of the working group on mackerel and horse mackerel egg surveys. *International Council for the Exploration of the Sea, Committee Meeting*, 1997/H: 4.

ICES (1997b) Report of the working group on the ecosystem effects of fishing activities. *International Council for the Exploration of the Sea, Committee Meeting*, 1998/ACFM/ACME:1 ref E.

ICES (1997c) Report of the multispecies assessment working group. *International Council for the Exploration of the Sea Committee Meeting*, 1997/Assess: 16.

ICES (1999a) Report of the working group of southern shelf demersal stocks. *International Council for the Exploration of the Sea, Committee Meeting*, 1999/ACFM: 4.

ICES (1999b) Report of the study group to evaluate the effects of multispecies interactions. *International Council for the Exploration of the Sea Committee Meeting*, 1999/D: 4.

Ihssen, P.E., Booke, H.E., Casselman, J.M., McGlade, J., Payne, N.R. & Utter, F.M. (1981) Stock identification: materials and methods. *Canadian Journal of Fisheries and Aquatic Sciences* **38**, 1838–1855.

Ikeda, T. & Kirkwood, R. (1989) Matabolism and body composition of two Euphausids (*E. superba* and *E. crystallorophias*) collected from under the pack ice of Enderby Land, Antarctica. *Marine Biology* **100**, 301–308.

IUCN (1996) *1996 Red List of Threatened Animals*. IUCN, Gland, Switzerland.

Iverson, R.L. (1990) Control of marine fish production. *Limnology and Oceanography* **35**, 1593–1604.

IWC (1995) Report of the Scientific Committee, Annex F. Report of the Sub-Committee on Aboriginal Subsistence Whaling. *Reports of the International Whaling Commission* **45**, 142–164.

Jakobsen, T. (1993) The behaviour of F_{low}, F_{med} and F_{high} in response to variation in parameters used for their estimation. *Canadian Special Publications of Fisheries and Aquatic Sciences* **120**, 119–125.

Jakobsson, J. & Stefánson, G. (1998) Rational harvesting of the cod–capelin–shrimp complex in the Icelandic marine ecosystem. *Fisheries Research* **37**, 7–21.

James, A.G. & Findlay, K.P. (1989) Effect of particle size and concentration on feeding behaviour, selectivity and rates of food ingestion by the Cape anchovy *Engraulis capensis*. *Marine Ecology Progress Series* **50**, 279–294.

James, A.G. & Probyn, T. (1989) The relationship between respiration rate, swimming speed and feeding behaviour in the cape anchovy, *Engraulis capensis* Gilchrist. *Journal of Experimental Marine Biology and Ecology* **131**, 81–100.

Jamieson, A. (1974) Genetic 'tags' for marine fish stocks. In: *Sea Fisheries Research* (ed. F.R. Harden-Jones). Elek Science, London.

Jamieson, G.S. & Francis, K. (1986) Invertebrate and marine plant resources of British Columbia. *Canadian Special Publications of Fisheries and Aquatic Sciences* **91**, 1–89.

Jefferson, T.A., Letherwood, S. & Webber, M.P. (1993) *Marine Mammals of the World* (FAO species identification guide). FAO, Rome.

Jennings, S. (1991) The effects of capture, net retention and preservation upon the lengths of larval and juvenile bass, *Dicentrarchus labrax* (L.). *Journal of Fish Biology* **38**, 349–357.

Jennings, S. & Beverton, R.J.H. (1991) Intraspecific variation in the life history tactics of Atlantic herring (*Clupea harengus* L.) stocks. *ICES Journal of Marine Science* **48**, 117–125.

Jennings, S. & Kaiser, M.J. (1998) The effects of fishing on marine ecosystems. *Advances in Marine Biology* **34**, 201–352.

Jennings, S. & Pawson, M.G. (1991) The development of bass, *Dicentrarchus labrax*, eggs in relation to temperature. *Journal of the Marine Biological Association of the United Kingdom* **71**, 107–116.

Jennings, S. & Polunin, N.V.C. (1995a) Comparative size and composition of yield from six Fijian reef fisheries. *Journal of Fish Biology* **46**, 28–46.

Jennings, S. & Polunin, N.V.C. (1995b) Biased underwater visual census biomass estimates for target species in tropical reef fisheries. *Journal of Fish Biology* **47**, 733–736.

Jennings, S. & Polunin, N.V.C. (1996a) Fishing strategies, fishery development and socioeconomics in traditionally managed Fijian fishing grounds. *Fisheries Management and Ecology* **3**, 335–347.

Jennings, S. & Polunin, N.V.C. (1996b) Effects of fishing effort and catch rate upon the structure and biomass of Fijian reef fish communities. *Journal of Applied Ecology* **33**, 400–412.

Jennings, S. & Polunin, N.V.C. (1996c) Impacts of fishing on tropical reef ecosystems. *Ambio* **25**, 44–49.

Jennings, S. & Polunin, N.V.C. (1997) Impacts of predator depletion by fishing on the biomass and diversity of non-target reef fish communities. *Coral Reefs* **16**, 71–82.

Jennings, S. & Reynolds, J.D. (2000) Impacts of fishing on diversity: from pattern to process. In: *Effects of Fishing on Non-Target Species and Habitats: Biological Conservation Socio-Economic Issues* (eds M.J. Kaiser & S.J. de Groot), pp. 235–250. Blackwell Science, Oxford.

Jennings, S., Grandcourt, E.M. & Polunin, N.V.C. (1995) The effects of fishing on the diversity, biomass and trophic struc-

ture of Seychelles' reef fish communities. *Coral Reefs* **14**, 225–235.

Jennings, S., Lancaster, J.E., Ryland, J.S. & Shackley, S.E. (1991) The age structure and growth dynamics of young-of-the-year bass, *Dicentrarchus labrax*, populations. *Journal of the Marine Biological Association of the United Kingdom* **71**, 799–810.

Jennings, S., Marshall, S.S. & Polunin, N.V.C. (1996) Seychelles' marine protected areas: comparative structure and status of reef fish communities. *Biological Conservation* **75**, 201–209.

Jennings, S., Reynolds, J.D. & Mills, S.C. (1998) Life history correlates of responses to fisheries exploitation. *Proceedings of the Royal Society: Biological Sciences* **265**, 333–339.

Jennings, S., Reynolds, J.D. & Greenstreet, S.P.R. (1999a) Structural change in an exploited fish community: a consequence of differential fishing effects on species with contrasting life histories. *Journal of Animal Ecology* **68**, 617–627.

Jennings, S., Reynolds, J.D. & Polunin, N.V.C. (1999b) Predicting the vulnerability of tropical reef fishes to exploitation using phylogenies and life histories. *Conservation Biology* **13**, 1466–1475.

Jennings, S., Alvsvåg, J., Cotter, A.J., Ehrich, S., Greenstreet, S.P.R., Jarre-Teichmann, A., Mergardt, N., Rijnsdorp, A.D. & Smedstad, O. (1999c) Fishing effects in northeast Atlantic shelf seas: patterns in fishing effort, diversity and community structure. III. International fishing effort in the North Sea: an analysis of temporal and spatial trends. *Fisheries Research* **40**, 125–134.

Jensen, A.C. (1972) *The Cod: an Uncommon History of a Common Fish and its Impact on American Life from Viking Times to the Present*. Thomas Crowell Company, New York.

Jensen, A.C., Collins, K.J., Free, E.K. & Bannister, R.C.A. (1994) Lobster (*Homarus gammarus*) movement on an artificial reef: the potential use of artificial reefs for stock enhancement. *Crustaceana* **67**, 198–211.

Johannes, R.E. (1978a) Traditional marine conservation methods in Oceania and their demise. *Annual Review of Ecology and Systematics* **9**, 349–364.

Johannes, R.E. (1978b) Reproductive strategies of coastal marine fishes in the tropics. *Environmental Biology of Fishes* **3**, 65–84.

Johannes, R.E. (1980) Using knowledge of reproductive behaviour of reef and lagoon fishes to improve yields. In: *Fish Behaviour and Fisheries Management* (eds J. Bardach, J. Magnuson, R. May & J. Reinhart), pp. 247–270. ICLARM, Manila.

Johannes, R.E. (1994) *Government-supported, Village-Based Management of Marine Resources in Vanuatu*. Forum Fisheries Agency, Honiara, Solomon Islands.

Johannes, R.E. (1998) The case for data-less marine resource management: examples from tropical nearshore finfisheries. *Trends in Ecology and Evolution* **13**, 243–246.

Jones, G.P., Millicich, M.J., Emslie, M.J. & Lunow, C. (1999) Self-recruitment in a coral reef fish population. *Nature* **402**, 802–804.

Jones, R. (1981) The use of length composition in fish stock assessments (with notes on VPA and cohort analysis). *FAO Fisheries Circular* **734**.

Joseph, J. (1994) The tuna–dolphin controversy in the eastern tropical Pacific Ocean: biological, economic and political impacts. *Ocean Development and International Law* **25**, 1–30.

Kaiser, M.J. (1996) Starfish damage as an indicator of trawling intensity. *Marine Ecology Progress Series* **134**, 303–307.

Kaiser, M.J. & de Groot, S.J. (eds) (2000) *Effects of Fishing on Non-Target Species and Habitats: Biological, Conservation and Socio-Economic Issues*. Blackwell Science, Oxford.

Kaiser, M.J. & Ramsay, K. (1997) Opportunistic feeding by dabs within areas of trawl disturbance: possible implications for increased survival. *Marine Ecology Progress Series* **152**, 307–310.

Kaiser, M.J. & Spencer, B.E. (1994) Fish scavenging behaviour in recently trawled areas. *Marine Ecology Progress Series* **112**, 41–49.

Kaiser, M.J. & Spencer, B.E. (1995) Survival of by-catch from a beam trawl. *Marine Ecology Progress Series* **126**, 31–38.

Kaiser, M.J. & Spencer, B.E. (1996a) The behavioural response of scavengers to beam-trawl disturbance. In: *Aquatic Predators and Their Prey* (eds S. Greenstreet & M. Tasker), pp. 116–123. Blackwell Scientific Publications, Oxford.

Kaiser, M.J. & Spencer, B.E. (1996b) The effects of beam-trawl disturbance on infaunal communities in different habitats. *Journal of Animal Ecology* **65**, 348–358.

Kaiser, M.J., Bullimore, B., Newman, P., Lock, K. & Gilbert, S. (1996a) Catches in 'ghost-fishing' set nets. *Marine Ecology Progress Series* **145**, 11–16.

Kaiser, M.J., Hill, A.S., Ramsay, K., Spencer, B.E., Brand, A.R., Veale, L.O., Prudden, K., Rees, E.I.S., Munday, B.W., Ball, B. & Hawkins, S.J. (1996b) Benthic disturbance by fishing gear in the Irish Sea: a comparison of beam trawling and scallop dredging. *Aquatic Conservation: Marine and Freshwater Ecosystems* **6**, 269–285.

Kaiser, M.J., Armstrong, P.J., Dare, P.J. & Flatt, R.P. (1998a) Benthic communities associated with a heavily fished scallop ground in the English Channel. *Journal of the Marine Biological Association of the United Kingdom* **78**, 1045–1059.

Kaiser, M.J., Edwards, D., Armstrong, P., Radford, K., Lough, N., Flatt, R. & Jones, H. (1998b) Changes in megafaunal benthic communities in different habitats after trawling disturbance. *ICES Journal of Marine Science* **55**, 353–361.

Kawasaki, T., Tanaka, S., Toba, Y. & Taniguchi, A. (eds) (1991) *Long-Term Variabilty of Pelagic Fish Populations and Their Environment*. Pergamon Press, Oxford.

Kent, G. (1998) Fisheries, food security and the poor. *Food Policy* **22**, 393–404.

Kerfoot, W.C. & Sih, A. (eds) (1987) *Predation: Direct and Indirect Effects on Aquatic Communities*. University Press of New England, Hanover.

King, J.E. (1964) *Seals of the World*. British Museum (Natural History), London.

King, M.G. (1995) *Fisheries Biology, Assessment and Management*. Fishing News Books, Oxford.

Kirk, J.T.O. (1992) The nature and measurement of the light environment in the ocean. In: *Primary Productivity and Biogeochemical Cycles in the Sea* (eds P.G. Falkowski & A.D. Woodhead), pp. 9–29. Plenum Press, New York.

Kirkwood, G.P., Beddington, J.R. & Rossouw, J.A. (1994) Harvesting species of different lifespans. In: *Large-Scale Ecology and Conservation Biology* (eds P.J. Edwards, R.M. May & N.R. Webb), pp. 199–227. Blackwell Scientific, Oxford.

Kjesbu, O.S. (1989) The spawning activity of cod, *Gadus morhua* L. *Journal of Fish Biology* **34**, 195–206.

Koslow, J.A., Bell, J., Virtue, P. & Smith, D.C. (1995) Fecundity and its variability in orange roughy- effects of population density, condition, egg size and senescence. *Journal of Fish Biology* **47**, 1063–1080.

Koslow, J.A., Hanley, F. & Wicklaud, R. (1988) Effects of fishing on reef fish communities at Pedro Bank and Port Royal Cays, Jamaica. *Marine Ecology Progress Series* **43**, 201–212.

Koslow, J.A., Aiken, K., Auil, S. & Clementson, A. (1994) Catch and effort analysis of the reef fisheries of Jamaica and Belize. *Fishery Bulletin* **92**, 737–747.

Krost, P., Bernhard, M., Werner, F. & Hukriede, W. (1990) Otter trawl tracks in Kiel Bay (Western Baltic) mapped by side-scan sonar. *Meeresforschung* **32**, 344–353.

Kruse, G.H. & Kimker, A. (eds) (1993) *Degradable Escape Mechanisms for Pot Gear: a summary report to the Alaska Board of Fisheries*. Alaska Department of Fisheries and Game, Juneau, Alaska.

Kulbicki, M. (1998) How the acquired behaviour of commercial fishes may influence the results obtained from visual censuses. *Journal of Experimental Marine Biology and Ecology* **222**, 11–30.

Kunin, W.E. & Lawton, J.H. (1996) Does biodiversity matter? Evaluating the case for conserving species. In: *Biodiversity: a Biology of Numbers and Difference* (ed. K.J. Gaston), pp. 283–308. Blackwell Science, Oxford.

Kurland, J.M. (1998) Implications of the essential fish habitat provisions of the Magnuson–Stevens Act. In: *Effects of Fishing Gear on the Sea Floor of New England* (eds E.M. Dorsey & J. Pederson), pp. 104–106. Conservation Law Foundation, Boston.

Kurlansky, M. (1997) *Cod: a Biography of the Fish that Changed the World*. Walker, New York.

Laing, I. & Spencer, B. (1997) *Bivalve Cultivation: Criteria for Selecting a Site*. The Centre for Environment, Fisheries and Aquaculture Science, Lowestoft.

Lande, R., Engen, S. & Saether, B.-E. (1994) Optimal harvesting, economic discounting and extinction risk in fluctuating populations. *Nature* **372**, 88–89.

Lapointe, B., Niell, F. & Fuentes, J. (1981) Community structure, succession and production of seaweeds associated with mussel-rafts in the Ria de Arosa, N.W. Spain. *Marine Ecology Progress Series* **5**, 243–253.

Lasker, R. (1975) Field criteria for survival of anchovy larvae: the relation between inshore chlorophyll maximum layers and successful first feeding. *Fishery Bulletin* **73**, 453–462.

Lasker, R. (1978) The relationship between oceanographic conditions and larval anchovy food in the California current: identification of factors contributing to recruitment failure. *Rapports et Procès-Verbaux des Réunions, Conseil International pour l'Exploration de la Mer* **173**, 212–230.

Lasker, R. (ed.) (1981) *Marine Fish Larvae: Morphology, Ecology and Relation to Fisheries.* University of Washington Press, Seattle.

Lasker, R. (1988) Food chains and fisheries: an assessment after 20 years. In: *Toward a Theory on Biological–Physical Interactions in the World Ocean* (ed. B.J. Rothschild), pp. 173–182. Kluwer Academic Publishers, Berlin.

Law, R. & Grey, D.R. (1989) Evolution of yields from populations with age-specific cropping. *Evolutionary Ecology* **3**, 343–359.

Law, R. & Rowell, C.A. (1993) Cohort structured populations, selection responses, and exploitation of the North Sea cod. In: *The Exploitation of Evolving Resources* (eds T.K. Stokes, J.M. McGlade & R. Law), pp. 155–173. Springer-Verlag, Berlin.

Law, R. & Stokes, T.K. (2000) Evolutionary impact of fishing on target populations. In: *Marine Conservation Biology: the Science of Maintaining Biodiversity.* Island Press (in press).

Laws, E.A. (1997) *El Niño and the Peruvian Anchovy Fishery.* University Science Books, Ausalito, California.

Laws, R.M. (1977) Seals and whales of the Southern Ocean. *Philosophical Transactions of the Royal Society* **B279**, 81–96.

Leakey, R.J.G., Burkill, P.H. & Sleigh, M.A. (1992) Planktonic ciliates in Southampton water: abundance, biomass, production, and role in pelagic carbon flow. *Marine Biology* **114**, 67–83.

Leakey, R.J.G., Burkhill, P.H. & Sleigh, M.A. (1994) Ciliate growth rates from Plymouth Sound: comparison of direct and indirect estimates. *Journal of the Marine Biological Association of the United Kingdom* **74**, 849–861.

LeClus, F. (1977) A comparison of four methods used in fecundity determination of the pilchard. *Fishery Bulletin* **78**, 603–618.

Leggett, W.C. & Carscadden, J.F. (1978) Latitudinal variation in the reproductive characteristics of American Shad *Alosa sapidissima*: evidence for population specific life history strategies in fish. *Journal of the Fisheries Research Board of Canada* **35**, 1469–1478.

Leis, J.M. (1983) Coral reef fish larvae (Labridae) in the east Pacific barrier. *Copeia* **1983**, 826–828.

Leis, J.M. (1991) The pelagic stage of reef fishes: the larval biology of coral reef fishes. In: *The Ecology of Fishes on Coral Reefs* (ed. P.F. Sale), pp. 183–230. Academic Press, San Diego.

Lequesne, C. (1999) The Common Fisheries Policy. In: *Policy Making in the European Union* (eds H. Wallace & W. Wallace), pp. 345–372. Oxford University Press, Oxford.

Leslie, P.H. & Davis, D.H.S. (1939) An attempt to determine the absolute number of rats on a given area. *Journal of Animal Ecology* **8**, 94–113.

Leslie, R.W. & Grant, W.S. (1990) Lack of congruence between genetic and morphological stock structure of the southern African anglerfish *Lophius vomerinus*. *South African Journal of Marine Science* **9**, 379–398.

Lessios, H.A. (1988) Mass mortality of *Diadema antillarum* in the Caribbean: what have we learned? *Annual Review of Ecology and Systematics* **19**, 371–393.

Lester, R.J.G. (1990) Reappraisal of the use of parasites for fish stock identification. *Australian Journal of Marine and Freshwater Research* **41**, 855–864.

Levin, L.A. (1984) Life history and dispersal patterns in a dense infaunal polychaete assemblage: community structure and response to disturbance. *Ecology* **65**, 1185–1200.

Levitan, D.R. (1991) Influence of body size and population density on fertilization success and reproductive output in a free-spawning invertebrate. *Biological Bulletin* **181**, 261–268.

Levitan, D.R. (1993) The importance of sperm limitation to the evolution of egg size in marine invertebrates. *American Naturalist* **141**, 517–536.

Levitan, D.R. (1995) Sperm limitation in the sea. *Trends in Ecology and Evolution* **10**, 228–231.

Levitan, D.R., Sewell, M.A. & Chia, F.S. (1992) How distribution and abundance influence fertilization success in the sea-urchin *Strongylocentrotus franciscanus*. *Ecology* **73**, 248–254.

Li, W.K.W. & Maestrini, S.Y. (eds) (1993) Measurement of primary production from the molecular to the global scale. *ICES Marine Science Symposia* **197**.

Lien, J., Hood, C., Pittman, D., Ruel, P., Borggaard, D., Chisholm, C., Wiesner, L., Mahon, T. & Mitchell, D. (1995) Field tests of acoustic devices on ground-fish gillnets: assessment of effectiveness in reducing harbour porpose by-catch. In: *Sensory Systems of Aquatic Mammals* (eds R. Kastelein, J.A. Thomas & P.E. Nachtigal), pp. 349–364. De Spil Publishers, Woerden, The Netherlands.

Liermann, M. & Hilborn, R. (1997) Depensation in fish stocks: a hierarchic Bayesian meta-analysis. *Canadian Journal of Fisheries and Aquatic Sciences* **54**, 1976–1984.

Lincoln-Smith, M.P. (1988) Effects of observer swimming speed on sample counts of temperate rocky reef fish assemblages. *Marine Ecology Progress Series* **43**, 223–231.

Lindley, J., Gamble, J. & Hunt, H. (1995) A change in the zooplankton of the central North Sea (55° to 58°N): a possible consequence of changes in the benthos. *Marine Ecology Progress Series* **119**, 299–303.

Lo, N.C.H. (1985) A model for temperature dependent northern anchovy egg development and an automated procedure for the assignment of age to staged eggs. *NOAA Technical Reports NMFS* **36**, 43–50.

Lockwood, S.J. (1988) *The Mackerel: its Biology Assessment and the Management of a Fishery.* Fishing News Books, Farnham.

Lockwood, S.J., Nichols, J.H. & Dawson, W.A. (1981) The estimation of mackerel (*Scomber scombrus* L.) spawning stock size by plankton survey. *Journal of Plankton Research* **3**, 217–233.

Løkkeborg, S. (1998) Seabird by-catch and bait loss in long-lining using different setting methods. *ICES Journal of Marine Science* **55**, 145–149.

Longhurst, A.R. & Harrison, W.G. (1989) The biological pump: profiles of plankton production and consumption in the upper ocean. *Progress in Oceanography* **22**, 47–123.

Longhurst, A.R., Sathyendranath, S., Platt, T. & Caverhill, C. (1995) An estimate of global primary production in the ocean from satellite radiometer data. *Journal of Plankton Research* **17**, 1245–1271.

Lopez-Jamar, E., Iglesias, J. & Otero, J. (1984) Contribution of infauna and mussel-raft epifauna to demersal fish diets. *Marine Ecology Progress Series* **15**, 13–28.

Love, M.S., Morris, P., McCrae, M. & Collins, R. (1990) Life history aspects of 19 rockfish species (Scorpaenidae: *Sebastes*) from the southern California Bight. *NOAA Technical Report NMFS* **87**, 38.

Lutz, R.A. & Kennish, M.J. (1993) Ecology of deep-sea hydrothermal vent communities: a review. *Review of Geophysics* **31**, 211–242.

Lynge, F. (1992) *Ethics of a Killer Whale.* University of Iceland, Sudurgata, Iceland.

Lyster, S. (1985) *International Wildlife Law.* Grotius Publications, Cambridge.

MacArthur, R.H. & Pianka, E. (1966) On the optimal use of a patchy environment. *American Naturalist* **100**, 603–609.

MacCall, A.D. (1986a) Changes in the biomass of the California current system. In: *Variability and Management of Large Marine Ecosystems* (eds K. Sherman & L.M. Alexander), pp. 33–54. Westview Press, Boulder.

MacCall, A.D. (1986b) Virtual population analysis (VPA) equations for nonhomogeneous populations, and a family of approximations including improvements on Pope's cohort analysis. *Canadian Journal of Fisheries and Aquatic Sciences* **43**, 2406–2409.

MacCall, A.D. (1990) *Dynamic Geography of Marine Fish Populations.* University of Washington Press, Washington.

MacDonald, D.S., Little, M., Eno, N.C. & Hiscock, K. (1996) Towards assessing the sensitivity of benthic species and biotopes in relation to fishing activities. *Aquatic Conservation: Marine and Freshwater Ecosystems* **6**, 257–268.

MacDonald, P.D.M. (1969) FORTRAN programs for statistical estimation of distribution mixtures: some techniques for statistical analysis of length-frequency data. *Fisheries Research Board of Canada, Technical Report* **29**.

Mace, G.M. & Hudson, E.J. (1999) Attitudes towards sustainability and extinction. *Conservation Biology* **13**, 242–246.

MacIntyre, H.L., Geider, R.J. & Miller, D.C. (1996) Microphytobenthos: the ecological role of the 'secret garden' of unvegetated, shallow-water marine habitats. 1. Distribution, abundance and primary production. *Estuaries* **19**, 186–201.

MacKenzie, K. (1982) Fish parasites as biological tags. *Scottish Fishery Bulletin* **47**, 27–32.

MacKenzie, K. & Abaunza, P. (1998) Parasites as biological tags for stock discrimination of marine fish: a guide to procedures and methods. *Fisheries Research* **38**, 45–56.

Mackinson, S., Sumaila, U.R. & Pitcher, T.J. (1997) Bioeconomics and catchability: fish and fishers behaviour during stock collapse. *Fisheries Research* **31**, 11–17.

MacLennan, D.N. & Simmonds, E.J. (1992) *Fisheries Acoustics.* Chapman & Hall, London.

Magnússon, K.G. (1995) An overview of multispecies VPA—theory and applications. *Reviews in Fish Biology and Fisheries* **5**, 195–212.

Magurran, A.E. (1988) *Ecological Diversity and its Measurement.* Croom-Helm, London.

Main, G.I. & Sangster, G.I. (1982a) A study of multi-level bottom trawl for species separation using direct observation techniques. *Scottish Fisheries Research Reports* **26**, 1–17.

Main, J. & Sangster, G.I. (1982b) A study of separating fish from *Nephrops norvegicus* L. in a bottom trawl. *Scottish Fisheries Research Reports* **24**, 1–9.

Mallin, M.A., Burkholder, J.M. & Sullivan, M.J. (1992) Contributions of benthic microalgae to coastal fishery yield. *Transactions of the American Fisheries Society* **121**, 691–695.

Mangel, M. (1985) *Decision and Control in Uncertain Resource Systems.* Academic Press, New York.

Mangel, M. & Clark, C.W. (1988) *Dynamic Modelling in Behavioural Ecology.* Princeton University Press, Princeton, NJ.

Mangel, M. & Switzer, P.V. (1998) A model at the level of the foraging trip for the indirect effects of krill (*Euphausia superba*) fisheries on krill predators. *Ecological Modelling* **105**, 235–256.

Mann, K.H. (1982) *Ecology of Coastal Waters: a Systems Approach.* Blackwell Scientific Publications, Oxford.

Mann, K.H. (1993) Physical oceanography, food chains and fish stocks: a review. *ICES Journal of Marine Science* **50**, 105–119.

Mann, K.H. & Breen, P.A. (1972) The relation between lobster abundance, sea urchins and kelp beds. *Journal of the Fisheries Research Board of Canada* **29**, 603–605.

Mann, K.H. & Lazier, J.R.N. (1996) *Dynamics of Marine Ecosystems.* Blackwell Science, London.

Manooch, C.S.I. & Potts, J.C. (1997) Age, growth and mortality of greater amberjack from the southeastern United States. *Fisheries Research* **30**, 229–240.

Mardle, S. & Pascoe, S. (1999) A review of applications of multiple-criteria decision-making techniques to fisheries. *Marine Resource Economics* **14**, 41–63.

Mariojouls, C. & Kusuki, Y. (1987) Appréciation des quantités de biodépôts émis par les huîtres en élevage suspendu dans la baie d'Hiroshima. *Haliotis* **16**, 221–231.

Martin, J.H. (1992) Iron as a limiting factor in ocean productivity. In: *Primary Productivity and Biogeochemical Cycles in the Sea* (eds P.G. Falkowski & A.D. Woodhead), pp. 123–137. Plenum Press, New York.

Martinez, N.D. (1996) Defining and measuring functional aspects of biodiversity. In: *Biodiversity: a Biology of*

Numbers and Difference (ed. K.J. Gaston), pp. 114–148. Blackwell Science, Oxford.

Matsuda, H., Yahava, T. & Uozumi, Y. (1997) Is tuna critically endangered? Extinction risk of a large and over-exploited population. *Ecological Research* 12, 345–356.

May, R.M., Beddington, J.R., Clark, C.W., Holt, S.J. & Laws, R.M. (1979) Management of multispecies fisheries. *Science* 205, 267–277.

Maynard-Smith, J. (1982) *Evolution and the Theory of Games.* Cambridge University Press, Cambridge.

McAllister, D.E. (1988) Environmental, economic and social costs of coral reef destruction in the Philippines. *Galaxea* 7, 161–178.

McAllister, M.K., Pikitch, E.K., Punt, A.E. & Hilborn, R. (1994) A Bayesian approach to stock assessment and harvest decisions using the sampling/importance resampling algorithm. *Canadian Journal of Fisheries and Aquatic Sciences* 51, 2673–2687.

McClanahan, T.R. (1992) Resource utilisation, competition and predation: a model and example from coral reef grazers. *Ecological Modelling* 61, 195–215.

McClanahan, T.R. (1995) A coral-reef ecosystem-fisheries model—impacts of fishing intensity and catch selection on reef structure and processes. *Ecological Modelling* 80, 1–19.

McClanahan, T.R. & Kaundaarara, B. (1996) Fishery recovery in a coral reef marine park and its effect on the adjacent fishery. *Conservation Biology* 10, 1187–1199.

McClanahan, T.R. & Shafir, S.H. (1990) Causes and consequences of sea urchin abundance and diversity in Kenyan coral reef lagoons. *Oecologia* 83, 362–370.

McClanahan, T.R., Kakamura, A.T., Muthiga, N.A., Gilagabher Yebio, M. & Obura, D. (1996) Effects of sea-urchin reductions on algae, coral and fish populations. *Conservation Biology* 10, 136–154.

McManus, J. (1996) Social and economic aspects of reef fisheries and their management. In: *Reef Fisheries* (eds N.V.C. Polunin & C.M. Roberts), pp. 249–281. Chapman & Hall, London.

Megrey, B.A. (1989) Review and comparison of age-structured stock assessment models from theoretical and applied points of view. *American Fisheries Society Symposium* 6, 8–48.

Mehl, S. & Sunnanå, K. (1991) Changes in the growth of northeast Arctic cod in relation to food consumption in 1984–88. *ICES Marine Science Symposia* 193, 109–112.

Melillo, J.M., McGuire, A.D., Kicklighter, D.W., Moore, B., Vorosmarty, C.J. & Schloss, A.L. (1993) Global climate change and terrestrial net primary production. *Nature* 363, 234–240.

Mensink, B.P., Boon, J.P., ten Hallers Tjabbes, C.C., Van Huttum, B., & Koeman, J.H. (1997) Bioaccumulation of organotin compounds and imposex occurrence in a marine food chain (Eastern Scheldt, The Netherlands) *Environmental Technology* 18, 1235–1244.

Mensink, B.P. Fischer, C.V., Cadée, G.C., Fonds, M., Hallers–Tjabbs, C.C. & Boon, J.P. (2000) Shell damage and mortality in the common whelk *Buccinum undatum* caused by beam trawl fishery. *Journal of Sea Research* 43, 53–64.

Mercer, M.C. (ed.) (1982) Multispecies approaches to fisheries management advice. *Canadian Special Publication of Fisheries and Aquatic Sciences* 59.

Merrett, N.R. & Haedrich, R.L. (1997) *Deep-Sea Demersal Fish and Fisheries.* Chapman & Hall, London.

Mertz, G. & Myers, R.A. (1994) Match/mismatch predictions of spawning duration versus recruitment variability. *Fisheries Oceanography* 4, 236–245.

Metcalfe, J.D. & Arnold, G.P. (1997) Tracking fish with electronic tags. *Nature* 387, 665–666.

Metcalfe, J.D., Fulcher, M. & Storeton-West, T.J. (1991) Progress and developments in telemetry for monitoring the migratory behaviour of plaice in the North Sea. In: *Wildlife Telemetry: Remote Monitoring and Tracking of Animals* (eds Priede, I.G. & Swift, S.M.), pp. 361–366. Ellis Horwood, New York.

Metcalfe, J.D., Holford, B.H. & Arnold, G.P. (1993) Orientation of plaice (*Pleuronectes platessa*) in the open sea: evidence for the use of external directional cues. *Marine Biology* 117, 559–556.

Meyer, M.A., Kotze, P.G.H. & Brill, G.W. (1992) Consumption of catch and interference with linefishing by South African (Cape) fur seals, *Arctocephalus pusillus pusillus. South African Journal of Marine Science* 12, 835–842.

Michod, R.E. & Levin, B.R. (1988) *The Evolution of Sex.* Sinauer Associates, Sunderland, Massachusetts.

Miller, B.A. & Emlet, R.B. (1997) Influence of nearshore hydrodynamics on larval abundance and settlement of sea urchins *Strongylocentrotus franciscanus* and *S. purpuratus* in the Oregon upwelling zone. *Marine Ecology Progress Series* 148, 83–94.

Miller, D.C., Geider, R.J. & MacIntyre, H.L. (1996) Microphytobenthos: the ecological role of the 'secret garden' of unvegetated, shallow-water marine hab itats. II. Role in sediment stability and shallow water food webs. *Estuaries* 19, 202–212.

Miller, R.J. (1977) Resource underutilization in a spider crab industry. *Fisheries* 2, 9–13.

Miller, R.J. (1985) Seaweeds, sea urchins and lobsters: a reappraisal. *Canadian Journal of Fisheries and Aquatic Sciences* 42, 2061–2072.

Millner, R. & Whiting, C. (1996) Long-term changes in growth and population abundance of sole in the North Sea from 1940 to the present. *ICES Journal of Marine Science* 53, 1185–1195.

Milner-Gulland, E.J. & Mace, R. (1998) *Conservation of Biological Resources.* Blackwell Science, Oxford.

Misund, O.A. (1997) Underwater acoustics in marine and fisheries research. *Reviews in Fish Biology and Fisheries* 7, 1–34.

Monaghan, P., Uttley, J.D. & Okill, J.D. (1989) Terns and sandeels—seabirds as indicators of changes in marine fish populations. *Journal of Fish Biology* 35, 339–340.

Moore, P.G. & Howarth, J. (1996) Foraging by marine scavengers: effects of relatedness, bait damage and hunger. *Journal of Sea Research* 36, 267–273.

Moreau, G. & Barbeau, C. (1982) Heavy metals as indicators of the geographic origin of the American eel (*Anguilla rostrata*). *Canadian Journal of Fisheries and Aquatic Sciences* 39, 1004–1011.

Morgan, G.R. (1997) Individual quota management in fisheries. *FAO Fisheries Technical Paper*, 371.

Morris, D.J., Watkins, J.L., Ricketts, C., Bucholz, F. & Priddle, J. (1988) An assessment of the merits of length and weight measurements of Antarctic krill *Euphausia superba*. *British Antarctic Survey Bulletin* 79, 27–50.

Munro, G.R. (1992) Mathematical bioeconomics and the evolution of modern fisheries economics. *Bulletin of Mathematical Biology* 54, 163–184.

Munro, G.R. (1998) The economics of overcapitalisation and fishery resource management: a review. In: *Overcapacity, Overcapitalisation and Subsidies in European Fisheries* (eds A. Hatcher & K. Robinson), pp. 7–23. Centre for the Economics and Management of Aquatic Resources, Portsmouth.

Munro, J.L. (1993) Giant Clams. In: *Nearshore Marine Resources of the South Pacific* (eds A. Wright & L. Hill), pp. 431–449. Institute of Pacific Studies, Suva.

Munro, J.L. & Fakahau, S.T. (1993) Appraisal, assessment and monitoring of small-scale coastal fisheries in the South Pacific region. In: *Nearshore Marine Resources of the South Pacific* (eds A. Wright & L. Hill), pp. 15–53. Institute of Pacific Studies, Suva.

Munro, J.L., Parrish, J.D. & Talbot, F.H. (1987) The biological effects of intensive fishing upon reef fish communities. In: *Human Impacts on Coral Reefs: Facts and Recommendations* (ed. B. Salvat), pp. 41–49. Antenne Museum EPHE, French Polynesia.

Murata, M. (1989) Population assessment, management and fishery forecasting for the Japanese common squid, *Todarodes pacificus*. In: *Marine Invertebrate Fisheries: their Assessment and Management* (ed. J.F. Caddy), pp. 613–636. John Wiley & Sons, New York.

Murawski, S.A. (1984) Mixed-species yield-per-recruitment analysis accounting for technological interactions. *Canadian Journal of Fisheries and Aquatic Sciences* 41, 897–916.

Murawski, S.A. & Idoine, J.S. (1992) Multispecies size composition: a conservative property of exploited fishery systems. *Journal of Northwest Atlantic Fishery Science* 14, 79–85.

Murphy, E.J. (1995) Spatial structure of the Southern Ocean ecosystem: predator–prey linkages in Southern Ocean food webs. *Journal of Animal Ecology* 64, 333–347.

Murphy, E.J., Watkins, J.L., Reid, K., Trathan, P.N., Everson, I., Croxall, J.P., Priddle, J., Brandon, M.A., Brierley, A.S. & Hofmann, E. (1998) Interannual variability of the South Georgia marine ecosystem: biological and physical sources of variation in the abundance of krill. *Fisheries Oceanography* 7, 381–390.

Murphy, M.D. (1997) Bias in Chapman-Robson least squares estimators of mortality rates for steady-state populations. *Fishery Bulletin* 95, 863–868.

Murray, J.W., Young, J., Newton, J., Dunne, J., Chapin, T., Paul, B. & McCarthy, J.J. (1996) Export flux of particulate organic carbon from the central equatorial Pacific using a combined drifting trap Th^{234} approach. *Deep Sea Research* (Part 2), 1095–1132.

Muscatine, L. (1990) The role of symbiotic algae in carbon and energy flux in reef corals. In: *Ecosystems of the World: Coral Reefs* (ed. Z. Dubinsky), pp. 75–87. Elsevier Science, New York.

Muscatine, L. & Weis, W. (1991) Productivity of zooxanthellae and biogeochemical cycles. In: *Primary Productivity and Biogeochemical Cycles in the Sea* (eds P.G. Falkowski & A.D. Woodhead), pp. 257–271. Plenum Press, New York.

Musick, J.A. (1999) Criteria to define extinction risk in marine fishes. *Fisheries* 24, 6–14.

Myers, R.A. & Stokes, K. (1989) Density-dependent habitat utilisation of groundfish and the improvement of research surveys. *International Council for the Exploration of the Seas, Committee Meeting*, D: 15.

Myers, R.A., Barrowman, N.J., Hutchings, J.A. & Rosenberg, A.A. (1995a) Population dynamics of exploited fish stocks at low population levels. *Science* 269, 1106–1108.

Myers, R.A., Bridson, J. & Barrowman, N.J. (1995b) Summary of worldwide spawner and recruitment data. *Canadian Technical Report of Fisheries and Aquatic Sciences* 2024.

Myers, R.A., Hutchings, J.A. & Barrowman, N.J. (1996) Hypothesis for the decline of cod in the North Atlantic. *Marine Ecology Progress Series* 138, 293–308.

Nakano, H., Koube, H., Umezawa, S., Momoyama, K., Hiraoka, M., Inouye, K. & Oseko, N. (1994) Mass mortalities of cultured Kuruma shrimp, *Penaeus japonicus*, in Japan in 1993: epizoological survey and infection trials. *Fish Pathology* 29, 135–139.

Nash, W.J. (1993) Trochus. In: *Nearshore Marine Resources of the South Pacific* (eds A. Wright & A. Hill), pp. 451–495. Institute of Pacific Studies, Suva.

Neill, W.M. (1994) Spatial and temporal scaling and the organisation of limnetic communities. In: *Aquatic Ecology: Scale, Pattern and Process* (eds P.S. Giller, A.G. Hildrew & D.G. Rafaelli), pp. 189–231. Blackwell Scientific Publications, Oxford.

Nelson, J.S. (1994) *Fishes of the World*. John Wiley and Sons, New York.

Nelson, C.S., Northcote, T.G. & Hendry, C.H. (1989) Potential use of oxygen and carbon isotopic composition of otoliths to identify migratory and non-migratory stocks of New Zealand common smelt: a pilot study. *New Zealand Journal of Marine and Freshwater Research* 23, 337–344.

New, M.B., Tacon, A.G.J. & Csavas, I. (eds) (1993) *Farm-Made Aquafeeds*. FAO (Food and Agriculture Organisation of the United Nations) RAPA (Regional Office for Asia and the Pacific) ASEAN (Association of Southeast Asian Nations) Commission of European Communities, Rome.

Newell, R. (1988) *Ecological Changes in Chesapeake Bay: are they the Result of Over-Harvesting the American Oyster* Crassostrea virginica? Understanding the Estuary; Advances in Chesapeake Bay Research. Chesapeake Research Consortium Publication, Baltimore.

Nichols, J.H. (1976) *Soleidae of the Eastern North Atlantic.* ICES Fiches d'identification du zooplancton (150/151). ICES, Charlottenlund, Denmark.

Nickell, L.A. & Atkinson, R.J.A. (1995) Functional morphology of burrows and trophic modes of three thalassinidean shrimp species, and a new approach to the classification of thalassinidean burrow morphology. *Marine Ecology Progress Series* **128**, 181–197.

Nickell, T.D. & Moore, P.G. (1992) The behavioural ecology of epibenthic scavenging invertebrates in the Clyde Sea area: laboratory experiments on attractions to bait in moving water, underwater TV observations in situ and general conclusions. *Journal of Experimental Marine Biology and Ecology* **159**, 15–35.

Nielsen, L.A. (1992) *Methods of Marking Fish and Shellfish.* American Fisheries Society Special Publication, 32. Bethesda, Maryland.

Nielsen, T.G. & Kiørboe, T. (1994) Regulation of zooplankton biomass and production in a temperate, coastal ecosystem. 2. Ciliates. *Limnology and Oceanography* **39**, 508–519.

Norse, E.A. (ed.) (1993) *Global Marine Biological Diversity: a Strategy for Building Conservation into Decision Making.* Island Press, Washington, DC.

Northridge, S. (1991) *Driftnet fisheries and their impacts on non-target species: a worldwide review.* Food and Agriculture Organization of the United Nations, Rome.

Nugues, M., Kaiser, M.J., Spencer, B.E. & Edwards, D.B. (1996) Benthic community changes associated with intertidal oyster cultivation. *Aquaculture Research* **27**, 913–924.

O'Brien, C.M., Fox, C.J., Planque, B. & Casey, J. (2000) Climate variability and North Sea cod. *Nature* **404**, 142.

O'Dor, R.K. & Wells, M.J. (1987) Energy and nutrient flow. In: *Cephalopod Life Cycles, 2 Comparative Review* (ed. P.R. Boyle), pp. 109–133. Academic Press, London.

OECD (1997) *Towards Sustainable Fisheries: Economic Aspects of the Management of Living Marine Resources.* Organisation for Economic Co-operation and Development, Paris.

Olla, B.L. & Davis, M.W. (1989) The role of learning and stress in predator avoidance of hatchery-reared coho salmon (*Oncorhynchus kisutch*) juveniles. *Aquaculture* **76**, 209–214.

Olla, B.L., Davis, M.W. & Schreck, C.B. (1997) Effects of simulated trawling on sablefish and walleye pollock: the role of light intensity, net velocity and towing duration. *Journal of Fish Biology* **50**, 1181–1194.

Ona, E. & Godø, O.R. (1990) Fish reaction to trawling noise: the significance of trawl sampling. *Rapports et Procès-Verbaux des Réunions, Conseil International pour l'Exploration de la Mer* **189**, 159–166.

Orensanz, J.M. & Jamieson, G.S. (1998) The assessment and management of spatially structured stocks: an overview of the North Pacific Symposium on invertebrate stock assessment and management. *Canadian Special Publications of Fisheries and Aquatic Sciences* **125**, 441–459.

Oro, D., Genovart, X., Ruiz, X., Jimenez, J. & Garcia-Gans, J. (1996) Differences in diet, population size and reproductive performance between two colonies of Audouin's Gull *Larus audouinii* affected by a trawling moratorium. *Journal of Avian Biology* **27**, 245–251.

Overholtz, W.J., Marawski, S.A. & Foster, K.L. (1991) Impact of predatory fish, marine mammals and seabirds on the pelagic fish ecosystem of the northeastern USA. *ICES Marine Science Symposia* **193**, 198–208.

Owens, N.J.P. (1987) Natural variations in ^{15}N in the marine environment. *Advances in Marine Biology* **24**, 389–451.

Pakhomov, E.A., Perissinotto, R., Froneman, P.W. & Miller, D.G.M. (1997) Energetics and feeding dynamics of *Euphausia superba* in the South Georgia region during the summer of 1994. *Journal of Plankton Research* **19**, 399–423.

Paloheimo, J.E. (1980) Estimation of mortality rates in fish populations. *Transactions of the American Fisheries Society* **109**, 378–386.

Panella, G. (1971) Fish otoliths: daily growth layers and periodical patterns. *Science* **173**, 1124–1127.

Parker, K. (1980) A direct method for estimating northern anchovy, *Engraulis mordax*, spawning biomass. *Fishery Bulletin* **78**, 541–544.

Parker, N.C., Giorgi, A.E., Heidinger, R.C., Jester, D.B., Prince, E.D. & Winans, G.A. (eds) (1990) *Fish Marking Techniques.* American Fisheries Society, Bethesda.

Parrish, F.A. & Kazama, T.K. (1992) Evaluation of ghost fishing in the Hawaiian lobster fishery. *Fisheries Bulletin* **90**, 720–725.

Parrish, J.D., Norris, J.E., Callahan, M.W., Magarifugi, E.J. & Schroeder, R.E. (1986) Piscivory in a coral reef community. In: *Gutshop '81: Fish Food Habits and Studies* (eds G.M. Caillet & C.A. Simenstad), pp. 73–78. University of Washington, Seattle.

Parsons, D.G. & Frechétte, J. (1989) Fisheries for northern shrimp (*Pandalus borealis*) in the northwest Atlantic from Greenland to the Gulf of Maine. In: *Marine Inveretebrate Fisheries: their Assessment and Management* (ed. J.F. Caddy), pp. 63–86. John Wiley & Sons, New York.

Parsons, T.R. & Lee Chen, Y.-L. (1994) Estimates of trophic efficiency, based on the size distribution of phytoplankton and fish in different environments. *Zoological Studies* **33**, 296–301.

Parsons, T.R. & Takahashi, M. (1973) *Biological Oceanographic Processes.* Pergamon Press, Oxford.

Parsons, T.R., LeBresseur, J.D., Fulton, J.D. & Kennedy, O.D. (1969) Production studies in the Strait of Georgia. II. Secondary production under the Fraser River plume, February to May, 1967. *Journal of Experimental Marine Biology and Ecology* **3**, 39–50.

Pascoe, S., Tamiz, M. & Jones, D.F. (1997) *Multi-objective modelling of the UK fisheries of the English Channel.* CEMARE Research Paper P113, University of Portsmouth.

Paulik, G.J. (1973) Studies of the possible form of the stock-recruitment curve. *Rapports et Procès-Verbaux des*

Réunions, Conseil International pour l'Exploration de la Mer **164**, 302–315.

Pauly, D. (1980) On the interrelationships between natural mortality, growth parameters and mean environmental temperature in 175 fish stocks. *Journal du Conseil, Conseil International pour l'Exploration de la Mer* **39**, 175–192.

Pauly, D. (1982) Studying single-species dynamics in a tropical multispecies context. *ICLARM Conference Proceedings* **9**, 33–70.

Pauly, D. (1985) Population dynamics of short lived species with emphasis on squids. *NAFO Science Council Studies* **9**, 143–154.

Pauly, D. & Christensen, V. (1993) Stratified models of large marine ecosystems: a general approach, and an application to the South China Sea. In: *Stress, Mitigation and Sustainability of Large Marine Ecosystems* (eds K. Sherman, L.M. Alexander & B.D. Gold) pp. 148–174. AAAS Press, Washington, D.C.

Pauly, D. & Christensen, V. (1995) Primary production required to sustain global fisheries. *Nature* **374**, 255–257.

Pauly, D. & David, N. (1981) ELEFAN 1, a BASIC program for objective extraction of growth parameters from length-frequency data. *Meeresforschung* **28**, 205–211.

Pauly, D. & Murphy, G.I. (eds) (1982) Theory and management of tropical fisheries. *ICLARM Conference Proceedings*, 9

Pauly, D., Silvestre, G. & Smith, I.R. (1989) On development, fisheries and dynamite: a brief review of tropical fisheries management. *Natural Resource Modeling* **3**, 307–329.

Pauly, D., Christensen, V., Dalsgaard, J., Froese, R. & Torres, F. (1998) Fishing down marine food webs. *Science* **279**, 860–863.

Pauly, D., Christensen, V. & Walters, C. (2000) Ecopath, Ecosim and Ecospace as tools for evaluating ecosystem impact of fisheries. *ICES Journal of Marine Science* **57**, 697–706.

Pawson, M.G. & Eaton, D.R. (1999) The influence of a power station on the survival of juvenile sea bass in an estuarine nursery area. *Journal of Fish Biology* **54**, 1143–1160.

Pawson, M.G. & Jennings, S. (1996) A critique of methods for stock identification in marine capture fisheries. *Fisheries Research* **25**, 203–217.

Pawson, M.G., Kelley, D.F. & Pickett, G.D. (1987) The distribution and migrations of bass *Dicentrarchus labrax* (L.) in waters around England and Wales as shown by tagging. *Journal of the Marine Biological Association of the United Kingdom* **67**, 183–217.

Pearson, T. & Rosenberg, R. (1978) Macrobenthic succession in relation to organic enrichment and pollution of the marine environment. *Oceanography and Marine Biology: an Annual Review* **16**, 229–311.

Pederson, B.H. (1997) The cost of growth in young fish larvae: a review of new hypotheses. *Aquaculture* **155**, 259–269.

Pella, J.J. & Tomlinson, P.K. (1969) A generalized stock production model. *Bulletin of the Inter-American Tropical Tuna Commission* **13**, 419–496.

Pemberton, D.J. & Shaughnessy, P.D. (1993) Interaction between seals and marine fish-farms in Tasmania, and management of the problem. *Aquatic Conservation: Marine and Freshwater Ecosystems* **3**, 149–158.

Pentilla, J. & Dery, L.M. (eds) (1988) Age determination methods for northwest Atlantic species. *NOAA Technical Report NMFS* **72**.

Pepin, P., Dower, J.F. & Leggett, W.C. (1998) Changes in the probability density function of larval fish body length following preservation. *Fishery Bulletin* **96**, 633–640.

Pergent, G., RicoRaimondino, V. & PergentMartini, C. (1997) Fate of primary production in *Posidonia oceanica* meadows of the Mediterranean. *Aquatic Botany* **59**, 307–321.

Perrins, C. (1987) *Birds of Britain and Europe*. Collins, London.

Peterson, C.H. & Summerson, H.C. (1992) Basin scale coherence of population dynamics of an exploited marine invertebrate, the bay scallop: implications of recruitment limitation. *Marine Ecology Progress Series* **90**, 257–272.

Petrakis, G. & Stergiou, K.I. (1995) Gill net selectivity for *Diplodus annularis* and *Mullus surmuletus* in Greek waters. *Fisheries Research* **21**, 455–464.

Philippart, C.J.M. (1998) Long-term impact of bottom fisheries on several by-catch species of demersal fish and benthic invertebrates in the south-eastern North Sea. *ICES Journal of Marine Science* **55**, 342–352.

Phillips, M.J., Lin, C.K. & Beveridge, M.C.M. (eds) (1993) Shrimp culture and the environment: lessons from the world's most rapidly expanding warm water aquaculture sector. *ICLARM Conference Proceedings* **31**.

Piatt, J.F. (1990) Aggregative response of common murres and Atlantic puffins to their prey. *Studies in Avian Biology* **14**, 36–51.

Piatt, J.F. (1997) *Behavioural ecology of common murre and Atlantic puffin predation on capelin: implications for population biology*. PhD Thesis, Department of Biology, Memorial University of Newfoundland, St. Johns.

Piatt, J.F. & Nettleship, D.N. (1985) Diving depths of four alcids. *Auk* **102**, 293–297.

Piatt, J.F. & Nettleship, D.N. (1987) Incidental catch of marine birds and mammals in fishing nets off Newfoundland, Canada. *Marine Pollution Bulletin* **18**, 344–349.

Pickering, H. & Whitmarsh, D. (1997) Artificial reefs and fisheries exploitation: a review of the 'attraction versus production' debate, the influence of design and its significance for policy. *Fisheries Research* **31**, 39–59.

Pickett, G.D. & Pawson, M.G. (1994) *Sea Bass: Biology, Exploitation and Conservation*. Chapman & Hall, London.

Pickett, S.T.A. & White, P.S. (1985) *The Ecology of Natural Disturbance and Patch Dynamics*. Academic Press, London.

Picquelle, S. (1985) Sampling requirements for the adult fish survey. *NOAA Technical Report NMFS* **36**, 55–57.

Picquelle, S. & Stauffer, G. (1985) Parameter estimation for an egg production method of northern anchovy biomass assessment. *NOAA Technical Report NMFS* **36**, 7–15.

Pierce, G.J. & Boyle, P.R. (1991) A review of methods for diet analysis in piscivorous marine mammals. *Oceanography and Marine Biology: an Annual Review* **29**, 409–486.

Pierce, G.J. & Guerra, A. (1994) Stock assessment methods used for cephalopod fisheries. *Fisheries Research* **21**, 255–285.

Pierce, R.W. & Turner, J.T. (1992) Ecology of planktonic ciliates in marine food webs. *Reviews in Aquatic Science* **6**, 139–181.

Piet, G.J. & Rijnsdorp, A.D. (1998) Changes in the demersal fish assemblage in the south-eastern North Sea following the establishment of a protected area ('plaice box'). *ICES Journal of Marine Science* **55**, 420–429.

Pimm, S.L. (1982) *Food Webs*. Chapman & Hall, London.

Pingree, R.D., Pugh, P.R., Holligan, P.M. & Forster, G.R. (1975) Summer phytoplankton blooms and red tides along tidal fronts in the approaches to the English Channel. *Nature* **258**, 672–677.

Pinkerton, E.W. (1994) Summary and conclusions. In: *Folk Management in the World's Fisheries* (eds C.L. Dyer & J.R. McGoodwin), pp. 317–337. University Press of Colorado, Colorado.

Pitcher, T.J. (ed.) (1986) *The Behaviour of Teleost Fishes*. Croom Helm, Beckenham.

Pitcher, T.J. (1995) The impact of pelagic fish behaviour on fisheries. *Scientia Marina* **59**, 295–306.

Pitcher, T.J. & Hart, P.J.B. (1982) *Fisheries Ecology*. Croom Helm, Beckenham.

Pitcher, T.J. & Parrish, J.K. (1993) Functions of shoaling behaviour in teleosts. In: *Behaviour of Teleost Fishes*, 2nd edn (ed. T.J. Pitcher), pp. 363–439. Chapman & Hall, London.

Pitcher, T.J., Hart, P.J.B. & Pauly, D. (eds) (1998) *Reinventing Fisheries Management*. Kluwer Academic Publishers, Dordrecht.

Planes, S. (1993) Genetic differentiation in relation to restricted larval dispersal of the convict surgeonfish *Acanthurus triostegus* in French Polynesia. *Marine Ecology Progress Series* **98**, 237–246.

Poiner, I.R. & Catterall, C.P. (1988) The effects of traditional gathering on populations of the marine gastropod *Strombus luhuanus* Linne 1758, in southern Papua New Guinea. *Oecologia* **76**, 191–199.

Poiner, I., Buckworth, R. & Harris, A. (1990) Incidental capture and mortality of sea turtles in Australia's northern prawn fishery. *Australian Journal of Marine and Freshwater Research* **41**, 97–110.

Polacheck, T., Mountain, D., McMillan, D., Smith, W. & Berrien, P. (1992) Recruitment of the 1987 year class of Georges Bank haddock (*Melanogrammus aeglefinus*): the influence of unusual larval transport. *Canadian Journal of Fisheries and Aquatic Sciences* **49**, 484–496.

Polacheck, T., Hilborn, R. & Punt, A.E. (1993) Fitting surplus production models: comparing methods and measuring uncertainty. *Canadian Journal of Fisheries and Aquatic Sciences* **50**, 2597–2607.

Policansky, D. (1993) Fishing as a cause of evolution in fishes. In: *the Exploitation of Evolving Resources* (eds T.K. Stokes, J.M. McGlade & R. Law), pp. 1–18. Springer-Verlag, Berlin.

Pollock, D.E. & Melville-Smith, R. (1993) Decapod life histories and reproductive dynamics in relation to oceanography off southern Africa. *South African Journal of Marine Science* **13**, 205–212.

Polovina, J.J. (1984) Model of a coral reef ecosystem: the ECOPATH model and its application to French Frigate Shoals. *Coral Reefs* **3**, 1–11.

Polovina, J.J. (1994) The lobster fishery in the north-western Hawaiian islands. In: *Spiny Lobster Management* (eds B.F. Phillips, J.S. Cobb & J. Kittaka), pp. 83–90. Blackwell Scientific Publications, Oxford.

Polunin, N.V.C. (1983) The marine resources of Indonesia. *Oceanography and Marine Biology: an Annual Review* **21**, 455–531.

Polunin, N.V.C. (1984) Do traditional marine reserves conserve? a review of Indonesian and New Guinean evidence. *Senri Ethnological Studies* **17**, 267–283.

Polunin, N.V.C. (1996) Trophodynamics of reef fisheries productivity. In: *Reef Fisheries* (eds N.V.C. Polunin & C.M. Roberts), pp. 113–135. Chapman & Hall, London.

Polunin, N.V.C. & Roberts, C.M. (eds) (1996) *Reef Fisheries*. Chapman & Hall, London.

Pongthanapanich, T. (1996) Economic study suggests management guidelines for mangroves to derive optimal economic and social benefits. *Aquaculture Asia* **1**, 16–17.

Pope, J.G. (1972) An investigation of the accuracy of virtual population analysis. *ICNAF Research Bulletin* **9**, 65–74.

Pope, J.G. (1979) A modified cohort analysis in which constant natural mortality is replaced by estimates of predation levels. *International Council for the Exploration of the Sea Committee Meeting*, 1979/H: 16.

Pope, J.G. (1991) The ICES multispecies assessment working group: evolution, insights, and future problems. *ICES Marine Science Symposia* **193**, 22–33.

Pope, J.G. (1988) Collecting fisheries assessment data. In: *Fish Population Dynamics* (ed. J.A. Gulland), pp. 63–82. John Wiley & Sons, Chichester.

Pope, J.G. & Gray, D. (1983) An investigation of the relationship between precision of assessment and precision of total allowable catches. *Canadian Special Publications of Fisheries and Aquatic Sciences*, **66**, 151–157.

Pope, J.G. & Macer, C.T. (1996) An evaluation of the stock structure of North Sea cod, haddock, and whiting since 1920, together with a consideration of the impacts of fisheries and predation effects on their biomass and recruitment. *ICES Journal of Marine Science* **53**, 1157–1169.

Pope, J.G. & Shepherd, J.G. (1985) A comparison of the performance of various methods for tuning VPAs using effort data. *Journal du Conseil, Conseil International pour l'Exploration de la Mer* **42**, 129–151.

Pope, J.G., Stokes, T.K., Murawski, S.A. & Iodoine, S.I. (1988) A comparison of fish size composition in the North Sea and on Georges Bank. In: *Ecodynamics: Contributions to Theoretical Ecology* (eds W. Wolff, C.J. Soeder & F.R. Drepper). pp. 146–152. Springer-Verlag, Berlin.

Pope, J.G., Shepherd, J.G. & Webb, J. (1994) Successful surf-riding on size spectra: the secret of survival in the sea. *Philosophical Transactions of the Royal Society* **343**, 41–49.

Pope, J.G., MacDonald, D.S., Daan, N., Reynolds, J.D. & Jennings, S. (2000) Gauging the vulnerability of non-target species to fishing. *ICES Marine Science Symposia* **57**, 689–696.

Posey, M., Lindberg, W., Alphin, T. & Vose, F. (1996) Influence of storm disturbance on an offshore benthic community. *Bulletin of Marine Science* **59**, 523–529.

Potin, P., Floc'h, J.Y., Augris, C. & Cabioch, J. (1990) Annual growth rate of the calcareous red alga *Lithothamnion corallioides* (Corallinales, Rhodophyta) in the Bay of Brest, France. *Hydrobiologia* **204**, 263–267.

Potter, E.C.E. & Pawson, M.G. (1991) Gill netting. *MAFF Directorate of Fisheries Research Laboratory Leaflet* 69.

Powers, K.D. & Brown, R.G.B. (1987) Seabirds. In: *Georges Bank* (ed. R.H. Backus), pp. 359–371. MIT Press, Cambridge, MA.

Pratt, H.L. & Casey, J.G. (1983) Age and growth of the shortfin Mako, *Isurus oxyrinchus*, using four methods. *Canadian Journal of Fisheries and Aquatic Sciences* **40**, 1944–1957.

Prendergast, A.F. (1994) Searching for substitutes-canola. *Northern Aquaculture*, 15–20 May.

Priddle, J., Croxall, J.P., Everson, I., Heywood, R.B., Murphy, E.J., Prince, P.A. & Sear, C.B. (1988) Large-scale fluctuations in distribution and abundance of krill—a discussion of possible causes. In: *Antarctic Ocean and Resources Variability* (ed. D. Sahrhage), pp. 169–182. Springer-Verlag, Berlin.

Priede, I.G. & Watson, J.J. (1993) An evaluation of the daily egg production method for estimating biomass of Atlantic mackerel (*Scomber scombrus*). *Bulletin of Marine Science* **53**, 981–911.

Puente, E. (1997) Incidental impacts of gill nets. *Report to the European Commission*, No. 94/095, 152.

Punt, A.E. (1992) Selecting management methodologies for marine resources, with an illustration for southern African hake. *Canadian Journal of Fisheries and Aquatic Sciences* **12**, 943–958.

Punt, A.E. (1999) A full description of the standard Baleen II model and some variants thereof. *Journal of Cetacean Research and Management* **1** (Suppl.), 267–276.

Punt, A.E. & Hilborn, R. (1997) Fisheries stock assessment and decision analysis: the Bayesian approach. *Reviews in Fish Biology and Fisheries* **7**, 35–63.

Punt, A.E. & Smith A.D.M. (2001) The gospel of maximum sustainable yield in fisheries management: birth, crucifixion and reincarnation. In: *Conservation of Exploited Species* (eds J.D. Reynolds, G.M. Mace, K.H. Redford & J.G. Robinson) Cambridge University Press, Cambridge.

Quinn, T.J.II. & Deriso, R.B. (1999) *Quantitative Fish Dynamics*. Oxford University Press, New York.

Quinn, T.J.II, Funk, F., Heifetz, J.N., Ianelli, J.E., Schweigert, J.F., Sullivan, P.J. & Zhang, C.I. (eds) (1998) *Fishery Stock Assessment Models*. University of Alaska, Fairbanks.

Ralston, S. & Polovina, J.J. (1982) A multispecies analysis of the commercial deep-sea handline fishery in Hawaii. *Fishery Bulletin* **80**, 435–448.

Ralston, S., Gooding, R.M. & Ludwig, G.M. (1986) An ecological survey and comparison of bottom fish resource assessments (submersible versus handline fishing) at Johnston Atoll. *Fishery Bulletin* **84**, 141–155.

Ramsay, K. & Kaiser, M.J. (1998) Demersal fishing increases predation risk for whelks *Buccinum undatum* (L.). *Journal of Sea Research* **39**, 299–304.

Ramsay, K., Kaiser, M.J. & Hughes, R.N. (1996) Changes in hermit crab feeding patterns in response to trawling disturbance. *Marine Ecology Progress Series* **144**, 63–72.

Ramsay, K., Kaiser, M.J., Moore, P.G. & Hughes, R.N. (1997) Consumption of fisheries discards by benthic scavengers: utilisation of energy subsidies in different marine habitats. *Journal of Animal Ecology* **66**, 884–896.

Ramsay, K., Kaiser, M.J., Rijnsdorp, A.D., Craeymeersch, J.A. & Ellis, J. (2000) Impacts of trawling on populations of the invertebrate scavenger *Asterias rubens*. In: *The effects of fishing on non-target species and habitats: biological, conservation and socioeconomic issues* (ed. M.J. Kaiser & S.J. de Groot), pp. 151–162. Blackwell Science, Oxford.

Rawlinson, N.J.F., Milton, D.A., Blaber, S.J.M., Sesewa, A. & Sharma, S.P. (1994) A survey of the subsistence and artisanal fisheries in rural areas of Viti Levu, Fiji. *ACIAR Mongraph* **35**, 1–138.

Reise, K., Gollasch, S. & Wolff, W.J. (1999) Introduced marine species of the North Sea coasts. *Helgolander Meeresuntersuchungen* **52**, 219–234.

Réyni, A. (1961) On measures of entropy and information. In: *Proceedings of the 4th Berkeley Symposium on Mathematical Statistics and Probability* (ed. J. Neyman), pp. 547–561. University of California Press, Berkeley.

Reynolds, J.D. (1996) Animal breeding systems. *Trends in Ecology and Evolution* **11**, 68–72.

Reynolds, J.D. & Jennings, S. (2000) The role of animal behaviour in marine conservation. In: *Behaviour and Conservation* (eds M.L. Gosling & W.J. Sutherland). pp. 238–257. Cambridge University Press, Cambridge.

Reynolds, J.D. & Mace, G.M. (1999) Risk assessment of threatened species. *Trends in Ecology and Evolution* **14**, 215–217.

Reynolds, J.D. Mace, G.M., Redford, K.H. & Robinson, J.G. (eds) (2001) *Conservation of Exploited Species*. Cambridge University Press, Cambridge.

Reznick, D.N., Bryga, H. & Endler, J.A. (1990). Experimentally induced life-history evolution in a natural population. *Nature* **346**, 357–359.

Rice, J. & Gislason, H. (1996) Patterns of change in the size spectra of numbers and diversity of the North Sea fish assemblage, as reflected in surveys and models. *ICES Journal of Marine Science* **53**, 1214–1225.

Rice, J.C., Daan, N., Pope, J.G. & Gislason, H. (1991) The stability of estimates of suitabilities in MSVPA over four years of data from predator stomachs. *ICES Marine Science Symposia* **193**, 34–45.

Richards, R.A., Cobb, J.S. & Fogarty, M.J. (1983) Effects of behavioural interactions on the catchability of American lobster, *Homarus americanus*, and two species of *Cancer* crab. *Fishery Bulletin* **81**, 51–60.

Richardson, K. (2000) Integrating environmental and fisheries objectives in the ICES area. *ICES Journal of Marine Science* 57, 766–770.

Ricker, W.E. (1954) Stock and recruitment. *Journal of the Fisheries Research Board of Canada* 11, 559–623.

Ricker, W.E. (1975) Computation and interpretation of biological statistics of fish populations. *Bulletin Fisheries Research Board of Canada* 191.

Rickman, S.J., Dulvy, N.K., Jennings, S. & Reynolds, J.D. (2000) Recruitment variation related to fecundity in marine fishes. *Canadian Journal of Fisheries and Aquatic Sciencs* 57, 116–124.

Riesen, W. & Reise, K. (1982) Macrobenthos of the subtidal Wadden Sea: revisited after 55 years. *Helgolander Meeresuntersuchungen* 35, 409–423.

Rijnsdorp, A.D. (1993) Fisheries as a large scale experiment on life history evolution: disentangling phenotypic and genetic effects in changes in maturation and reproduction of North Sea plaice, *Pleuronectes platessa* L. *Oecologia* 96, 391–401.

Rijnsdorp, A.D. & Leeuwen, P.I.V. (1996) Changes in growth of North Sea plaice since 1950 in relation to density, eutrophication, beam-trawl effort, and temperature. *ICES Journal of Marine Science* 53, 1199–1213.

Rijnsdorp, A.D., Daan, N., van Beek, F.A. & Heessen, H.J.L. (1991) Reproductive variability in North Sea plaice, sole and cod. *Journal du Conseil, Conseil International pour l'Exploration de la Mer* 47, 352–375.

Rijnsdorp, A.D., Buijs, A.M., Storbeck, F. & Visser, E. (1998) Micro-scale distribution of beam trawl effort in the southern North Sea between 1993 and 1996 in relation to the trawling frequency of the sea bed and the impact on benthic organisms. *ICES Journal of Marine Science* 55, 403–419.

Rijnsdorp, A.D., Dol, W., Hooyer, M. & Pastoors, M.A. (2000a) Effects of fishing power and competitive interactions among vessels on the effort allocation on the trip level of the Dutch beam trawl fleet. *ICES Journal of Marine Science* 57, 927–937.

Rijnsdorp, A.D., van Maurik Broekman, P.L. & Visser, E.G. (2000b) Competitive interactions among beam trawlers exploiting local patches of flatfish in the North Sea. *ICES Journal of Marine Science* 57, 894–902.

Roberts, C.M. & Hawkins, J.P. (1999) Extinction risk in the sea. *Trends in Ecology and Evolution* 14, 241–246.

Roberts, C.M. & Ormond, R.F.G. (1987) Habitat complexity and coral reef fish diversity and abundance on Red Sea fringing reefs. *Marine Ecology Progress Series* 41, 1–8.

Roberts, C.M. & Polunin, N.V.C. (1991) Are marine reserves effective in management of reef fisheries. *Reviews in Fish Biology and Fisheries* 1, 65–91.

Roberts, C.M. & Polunin, N.V.C. (1993) Marine reserves: simple solutions to managing complex fisheries. *Ambio* 22, 363–368.

Robertson, A.I. & Philips, M.J. (1995) Mangroves as filters of shrimp pond effluent: predictions and biogeochemical research needs. *Hydrobiologia* 295, 311–321.

Robins-Troeger, J.B. (1994) Evaluation of the Morrison soft turtle excluder device: prawn and bycatch variation in Moreton Bay, Queensland. *Fisheries Research* 19, 205–217.

Robins-Troeger, J.B. (1995) Estimated catch and mortality of sea turtles from the east coast otter trawl fishery of Queensland, Australia. *Biological Conservation* 74, 157–167.

Robins-Troeger, J.B., Buckworth, R.C. & Dredge, M.C.L. (1995) Development of a trawl efficiency device (TED) for Australian prawn fisheries. II. Field evaluations of the AusTED. *Fisheries Research* 22, 107–117.

Rochette, R. & Himmelman, J.H. (1996) Does vulnerability influence trade-offs made by whelks between predation risk and feeding opportunities? *Animal Behaviour* 52, 783–794.

Rodhouse, P.G. & Hatfield, E.M.C. (1990) Age determination in squid using statolith growth increments. *Fisheries Research* 8, 323–334.

Rødseth, T., ed. (1998) *Models for Multispecies Management.* Physica-Verlag. Heidelberg.

Roff, D.A. (1984) The evolution of life history parameters in teleosts. *Canadian Journal of Fisheries and Aquatic Sciences* 41, 989–1000.

Roff, D.A. (1992) *The Evolution of Life Histories: Theory and Analysis.* Chapman & Hall, New York.

Rogers, S.I., Rijnsdorp, A.D., Damm, U. & Vanhee, W. (1998) Demersal fish populations in the coastal waters of the UK and continental NW Europe from beam trawl survey data collected from 1990 to 1995. *Journal of Sea Research* 39, 79–102.

Rogers, S.I., Clarke, K.R. & Reynolds, J.D. (1999a) The taxonomic distinctness of coastal bottom-dwelling fish communities in the northeast Atlantic. *Journal of Animal Ecology* 68, 769–782.

Rogers, S.I., Maxwell, D., Rijnsdorp, A.D., Damm, U. & Vanhee, W. (1999b) Fishing effects in northeast Atlantic shelf seas: patterns in fishing effort, diversity and community structure. IV. Can comparisons of species diversity be used to assess human impacts on demersal fish faunas. *Fisheries Research* 40, 135–152.

Rohde, K., Heap, M. & Heap, D. (1993) Rapoport's rule does not apply to marine teleosts and cannot explain latitudinal gradients in species richness. *American Naturalist* 142, 1–16.

Romero, J., Pérez, M., Mateo, M.A. & Sala, E. (1994) The below ground organs of the Mediterranean seagrass *Posidonia oceanica* as a biogeochemical sink. *Aquatic Botany* 47, 13–19.

Ropes, J.W. & O'Brien, L. (1979) A unique method of ageing surf clams. *Bulletin American Malacology Union* 1978, 58–61.

Ropes, J.W. & Shepherd, G.R. (1988) Surf clam *Spisula solidissima. NOAA Technical Report NMFS* 72, 125–128.

Rose, G.A. (1993) Cod spawning on a migration highway in the north-west Atlantic. *Nature* 366, 458–461.

Rosenberg, A.A. (1998) Controlling marine fisheries 50 years from now. *Journal of Northwest Atlantic Fishery Science* 23, 95–103.

Rosenberg, A.A. & Beddington, J.R. (1987) Monte-Carlo testing of two methods of estimating growth from length-frequency data with general conditions for their applicability. In: *The Theory and Application of Length Based Methods of Stock Assessment* (eds D. Pauly & G.P. Morgan). ICLARM, Manila.

Rosenberg, A.A. & Beddington, J.R. (1988) Length-based methods of fish stock assessment. In: *Fish Population Dynamics* (ed. J.A. Gulland), pp. 83–103. John Wiley & Sons, Chichester.

Rosenberg, A.A., Kirkwood, G.P., Crombie, J.A. & Beddington, J.R. (1990) The assessment of stocks of annual squid species. *Fisheries Research* 8, 335–350.

Rosenberry, R. (1994) *World Shrimp Farming 1994*. Annual Report, Shrimp News International, San Diego, California.

Rosenberry, R. (1996) *World Shrimp Farming 1996*. Annual Report, Shrimp News International, San Diego, California.

Rosenthal, H. & Hempel, G. (1970) Experimental studies in feeding and food requirements of herring larvae (*Clupea harengus* L.). In: *Marine Food Chains* (ed. J.H. Steele), pp. 344–364. Oliver and Boyd, Edinburgh.

Rothschild, B.J. (1986) *Dynamics of Marine Fish Populations*. Harvard University Press, Cambridge, Massachusetts.

Rothschild, B.J. (1991) Multispecies interactions on Georges Bank. *ICES Marine Science Symposia* 193, 86–92.

Rowden, A.A. & Jones, M.B. (1993) Critical evaluation of sediment turnover estimates for Callianassidae (Decapoda: Thalassinidea). *Journal of Experimental Marine Biology and Ecology* 173, 265–272.

Rowe, P.M. & Epifanio, C.E. (1994) Flux and transport of larval weakfish in Delaware Bay, USA. *Marine Ecology Progress Series* 110, 115–120.

Rowell, C.A. (1993) The effects of fishing on the timing of maturity in North Sea cod (*Gadus morhua* L.). In: *The Exploitation of Evolving Resources* (eds T.K. Stokes, J.M. McGlade & R. Law), pp. 44–61. Springer-Verlag, Berlin.

Rowley, R.J. (1994) Marine reserves in fisheries management. *Aquatic Conservation: Marine and Freshwater Ecosystems* 4, 233–254.

Ruddle, K. (1989) The continuity of traditional management practices: the case of Japanese coastal fisheries. In: *Traditional Marine Resource Management in the Pacific Basin: an Anthology* (eds K. Ruddle & R.E. Johannes), pp. 263–285. CSIRO, Hobart.

Ruddle, K. (1996) Traditional management of reef fishing. In: *Reef Fisheries* (eds N.V.C. Polunin & C.M. Roberts), pp. 315–335. Chapman & Hall, London.

Ruddle, K., Hviding, E. & Johannes, R.E. (1992) Marine resources management in the context of customary tenure. *Marine Resources Economics* 7, 249–273.

Rueness, J. (1989) *Sargassum muticum* and other introduced Japanese macroalgae: biological pollution of European coasts. *Marine Pollution Bulletin* 4, 173–176.

Rulifson, R.A., Murray, J.D. & Bahen, J.J. (1992) Finfish catch reduction in south-Atlantic shrimp trawls using 3 designs of by-catch reduction devices. *Fisheries* 17, 9–20.

Russ, G.R. (1985) Effects of protective management on coral reef fishes in the central Philippines. *Proceedings of the Fifth International Coral Reef Symposium* 4, 219–224.

Russ, G.R. & Alcala, A.C. (1989) Effects of intense fishing pressure on an assemblage of coral reef fishes. *Marine Ecology Progress Series* 56, 13–27.

Russ, G.R. & Alcala, A.C. (1994) Sumilon Island Reserve: 20 years of hopes and frustrations. *Naga* 7, 8–12.

Russ, G.R. & Alcala, A.C. (1996a) Marine reserves—rates and patterns of recovery and decline of large predatory fish. *Ecological Applications* 6, 947–961.

Russ, G.R. & Alcala, A.C. (1996b) Do marine reserves export adult fish biomass—evidence from Apo Island, Central Philippines. *Marine Ecology Progress Series* 132, 1–9.

Russ, G.R. & Alcala, A.C. (1998a) Natural fishing experiments in marine reserves 1983–93: roles of life history and fishing intensity in family responses. *Coral Reefs* 17, 399–416.

Russ, G.R. & Alcala, A.C. (1998b) Natural fishing experiments in marine reserves 1983–93: community and trophic responses. *Coral Reefs* 17, 383–397.

Russell, E.S. (1931) Some theoretical considerations on the overfishing problem. *Journal du Conseil, Conseil International pour l'Exploration de la Mer* 6, 3–20.

Russell, G., Hawkins, S.J., Evans, L.C., Jones, H.D. & Holmes, G.D. (1983) Restoration of a disused dock basin as a habitat for marine benthos and fish. *Journal of Applied Ecology* 20, 43–58.

Russell, M.P., Ebert, T.A. & Petraitis, P.S. (1998) Field estimates of growth and mortality of the green sea urchin *Strongylocentrotus droebachiensis*. *Ophelia* 48, 137–153.

Ryman, N., Utter, F. & Laikre, L. (1995) Protection of intraspecific diversity of exploited fishes. *Reviews in Fish Biology and Fisheries* 5, 417–446.

Ryther, J.H. (1969) Relationships of photosynthesis to fish production in the sea. *Science* 166, 72–76.

Sadovy, Y.J. (1996) Reproduction of reef fishery species. In: *Reef Fisheries* (eds N.V.C. Polunin & C.M. Roberts), pp. 15–59. Chapman & Hall, London.

Sainsbury, J.C. (1996) *Commercial Fishing Methods: an Introduction to Vessels and Gears*. Fishing News Books, Oxford.

Sainsbury, K.J. (1991) Application of an experimental approach to management of a tropical multispecies fishery with highly uncertain dynamics. *ICES Marine Science Symposia* 193, 301–320.

Sainsbury, K.J. (1987) Assessment and management of the demersal fishery on the continental shelf of northwestern Australia. In: *Tropical Snappers and Groupers—Biology and Fisheries Management* (eds J.J. Polovina & S. Ralston), pp. 465–503. Westview Press, Boulder, Colorado.

Sainsbury, K.J., Campbell, R.A., Lindholm, R. & Whitlaw, A.W. (1998) Experimental management of an Australian multispecies fishery: examining the possibility of trawl-induced habitat modification. In: *Global Trends: Fisheries Management* (eds E.K. Pikitch, D.D. Huppert & M.P. Sissenwine), pp. 107–112. American Fisheries Society, Bethesda, Maryland.

Sainte-Marie, B. (1986) Effect of bait size and sampling time on the attraction of the lysianassid amphipods *Anonyx sarsi* Steele and Brunel and *Orchomenella pinguis* (Boeck). *Journal of Experimental Marine Biology and Ecology* **99**, 63–77.

Sainte-Marie, B. & Hargrave, B.T. (1987) Estimation of scavenger abundance and distance of attraction to bait. *Marine Biology* **94**, 431–443.

Sammarco, P.W. (1980) *Diadema* and its relationship with coral spat mortality: grazing, competition and biological disturbance. *Journal of Experimental Marine Biology and Ecology* **45**, 245–272.

Samoilys, M.A. (1997) Movement of a large predatory fish: the coral trout, *Plectropomus leopardus* (Pisces: Serranidae), on Heron Reef, Australia. *Coral Reefs* **16**, 151–158.

Samoilys, M.A. & Carlos, G.M. (1992) *Development of an Underwater Visual Census Method for Assessing Shallow Water Reef Fish Stocks in the South-West Pacific.* Queensland Department of Primary Industries, Cairns.

Sanders, M.J. & Beinssen, K.H.H. (1996) Bioeconomic modelling of a fishery under individual transferable quota management: a case study of the fishery for blacklip abalone *Haliotis rubra* in the western zone of Victoria (Australia). *Fisheries Research* **27**, 179–201.

Sanders, M.J. & Beinssen, K.H.H. (1998) A comparison of man-agement strategies for the rehabilitation of a fishery: applied to the blacklip abalone *Haliotis rubra* in the western zone of Victoria (Australia). *Fisheries Research* **38**, 283–301.

Sardà, F. & Maynou, F. (1998) Assessing perceptions: Do Catalan fishermen catch more shrimp on Fridays? *Fisheries Research* **36**, 149–157.

Saville, A. (1964) Estimation of the abundance of a fish stock from egg and larval surveys. *Rapports et Procès-Verbaux des Réunions, Conseil International pour l'Exploration de la Mer* **155**, 164–170.

Schaefer, M.B. (1954) Some aspects of the dynamics of populations important to the management of commercial marine fisheries. *Inter-American Tropical Tuna Commission Bulletin* **1**, 27–56.

Schmidt, K.F. (1997) 'No-Take' zones spark fisheries debate. *Science* **277**, 489–491.

Schnute, J. (1977) Improved estimates from the Schaefer production model: theoretical considerations. *Journal of the Fisheries Research Board of Canada* **34**, 583–603.

Schnute, J. (1985) A general theory for analysis of catch and effort data. *Canadian Journal of Fisheries and Aquatic Sciences* **42**, 414–429.

Schwartz, C.S. & Seber, G. (1999) Estimating animal abundance: Review III. *Statistical Science* **14**, 427–456.

Schwartzlose, R., Alheit, J., Bakun, A. *et al.* (1999) Worldwide large-scale fluctuations of sardine and anchovy populations. *South African Journal of Marine Science* **21**, 289–347.

Schweigert, J.F. (1993) A review and evaluation of methodology for estimating Pacific herring egg deposition. *Bulletin of Marine Science* **53**, 818–841.

Schwinghamer, P., Guigné, J. & Siu, W. (1996) Quantifying the impact of trawling on benthic habitat structure using high resolution acoustics and chaos theory. *Canadian Journal of Fisheries and Aquatic Sciences* **53**, 288–296.

Seber, G.A.F. (1982) *The Estimation of Animal Abundance*, Second Edn. Griffin, London.

Secor, D.H., Dean, J.M. & Campana, S.E. (eds) (1995) *Recent Developments in Fish Otolith Research*. University of South Carolina Press. Columbia, South Carolina.

Shelbourne, J.E. (1964) The artificial propagation of marine fish. *Advances in Marine Biology* **2**, 1–83.

Shelton, P.A. (1992) Detecting and incorporating multispecies effects into fisheries management in the north-west and south-east Atlantic. *South African Journal of Marine Science* **12**, 723–737.

Shepherd, J.G. (1982) A versatile new stock–recruitment relationship for fisheries and the construction of sustainable yield curves. *Journal du Conseil, Conseil International pour l'Exploration de la Mer* **40**, 67–75.

Shepherd, J.G. (1987) A weakly parametric method for the analysis of length composition data. In: *The Theory and Application of Length Based Methods of Stock Assessment* (eds D. Pauly & G.P. Morgan). ICLARM, Manila.

Shepherd, J.G. (1988) Fish stock assessments and their data requirements. In: *Fish Population Dynamics* (ed. J.A. Gulland), pp. 35–62. John Wiley & Sons, Chichester.

Shepherd, J.G. & Cushing, D.H. (1982) A mechanism for density-dependent survival of larval fish as the basis of a stock-recruitment relationship. *Journal du Conseil, Conseil International pour l'Exploration de la Mer* **39**, 160–167.

Shepherd, J.G. & Cushing, D.H. (1990) Regulation in fish populations: myth or mirage. *Philosophical Transactions of the Royal Society* **B330**, 151–164.

Shepherd, S.A. (1983) The epifauna of megaripples: species' adaptations and population responses to disturbance. *Australian Journal of Ecology* **8**, 3–8.

Short, F.T. & Wyllie-Echeverria, S. (1996) Natural and human-induced disturbance of seagrasses. *Environmental Conservation* **23**, 17–28.

Shpigel, M., Neori, A., Popper, D.M. & Gordin, H. (1993) A proposed model for 'environmentally clean' land-based culture of fish, bivalves and seaweeds. *Aquaculture* **117**, 115–128.

Siegel, V. & Loeb, V. (1995) Recruitment of Antarctic krill *Euphausia superba* and possible causes for its variability. *Marine Ecology Progress Series* **123**, 45–56.

Silliman, R. (1975) Selective and unselective exploitation of experimental populations of *Tilapia mossambica*. *Fishery Bulletin* **73**, 495–507.

Simenstad, C. & Fresh, K. (1995) Influence of intertidal aquaculture on benthic communities in Pacific northwest estuaries: scales of disturbance. *Estuaries* **18**, 1A, 43–70.

Simpson, A.C. (1951) The fecundity of the plaice. *Fishery Investigations London Series 2*, **17**(5).

Simpson, D.G. (1989) Codend selection of winter flounder *Pseudopleuronectes americanus*. *NOAA Technical Report NMFS* **75**, 10.

Simpson, J.H. (1981) The shelf-sea fronts: implications of their existence and behaviour. *Philosophical Transactions of the Royal Society* 302, 531–546.

Sims, N.A. (1993) Pearl oysters. In: *Nearshore Marine Resources of the South Pacific* (eds A. Wright & L. Hill), pp. 414–423. Institute of Pacific Studies, Suva.

Sinclair, A.R.E. (1989) Population regulation in animals. In: *Ecological Concepts: the Contribution of Ecology to the Understanding of the Natural World* (ed. J.M. Cherrett), pp. 197–241. Blackwell Scientific Publications, London.

Sinclair, M. (1988) *Marine Populations: an Essay on Population Regulation and Speciation.* University of Washington Press, Seattle.

Sissenwine, M.P. & Rosenberg, A.A. (1993) Marine fisheries at a critical juncture. *Fisheries* 18, 6–14.

Sissenwine, M.P. & Shepherd, J.G. (1987) An alternative perspective on recruitment overfishing and biological reference points. *Canadian Journal of Fisheries and Aquatic Sciences* 44, 913–918.

Smayda, T.J. (1970) The suspension and sinking of phytoplankton in the sea. *Oceanography and Marine Biology: an Annual Review* 8, 353–414.

Smith, A. & Dalzell, P. (1993) Fisheries resources and management investigations in Woleai Atoll, Yap State, Federated States of Micronesia. *South Pacific Commission, Inshore Fisheries Research Project Technical Document*, 4.

Smith, A.D.M. & Punt, A.E. (1998) Stock assessment of the gemfish (Rexea solandri) in eastern Australia using maximum likelihood and Bayesian methods. In: *Fishery Stock Assessment Models* (eds T.J. Quinn, I.I.F. Funk, J.N. Heifetz, J.E. Ianelli, J.F. Schweigert, P.J. Sullivan & C.I. Zhang), pp. 245–286. Alaska Sea Grant College Program Report no, AK-SG-98–01, University of Alaska, Fairbanks.

Smith, B.D. & Jamieson, G.S. (1991) Possible consequences of intensive fishing for males on the mating opportunities of Dungeness crabs. *Transactions of the American Fisheries Society* 120, 650–653.

Smith, P.E. & Hewitt, R.P. (1985) Sea survey design and analysis for an egg production method of anchovy biomass assessment. *NOAA Technical Report NMFS* 36, 17–26.

Smith, P.E., Flerx, W. & Hewitt, R.P. (1985) The CalCOFI vertical egg tow (CalVET) net. *NOAA Technical Report NMFS* 36, 27–35.

Smith, P.J., Francis, R.I.C.C. & McVeagh, M. (1991) Loss of genetic diversity due to fishing pressure. *Fisheries Research* 10, 309–316.

Smith, S.V. & Hollibaugh, J.T. (1993) Coastal metabolism and the oceanic organic carbon balance. *Reviews in Geophysics* 31, 75–89.

Smith, S.J., Hunt, J.J. & Rivard, D. (eds) (1993) Risk evaluation and biological reference points for fisheries management. *Canadian Special Publication of Fisheries and Aquatic Sciences*, 120.

Smith, T.D. (1994) *Scaling Fisheries: the Science of Measuring the Effects of Fishing, 1855–1955.* Cambridge University Press, Cambridge.

Smolowitz, R.J. (1978) Trap design and ghost fishing: a review. *Marine Fisheries Review* 40, 2–8.

Snedaker, S.C., Dickinson, J.C., Brown, M.S. & Lahmann, E.J. (1986) *Shrimp Pond Siting and Management Alternatives in Mangrove Ecosystems in Ecuador.* US Agency for International Development, Washington, DC.

Sogard, S.M. (1997) Size selective mortality in the juvenile stages of teleost fishes: a review. *Bulletin of Marine Science* 60, 1129–1157.

Solemdal, P. (1997) Maternal effects: a link between the past and the future. *Journal of Sea Research* 37, 213–227.

Solemdal, P., Kjesbu, O.S. & Fonn, M. (1995) Egg mortality in recruit and repeat spawning cod: an experimental study. *International Council for the Exploration of the Seas, Committee Meeting*, G: 35.

Sotheran, I.S., Foster-Smith, R.L. & Davies, J. (1997) Image processing techniques within a raster-based geographic information system. *Esturarine, Coastal and Shelf Science* 44 (Suppl. A), 25–32.

Soutar, A. & Isaacs, J.D. (1974) Abundance of pelagic fish during the 19th and 20th centuries as recorded in anaerobic sediment off the Californias. *Fishery Bulletin* 72, 257–273.

South, G.R. (1993) Seaweeds. In: *Nearshore Marine Resources of the South Pacific* (eds A. Wright & L. Hill), pp. 683–710. Institute of Pacific Studies, Suva.

South, G.R. & Whittick, A. (1987) *Introduction to Phycology.* Blackwell Scientific, Oxford.

Southward, A.J., Boalch, G.T. & Maddock, L. (1988) Fluctuations in the herring and pilchard fisheries of Devon and Cornwall linked to changes in climate since the 16th century. *Journal of the Marine Biological Association of the United Kingdom* 68, 423–445.

Sparholt, H. (1990) An estimate of the total biomass of fish in the North Sea. *Journal du Conseil, Conseil International pour l'Exploration de la Mer* 46, 200–210.

Sparre, P. (1991) Introduction to multispecies virtual population analysis. *ICES Marine Science Symposia* 193, 12–21.

Sparre, P. & Venema, S.C. (1998) Introduction to tropical fish stock assessment. Part 1: Manual. *FAO Fisheries Technical Paper*, 306/1 (Rev. 2).

Sprent, P. & Dolby, G.R. (1980) The geometric mean functional relationship. *Biometrics* 36, 547–550.

Spurgeon, J.P.G. (1992) The economic valuation of coral reefs. *Marine Pollution Bulletin* 24, 529–536.

Stearns, S.C. (1992) *The Evolution of Life Histories.* Oxford University Press, Oxford.

Stearns, S.C. & Crandall, R.E. (1984) Plasticity for age and size at sexual maturity: a life history response to unavoidable stress. In: *Fish Reproduction: Strategies and Tactics* (eds G.W. Potts & R.J. Wootton), pp. 13–34. Academic Press, London.

Steele, J.H. (ed.) (1978) *Spatial Pattern in Plankton Communities.* Plenum Press, New York.

Stephens, D.W. & Krebs, J.R. (1986) *Foraging Theory.* Princeton University Press, Princeton.

Stevens, J.D. (1975) Vertebral rings as a means of age determination in the blue shark (*Prionace glauca* L.). *Journal of* the Marine Biological Association of the United Kingdom 55, 657–665.

Stevenson, D.K. & Campana, S.E. (eds) (1992) Otolith microstructure examination and analysis. *Canadian Special Publications of Fisheries and Aquatic Sciences* 117.

Stillman, R.A.D.G.-C.J., McGrorty, S., West, A.D., dit-Durell, S.E.A.L., Clarke, R.T., Caldow, R.W.G., Norris, K.J., Johnstone, I.G., Ens, B.J., Bunskoeke, E.J., Merwe, A.V.-d., van-der-Meer, J., Triplet, P., Odoni, N., Swinfen, R. & Cayford, J.T. (1996) *Models of Shellfish Populations and Shorebirds.* Report to Commission of the European Communities by the Institute of Terrestrial Ecology, Furzebrook.

Stockton, W.L. & DeLaca, T.E. (1982) Food falls in the deep sea: occurrence, quality and significance. *Deep-Sea Research* 29, 157–169.

Stokes, T.K., McGlade, J.M. & Law, R. (eds) (1993) *The Exploitation of Evolving Resources.* Springer-Verlag, Berlin.

Stokesbury, K.D.E. & Himmelman, J.H. (1993) Spatial distribution of the giant scallop *Placopecten magellanicus* in unharvested beds in the Baie des Chaleurs, Quebec. *Marine Ecology Progress Series* 96, 159–168.

Stotz, W.B. & González, S.A. (1997) Abundance, growth, and production of the sea scallop *Argopecten purpuratus* (Lamarck, 1819): basis for sustainable exploitation of natural scallop beds in north-central Chile. *Fisheries Research* 32, 173–183.

Straile, D. (1997) Gross growth efficiencies of protozoan and metazoan zooplankton and their dependence on food concentration, predator–prey weight ratio and taxonomic group. *Limnology and Oceanography* 42, 1375–1385.

Suchanek, T.H., Williams, S.L., Ogden, J.C., Hubbard, D.K. & Gill, I.P. (1985) Utilization of shallow-water seagrass detritus by Caribbean deep-sea macrofauna: ^{13}C evidence. *Deep-Sea Research* 32, 201–214.

Sugihara, G., Garcia, S., Gulland, J.A., Lawton, J.H., Maske, H., Paine, R.T., Platt, T., Rachor, E., Rothschild, B.J., Ursin, E.A. & Zeitzschel, B.F.K. (1984) Ecosystem dynamics. In: *Exploitation of Marine Communities* (ed. R.M. May), pp. 131–153. Springer-Verlag, Berlin.

Summerfelt, R.C. & Hall, G.E. (ed.) (1987) *Age and Growth of Fish.* Iowa State University Press, Iowa.

Sünderman, J. (ed.) (1986) *Oceanography.* Springer-Verlag, Berlin.

Sutherland, W.J. (1996) *From Individual Behaviour to Population Ecology.* Oxford University Press, Oxford.

Svånstad, T. & Kristiansen, T.S. (1990) Enhancement studies of coastal cod in western Norway. Part II. Migration of reared coastal cod. *Journal du Conseil International pour l'Exploration de la Mer* 47, 13–22.

Sverdrup, H.U. (1953) On conditions for the vernal blooming of phytoplankton. *Journal du Conseil, Conseil International pour l'Exploration de la Mer* 18, 287–295.

Swain, D.P. & Morin, R. (1996) Relationships between geographic distribution and abundance of American plaice (*Hippoglossoides platessoides*) in the southern Gulf of St. Lawrence. *Canadian Journal of Fisheries and Aquatic Sciences* 53, 106–119.

Swain, D.P. & Wade, E.J. (1993) Density-dependent geographic distribution of Atlantic cod (*Gadus morhua*) in the southern Gulf of St. Lawrence. *Canadian Journal of Fisheries and Aquatic Sciences* 50, 725–733.

Swearer, S.E., Caselle, J.E., Lea, D.W. & Warner, R.R. (1999) Larval retention and recruitment in an island population of a coral-reef fish. *Nature* 402, 799–802.

Tegner, M.J. & Dayton, P.K. (2000) Ecosystem effects of fishing in kelp forest communities. *ICES Journal of Marine Science* 57, 579–589.

Tegner, M.J., DeMartini, J.D. & Karpov, K.A. (1992) The California red abalone fishery: a case study in complexity. In: *Abalone of the World: Biology, Fisheries and Culture* (eds S.A. Shepherd, M.J. Tegner & S.A. Guzman del Proo), pp. 370–383. Fishing News Books, Oxford.

Tegner, M.J., Basch, L.V. & Dayton, P.K. (1996) Near extinction of an exploited marine invertebrate. *Trends in Ecology and Evolution* 11, 278–280.

Templeman, W. (1981) Vertebral numbers in Atlantic cod, *Gadus morhua*, of the Newfoundland and adjacent areas. 1947–71, and their use for delineating cod stocks. *Journal of Northwest Atlantic Fishery Science* 2, 21–45.

Tenore, K. & Gonzalez, N. (1976) Food chain patterns in the Ria de Arosa, Spain, an area of intense mussel aquaculture. In: *Population Dynamics* (eds G. Persoone & E. Jaspers), pp. 601–609. Ostend, Belgium.

Tenore, K., Boyer, L., Cal, R., Corral, J., Garcia, F., Gonzalez, M., Gonzalez, E., Hanson, R., Iglesias, J. & Krom, M. (1982) Coastal upwelling in the Rias Bajas, NW Spain: constrasting the benthic regimes of the Rias de Arosa and Muros. *Journal of Marine Research* 40, 701–772.

Tenore, K.R., Corral, J. & Gonzales, N. (1985) Effects of intense mussel culture on food-chain patterns and production in coastal Galicia. NW Spain. *International Council for the Exploration of the Seas, Committee Meeting*, 1985/F: 62, 1–9.

Theil, H. & Schriever, G. (1990) Deep-sea mining, environmental impact and the DISCOL project. *Ambio* 19, 245–250.

Thiebaux, M.L. & Dickie, L.M. (1992) Models of aquatic biomass size spectra and the common structure of their solutions. *Journal of Theoretical Biology* 159, 147–161.

Thiebaux, M.L. & Dickie, L.M. (1993) Structure of the body size spectrum of the biomass in aquatic ecosystems: a consequence of allometry in predator–prey interactions. *Canadian Journal of Fisheries and Aquatic Sciences* 50, 1308–1317.

Thomas, C., Cauwet, G. & Minster, J.F. (1995) Dissolved organic carbon in the equatorial Atlantic Ocean. *Marine Chemistry* 49, 15–169.

Thompson, R. & Munro, J.L. (1983) The biology, ecology and bionomics of the hinds and groupers, Serranidae. *ICLARM Studies and Reviews* 7, 59–81.

Thorpe, S., Landeghem, K.V., Hogan, L., Holland, P. & Landegham, K.V. (1997) *Ecomomic Effects on Australian Southern Bluefin Tuna Farming of a Quarantine Ban on Imported Pilchards.* Australian Bureau of Agricultural and Resource Economics, Canberra, Australia.

Thrush, S.F., Hewitt, J.E., Cummings, V.J. & Dayton, P.K. (1995) The impact of habitat disturbance by scallop dredging on marine benthic communities: what can be predicted from the results of experiments? *Marine Ecology Progress Series* 129, 141–150.

Thrush, S.F., Whitlatch, R.B., Pridmore, R.D., Hewitt, J.E., Cummings, V.J. & Wilkinson, M.R. (1996) Scale-dependent recolonization: the role of sediment stability in a dynamic sandflat habitat. *Ecology* 77, 2472–2487.

Toresen, R. (1986) Length and age at maturity of Norwegian spring-spawning herring for the year classes 1959–1961 and 1973–1978 *International Council for the Exploration of the Seas, Committee Meeting* 1986/ H:42.

Toresen, R. (1990) Long-term changes in growth of Norwegian spring spawning herring. *Journal du Conseil, Conseil International pour l'Exploration de la Mer* 47, 48–56.

Toresen, R. (1991) Absorption of acoustic energy in dense schools of herring studies by the attenuation in the bottom echo signal. *Fisheries Research* 10, 317–327.

Townsend, D.W. & Pettigrew, N.R. (1996) The role of frontal currents in larval fish transport on Georges Bank. *Deep-Sea Research* (Part 2) 43, 1773–1792.

Trathan, P.N., Everson, I., Miller, D.G.M., Watkins, J.L. & Murphy, E.J. (1995) Krill biomass in the Atlantic. *Nature* 373, 201–202.

Tregenza, N., Berrow, S., Hammond, P. & Leaper, R. (1997) Harbour porpoise (*Phocoena phocoena* L.) by-catch in set gillnets in the Celtic Sea. *ICES Journal of Marine Science* 54, 896–904.

Trella, K. (1998) Fecundity of the blue whiting (*Micromesistius australis* Norman, 1937) from the Falkland fishing grounds in the years 1983, 1984 and 1986. *Bulletin of the Sea Fisheries Institute* 2(144), 25–34.

Trippel, E.A. (1995) Age at maturity as a stress indicator in fishes. *Bioscience* 45, 759–771.

Trippel, E.A., Kjesbu, O.S. & Solemdal, P. (1997) Effects of adult size and age structure on reproductive output in marine fishes. In: *Early Life History and Recruitment in Fish Populations* (eds R.C. Chambers & E.A. Trippel), pp. 31–62. Chapman & Hall, New York.

Tuck, I., Hall, S.J., Roberston, M., Armstrong, E. & Basford, D. (1998) Effects of physical trawling disturbance in a previously unfished sheltered Scottish sea loch. *Marine Ecology Progress Series* 162, 227–242.

Tucker, A.D., Robins, J.B. & McPhee, D.P. (1997) Adopting turtle excluder devices in Australia and the United States: What are the differences in technology transfer, promotion and acceptance? *Coastal Management* 25, 405–421.

Twichell, D., McClennen, C. & Butman, B. (1981) Morphology and processes associated with the accumulation of fine grain sediment deposit on the southern New England shelf. *Journal of Sedimentary Petrology* 51, 269–280.

Tyler, J.E. & Smith, R.C. (1970) *Measurements of Spectral Irradiance Underwater*. Gordon and Breach, New York.

UNDP (1999) *Human Development Report 1999*. UNDP, London.

Ursin, E. (1982) Stability and variability in the marine ecosystem. *Dana* 2, 51–67.

Vadas, R.L. & Steneck, R.W. (1995) Overfishing and inferences in kelp–sea urchin interactions. In: *Ecology of Fjords and Coastal Waters* (eds H.R. Skjoldal, C. Hopkins, K.E. Erickstad & H.P. Leinaas) pp. 509–524. Elsevier Science, Amsterdam.

Van Blaricom, G.R. (1982) Experimental analysis of structural regulation in a marine sand community exposed to oceanic swell. *Ecological Monographs* 52, 283–305.

Victor, B.C. (1986) Duration of the planktonic larval stage of one hundred species of Pacific and Atlantic wrasses (family Labridae). *Marine Biology* 90, 317–327.

Victor, B.C. (1991) Settlement strategies and biogeography of reef fishes. In: *The Ecology of Fishes on Coral Reefs* (ed. P.F. Sale), pp. 231–260. Academic Press, San Diego.

Videler, J.J. (1993) *Fish Swimming*. Chapman & Hall, London.

Vincent, A. & Sadovy, Y. (1998) Reproductive ecology in the conservation and management of fishes. In: *Behavioural Ecology and Conservation Biology* (ed. T. Caro), pp. 209–245. Oxford University Press, Oxford.

Vinogradov, A.P. (1953) The elementary composition of marine organisms. *Memoirs of the Sears Foundation for Marine Research* 2, 647.

Vinogradov, M.E. (1981) Ecosystems of equatorial up-well-ings. In: *Analysis of Marine Ecosystems* (ed. A.R. Longhurst), pp. 69–93. Academic Press, London.

Vinther, M., Lewy, P., Thomsen, L. & Petersen, U. (1998) *Specification and Documentation of the 4M Package containing Multi-species, Multi-Fleet, and Multi-Area Models*. Danish Institute for Fisheries Research, Charlottenlund.

Von Brandt, A. (1984) *Fish Catching Methods of the World*. Fishing News Books at Blackwell Science, Oxford.

Wade, P.R. (1998) Calculating limits to the allowable human-caused mortality of cetaceans and pinnipeds. *Marine Mammal Science* 14, 1–37.

Wafar, S., Untawale, A.G. & Wafar, M. (1997) Litter fall and energy flux in a mangrove ecosystem. *Estuarine Coastal and Shelf Science* 44, 111–124.

Walker, P.A. & Heessen, H.J.L. (1996) Long-term changes in ray populations in the North Sea. *ICES Journal of Marine Science* 53, 1085–1093.

Walker, P.A. & Hislop, J.R.G. (1998) Sensitive skates or resilient rays? Spatial and temporal shifts in ray species composition in the central and north-western North Sea between 1930 and the present day. *ICES Journal of Marine Science* 55, 392–402.

Walsh, S.J. (1991) Diel variation in availability and vulnerability of fish to a survey trawl. *Journal of Applied Ichthyology* 7, 147–159.

Walsh, S.J. (1996) Efficiency of bottom sampling trawls in deriving survey abundance indices. In: *Assessment of Groundfish Stocks Based on Bottom Trawl Results* (ed. Northwest Atlantic Fisheries Organisation). St. Petersburg, Russia.

Walters, C.J. (1975) Optimal harvest strategies for salmon in relation to environmental variability and uncertain production parameters. *Journal of the Fisheries Research Board of Canada* **32**, 1777–1784.

Walters, C.J. (1984) Managing fisheries under biological uncertainty. In: *Exploitation of Marine Communities* (ed. R.M. May), pp. 263–274. Springer-Verlag, Berlin.

Walters, C.J. (1986) *Adaptive Management of Renewable Resources*. Macmillan, New York.

Walters, C.J. (1987) Nonstationarity of production relationships in exploited populations. *Canadian Journal of Fisheries and Aquatic Sciences* **44** (Suppl. 2), 156–165.

Walters, C.J. & Hilborn, R. (1976) Adaptive control of fishing systems. *Journal of the Fisheries Research Board of Canada* **33**, 145–159.

Walters, C.J. & Ludwig, D. (1994) Calculation of Bayes posterior probability distributions for key population parameters. *Canadian Journal of Fisheries and Aquatic Sciences* **51**, 946–958.

Walters, C.J. & Maguire, J.-J. (1996) Lessons for stock assessment from the northern cod collapse. *Reviews in Fish Biology and Fisheries* **6**, 125–137.

Walters, C.J., Christensen, V. & Pauly, D. (1997) Structuring dynamic models of exploited ecosystems from trophic mass-balance assessments. *Reviews in Fish Biology and Fisheries* **7**, 139–172.

Waples, R.S. (1998) Separating the wheat from the chaff: patterns of genetic differentiation in high gene flow species. *Journal of Heredity* **89**, 438–450.

Ward, T.M. (1996) Sea snake by-catch of prawn trawlers on the northern Australian continental shelf. *Marine and Freshwater Research* **47**, 625–630.

Wardle, C.S. (1986) Fish behaviour and fishing gear. In: *The Behaviour of Teleost Fishes* (ed. T.J. Pitcher), pp. 463–495. Croom-Helm, Beckenham.

Warner, R.R. (1975) The adaptive significance of sequential hermaphroditism in animals. *American Naturalist* **109**, 61–89.

Warner, R.R. (1988a) Traditionality of mating-site preference in a coral reef fish. *Nature* **335**, 719–721.

Warner, R.R. (1988b) Sex change in fishes: hypotheses, evidence and objections. *Environmental Biology of Fishes* **22**, 81–90.

Warwick, R.M. & Clarke, K.R. (1995) New 'biodiversity' measures reveal a decrease in taxonomic distinctness with increasing stress. *Marine Ecology Progress Series* **129**, 301–305.

Watanabe, T. & Nomura, M. (1990) Current status of aquaculture in Japan. In: *Aquaculture in Asia* (ed. M. Mohan Joseph), pp. 223–253. Asian Fisheries Society, Indian Branch.

Watling, L. & Norse, E.A. (1998) Disturbance of the seabed by mobile fishing gear: a comparison to forest clearcutting. *Conservation Biology* **12**, 1180–1197.

Watson, R.A., Carlos, G.M. & Samoilys, M.A. (1995) Bias introduced by the non-random movement of fish in visual transect surveys. *Ecological Modelling* **77**, 205–214.

Weber, D.D. & Ridgeway, G.J. (1962) The deposition of tetracycline drugs in bones and scales of fish and its possible use for marking. *Progressive Fish Culturist* **24**, 150–155.

Weimerskirch, H. & Jouventin, P. (1987) Population dynamics of the wandering albatross, *Diomedia exulans*, of the Crozet Islands: causes and consequences of the population decline. *Oikos* **49**, 195–213.

Weimerskirch, H., Brothers, N. & Jouventin, P. (1997) Population dynamics of wandering albatross *Diomedea exulans* and Amsterdam albatross *D. amsterdamensis* in the Indian Ocean and their relationships with long-line fisheries: conservation implications. *Biological Conservation* **79**, 257–270.

Weinstein, M.P. (ed.) (1988) Fish and shellfish transport through inlets. *American Fisheries Society Symposia* **3**.

Wellington, G.M. & Victor, B.C. (1989) Planktonic larval duration of one hundred species of Pacific and Atlantic damselfishes (Pomacentridae). *Marine Biology* **101**, 557–567.

Wells, M.J. & Clarke, A. (1996) Energetics: the costs of living and reproducing for an individual cephalopod. *Philosophical Transactions of the Royal Society* **351**, 1083–1104.

Werner, E.E. & Gilliam, J.F. (1984) The ontogenetic niche shift and species interactions in size-structured populations. *Annual Review of Ecology and Systematics* **15**, 393–426.

Whittle, P. & Horwood, J.W. (1995) Population extinction and optimal resource management. *Philosophical Transactions of the Royal Society, London, B* **350**, 179–188.

Whitney, R.R. & Carlander, K.D. (1956) Interpretation of body-scale regression for computing body length of fish. *Journal of Wildlife Management* **20**, 21–27.

Williams, G.C. (1975) *Sex and Evolution*. Princeton University Press, Princeton.

Williams, T. & Bedford, B.C. (1974) The use of otoliths for age determination. In: *Ageing of Fish* (ed. T.B. Bagenal), pp. 114–123. Unwin Brothers Ltd, Woking.

Winemiller, K.O. & Rose, K.A. (1993) Why do most fish produce so many tiny offspring? *American Naturalist* **142**, 585–603.

Winterhalder, B. (1981). Foraging strategies in the boreal environment: an analysis of Cree hunting and gathering. In: *Hunter–Gatherer Foraging Strategies* (eds. A.E. Smith & B. Winterhalder) pp. 13–35. Chicago University Press, Chicago.

Witbaard, R. & Klein, R. (1994) Long-term trends on the effects of the southern North Sea beamtrawl fishery on the bivalve mollusc *Arctica islandica* L. (Mollusca, bivalvia). *ICES Journal of Marine Science* **51**, 99–105.

Witthames, P.R. & Greer-Walker, M.G. (1987) An automated method for counting and sizing fish eggs. *Journal of Fish Biology* **30**, 225–235.

Wongteerasupaya, C., Sriurairatana, S., Vickers, J.E., Akrajamorn, A., Boonsaeng, V., Panyim, S., Tassanakajon, A., Withyachumnarnkul, B. & Flegel, W. (1995) Yellowhead virus of *Penaeus monodon* is an RNA virus. *Diseases in Aquatic Organisms* **22**, 45–50.

Woodhouse, C. (1996) Farms avoid new U.S. turtle curb on shrimp imports. *Fish Farming International* **23**, 24.

Wright, P.J. (1996) Is there a conflict between sandeel fisheries and seabirds? a case study at Shetland. In: *Aquatic Predators and Their Prey* (eds S.P.R. Greenstreet & M. Tasker), pp. 154–165. Blackwell Scientific Publications, Oxford.

Xie, L. & Hsieh, W.W. (1995) The global distribution of wind-induced upwelling. *Fisheries Oceanography* **4**, 52–67.

Yamaguchi, M. (1993) Green snail. In: *Nearshore Marine Resources of the South Pacific* (eds A. Wright & L. Hill), pp. 497–511. Institute of Pacific Studies, Suva.

Yellen, J.E., Brooks, A.S., Cornelissen, E., Mehlman, M.J. & Stewart, K. (1995) A middle stone-age worked bone industry from katanda, Upper Semliki Valley, Zaire. *Science* **268**, 553–556.

Yodzis, P. (1998) Local trophodynamics and the interaction of marine mammals and fisheries in the Benguela ecosystem. *Journal of Animal Ecology* **67**, 635–658.

Youngs, W.D. & Robson, D.S. (1978) Estimation of population number and mortality rates. In: *Methods for Assessment of Fish Production in Freshwaters* (ed. T. Bagenal), pp. 137–164. Blackwell, Oxford.

Zann, L.P. (1989) Traditional management and conservation of fisheries in Kiribati and Tuvalu atolls. In: *Traditional Marine Resource Management in the Pacific Basin: an Anthology* (eds K. Ruddle & R.E. Johannes), pp. 78–101. CSIRO, Hobart.

Zar, J.H. (1996) *Biostatistical Analysis*. Prentice-Hall, London.

Zhou, S.J. & Shirley, T.C. (1997) Behavioural responses of red king crab to crab pots. *Fisheries Research* **30**, 177–189.

Zimmer-Faust, R.K. (1993) ATP: a potent prey attractant evoking carnivory. *Limnology and Oceanography* **38**, 1271–1275.

Appendix 1: List of symbols

This is a key to symbols used in this book. We have excluded symbols such as *a*, *b*, α or β that are used throughout the text as generic parameters in linear regressions, spawner–recruitment relationships etc. We have also excluded standard mathematical symbols such as *d*, *e* or Σ. Their use is explained in context. Where possible we have used conventional symbols rather than changing them to ensure that there is a one-to-one relationship between symbols and meanings. Conventional symbols are most likely to be encountered in other publications (*F* for example is widely used to denote fishing mortality and fecundity) and there is little scope for confusion (we hope!). To avoid ambiguity, we explicitly state the meanings of symbols at first use in every section of the text, and use within sections is internally consistent. 'Example' refers to a section or box where the symbol is used, usually as part of an example.

Symbol	Meaning	Example
a	age	7.6
A	area	10.2.4
B	biomass	7.3.1
B^*	optimal biomass	11.3.1
B_{lim}	lowest acceptable biomass (biological reference point)	7.9
B_{LOSS}	lowest biomass ever observed (biological reference point)	7.9
B_{max}	maximum biomass	7.3.2
B_{MSY}	biomass corresponding to the maximum sustainable yield (biological reference point)	7.9
B_{MSY}	biomass of population at which MSY is taken	7.3.1
B_{pa}	target biomass set as a precautionary approach (biological reference point)	7.9
$B_{x\%R}$	biomass at which the average recruitment is $x\%$ of the maximum of the underlying spawner–recruitment relationship (biological reference point)	7.9
c	operating costs	11.3.2
c_y	cost per unit caught	11.3.2
C	amplitude of seasonal oscillation in seasonally modified von Bertalanffy growth equation	9.3.3
C	catch	9.3.1
C^*	prey consumption *d* level of depensation	4.2.3
D	density dependence: represents a stock-recruitment function	Box 7.1
D	diversity (e.g. Margalef, Menhinick's or Simpsons Dominance indices)	12.4.1
D	egg stage duration	10.2.7
D	density of fish	10.2.4
DC	diet composition: fraction of *i* in the diet of *j*	8.7

Continued

Symbol	Meaning	Example
Ee_i	ecotrophic efficiency	8.7.1
ER	expected revenue (net of operating costs)	11.3.2
f	fishing effort	10.2.4
f	fraction of females spawning in a given time interval	10.2.7
f_{MSY}	the level of fishing effort at which MSY is achieved	
F	fecundity	3.4.3
F	instantaneous rate of fishing mortality	7.5
F′	relative fecundity adjusted to allow for spawning frequency	10.2.7
$F_{0.1}$	F where slope of yield per recruit vs. F is one-tenth of its value near the origin (biological reference point)	7.9
F_{atr}	proportion of the eggs in ovary that are atretic	10.2.7
F_{crash}	F that would drive the population to extinction (biological reference point)	7.9
F_{high}	F in an equilibrium population where recruitment per spawning stock biomass in 10% of years has been above the replacement level (biological reference point)	7.9
F_{lim}	F set as the highest that is acceptable by some specified criterion (biological reference point)	7.9
F_{LOSS}	F that would drive the stock to the lowest observed spawning stock size (LOSS) (biological reference point)	7.9
F_{low}	F in an equilibrium population where recruitment per spawning stock biomass in 90% of years has been above the replacement level (biological reference point)	7.9
F_{max}	F where total yield or yield per recruit is highest (biological reference point)	7.9
F_{MCY}	F giving maximum consistent yield, i.e. largest long-term yield without reducing population below a predetermined level (biological reference point)	7.9
F_{med}	F in an equilibrium population where recruitment per spawning stock biomass in half of the years has been above the replacement level (biological reference point)	7.9
F_{MSY}	F giving maximum sustainable yield (also called F_m) (biological reference point)	7.9
F_{pa}	F that is targeted according to a specific precautionary approach (biological reference point)	7.9
F_{pot}	potential fecundity (the stock of eggs in the ovary before spawning)	10.2.7
F_{real}	realized fecundity (eggs released into the water)	10.2.7
F_{trial}	trial value of fishing mortality	7.7.1
$F_{x\%}$	F in an equilibrium population where recruitment per spawning stock biomass is x% of the corresponding unfished population (biological reference point)	7.9
G	sea area over which egg production is assessed	10.2.7
H′	eveness	12.4.1
H′	Shannon–Wiener diversity	12.4.1
H_1	alternative hypotheses in Bayesian statistics	12.4.1
H_2	alternative hypotheses in Bayesian statistics	12.4.1
H_α	Reyni diversity index	12.4.1
HB	Brillouin or Pielou eveness	12.4.1
I	index of relative abundance	7.3.3

Continued

Symbol	Meaning	Example
k_1	catch per unit capital invested	11.3.2
K	cumulative catch	10.2.5
K	Brody growth coefficient (von Bertalanffy growth equation)	9.3.3
K	fleet capacity represented by amount of money invested in fleet	11.3.2
K_{opt}	optimal fleet capacity (investment) to maximize ER	11.3.2
l_t	annual survival rate after natural mortality	Box 7.1
L	length	3.4.2
L	depth	10.2.7
L_∞	asymptotic length (von Bertalanffy growth equation)	3.4.2
L_c	mean length at capture	9.3.1
L_i	length at time when ith scale annulus was formed	9.3.3
L_{mat}	length at maturity	9.3.4
m	mesh size	9.3.1
m	maintenance and depreciation	11.3.2
M	instantaneous rate of natural mortality	7.5
M^*	instantaneous natural mortality due to predation by marine mammal	15.3.1
M_1	residual mortality (MSVPA)	8.4
M_2	predation due to predators included in model (MSVPA)	8.4
N	number, population size	Box 7.1
N^*	equilibrium population size	7.4.1
N_0	Hill diversity: number of species	12.4.1
N_1	Hill diversity: number of 'abundant' species	12.4.1
N_2	Hill diversity: number of 'very abundant' species	12.4.1
N_{max}	maximum population number (size)	Box 7.1
p	price	11.3.2
p	proportion	7.3.1
P	egg production (numbers)	10.2.7
P	production (biomass)	8.7
$P(.)$	probability	Box 7.7
$P(.|.)$	conditional probability (Bayesian statistics)	Box 7.7
P	prey biomass	15.3.1
PV	present value	11.3.1
q	catchability coefficient	10.2.4
q	recruitment ratio of depensatory to standard recruitment models	4.2.3
q_a	availability to catchability (availability of stock in survey area)	10.2.4
q_e	catching efficiency (of net)	10.2.4
Q	consumption	8.7

Continued

Symbol	Meaning	Example
Q	quota (total allowable catch)	11.3.2
Q^*	upper limit to fleet's catching capacity	11.3.2
r	intrinsic rate of population increase	7.3.2
R	current observation (Bayesian statistics)	Box 7.7
R	number of tagged fish recaptured	10.2.6
R	recruit abundance	4.2.1
R	revenue (net of operating costs)	11.3.2
R	sex ratio or female fraction	10.2.7
R^*	level of recruitment at which standard model agrees with depensatory spawner–recruitment model	4.2.3
S	species richness (number of species)	12.4.1
S	survivorship	Box 7.1
S	scale radius	9.3.3
S	spawner abundance	4.2.1
S^*	maximum spawner abundance	4.2.3
S_c	scale radius at capture	9.3.3
S_i	scale radius at time when ith annulus was formed	9.3.3
S_m	optimal escapement (survivorship)	Box 7.1
S_t	number of adults that escape from the fishery	Box 7.1
t	time	3.4.2
t_0	parameter of von Bertalanffy growth equation	3.4.2
t_c	age at capture	7.7
t_{mat}	age at maturity	9.3.4
t_{max}	maximum age of cohort	7.7
t_s	phase shift of seasonal growth oscillations	9.3.3
T	number of tagged fish	10.2.6
T	temperature	10.2.7
v	vulnerability to the fishery	9.3.6
V	value of income	11.3.1
V	filtered volume	10.2.7
W	weight	9.3.2
W_∞	asymptotic weight (von Bertalanffy growth equation)	9.3.3
Y	yield (biomass caught)	7.3.1
Y^*	yield at which profit is maximized	11.3.1
$Z_{r=0}$	maximum level of mortality that population can withstand	12.4.1
δ	discount rate	11.3.1
Δ	taxonomic diversity	12.4.1

Continued

Symbol	Meaning	Example
Δ^*	taxonomic distinctness without species diversity	12.4.1
Δ^+	taxonomic distinctness for presence/absence data	12.4.1
ω_{ij}	weights the path length linking species i and j through a taxonomy (in taxonomic diversity measures)	12.4.1

Appendix 2: Fisheries websites

The internet is an important source of information on fish and fisheries. Data, publications and software are available from many sites, often without charge. Since the words 'fish' and 'fisheries' produce several thousand 'hits' with most search engines, we are not going to list all the sites where they feature. Rather, our list gives a selection of sites that we have found useful. Web addresses change rather frequently, but all these sites were functional in mid-2000!

Alaska Department of Fish and Game (ADFG), US
www.cf.adfg.state.ak.us/
Alaskan fisheries, data, research, regulations and publications.

American Fisheries Society (AFS).
www.fisheries.org/
News, activities and publications.

American Society of Ichthyologists and Herpetologists (ASIH).
www.utexas.edu/depts/asih/
Society news, research and publications.

Aquatic Network, US
www.aquanet.com/
Information service with links to sites dealing with fisheries and aquaculture. Publications and news.

Australian Institute of Marine Science (AIMS).
www.aims.gov.au/
Information on reef fish research projects in progress.

Australian Museum-Ichthyology
www.austmus.gov.au/fish/
Information on research collections and activities

Bernice P. Bishop Museum, Hawaii, US
www.bishop.hawaii.org/bishop/fish/
Information on research collections and publications

California Academy of Sciences-Ichthyology, US
www.calacademy.org/research/ichthyology/
Catalogues, research, publications and database of fish collections

California Department of Fish and Game (CFG), US
www.dfg.ca.gov/
Fish, fisheries and news from California. Data and maps.

Center for Aquatic Biology and Aquaculture, University of California, Davis, US
aquafishprog.ucdavis.edu/
Research and teaching activities, news.

Commonwealth Scientific and Industrial Research Organization (CSIRO), Australia
www.marine.csiro.au/
Wide ranging information on assessment, management of Australian fisheries. Links to CSIRO laboratories throughout Australia.

CRC Reef Research Centre (Great Barrier Reef, Australia).
www.reef.crc.org.au/
Description of current research on reef fisheries and fishing effects.

Department of Fisheries and Oceans (DFO), Canada
www.dfo-mpo.gc.ca/
Information, publications and statistics relating to Canadian fisheries. Links to individual DFO laboratories.

Department of Fisheries, Malaysia
agrolink.moa.my/dof/
Fisheries conservation and management, research, aquaculture, news and activities.

Department of Oceanography, Dalhousie University, Canada

www.phys.ocean.dal.ca/
Research, publications, news and activities.

Department of Primary Industry and Fisheries (DPI), Northern Territory, Australia
www.nt.gov.au/dpif/
Fisheries and aquaculture research, statistics and news.

Ecopath with Ecosim
www.ecopath.org/
Information on Ecopath software, downloadable software and presentations.

FishBase
www.fishbase.org/
A database that gives information on the taxonomy, life history and distribution of many fish. Downloadable data, analyses, pictures and references.

Fisheries Net, US
www.fis-net.com/
Information on commercial fisheries, fishing equipment, fish processing and marketing

Fisheries Society of the British Isles (FSBI).
www.le.ac.uk/biology/fsbi/
Society news, research and publications.

Food and Agriculture Organization of the United Nations (FAO), Italy
www.fao.org/
Excellent site with huge range of information on global fisheries. Downloadable data, publications, assessments, pictures and analyses.

Forum Fisheries Agency (FFA), Solomon Islands
www.bscene.com.au/ffa/
South Pacific fisheries management and development. News and publications.

Global Change Research Information Office (GCRIO), US
gcrio.gcrio.org/
Publications and data relating to the impacts of climate change on the oceans and their fisheries.

Greenpeace
www.greenpeace.org/
Information on a range of fisheries and environmental campaigns

GSFC Earth Sciences (GES) Distributed Archive Center
daac.gsfc.nasa.gov/

Maps of ocean currents, ocean productivity and remote sensing images of the oceans

Indian Centre for Aquaculture and Fisheries Trade, India
www.fishindia.com/
News and information on commercial fisheries and aquaculture

International Baltic Sea Fishery Commission (IBSFC), Poland
www.ibsfc.org/
Baltic Fisheries management, assessment and regulation.

International Center for Living Aquatic Resources Management (ICLARM).
www.cgiar.org/iclarm/
Extensive information on fish biology and fisheries, news, and research activities.

International Council for the Exploration of the Sea (ICES), Denmark
www.ices.dk/
News, activities, research, publications on North Atlantic fisheries

International Pacific Halibut Commission, US
www.iphc.washington.edu/
Information on the Pacific halibut fishery including catch data, research and recipes.

Lobster Conservancy, Maine, US
www.lobsters.org/
News on the management and protection of Maine lobsters, lobster biology and research.

Lobster Institute, University of Maine, US
www.lobster.um.maine.edu/
Information on lobster biology, lobster fisheries and publications.

Marine Institute (MI), Memorial University of Newfoundland, Canada
www.mi.mun.ca/
Research and teaching activities, news.

Marine Stewardship Council, US
www.msc.org/
Information on the work of MSC, a charity trying to promote sustainable fishing.

Milton Love Fisheries Research Laboratory, University of California, Santa Barbara (UCSB), US

www.ucsb.edu/lovelab/
Taxonomy and biology of fishes on Western coast of central and North America

Ministry of Agriculture, Fisheries and Food (MAFF), UK
www.maff.gov.uk/fish
Description of the UK fishing industry and fishery conservation policies

Ministry of Agriculture, Forestry and Fisheries of Japan
www.maff.go.jp/
News, statistics, fishery policy and publications

Ministry of Fisheries, Iceland
www.fisheries.is/
Fisheries and their management, seafood utilization and markets.

Ministry of Fisheries, New Zealand
www.fish.govt.nz
News, research, publications, statistics and policy.

Nagasaki University Faculty of Fisheries, Japan
www.fish.nagasaki-u.ac.jp/
Research, news, publications, data and photographs.

National Marine Fisheries Service (NMFS) of the National Oceanic and Atmospheric Administration (NOAA), US
www.nmfs.gov/
Excellent site with links to all the NMFS laboratories. Information on fisheries assessment and management, downloadable data and publications, US Government policy documents and news.

National Shellfisheries Association, US
www.shellfish.org/
The ecology, production, economics and management of shellfish resources.

North Atlantic Fishery College, Shetland Islands
www.nafc.ac.uk/
Research and teaching activities, local fishery data.

North Carolina Marine Fisheries, US
www.ncfisheries.net/
News, fishery statistics, biology and photographs

Pacific States Marine Fisheries Commission (PSMFC), US
www.psmfc.org/

Projects, publications and news on fishery resources on the west coast of the US.

Queensland Department of Primary Industries (DPI), Australia
www.dpi.qld.gov.au/
Fisheries and aquaculture research, statistics and news.

Ransom Myer's stock recruitment database, Canada
www.mscs.dal.ca/~myers/
Downloadable stock and recruitment data for many fish stocks, publications.

ReefBase
www.reefbase.org/
Downloadable data and maps describing reef distributions and reef related research.

School of Fisheries, University of Washington, US
www.fish.washington.edu/
Information on the School, research activities, publications and events

Scripps Institute of Oceanography (SIO), San Diego, US
www.sio.ucsd.edu/scripps/
Research activities, publications.

Sea Fisheries Research Institute, Cape Town, South Africa
www.gov.za/sfri/
Wideranging information on South African fishery resources, their assessment and management. Satellite images, publications and news.

South-east Asian Fisheries Development Center (SEAFDEC).
www.seafdec.org/
Fisheries conservation and management, gear technology, aquaculture.

United Nations Environment Programme (UNEP), Switzerland
www.unep.ch/
Details of a range of UNEP conventions that relate to marine life and protection of the marine environment

United States Fish and Wildlife Service (FWS).
www.fws.gov/
News and activities, information on management and research.

University of British Colombia (UBC) Fisheries Centre, Canada

www.fisheries.com/
Research activities, publications and news relating to many fisheries issues.

Western Australian Fisheries, Australia
www.wa.gov.au/
Information on recreational and commercial fisheries in Western Australia

World Conservation Monitoring Centre (WCMC), UK
www.wcmc.org.uk/
Wealth of information on distribution of marine habitats, marine reserves. Downloadable maps and data.

Appendix 3: Geographic index

The numbers in this index refer to the locations of sites, seas and countries that are shown on the accompanying map (page 391).

Location	Code	Location	Code
Aberdeen, Scotland	3	Chatham Rise, New Zealand	86
Alaska, USA	104	Chesapeake Bay, USA	113
Aleutian Islands	101	Chile	74
Amercian Samoa	92	China	28
Antarctic	64	Christmas Island (Pacific Ocean)	84
Antarctic Ocean (Southern Ocean)	64	Cornwall, England	5
Apo Island, Philippines	33	Denmark	11
Arcachon Bay, France	125	Devon, England	5
Arctic	129	Dublin, Eire	126
Atlantic Bight, USA	113	Dutch Wadden Sea, Holland	9
Atlantic Ocean	67	Ebro Delta, Spain	122
Australia	43	Ecuador	77
Aveiro, Portugal	121	Egypt	29
Baja California, Mexico	82	Eire (southern Ireland)	126
Baltic Sea	17	El Quisco, Chile	75
Barbados	70	Enewetak Atoll, Marshall Islands	96
Barents Sea	15	England	5
Basque Country	124	English Channel	20
Belgium	8	Euboikos Gulf, Greece	27
Belize	78	Exe estuary, England	5
Black Sea	23	Exmouth Gulf, Australia	49
Bradda Ground, Isle of Man	127	Falkland Islands	71
Brazil	68	Faroe Bank	2
Breton coast, France	125	Faroe Islands	2
Bristol Bay, Alaska, USA	103	Federated States of Micronesia	34
British Columbia, Canada	107	Finland	14
Brixham, Devon, England	5	Fleetwood, England	5
Burry Inlet, Wales	4	Florida Bay, USA	111
California Bight	100	France	125
California, United States	100	French Polynesia	83
Canada	118	Galápagos Islands	79
Canary Islands	119	Galicia, Spain	122
Cape Cod Bay, Massachusetts, USA	114	Gareloch, Scotland	3
Cape Hatteras, USA	112	Georges Bank, USA	115
Cape Town, South Africa	61	Grand Banks, Newfoundland, Canada	118
Cape Verde	65	Great Astrolabe Reef, Fiji	90
Caribbean Sea	72	Great Barrier Reef, Australia	41
Caspian Sea	22	Great Britain (England, Scotland, Wales)	3–5
Celtic Sea	21	Greece	27
Challenger Plateau, New Zealand	88	Greenland	128

Continued

Location	Code	Location	Code
Guayaquil, Ecuador	77	Pacific Ocean	99
Guayas River, Ecuador	77	Palau	36
Gulf of Alaska, USA	106	Papua New Guinea	38
Gulf of Carpentaria, Australia	46	Peru	75
Gulf of Maine, Maine, USA	115	Peterhead, Scotland	3
Gulf of Mexico	80	Philippines	32
Gulf of St. Lawrence, Canada and USA	116	Point Barrow, Alaska	105
Gulf of Thailand	50	Poland	19
Halifax, Nova Scotia, Canada	117	Portugal	121
Hawaii	98	Queensland, Australia	41
Hokkaido, Japan	24	Red Sea	31
Holland	9	Rome, Italy	25
Iceland	1	Russia	16
Indian Ocean	55	Salcombe, Devon, England	5
Ireland	126	Santa Barbara Basin	100
Irish Sea	127	Santa Barbara, California, USA	100
Isle of Man	127	Sarawak	37
Italy	25	Scotland	3
Ivory Coast	63	Seychelles Islands	56
Jamaica	76	Shetland Islands	6
Japan	24	Solomon Islands	40
Jeffreys Bank, Gulf of Maine, Maine, USA	115	Solway Firth, Scotland	3
Johnston Atoll	97	Somalia	57
Kanoehe Bay, Hawaii	98	South Africa	60
Kattegat, Baltic Sea	18	South Australia, Australia	47
Kenya	59	South China Sea	35
Kiribati	95	South Georgia	66
Ko Ono, Fiji	90	Southern Ocean (Antarctic Ocean)	64
Lofoten Islands, Norway	12	Spain	122
Long Island Sound, Connecticut, USA	113	Sri Lanka	53
Loughor estuary, Wales	4	St. Lawrence River, Canada	116
Louisiana, United States	110	St. John's, Newfoundland, Canada	118
Lupar Estuary, Sarawak (state), Malaysia	37	Stellwagen Bank, Gulf of Maine, Maine, USA	115
Maine, USA	115	Sumilon Island, Philippines	33
Maldive Islands	54	Swans Island, Gulf of Maine, Maine, USA	115
Mallaig, Scotland	3	Sweden	13
Mediterranean Sea	123	Taiwan	30
Medway Estuary, England	5	Tasmania	45
Mexico	82	Thailand	52
Moorea Atoll, French Polynesia	83	Tokelau	94
Morocco	120	Totoya Island, Fiji	90
Mukah, Sarawak	37	Turkey	26
Namibia	62	United Kingdom (England, Scotland,	
Naragansett Bay, Rhode Island	113	Wales and Northern Ireland)	3–5, 126
Netherlands	9	United States	110
New England, USA	115	Urk, Holland	9
New York Bight	113	Vanua Levu Islands, Fiji	90
New Zealand	87	Vanuatu	91
Newfoundland, Canada	118	Victoria, Australia	44
Niue	85	Virgin Islands (USA)	69
North Carolina, United States	112	Viti Levu Island, Fiji	90
North Sea	7	Wadden Sea, Holland	9
Norway	10	Wales	4
Norwegian Sea	12	Western Samoa	92
Nova Scotia	117	Willapa Bay, Washington State, USA	109
Ontong Java Atoll, Solomon Islands	39	Woleai Atoll, Federated States of Micronesia	34
Oregon, USA	108	Yap, Federated States of Micronesia	34

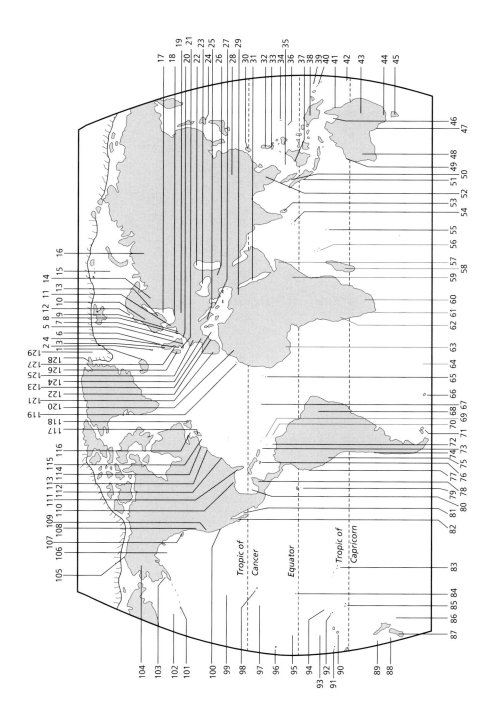

Tropic of Cancer

Equator

Tropic of Capricorn

Index

Note: Page numbers in **_bold italics_** refer to pages where a terms is printed in bold and defined. Page numbers in _italics_ refer to figures/boxes; pages in **bold** refer to tables.